# Complementary and Alternative Approaches to Biomedicine

# ADVANCES IN EXPERIMENTAL MEDICINE AND BIOLOGY

Editorial Board:

NATHAN BACK, *State University of New York at Buffalo*

IRUN R. COHEN, *The Weizmann Institute of Science*

DAVID KRITCHEVSKY, *Wistar Institute*

ABEL LAJTHA, *N. S. Kline Institute for Psychiatric Research*

RODOLFO PAOLETTI, *University of Milan*

---

Recent Volumes in this Series

Volume 539
BLADDER DISEASE, Part A and Part B: Research Concepts and Clinical Applications
Edited by Anthony Atala and Debra Slade

Volume 540
OXYGEN TRANSPORT TO TISSUE, VOLUME XXV
Edited by Maureen Thorniley, David K. Harrison, and Philip E. James

Volume 541
FRONTIERS IN CLINICAL NEUROSCIENCE: Neurodegeneration and Neuroprotection
Edited by László Vécsei

Volume 542
QUALITY OF FRESH AND PROCESSED FOODS
Edited by Fereidoon Shahidi, Arthur M. Spanier, Chi-Tang Ho, and Terry Braggins

Volume 543
HYPOXIA: Through the Lifecycle
Edited by Robert C. Roach, Peter D. Wagner, and Peter H. Hackett

Volume 544
PEROXISOMAL DISORDERS AND REGULATION OF GENES
Edited by Frank Roels, Myriam Baes, and Sylvia De Bie

Volume 545
HYPOSPADIAS AND GENITAL DEVELOPMENT
Edited by Laurence S. Baskin

Volume 546
COMPLEMENTARY AND ALTERNATIVE APPROACHES TO BIOMEDICINE
Edited by Edwin L. Cooper and Nobuo Yamaguchi

Volume 547
ADVANCES IN SYSTEMS BIOLOGY
Edited by Lee K. Opresko, Julie M. Gephart, and Michaela B. Mann

Volume 548
RECENT ADVANCES IN EPILEPSY RESEARCH
Edited by Devin K. Binder and Helen E. Scharfman

Volume 549
HOT TOPICS IN INFECTION AND IMMUNITY IN CHILDREN
Edited by Andrew J. Pollard, George H. McCracken, Jr., and Adam Finn

---

A Continuation Order Plan is available for this series. A continuation order will bring delivery of each new volume immediately upon publication. Volumes are billed only upon actual shipment. For further information please contact the publisher.

# Complementary and Alternative Approaches to Biomedicine

Edited by

**Edwin L. Cooper**
*University of California, Los Angeles*
*Los Angeles, California*

and

**Nobuo Yamaguchi**
*Kanazawa Medical University*
*Uchinada, Ishikawa, Japan*

Springer Science+Business Media, LLC

Library of Congress Cataloging-in-Publication Data

Congress on Alternative and Complementary Medicine (5th: 2002: Kanazawa-shi Japan)
  Complementary and alternative appoaches to biomedicine/edited by Edwin L. Cooper and Nobuo Yamaguchi.
    p. ; cm. — (Advances in experimental medicine and biology; v. 546)
  Includes bibliographical references and index.

  1. Alternative medicine—Congresses.   2. Medicine, Chinese—Congresses.   I. Cooper, Edwin L. (Edwin Lowell), 1936–   II. Yamaguchi, Nobuo.   III. Title.   IV. Series.
  [DNLM: 1. Complementary Therapies—Congresses.   2. Medicine, Chinese Traditional—Congresses.  3. Phytotherapy—Congresses.   WB 890 C749c 2004]
  R733.C664   2002
  615.5—dc22

2003068658

Proceedings of the 5th Annual Japanese Congress on Alternative and Complementary Medicine, held November 8–10, 2002, in Kanazawa, Japan

ISSN 0065-2598

ISBN 978-1-4419-3441-3      ISBN 978-1-4757-4820-8 (eBook)
DOI 10.1007/978-1-4757-4820-8

© Springer Science+Business Media New York 2004
Originally published by Kluwer Academic Publishers/Plenum Publishers, New York in 2004
Softcover reprint of the hardcover 1st edition 2004

http://www.wkap.nl

10  9  8  7  6  5  4  3  2  1

A C.I.P. record for this book is available from the Library of Congress

All rights reserved

No part of this book may be reproduced, stored in a retrieval system, or transmitted in any form or by any means, electronic, mechanical, photocopying, microfilming, recording, or otherwise, without written permission from the Publisher, with the exception of any material supplied specifically for the purpose of being entered and executed on a computer system, for exclusive use by the purchaser of the work.

Permissions for books published in Europe: *permissions@wkap.nl*
Permissions for books published in the United States of America: *permissions@wkap.com*

Foreground: G. Talwar, E. L. Cooper, P. C. Willis. Left: M. Irwin, A. Cossarizza

**THE LOGO: International Symposium on Alternative and Complementary Medicine**
There are several characters that express a concept. **ISACM:** International Symposium on Alternative and Complementary Medicine. **INMPRC:** Ishikawa Natural Medicinal Products Research Center. **THE WORLD**: alternative and complementary medicine (ACM) has its source internationally. **MEDITATING HUMAN:** 1) ACM is involved in human health; 2) meditation or Yoga is one form of ACM; **SNAKE (LEFT) AND YIN & YANG (RIGHT):** The idea is to combine Western medicine and Eastern medicine. Snake is the symbol of Hippocrates representing Western medicine, while Yin & Yang represent the philosophy of Eastern (Chinese) medicine; **GINKO:** Ginko is used in Chinese medicine especially for the improvement of brain functions; **MAGNET:** Magnetism is used to improve various physical problems; **BEE:** Proporis, which is the mixture of sap and bee secretion, has strong antimicrobial activity, a well-known potion; **HERBS:** Various herbs have been used historically; **EARTHWORM:** Used for many diseases in Chinese medicine; **MUSHROOM:** Mushrooms known to have immune enhancing effect, especially for cancer; **MUSIC NOTE:** Music therapy, which is also getting popularity as ACM. **Designed by Edwin L. Cooper, Atsue Kamata, Shinji Kasahara.**

# Contents

**Preface** ................................................ *page* xi

**What Happened in Kanazawa? The Birth of *eCAM*** ......... xi
**National Approaches to CAM** ........................... xi
**CAM Approaches to Specific Diseases and Biomedical Conditions** xiii

## INTRODUCTION

**The Urgent Need for Evidence-Based CAM in the World** ...... 3
*Nobuo Yamaguchi*

**Beginnings** ............................................. 5
*Patty Christiena Willis*

## I. NATIONAL APPROACHES TO CAM

**1. Complementary and Alternative Medicine in Japan** ......... 9
*Nobutaka Suzuki, Satoshi Ohno, Tsutomu Kamei, Yumiko Yoshiki, Yuji Kikuchi, Kazuyoshi Okubo, Tomihisa Ohta, Shoji Shimizu, Saburo Koshimura, Atsufumi Taru, and Masaki Inoue*

**2. New Scientific Approach for Natural Medicine: Examples of Kampo Medicine** ...................................... 27
*Haruki Yamada*

**3. The Modernization of Traditional Chinese Medicine in Taiwan—Past, Present and Future** ..................... 35
*Chieh Fu Chen, Yau Chik Shum and Sze Piao Yang*

4. Quality Requirements for Herbal Medicinal Products
   in Europe .................................................... 43
   G. Franz

5. Integrating Medical Knowledge: Old Roots and a Modern
   Science-Based Approach at the University of Milan .......... 57
   E. Minelli, F. Marotta, U. Solimene

## II. CAM APPROACHES TO SPECIFIC DISEASES AND BIOMEDICAL CONDITIONS

6. Identification of Diseases that may be Targets for
   Complementary and Alternative Medicine (CAM) ............ 75
   Aristo Vojdani and Edwin L. Cooper

7. Complementary and Alternative Medicine during
   HIV Infection ................................................. 105
   Milena Nasi, Marcello Pinti, Leonarda Troiano,
   Andrea Cossarizza

8. A Polyherbal Formulation for a Wide Spectrum of Reproductive
   Tract and Sexually Transmitted Infections ................ 111
   G.P. Talwar

9. Anticancer Therapeutic Potential of Soy Isoflavone, Genistein ... 121
   Mepur H. Ravindranath, Sakunthala Muthugounder, Naftali Presser,
   and Subramanian Viswanathan

10. Chinese Medicine and Immunity ........................... 167
    Haruhisa Wago and Hong Deng

11. Testing Efficacy of Natural Anxiolytic Compounds ........... 181
    A.A. Roberts

12. Alternative Approaches to Pain Relief during Labor
    and Delivery ................................................. 193
    Michel Tournaire

13. **Alternative and Comparative Medicine in Dentistry** .......... 207
   *Kazuo Komiyama, Kazuyosi Koike, Masahiro Okaue, Takahiro Kaneko, Mitsuhiko Matsumoto*

## III. PHYSICAL INTERVENTION: TOUCH, HYDROTHERAPY, SOUND

14. **Methodological Concerns when Designing Trials for the Efficacy of Acupuncture for the Treatment of Pain** .......... 217
   *P.J. White*

15. **Treatment for Atopic Dermatitis by Acupuncture** .......... 229
   *M. Fukuda, N. Kawada, H. Kawamura and T. Abo*

16. **Hydrotherapy can Modulate Peripheral Leukocytes: An Approach to Alternative Medicine** .......... 239
   *N. Yamaguchi, S. ShimizuS and H. Izumi*

17. **Music Therapy, Wellness, and Stress Reduction** .......... 253
   *Andrea M. Scheve*

18. **Music Therapy, a Future Alternative Intervention against Diseases** .......... 265
   *Haruhisa Wago and Shinji Kasahara*

## IV. DIETARY INTERVENTION IN SPECIFIC DISEASES

19. **Diet Instruction for Japanese Traditional Food in Therapy for Atopic Dermatitis** .......... 281
   *Hiromi Kobayashi, Nobuyuki Mizuno, Hiroyuki Teramae, Haruo Kutsuna, Mika Nanatsue, Kazuko Hirai, and Masamitsu Ishii*

20. **Kampo Therapy for Adult Atopic Dermatitis by Dieting and Herbal Medications: Evaluating the Disappearance of Disease Phases** .......... 297
   *Masamitsu Ishii, Hiromi Kobayashi, Nobuyuki Mizuno, Hiroyuki Teramae, Takeshi Nakanishi, Haruo Kutsuna and Iwao Yamamoto*

21. Management of Nutritional and Health Needs of Malnourished
    and Vegetarian People in India .......................... 311
    H.D. Kumar

## V. BASIC SCIENCE: FUTURE APPROACHES TO NOVEL MOLECULES FOR CAM

22. Cultural Heritage: Porifera (Sponges), A Taxon Successfully
    Progressing Paleontology, Biology, Biochemistry, Biotechnology
    and Biomedicine ...................................... 325
    Werner E.G. Müller, Renato Batel, Isabel M. Müller
    and Heinz C. Schröder

23. Earthworms: Sources of Antimicrobial and Anticancer
    Molecules ........................................... 359
    Edwin L. Cooper, Binggen Ru and Ning Weng

24. What Can We Learn from Marine Invertebrates to be Used as
    Complementary Antibiotics? ............................ 391
    Philippe Roch

25. Nervous, Endocrine, Immune Systems as a Target for
    Complementary and Alternative Medicine ................. 405
    Shinji Kasahara and Edwin L. Cooper

## VI. EDUCATION AND PHILOSOPHY

26. Scientific Thinking: Its History, Methods, and Advantages ..... 427
    C.S. Wallis

27. Glycome: A Medical Paradigm .......................... 445
    Arnold Loel

## VII. ABSTRACTS

1. Exploring the scientific basis of complementary and alternative
   medicine: Plenary. .................................... 455
   Michael Irwin

2. Compliments to complementary medicine: Plenary . . . . . . . . . . . . . 455
   *Gerhard G. Uhlenbruck and Edwin L. Cooper*

3. Effects of Tai Chi Chih on varicella-zoster virus specific. Immunity
   and health functioning in older adults. . . . . . . . . . . . . . . . . . . . . . . 455
   *Michael Irwin, J.L. Pike and M.N. Oxman*

4. Safety and effectiveness of health strategies to eliminate behaviors
   detrimental to health. . . . . . . . . . . . . . . . . . . . . . . . . . . . . . . . . . . . . 456
   *Bruce S. Rabin*

5. Expression of matrix metalloproteinase in cervical squamous cell
   carcinoma tissue and its significance. . . . . . . . . . . . . . . . . . . . . . . . 456
   *Zhang Shulan, Lin Bei, CAI Wei, et al.*

6. Expression of surviving gene and its relation with the expression
   of Bcl-2 and Bax protein in epithelial ovarian cancer. . . . . . . . . . . . 457
   *Zhang Shulan, Zhao Chang, Qing Lin Pei, Li Yan, and Gao Hong*

   **Index** . . . . . . . . . . . . . . . . . . . . . . . . . . . . . . . . . . . . . . . . . . . . . . . 459

# Preface

## WHAT HAPPENED IN KANAZAWA? THE BIRTH OF eCAM

This book contains the proceedings of the International Symposium on Complementary and Alternative Medicine, (CAM) which was convened in Kanazawa Japan, November 8–10, 2002. The participants were mainly from Japan, USA, China, France, England, Germany, Taiwan, and India.

The world of western medicine is gradually opening its doors to new ways of approaching healing. Since many of these approaches began centuries and even millennia ago in Asia, it was entirely appropriate to open our symposium in Kanazawa, a beautiful, traditional city located on the Sea of Japan. Experts from Asia, Europe and the United States gathered together for true discussions on complementary and alternative medicine and its role developing all over the world. As scientists, we listened to historical perspectives from India, China and Japan, where CAM is still being practiced as it has been for centuries.

It is well to mention at the outset that this book will cover a rapidly growing field that has strong advocates but others who are less than enthusiastic. This should be evident by the presentation of chapters that aim to significantly dispel some of the criticisms of pseudoscience and myth that often surround the discipline. It is our purpose to present high quality peer reviewed chapters.

This book should have great appeal to a number of audiences with its wide range of contributions from fields such as psychiatry, neurology, and the several basic sciences to medicine, neurobiology, and the relatively new field of psychoneuroimmunology. There is to our knowledge no book of this kind composed of presentations by biomedical researchers who are already recognized in fields that may be somewhat peripheral to the book. Most have established records in neurobiology and immunology as closely related disciplines; some are active in the field of alternative and complementary medicine. Once published, the book can be displayed at relevant meetings attended by neuroscientists, psychiatrists and certain practitioners of alternative and complementary medicine.

## NATIONAL APPROACHES TO CAM

The five chapters making up the first section on national approaches to CAM from Japan, Taiwan, Germany and Italy focus on the various practices in these regions. Suzuki, et al. gives an excellent account of the history of CAM in Japan and USA. In 1990, a small research group of physicians interested in CAM was born at Kanazawa University and in 1997; this group established the Japanese Society for Complementary and Alternative Medicine (JCAM). The members of JCAM are doctors, nurses, dentists, veterinarians, acupuncturists, nutritionists, pharmacists and researchers of various fundamental sciences. After the foundation of JCAM, the Japanese Society for Alternative and Complementary Medicine the Japanese Association for Alternative, Complementary and Traditional Medicine (JACT) and the Japanese Society for Integrative Medicine (JIM) were successively initiated, making Japan a locale of intense activity.

According to Yamada, the origins of Kampo in traditional Chinese medicines give it access to the valuable experience acquired during more than 2000 years of practice. With this long history, Chinese medicines including Kampo are not considered alternative medicine for peoples in China, Japan and Korea. Kampo medicines have been often used for the treatments of hepatitis, menopausal disorder such as autonomic-nervous and hormonal manifestations, autonomic imbalance, bronchial asthma, cold syndrome, digestive disorder, atopy, dermatitis, eczema, sensitive to the cold, allergic rhinitis, and complaints of general malaise. Although these diseases are still difficult to treat by modern medicine, Kampo approaches are relatively effective by treating imbalances of the whole body. Kampo medicines are used in traditional formulations and each formula has a particular name. Although Kampo medicines have a long clinical experience that supports a belief in their efficacy and safety, Kampo medicine must be evaluated by scientific clarification of mechanisms of action and the active ingredients.

According to Chen et al., honed by clinical wisdom accumulated over thousands of years, a time-tested system encompassing theory, system, structure, education, examination, and practice has evolved for Traditional Chinese Medicine (TCM). Characteristic of Oriental tradition and heritage, Chinese Medicine is unassuming yet it has faithfully and steadfastly protected the health of the Chinese and neighboring populace over thousands of years. In Mainland China, Chinese medicine has never been relegated to the status of alternative medicine. Instead the government has spearheaded the modernization of Chinese Medicine since 1950. Instead of pitting the two major health protection systems, viz. Chinese medicine and the Western system, against each other by confrontations and competitions, the effort of recent years has been towards integrating the complementary aspects of the two systems into a comprehensive system, that may be referred to as integrative medicine. Young doctors, known as East-West Integrative Medical doctors ("New Generation" Chinese Medicine medical doctors in Taiwan) are trained in both disciplines to offer primary and secondary medical care to the community. The results are increasingly recognized and accepted by Western nations.

The increased use of herbal medicinal products (HMP) in Europe requires a thorough review of appropriate guidelines for evaluation of their safety and efficacy, i.e. therapeutic benefits. The generally accepted and well-structured quality management is formally laid down in the European Pharmacopoeia, where an increasing number of monographs for HMP and the products thereof are documented (see Chapter by Franz). The quality of HMP defines the optimal specifications for each compound by specifying one or more analytical procedures, which, in association with the limits of acceptance, define qualitative (identification) and quantitative (limits for impurities and content of active ingredient) characteristics for the HMP. Since it is known that HMPs adulterations are a common problem, sensitive analytical methods have been developed. Frequent contamination with organic compounds, heavy metals or microorganisms clearly demonstrates the need for highly sophisticated and sensitive procedures to guarantee a safe and non-toxic product for the patient.

The proper use of good quality products can reduce most of the risks linked to Traditional Phytotherapy (See Chapter by Minelli et al.). Intensive work is essential in order to increase the information about safety and the proper use of Traditional Phytotherapy. For example, adverse effects may arise as a consequence of the interaction between herbal medicines and chemical drugs. There are still many patients who do not inform their allopathic doctors that they are taking herbal medicines. By means of information, education and communication these problems can be overcome. The WHO Collaborating Center for Traditional Medicine of the University of Milan faces all these aspects of Traditional Phytotherapy with a number of activities such as research, educational training, and legal advice to several public institutions. Indeed, the Research Centre in Bioclimatology, Biotechnology and Natural Medicine is one of the oldest centers within the University of Milan. It was founded in 1969, and it refers to the Departments of Normal Human Anatomy, Biochemistry, Microbiology, Medical Chemistry and Thermal Medicine. Since its establishment the Center has dealt with issues concerning the QOL (quality of life), and it has developed research and training in this field. Particular attention has been given to natural and complementary medicines. In the US, there is a National Center for Alternative and Complementary Medicine, in Bethesda, Maryland. This Center that may become a full-fledged institute of the National Institutes of Health, actively funds peer reviewed proposals for research in the field.

## CAM APPROACHES TO SPECIFIC DISEASES AND BIOMEDICAL CONDITIONS

This second section on CAM approaches specific diseases and biomedical conditions taking a global approach that begins with environmental concerns and the causes of disease. Despite significant advancements in medicine and identifying the human genome, mainstream practice has paid less attention to early diagnosis and prevention of complex diseases (e.g. GI, cardiovascular, autoimmune) (See chapter by Vojdani and Cooper). These diseases cannot be ascribed to mutations in a single gene; rather they arise from the combined action of many genes, environmental factors and risk-conferring behaviors. Complementary medicine has an enormous challenge: to determine how environmental factors (infections and xenobiotics) interact. Effective strategies for diagnosis, prevention and therapy should then emerge. Understanding the mechanisms of environmental actions and role of laboratory examinations will demonstrate how: 1) introducing indigenous bacteria into the intestine of germ-free mice modulates gene function; 2) exposure to bacteria, yeasts, molds and viruses and release of exotoxins/ enzymes result in unfavorable mucosal immune function; 3) infectious agents, (e.g. oral bacteria), can induce cardiovascular and autoimmune disease by molecular mimicry or super-antigen induction; 4) xenobiotics, activate specific genes; 5) Lab tests can reveal neuroimmunotoxicity induction by metals.

According to the Joint United Nations Program of HIV/AIDS (UNAIDS) and the World Health Organization (WHO) (Joint United Nations Program of HIV/AIDS, 2001), as of the end of 2001, there were about 40 million adults and children living with human immunodeficiency virus (HIV) infection. (See chapter by Cossarizza and his students). This total does not include the 20 million people around the world who have already died of AIDS. Of the 40 million currently alive, 37.2 are adults, 17.6 are women, and more than 2.7 are children. In 2001, there were 5 million new cases of HIV infection in the world, and 3 million AIDS related deaths. The large majority (almost three-quarters) live in Sub-Saharan Africa where the prevalence rate of the infection among adults is 8.4%, with women comprising more than 55% of-infected individuals. The second major pocket of HIV infection is in South and Southeast Asia, with more than 6 million people infected. In North America, where the epidemics were first described, HIV+ individuals are 940,000 with 560,000 in Western Europe. Furthermore, South America, China, and Eastern Europe are characterized by a rapid increase in infection rates. These dramatic numbers clearly indicate that the fight against HIV/AIDS is an absolute health, social, economical and political priority in all parts of the world. In 1995, the introduction of highly active antiretroviral therapy (HAART) for HIV-infected patients had a terrific impact and changed the story of the infection. In a consistent percentage of cases HAART can efficiently suppress HIV production, by using the strategy of inhibiting different steps in the viral cycle and replication with the combination of drugs.

According to WHO over 300 million people contract sexually transmitted infections every year. In developing countries about 30% of all women suffer from abnormal vaginal discharge due to reproductive tract infections (RTIs) caused by various pathogens (see chapter by Talwar). Dr. Talwar has developed a polyherbal formulation (Praneem) dispensed as a tablet, pessary or cream for topical vaginal use; it acts on a wide spectrum of genital pathogens. It inhibits the growth of *N. gonorrhea* including penicillin resistant strains, *Candida albicans, Candida krusei, Candida tropicalis* and urinary tract *E. coli* (including multi-drug resistant strains). Phase I trials (4 centers, India and in Salvador Bahia Brazil) show that Praneem is safe and devoid of local or systemic side effects. With permission from the Drug Regulatory Agencies and Ethical Committees, Phase II efficacy trials have been conducted in women with abnormal discharge due to pathogens. A daily treatment of Praneem over 7 days cured symptoms in 95% of women irrespective of the causative organisms.

Ravindranath and his colleagues now present the results turning to the immune system and cancer. Genistein ($4'5$, 7-trihydroxyisoflavone) occurs as a glycoside (genistin) in the plant family Leguminosae, which includes the soybean (*Glycine max*). A significant correlation between the serum/plasma level of genistein and the incidence of gender-based cancers in Asian, European and American populations suggests that genistein may reduce the risk of tumor formation. Other evidence includes the mechanism of action of genistein in normal and cancer cells. Genistein inhibits protein tyrosine kinase (PTK), which is involved in phosphorylation of tyrosyl residues of membrane-bound receptors leading to signal transduction, and it inhibits topoisomerase II, which participates in DNA replication, transcription and repair. Furthermore, genistein can alter the expression of gangliosides and other carbohydrate antigens to facilitate their immune recognition. Genistein acts synergistically with drugs such as tamoxifen, cisplatin, 1,3-bis 2-chloroethyl-1-nitrosourea (BCNU), dexamethasone, daunorubicin and tiazofurin, and with bioflavonoid food supplements such as quercetin, green-tea catechins and black-tea thearubigins. Genistein can augment the efficacy of radiation for breast and prostate carcinomas. Because it increases melanin production and tyrosinase activity, genistein can protect melanocytes of the skin of Caucasians from UV-B radiation-induced melanoma. Genistein-induced antigenic alteration has the potential for improving active specific immunotherapy of melanoma and carcinomas.

Traditional Chinese Medicine (TCM) has so far been known to be useful for cancer patients as both an anti-tumor and an immuno-enhancing drug (chapter by Wago and Deng).

If some kinds of crude drugs of TCM possess such biological activities, they would be of great help in cancer treatment. For example, recently, *Astragal radix* has been shown to possess immune-enhancing and anti-tumor effects and *Angelica radix* has also shown anti-inflammatory, analgesic, interferon inducing, and immunopotentiating effects. In addition, *Zizith fructose* augments natural killer cell activity, while *Cervix arum* corn exerts anti-tumor effects by inhibiting monoamine oxidase. *Ammo semen* and *Romania radix* effect anti-tumor activity and enhance the function of T lymphocytes with the increase of IL-2. This work tested a mixture of these six kinds of crude drugs used for cancer patients in China named Ekki-Youketsu-Fusei-Zai (EYFZ), and its biological activities. They tested in vitro functional activation of macrophages and T lymphocytes, in vivo effects on natural killer (NK) cell activity and life-prolongation in tumor-bearing mice. This strategy was adopted in order to elucidate relationships between immuno-enhancement and anti-tumor influences from the viewpoints of alternative and complementary medicine.

Alternative and complementary medical therapies aim for a balanced state of wellness or homeostasis in patients that translates into optimal adaptation to physical and psychological environments (see chapter by Roberts). In mammals, stressor exposure results in rapid, reflexive central nervous system (CNS) activation as part of the sympathetically mediated "fight or flight" response. The inhibitory neurotransmitter, gamma-aminobutyric acid (GABA) is a key regulator of the integrated stress response, and low levels are implicated in the pathophysiology of diseases of adaptation: depression, bipolar disease, and schizophrenia. GABA acts on many levels of the CNS as a flexible gate governing neuronal responsiveness to external stimuli. For example, GABAergic neurons of the spinal cord function in pain stimulus gating and their function can be altered by cutaneous exposure to an agent that elicits inflammation such as environmental stimuli and behavior; therefore it is a powerful pathway for self-modification of individual pain and stress responses. The popular natural sedative and hormone, melatonin, interacts with GABA by inducing anxiolytic and analgesic effects, as well as setting circadian and seasonal rhythms. Acupuncture alters GABAergic function in the spinal cord; this suggests a mechanism for inducing pain relief by needle or electrostimulation. Interactions between GABA and other neural and endocrine players reveal insights into homeostatic mechanisms essential for optimal wellness in a changing environment.

Turning now to childbirth, Tournaire presents a chapter on alternative approaches to pain during labor and delivery. Even if delivery is a natural phenomenon it has been demonstrated that the associated pain is considered too severe in more than half of the cases. Beside medical technology such as epidural anesthesia many alternative or complementary methods have been described throughout history. These can be divided into psychological, physical and chemical. Examples of the first group: "delivery without pain" or Lamaze method, sophrology, birth without violence of Leboyer, haptonomy. In the second group we find acupuncture reflex-therapy, electrical anesthesia, delivery in water. Homeopathy encompasses the third group with its use of plants. For centuries, pain associated with delivery was denied or considered a part of god's creation that had to be accepted. Only recently has this pain like others been better analyzed and more or less efficient means are now proposed for its control.

This section ends by giving examples of alternative and complementary medicine in dentistry (see Komiyama). Alternative and complementary approaches are of enormous importance in all areas of human health. Each biomedical discipline can now combine technologies and ideas with those that are more traditional. It is well known that almost all dental treatments are alternative, such as dentures, where implants replace portions of missing teeth. In every instance there is a concern for minimizing the risks of tissue injury and plaque accumulation in accordance with modern concepts of preventive dentistry. Chronic oral pain experienced by patients suffering from oral disorders is associated with oral refractory lesions and psychological factors. As one approach, dentists have tried Chinese medicine for such intractable pain and achieved satisfactory results. There is another problem called dry mouth. This occurs in patients with Sjögren's syndrome after cancer surgery and in aged individuals. Sjögren's syndrome involves atrophy of salivary glands partially caused by lymphocyte infiltration (autoimmune disease) into them. Dentists are now able to introduce a treatment for patients that involve artificial saliva and galenicals.

## PHYSICAL INTERVENTION: TOUCH, HYDROTHERAPY AND SOUND

The section on physical intervention: touch, hydrotherapy and sound begin with acupuncture. Peter White's study aims at determining whether acupuncture treatment is superior to a placebo intervention for chronic mechanical neck pain. All patients were treated twice a week for 4 weeks, with 20-minute treatments. The acupuncture group receiving largely individualized treatment with an average of 6 needles each session. The placebo group received mock electrical stimulation to acupuncture points using a decommissioned electro acupuncture stimulation unit via TENS electrodes. This was used as a comparison group for mean pain at 1-week post treatment using a VAS. Both groups significantly improved from the baseline ($p < 0.0005$) but no significant difference was found between the groups at 1-week post treatment ($p = 0.106$). All secondary outcomes showed a similar pattern. Acupuncture was shown to be effective at reducing neck pain in the short and long term, but performed no better than the placebo treatment. The improvements in outcomes are therefore probably due to either the non-specific effects of treatment or regression to the mean, rather than the process of needling.

With respect to disease, the onset of atomic dermatitis usually begins at infancy or childhood and some cases experience spontaneous cure before adolescence (Chapter by Abo). Primarily, children show a high level of lymphocytes that gradually decreasing in adults. The gametocyte levels surpass those of lymphocytes at 15–20 years old in Japanese. Apparently, age-associated changes of the immune system are responsible for the spontaneous cure of atopic dermatitis. In other words, many cases of atopic dermatitis naturally subside due to an age-dependant decrease in the number of T cells. Such gametocytes are further activated by many inflammatory cytokines, as well as by sympathetic nerve stimulation after the interaction with resident bacteria. Although the precise reasons are not yet determined, granulocytosis seen in patients with severe atopic dermatitis and in the withdrawal syndrome during therapy is interesting. Several factors must be considered such as granulocyte function as well as the functions of IgE, T cells, and eosinophils, in severe cases of atopic dermatitis.

Along with treating illness, one aim of alternative medicine is to promote quality of life (QOL) in healthy people. In Japan, centuries of tradition have shown that alternative therapies like hot spring hydrotherapy, acupuncture, and herbal medicine enhance the QOL of empirically healthy individuals (Chapter by Yamaguchi). Accumulating evidence suggests immune system regulation. The scientific basis, however, has not yet been established.

Results reveal that CD positive cell numbers and cytokine producing cells changed before and after hot spring hydrotherapy. Thus these subsets could reflect the number and function of immuno-competent cells. For example, in an individual with a low CD16 cell number, the number increased after treatment while it decreased in another individual with a high CD16 cell number. Leukocyte subsets may be an interesting indicator for the evaluation of alternative therapy. Many systems are in place to evaluate Western therapies that aim at healing the symptoms of an illness. However, when the purpose of a therapy is to enhance the QOL of healthy people, such as some alternative medical therapies, no widely accepted evaluation system has been established. To fill this lack, it is essential to propose numbers and functions of leukocyte subsets as indicators for evaluating alternative therapies.

Music therapy (MT) is an established health profession using music and music activities or interventions to address physical, emotional, cognitive, social, psychological, and physiological needs of infants, children, adolescents, and adults including the elderly, with disabilities or illnesses (Chapter by Scheve). MT may also be used with healthy children and adults to maintain wellness throughout life. MT builds on the power of music, using music in a focused and concentrated way for healing and for change. It is a non-invasive medical treatment designed to prevent illness and disease, alleviate pain and stress, help people express feelings, promote physical rehabilitation, positively affect moods and emotional states, enhance memory recall, and provide unique opportunities for social interaction and emotional intimacy. Concerning history of music as medicine, one may only speculate how far back the use of music as medicine dates. Written history dating back to 6,000 B.C. from Iraq identifies music's role in magic, healing ceremonies, and medicine. Evidence of the use of music in other ancient cultures exists as well. Music was used in chant therapies in Egypt circa 5,000 B.C., and in Babylonian culture circa 1850 B.C. in religious ceremonies as treatment for sin, which caused illness. In Greece circa 600 B.C., music was prescribed for emotionally disturbed individuals. In recent history, the first acknowledgement of the use of music in general hospital treatment came in 1914 by the American Medical Association. Today, over 5,000 music therapists are employed throughout the United States in over 50 different settings including hospitals, clinics, day care facilities, schools, community health centers, substance abuse facilities, nursing homes, hospices, rehabilitation centers, correctional facilities, and private practices.

Bolstered by experimental evidence, music has been shown to modulate neuroendocrine molecules (Chapter by Wago and Kasahara). The hypothalamic-pituitary-adrenal (HPA) axis, melatonin, prolactin, and biogenic amines (serotonin and catecholamines) are involved in altering functions of neuroendocrines in the brain in response to stressful inputs. These neurohormones and neurotransmitters have also been implicated in regulating behavioral responses to various stimuli and in modulating aggression and depression. Hormones of the HPA axis as well as melatonin and serotonin modulate various physiological functions and behaviors including circadian rhythms, sleep, sexual behaviors, mood, and tumor growth. Neuroendocrines also modulate immune system as a consequence of stress or depression. A molecular basis for reciprocal communication between the immune and neuroendocrine systems has been described. It is now commonly known that neuro-molecules, such as ACTH, dynorphin, dopamine, growth hormone (GH), opioid peptides, prolactin (PRL), thyrotropin (TSH), and others can modulate immune functions. stress responses of neuroendocrine, such as an increase of plasma corticosterone concentrations through ACTH secretion, could have deteriorating effects on specific cells and tissues that are required for optimum immunological defense. Music has been found to palliate stress induced hyperactivity of HPA axis, involving ACTH and corticosteroid secretion as well as to modulate norepinephrine, epinephrine, GH, PRL, and endorphin secretion. With regard to hormonal changes and behavioral modifications, reduced levels of plasma cortisol among night shift nurses and obstetric patients accompany the anxiolytic effects of music. Thus, recent investigations on music therapy intervention for stress and behavioral modification is directed toward measurements of changes in stress-related neuroendocrine molecules such as cortisol. These studies suggest that the relaxing effects of music are critically involved in stimulation of neuroendocrine systems.

## DIETARY INTERVENTION IN SPECIFIC DISEASES

Clearly dietary intervention is involved in the amelioration of disease (Chapter by Kobayashi). Atopic dermatitis is a disease whose main symptom is chronic recurrent eczematous lesions. Increased incidence and progression to intractable cases have become problems in recent years. The first choice of treatment is the detection and elimination of allergens and aggravating factors, and external treatment, mainly using steroids and proper skin care. For cases that do not resolve despite these efforts, a concomitant Japanese traditional herb treatment has been used for more than 20 years. Japanese traditional medicine (JTM) has developed independently in Japan based on medicine introduced from China via Korea in the 7th century. It was in the main stream of public health care until the latter half of the 19th century, but only Western medicine was incorporated into the official medical system. After passing through an era in which only a limited number of physicians performed JTM, it was re-incorporated into the medical care provided by health insurance in the 1970's. Since then, JTM has been used as a treatment concomitantly with modern Western medicine in all fields. Even in Western medicine, the dietary effect on skin symptoms has been reported such as Lind's discovery regarding prevention of scurvy by ingestion of fruits in 1742. Later, skin lesions caused by deficiencies of vitamin C, vitamin B complex, vitamin $B_6$, pantothenic acid, and linolenic acid have been reported.

A considerable time has passed since the gradual increase of refractory atopic dermatitis and its emergence as a social issue in Japan (chapter by Ishii). Compared with the past, the disease appears to already be refractory in children and most refractory in the adults. Treatment of atopic dermatitis by Kampo (Japanese Herbal Medicine) therapies including dieting, has achieved better results than by Western treatments alone, but it is difficult to demonstrate the genuine effectiveness of Kampo therapies. Therefore, Kampo therapies including diet are effective in the most refractory cases, i.e. severe cases with persistent rash.

Turning now to India, about seventy percent of the population lives in rural areas and urban slums (see Kumar). Many are malnourished and suffer from various ailments. To address the nutritional and health needs of the poor (most of them vegetarian) in a sustainable manner, the following ten plants have proven particularly useful: *Spirulina* (cyanobacterium), neem (*Azadirachta indica*), turmeric (*Curcuma longa*), *Allium sativum, Aegle marmelos, Trigonella foenicum graecum, Piper nigrum, Eugenia jambolina, Momordica charantia* and *Zingiber officinale*. In different permutations and combinations, these plants relieve protein-calorie malnutrition, anemia, xerophthalmia, goiter, mild infections, cardiovascular disorders, gastrointestinal disorders, respiratory problems (cough, cold) and borderline cases of diabetes. Villagers in the vicinity of saline-alkaline wastelands can be trained to grow *Spirulina* themselves. Besides the above plants, regular daily consumption of three to five different-colored fruits/vegetables provides antioxidants and promotes all round good health.

## BASIC SCIENCE: FUTURE APPROACHES TO NOVEL MOLECULES FOR CAM

This section is fascinating and perhaps the most unique section in the book. This basic science approach offers a beginning and a long-range prediction of what might occur if we harness the countless number of substances from terrestrial and aquatic species as new therapeutic aids. We should not forget that there is a substantial literature concerned with phytotherapy. However, derivatives from animals require more attention. For example Muller proposes the term evochemistry as efficient natural strategy to develop bioactive compounds for medical use. In the last 20 years combinatorial chemistry, mainly aimed to increase molecular diversity, has discovered bioactive compounds for medical therapy after thorough screening. Now, after some fruitful approaches this approach is flanked by a new rational, evochemistry. The evochemical strategy utilizes the evolutionary pressure that exists in some animal, fungal and plant phyla and results in the development of highly bioactive compounds. As an example, sponges diverged first from a common metazoan ancestor, the Urmetazoa, approximately 800 million years ago. They are the richest animal phylum regarding bioactive compounds. Besides sponges, fungi have also gained interest for medical application. One example is the chitin-containing extract Neo-Immune, which turned out to be an effective immune enhancer. It is suggested that the evochemical approach will result in the discovery of powerful bioactive compounds acting against specific human diseases.

According to Cooper and his colleagues in Beijing, earthworms offer another rich source of molecules that have been characterized and shown to be strongly lytic against a variety of tumor cell targets grown in vitro. Apparently there is a need for molecules to be used in clinical situations, e.g. thrombosis. Lumbrokinase (LK) is a group of proteolytic enzymes, including plasminogen activator and plasmin, separated biochemically from earthworms. Few people know of the earthworm's long association with medicine and various remedies since 1340 A.D. For example, in Burma, earthworms have been used to treat a disease called ye se Kun byo, (e.g. symptoms of pyorrhea). These old remedies may be resurrected in the current climate of complementary and alternative medicine.

Roch poses the question: What can we learn from marine invertebrates to be used as complementary antibiotics? The main problem in infectious diseases is an alarming increase in the resistance of microorganisms to classical antibiotics. We constantly need new molecules but most of them are derived from already used compounds. A better approach consists of developing molecules that possess totally different modes of action. Due to their life in hostile environments, marine invertebrates may have developed particular antiinfection systems and/or molecules. For instance, DNA polymerase from abyssal hot spring bacteria functions up to 90°C, which is extremely valuable for biotechnology. Shrimp and mollusks possess antimicrobial peptides routinely used by their innate immune system to prevent infections. Dissecting peptides produces various fragments with interesting in vitro antibacterial (*Staphylococcus*), antifungal (*Fusarium*), antiprotozoan (*Trypanosoma, Leishmania*) and antiviral (HIV) properties. Introducing modifications during chemical synthesis modulates activities, without inducing cytotoxicity. As patented molecules, their use in complementary medicine is under consideration.

As linkages, Kasahara and Cooper present the nervous, endocrine, immune systems as a target for complementary and alternative medicine. Many aspects of complementary and alternative medicine connect the three integrative systems: nervous, immune and endocrine. Alternative and complementary medicine, as we know, although controversial, is becoming a "hot topic" and is clearly recognized and acceptable in many academic and professional circles. We are reminded that homeostasis is a fundamental process that is essential for life under various stressors. Human health in a highly variable environment is dependent upon the proper balance of physiological processes. According to emerging views, homeostasis may be achieved by the coordinated activities of the three major integrative systems: the nervous, endocrine, and immune systems. The immune system has for a long time been considered as independent, but recent numerous studies have revealed it to be in a delicate state of balance with neuro-endocrine system. There are a number of animal models that will allow us to dissect the gene control of modulation between neuro-endocrine and immune systems in relation to stress and diseases.

## EDUCATION AND PHILOSOPHY

As Wallis points out clearly there is a place for scientific thinking when considering the alternative and complementary approach to medicine. Many philosophers and medical researchers dismiss placebo effects as the result of false reporting due to subject expectations. Placebo effects may measure changes in socially learned behavior affected by a belief in the treatment. Still others claim that the placebo effect is due to an illness or injury taking its natural course. There are three different placebo debates: Freudian psychotherapy, pain perception, and depression. First, research suggests that a complex range of interactive causes are responsible for placebo effects that may constitute a spectrum from clinical chimeras to measurable neurobiological changes. Second, using this conception of placebo effects, objections do exist by mainstream scientists to research methods and results in several areas of alternative and complementary medicine. Third, controlling the range of possible sources of placebo effects is a challenge for mainstream science and alternative and complementary medicine.

Finally Loel introduces in his proposed new glycome, a medical paradigm. He rightly asserts that humans have the compassion, empathy, and inventiveness to bring homeostasis to unfortunate individuals who lose their equilibrium. Throughout history we have sought cures from the environment. Many therapies defy logic, science, and reality, but they provide relief to some who partake of them. Gods, herbs, laying-on of hands, bleeding, burning, sticking, starving and feeding have all been tried. Some are with us today in the form of acupuncture, chiropractic, homeopathy, and herbalist, spiritual and nutritional therapies.

Computer technology has opened a new frontier-glycobiology. The new alternative therapy, glyconutrient support of cell-to-cell communication is essential for a healthy immune response.

Edwin L. Cooper, Ph.D., Sc.D.
Professor
Laboratory of Comparative Neuroimmunology
Department of Neurobiology
David Geffen School Of Medicine at UCLA
University of California, Los Angeles
Los Angeles California 90095-1763
Tel: (310) 825-9567; Fax: (310) 825-2224
email: cooper@mednet.ucla.edu

# Introduction

# The Urgent Need for Evidence-Based CAM in the World

Nobuo Yamaguchi

## What is the Best Way to Measure the Efficacy of CAM?

Recent excessive commercialization is particularly confusing for patients and doctors who seek remedies for heretofore undefined symptoms. Furthermore, since these treatments have not undergone strict testing, they are not always safe and the same drug may have different effects according to the individual patient and dosage. Complicated considerations are necessary for the application of practices such as those found in Chinese traditional medicine.

In Eastern countries, such as China and Japan, the terms Alternative and Complementary do not accurately describe healing modalities that have been practiced for centuries, many of which have been authorized for treatment under national insurance systems. Even with differing points of reference, the growing global demand for these treatments has given scientists and practitioners in the East and the West a responsibility to scientifically assess CAM to ensure its safe and effective use in the future. This new demand includes aromatherapy and herbal medications, acupuncture, moxibustion and yoga, among others. These therapies, however, have not been well defined and often have been applied traditionally without scientific evidence. Each needs to be assessed by scientific methods especially developed by Western medicine. The close relationship that has been identified among the neuropsychological network, endocrine system and peripheral white blood cells points to the number and function of leukocytes as well as immunocompetent cells as one of the most positive scales for CAM.

## CAM in the East and the West

The Japanese Society for Alternative and Complementary Medicine defines alternative and complementary medicine as "the modern Western medicine which has not been

---

**Nobuo Yamaguchi** • Organizer for International Symposium on Alternative and Complementary Medicine, November 8–10, 2002, Kanazawa, Japan • President of IMPRC, Ishikawa Natural Medicinal Products Research Center

*Complementary and Alternative Approaches to Biomedicine*, edited by Edwin L. Cooper and Nobuo Yamaguchi. Kluwer Academic/Plenum Publishers, 2004.

verified scientifically and is not regularly practiced in clinics." In the U.S., it is called an alternative medicine or (CAM) complementary and alternative medicine and comprises healing modalities which are not traditionally included in the curricula of medical schools or regularly practiced in Western-medicine based clinics. Using different words, medicine can be classified as either conventional or unconventional. Conventional medicine is equivalent to modern Western medicine, scientific medicine or technical medicine, while unconventional medicine is alternative, complementary, natural or fringe medicine. In Japan, alternative medicine includes traditional Chinese medicine, acupuncture and *judo* reduction etc., therapies originating in the East, each with its own long, independent history. Since herbal medicine is often covered by Japanese health insurance, some insist that it is not an alternative medicine, even though it belongs to that category in the West. The World Health Organization classifies 65~85% of the medicine practiced in the world as "traditional". These traditional medicines become "alternative" when they are practiced in the West.

## Alternative and Complementary Medicine in the Future

In Japan, herbal medicine (Kampo) was the primary medical therapy until the Meiji Era (1867–1912) when the government decided to import Western medicine for conventional care. With this long experience in both Eastern and Western healing methods, Japan should take the initiative in standardizing CAM. How should this standardization be achieved? I would like to propose that most alternative medicine works through affecting the regulatory system of the body. Recent studies, for example, revealed that herbal medicine caused interactions among the immune system, the endocrine system and the nervous system. Thus, observing the immune system by sampling the peripheral blood could be an excellent indicator for the standardization of these alternative therapies. Without standardization, this realm of medicine has often been shut out of the serious journals of Western medicine. The symposium that spawned the research in this book was instrumental in launching a new journal for evidence based CAM. It is our desire that this journal encourages the publication of original scientific papers based on sound scientific guidelines, but without prejudice against the possible efficacy of these new and ancient treatments.

# Beginnings

## Patty Christiena Willis

The idea of an international symposium was first proposed to me on a cold day in March of 2002. Professor Nobuo Yamaguchi had invited Dr. Toyoshi Onji of Oxford University Press to our part of western Japan for a feast of oysters. Auspiciously, we were wearing bibs to facilitate the consumption of this delicacy of the Noto peninsula. This is Professor Yamaguchi's *modus operandi*: the best regional dishes and wine or sake always accompany momentous discussions. When Professor Edwin L. Cooper reminisces of their 25-year-old friendship, his stories are dotted with unusual and sometimes heavenly food and drink. It was at this point, our faces hot from the charcoal fire and pink with new rice wine that Dr. Onji announced that a successful symposium in November of that year was the only chance for the launching of a new international journal for complementary and alternative medicine.

The dream was born many years ago with Professor Tomio Tada, a world-renowned immunologist who as he aged grew closer to the traditional Asian therapies disdained by most Japanese trained in Western medical science. A playwright as well as a scientist, his sharp eyes and ears observed the growing need for Western medicine to loosen its vise-like grip on healing. With his wide circle of colleagues in and out of academia, a plan was generated to disseminate and gather information about new as well as old non-Western therapies. Talk of an international journal caught the attention of Toyoshi Onji at OUP who watched as the ideas progressed. Professor Yamaguchi offered the resources of his organization, the Ishikawa Natural Products Research Center (INMPRC), as a main source of funding for the effort. Yet, just as the plan was nearing fruition, illness and administrative difficulties drained the life energy from the project until the last hope was discussed over oysters. A symposium.

Almost immediately, Edwin Cooper was contacted for he was known not only as a symposium organizer extraordinaire but also as a friend and compatriot to many on the Asian side of the Pacific. In the following months we wrote many times a day as I moved between homes in Japan and the US and Edwin traveled from California to Germany and France and then back. Professor Yamaguchi egged us on, wishing that he could take us out for *foie gras* and salmon steaks with a young Burgundy and in the stead of food promising financial support as Edwin's Herculean efforts pushed the number of invitees from five to

---

**Patty Christiena Willis** • Founding International Administrator
*Complementary and Alternative Approaches to Biomedicine*, edited by Edwin L. Cooper and Nobuo Yamaguchi. Kluwer Academic/Plenum Publishers, 2004.

twenty-two. We all forged on, calling on the talents of many friends such as Lisa LeRose who constructed a beautiful website and pamphlet for the symposium and Izumi Kadoshima who designed the flower arrangements that adorned the halls of the symposium. Yasuko Fukamura, a kimono designer, agreed to ferry the accompanying persons to places rarely seen by first-time visitors and Mary Lou Prince gave her generous moral support and real talent as she acted as the banker to handle the growing funds.

November 7th arrived quickly and invitees gathered from India, China, Taiwan, Europe and the United States, each feeling like an old friend after our months of correspondence. They all brought their unique histories and expertise, coming together for the first time in the conference halls in Kanazawa. Preoccupations behind the scenes kept me from attending all the meetings but when I did, what moved me most was the energy and flow of ideas. Sometimes Edwin caught my eye and we smiled. In only six months, we had brought these people together and now they were doing their part by contributing their ideas and most importantly, their energy. In a favorite Native American legend, the energy of the movement of a dancing rabbit generates the creation of humankind. All those who participated in the symposium provided the energy that would gain speed and evolve into a new creation, an international journal dreamed of many years ago. Many thanks to all. Arigatou gozaimashita.

# National Approaches to CAM

# 1

# Complementary and Alternative Medicine in Japan

Nobutaka Suzuki, Satoshi Ohno, Tsutomu Kamei, Yumiko Yoshiki, Yuji Kikuchi, Kazuyoshi Okubo, Tomihisa Ohta, Shoji Shimizu, Saburo Koshimura, Atsufumi Taru, and Masaki Inoue

## 1. Introduction

In recent years, complementary and alternative medicine (CAM) has grown in popularity worldwide. The World Health Organization (WHO) classifies 65–80% of the world's health care services as "traditional medicine"[1]. Therefore, from the viewpoint of the population ratio, more people use CAM than modern Western medicine.

Although much attention has focused on CAM not only from physicians but also basic medical researchers, the scientific evidence for most of CAM is sparse. The emergence of CAM as a current of new medicine depends on whether precise scientific evidence can be accumulated or not.

Why has CAM become so popular among consumers? We think that the following characteristic factors of CAM are responsible:

- It is easy to understand and familiar
- Non-invasive, few side effects
- Helps one to maintain one's own health
- Western modern medicine does not fully correspond to the patients' demands.

In this chapter, the current status of CAM in Japan will be explained with special reference to a new CAM department.

## 2. Classification of CAM

CAM, as defined by The National Center for Complementary and Alternative Medicine (NCCAM) in the USA, is a group of diverse medical and health care systems,

---

**Nobutaka Suzuki** • Author. Director of Laboratories, Department of Complementary and Alternative Medicine, Kanazawa University Graduate School of Medical Science

*Complementary and Alternative Approaches to Biomedicine*, edited by Edwin L. Cooper and Nobuo Yamaguchi. Kluwer Academic/Plenum Publishers, 2004.

**Table 1.1.** NCCAM classifies CAM therapies into five categories, or domains

---
**1. Alternative Medical Systems**
Traditional Chinese medicine, Ayurveda, Homeopathic medicine, Chiropractic, Naturopathic medicine etc.
**2. Mind-Body Interventions**
Meditation, Prayer, Mental healing, Art, Dance , Music therapy etc.
**3. Biologically Based Therapies**
Foods, Herbs, Vitamins, Dietary supplements, Aromatherapy etc.
**4. Manipulative and Body-Based Methods**
Chiropractic or Osteopathic manipulation, and Massage etc.
**5. Energy Therapies**
**Biofield therapies**──────────── qi gong, therapeutic Touch
**Bioelectromagnetic-based therapies**── electromagnetic fields

---

practices, and products that are not presently considered to be a part of conventional medicine. NCCAM classifies CAM therapies into five categories, or domains (Table 1.1).

In Japan, a part of anti-aging medicine (growth hormone treatment etc), lifestyle drugs (remedies for alopecia, obesity, and impotence etc) and environmental medicine are involved in CAM treatment. For example, developing a functional food which absorbs dioxins and promotes discharge into facilities or developing a filter to trap the free radicals of automobile exhaust gas, etc, are included in the category of environmental medicine.

The CAM domain contains advanced medical treatments, which are not yet widely carried out, such as immunotherapy for cancer using tumor specific antigens, etc. In-vitro fertilization and embryo transfer (IVF/ET), one of the most advanced medical technologies, was regarded as CAM at the very beginning. Now, however, IVF/ET is widely used in usual hospitals, and it can be said that CAM had been taken into conventional medical treatment.

Unlike Western countries, in Japan, a part of Kampo medicine and acupuncture is covered by public health insurance. Therefore, some Japanese object to their inclusion in CAM. However, these treatments are categorized as CAM in Europe or the USA. Thus, the definition of CAM differs slightly from country to country.

In Japan, after the 3 years of education at a professional school of acupuncture, moxibustion, acupressure or massage, one can get a license after passing the national examination. However, any doctor can perform one of the above treatments even if he or she hasn't attended one of these schools.

Judo has been very popular from old times. Because injuries such as fracture or dislocation were frequent among judo wrestlers, the government recognizes the occupation of Judo-Orthopaedics. To get a national license for Judo-Orthopaedics, one must be educated for 3 years. Chiropractic and osteopathy, which are widely used in the USA, are not authorized by the Japanese government.

The patients who use CAM are not only human beings. Recently, the popularity of pet ownership is rising in Japan. Since a pet is usually considered to be a member of the family, the owner is concerned that the pet receives CAM treatment. In response to the owners' strong desires, CAM treatment by veterinarians is on the rise. Furthermore, many companies are researching and developing dietary supplements for animals.

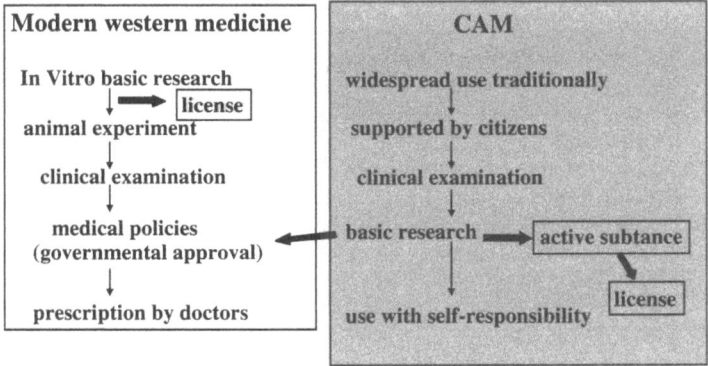

**Figure 1.1.** Flowchart of Scientific Method

## 3. How to Research CAM?

CAM domains should be studied using the scientific research methods of modern western medicine. However, the verification procedure for CAM differs somewhat from that of Western modern medicine.

Figure 1.1 shows a flowchart of the scientific method. Most CAM is already widely used without scientific endorsement. Therefore, the first step of CAM research should be focused on clinical testing to confirm its efficacy and usefulness. Of course, before performing a clinical trial, it is necessary to finish the required safety testing, such as toxicity tests on animals. If the results of a clinical test are desirable, fundamental research to identify and elucidate the action of the substance will be needed. CAM is usually done at the user's own expense, but a part of it may be able to get the approval of the government. In the case of modern Western medicine, the order of these steps is reversed.

The interesting feature of CAM research for the physician is to encounter new medical methods, which have high practicality, as well as high cost performance and safety. Furthermore, from the viewpoint of basic medicine, CAM offers unexpected opportunities to discover a new substance or a new mechanism.

## 4. History of CAM in Japan and the USA

First of all, we want to show the reason for our participation in this field. In 1990, Dr. Suzuki, who is a physician of obstetrics and gynecology and one of the authors of this chapter, worked at Keiju General Hospital, which is located in Noto Peninsula of Japan. One day, he met a curious case. He had planned to operate on an 18-year-old patient who suffered from condyloma acuminatum. Condyloma acuminatum is a troublesome disease for Ob/gyn physicians. It is hard to cure completely, and after surgery, the patient still feels much pain. On the day of the operation, he and his colleague saw that the lesions had disappeared completely. There should have been many condyloma lesions which had spread from the outer genitalia to the vaginal wall. Soon, the reason for her spontaneous recovery became clear. Her grandmother had given her a kind of food that consisted of

**Table 1.2.** History of CAM (Japan and USA)

| | | |
|---|---|---|
| 1990 | A small research group for CAM at Kanazawa University | (Japan) |
| 1992 | Office of Alternative Medicine (OAM) in NIH | (USA) |
| 1997 | Japanese Society for Complementary and Alternative Medicine (JCAM) | (Japan) |
| 1998 | National Center for Complementary and Alternative Medicine (NCCAM) | (USA) |
| | Japanese Association for Alternative, Complementary and Traditional Medicine (JACT) | (Japan) |
| 1999 | International Research Center for Traditional Medicine | (Japan) |
| 2001 | Japanese Society for Integrative Medicine (JIM) | (Japan) |
| 2002 | Department for Complementary and Alternative medicine, Kanazawa University Graduate School of medical Science | (Japan) |

**Table 1.3.** Japanese Societies in the field of CAM

The Japanese Society for Complementary and Alternative Medicine (JCAM)
The Japanese Association for Alternative, Complementary and Traditional Medicine (JACT)
The Japanese Society for Integrative Medicine (JIM)
Japan pre-symptomatic medical system society
Japanese Society of Aromatherapy
Japanese Association of Physical medicine, Balneology and Climatology
Japanese Society of Psychopathology of Expression and Arts Therapy
Japan Society of Acupuncture and Moxibution
Japan Traditional Acupuncture and Moxibution Society
Japan Dental Society of Oriental Medicine
Japan Society for Oriental Medicine
Medical and Pharmaceutical Society for WAKAN-YAKU
Japanese Society of Biofeedback Research
Society of Hemorheology and Related Research
Japan Medical Conference on Magnetism
Japan Music Therapy Association
Japan Biomusic Association
Japan Fasting Therapy Society
others

various natural products, and mainly contained the fermented coix seed (a kind of wheat). They got the coix seed from her grandmother, and they were sure that it had worked wonders on the condyloma. This was our first contact with traditional folk medicine. And now, we are investigating the fundamental aspects and clinical efficacy of this food.

Table 1.2 shows the history of CAM in Japan and USA. In 1990, soon after the above event, a small research group of physicians for CAM was born at Kanazawa University. In 1997, this group established the Japanese Society for Complementary and Alternative Medicine (JCAM). The members of JCAM are doctors, nurses, dentists, veterinarians, acupuncturists, nutritionists, pharmacists and researchers of various fundamental sciences etc. After the foundation of JCAM, the Japanese Association for Alternative, Complementary and Traditional Medicine (JACT) and the Japanese Society for Integrative Medicine (JIM) were successively initiated.

The medical societies in Japan relevant to CAM domains are shown in Table 1.3. Because most of these societies do not have a journal, they collaborate through discussion.

**Ayurveda**

**Figure 1.2.** International Research Center for Traditional Medicine

In 1999, the International Research Center for Traditional Medicine was founded in Toyama prefecture, about one hour from Kanazawa University. Research into Ayurveda or hot spring therapy, etc., is being conducted (Fig. 1.2).

In 2002, the CAM department was founded at Kanazawa National University. The details of this department are described later.

In the USA, to address the increasing need for research in CAM, Congress created the Office of Alternative Medicine (OAM) in 1992 at the National Institute of Health (NIH). Funds allocated for the OAM were 2 million dollars in 1992 and 1993, and have steadily grown to 12 million dollars in 1997[1]. In 1998, OAM came to be known as the National Center for Complementary and Alternative Medicine (NCCAM)[2]. Now, the funds for the NCCAM are over 100 million dollars. The NCCAM spends funds on CAM-related research, and the major part of this research goes toward areas already largely accepted by conventional scientists, such as antioxidants or dietary manipulation, etc.

## 5. CAM in the Field of Japanese Modern Western Medical Societies (2002–2003)

Table 1.4 shows CAM Symposia of modern Western medical societies in Japan from 2002 to 2003, showing a significant change in the acceptance of CAM by modern western medical societies.

In 1999, ASCO (American Society of Clinical Oncology) held the 35th Annual Meeting. In this congress the symposium named "Alternative and Complementary Therapies and Oncologic Care" was held. 3 years later, JSCO (Japan Society of Clinical Oncology) held the 40th Annual Meeting, and the symposium "Alternative Medicine and Oncologic Care" was held.

Alternative medicine appeared in the Journal of the American Medical Association (JAMA) as a " JAMA patient page. Alternative choices: what it means to use unconventional

**Table 1.4.** CAM Symposium in the Field of Modern Western Medical Societies

| | |
|---|---|
| Japanese Society for Palliative Medicine | (June, 2002) |
| Japan Society of Clinical Oncology | (October, 2002) |
| General Assembly of the Japan Medical Congress | (April, 2003) |
| Japan Pediatric Society | (April, 2003) |
| Japan Society for Occupational Health | (April, 2003) |
| Japanese College of Surgeons | (June, 2003) |

medicine"[3]. 5 years later, a lecture about CAM was done at the 26th General Assembly of the Japan Medical Congress. This is the biggest medical congress and is held every 4 years by the Japanese Medical Association.

Recently, The Japanese Society of Internal Medicine established a committee of specialists for CAM.

## 6. Attitude of Japanese Government Toward CAM Domains

In contrast to the USA, the attitude of the Japanese government toward CAM was slow to change. Recently, however, concerning the CAM of cancer, an Advisory Panel of the Health, Labor and Welfare Ministry was established. Dr I Hyodo of the National Shikoku Cancer Center is the chief of this committee. He stated that "there is a lack of systematic research concerning the evidence of CAM efficacy. Many patients use CAM without enough accurate information. Many oncologists tend to ignore CAM use. Such circumstances should be improved as soon as possible". In 2002, Dr Hyodo et al presented data of the cancer survey at the symposium of JSCO (Fig. 1.3).

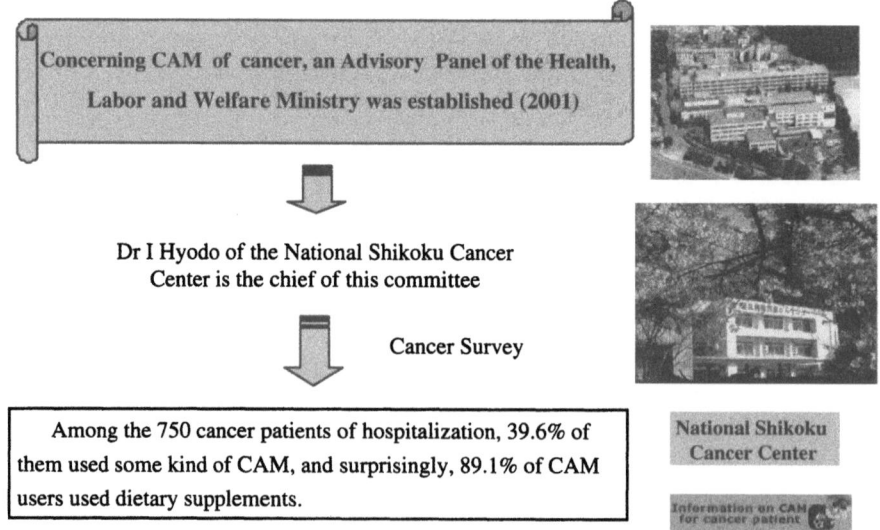

**Figure 1.3.** Cancer survey by the Health, Labour and Welfare Ministry of Japan

In 2003, the term "CAM" was adopted as a new key word of "the Grant-in-Aid for Scientific Research" of the Ministry of Education, Culture, Sports, Science and Technology. Because Japan does not yet have an organization like NCCAM, CAM Department of Kanazawa University, which is a National University, has played a role like that of NCCAM.

## 7. The Types of CAM in the USA and Japan

In the USA, CAM use increased from 33.8% in 1990[4] to 42.1% in 1997[5]. CAM was used by 20 to 50% of the population in many European countries[6-7] and 48.5% in Australia[8]. In Japan, 65.6% of adults used CAM [9] (Table 1.5). A telephone survey of 1000 Japanese respondents showed that the percentage of persons who used at least one CAM therapy during the past 12 months was greater than that of those who had used conventional western medicine (76.0% vs. 65.6%)[10].

Furthermore, in the United States, 61.5% of the CAM users did not tell their doctors[5] while 78.9% of the CAM users in Japan did not tell theirs[9].

Table 1.6 shows the type of CAM in the USA and Japan. In Japan, the domain of dietary supplements is most important, followed by aromatherapy, traditional Chinese medicine, Ayurveda, and electromagnetic fields. In addition to these domains, hot spring bathing, life style drugs, anti-aging medicine and environmental medicine are thought to be important.

A telephone survey revealed the percentage of use for each CAM therapy as follows: nutritional and tonic drinks (43.1%), dietary supplements (43.1%), health-related appliances (21.5%), herbs or over-the-counter Kampo (17.2%), massage or acupressure (14.8%), ethical Kampo (Kampo prescribed by medical doctors) (10.0%), aromatherapy (9.3%), chiropractic or osteopathy (7.1%), acupuncture and moxibustion (6.7%), homeopathy (0.3%), and other therapies (6.5%)[10]. Regarding the reason for the use of CAM, 60.4% responded

Table 1.5. Prevalence of CAM in USA and Japan

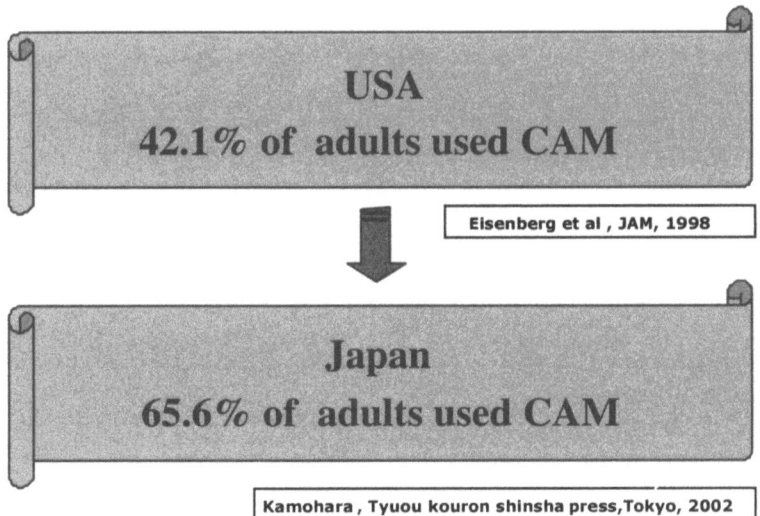

**Table 1.6.** Type of CAM in USA and Japan

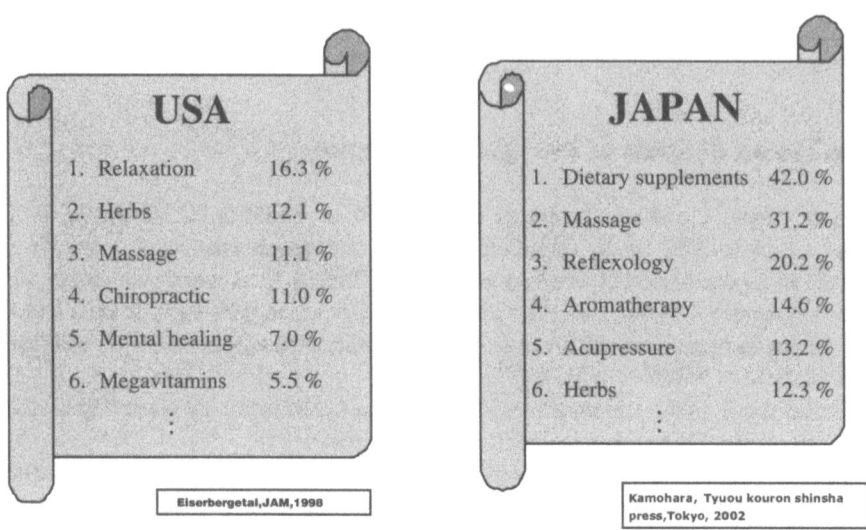

that 'the condition was not serious enough to warrant conventional western medicine', and 49.3% were 'expecting health enhancement or disease prevention'. The average annual out-of-pocket expenditures of all the 1000 respondents were half as much for CAM as for conventional western medicine (19,080 yen vs. 38,360 yen).

According to these data, CAM is very popular in Japan and the money spent is not negligible.

## 8. Dietary Supplement in USA and Japan

The popular dietary supplements in the USA are Gingko Biloba, Echinacea, Garlic, Ginseng, Soybean, Saw Palmetto, St. John's Wort, Valerian, Black Cohosh, Milk Thistle, Evening Primrose, Grape Seed Extract, Bilberry, Green Tea, Pycnogenol, Ginger, Feverfew and Dong quai etc.

In Japan, Agaricus blazei Murill is the most popular product. Many cancer patients take this mushroom. Beer yeast, Propolis, Cereals, Banaba, Japanese Plum, Chlorella, Barley verdure, vegetable juice, Collagen, Royal jelly, and Mulberry are also taken. These foods are quite different from the food supplements popular in the USA. For example, Dong quai is a food supplement in the USA, while it is classified as a medicine in Japan. Dong quai is one ingredient used in Japanese traditional medicine "Kampo".

## 9. CAM Department at Kanazawa University Graduate School of Medical Science

### 9.1. An outline of the Kanazawa CAM Department

On March 1, 2002, a new department for CAM was established at Kanazawa University Graduate School of Medical Science (Fig. 1.4). The distinctive characteristic

Kanazawa University Graduate School of Medical Science

March 1, 2002

The Department of Complementary and Alternative Medicine was established

**Figure 1.4.** A new CAM department at the Japan national university

April 20, 2002
A university extension lecture for citizens and medical students was held by the CAM Department of Kanazawa University.

The contents of the lectures were as follows;
● General statement of CAM
● Relation between mind and disease
● Reactive oxygen and antioxidant food
● Blood rheology
● Red photodiode light and immunity
● Development of the Healing Robots

**Figure 1.5.** An extension lecture by Kanazawa CAM department

of this department is that various kinds of joint research are conducted by the faculties of agriculture, science, engineering, medicine, pharmaceutical science, etc.

On April 20, 2002, this Department held a university extension lecture for citizens and medical students. About one thousand people gathered for the occasion, which shows the height of interest (Fig. 1.5).

## 9.2. Staff and Role

Table 1.7 shows the staff of Kanazawa CAM Department and their research areas.

**Table 1.7.** Staff of the Kanazawa CAM Department

| | |
|---|---|
| Nobutaka Suzuki, MD, PhD (Director) | Clinical research for CAM, Functional food |
| Kazuyoshi Okubo, PhD (professor) | Reactive oxygen and scavenging compounds |
| Saburo Koshimura, PhD (professor) | Anticancer actions of natural products |
| Shoji Shimizu, PhD (professor) | Natural products and immunity |
| Atuhumi Taru, MD, PhD (professor) | Acupuncture, Kampo medicine, Oriental medicine |
| Tsutomu Kamei, MD, PhD (associate professor) | Application of red photodiode light and immunity |
| Yuji Kikuchi, PhD (associate professor) | Hemorheology |
| Satoshi Ohno, MD, PhD (assistant professor) | Clinical research for CAM, Cancer immunotherapy |

The role of the Kanazawa CAM Department is as follows:

1. Accumulation and analysis of evidence for CAM
2. Investigation of CAM in hospitals
3. Research for reactive oxygen species and anti-oxidants to develop anti-oxidant foods or remedies
4. Identification and clinical application of natural products, such as coix seeds, mushrooms etc.
5. Clinical trials of dietary supplements etc.
6. Research for cancer chemoprevention
7. Forehead exposure to red photodiode light and immunity
8. Development of equipment for blood rheology
9. Investigation the side effects of CAM
10. Others

### 9.3. Curricula

In general, there is an increasing demand for CAM curricula in medical schools. In the USA, the number of medical schools where CAM instruction is given, increased from 33(34%) in 1995 to 75(64%) in 1997–1998[11]. In Canada, 12 of all 16 medical schools introduced CAM into the undergraduate medical curricula[12]. In contrast, there is little education in CAM at British medical schools, but it is an area of active curricular development[13].

According to a previous report, of 80 medical schools in Japan, CAM was officially taught in 16 schools (20%)[14]. Almost all the courses had oriental medicine titles such as acupuncture and Kampo.

The characteristic CAM lectures at Kanazawa University are as follows:

- General introduction of CAM
- Dietary supplements
- Antioxidant food
- Anticancer action of natural products
- Natural products and immunity
- Acupuncture, Kampo medicine
- Others

The contents of lectures will be modified according to the changes in CAM use among Japanese people.

As a point of interest, the Yamanashi nursing college incorporated CAM into their educational curriculum from 2003. Our department is helping to develop their curriculum. As far as we know, this will be the first time that any CAM lectures will be held at a Japanese nursing college.

An outpatient CAM department will be founded in the near future. This department will be defined as a place for clinical testing, patient consultation, database survey, and practical education for medical students.

### 9.4. Research

The contents of the research at Kanazawa CAM Department are as follows:

#### 9.4.1. Clinical Trials of Various Natural Products

The French maritime pine bark water-soluble extract (Pycnogenol) has several clinical effects[15-18]. Pycnogenol is considered to be a medicine in several countries in Europe, but in Japan and USA, it is classified as a food supplement (Fig. 1.6).

Recently, we performed an open clinical trial and found that pycnogenol has a curative effect on dysmenorrhea and endometriosis[19]. Now, we are trying a multicenter, double-blind, randomized placebo-controlled study of Pycnogenol for the treatment of dysmenorrhea. Other clinical trials of natural products or vitamins, such as Coix Seed, Nucleic Acid, Agaricus blazei Murill, Phellinus Linteus, Sasa Veitchii, Gingko Biloba, Folic acid, Vitamin H (Biotin) are being planned.

#### 9.4.2. Researching Natural Products to Treat or Prevent Cancer

Collaborating with the pharmaceutical faculty of Kanazawa University, the anti-tumor actions of various mushrooms are being investigated in relation to the attention being given by the public. Agaricus blazei murill and Phellinus linteus are the most popular mushrooms

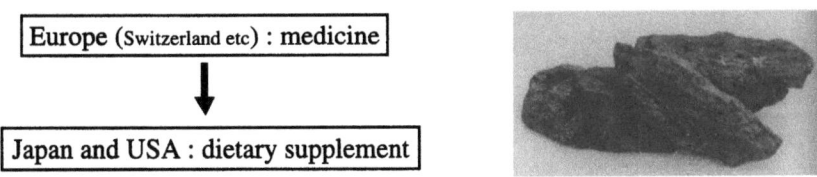

**Figure 1.6.** French maritime pine bark extract

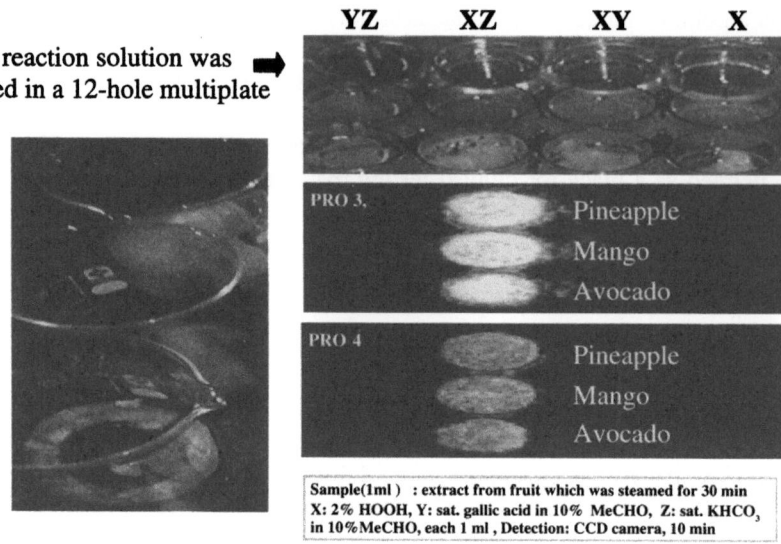

**Figure 1.7.** Hydrogen donor emission from fruits

in Japanese dietary supplements. Other mushrooms, such as Grifora frondosa or lentinula edodes, etc., are under experimentation.

### 9.4.3. Approach to a New Measurement of Reactive Oxygen and its Scavenging Compounds by a Chemiluminescence System

In recent years, Dr. Yoshiki and Dr. Okubo proposed a new chemiluminescence system[20–23] that measures the reactive oxygen species (ROS), hydrogen donor, and mediator. The occurrence and intensity of the photon emission were closely related to the partial structures of the hydrogen donor (phenolic compounds) and radical scavenging activity, and showed a high concentration dependence on reactive oxygen species (ROS), hydrogen donors and mediators.

Based on the results of the chemiluminescence system, we developed a photon system to quantify hydroperoxide, hydrogen donor or mediators of crude extracts of food or human plasma (Fig. 1.8, 1.9). The ROS/hydrogen donor/mediator system might be useful not only for the investigation of food function but also for clinical application.

Currently, we can catch the reactive oxygen species from various kinds of gases, such as automobile exhaust or smoke from a cigarette. From the viewpoint of environmental medicine, this system may be useful for the development of a new anti-oxidant filter, food or remedy.

### 9.4.4. Hemorheology

Dr. Kikuchi developed a piece of equipment that measures the human blood flow[24–27]. This equipment named MCFAN, and is useful for research of blood rheology.

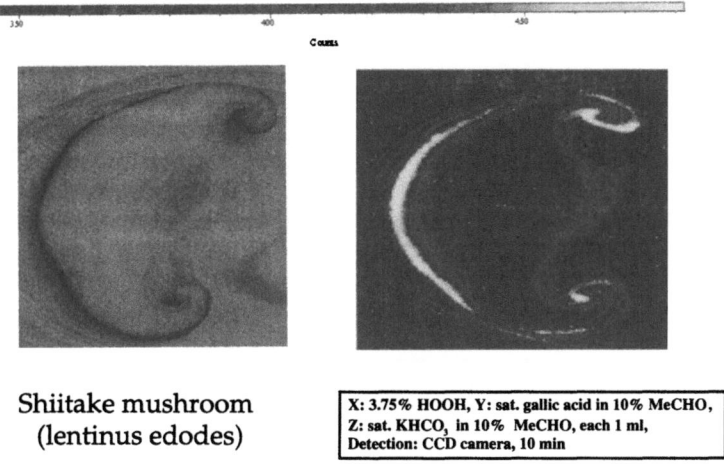

Shiitake mushroom
(lentinus edodes)

X: 3.75% HOOH, Y: sat. gallic acid in 10% MeCHO,
Z: sat. KHCO$_3$ in 10% MeCHO, each 1 ml,
Detection: CCD camera, 10 min

**Figure 1.8.** Localization of Hydrogen donor emission from a Shiitake mushroom

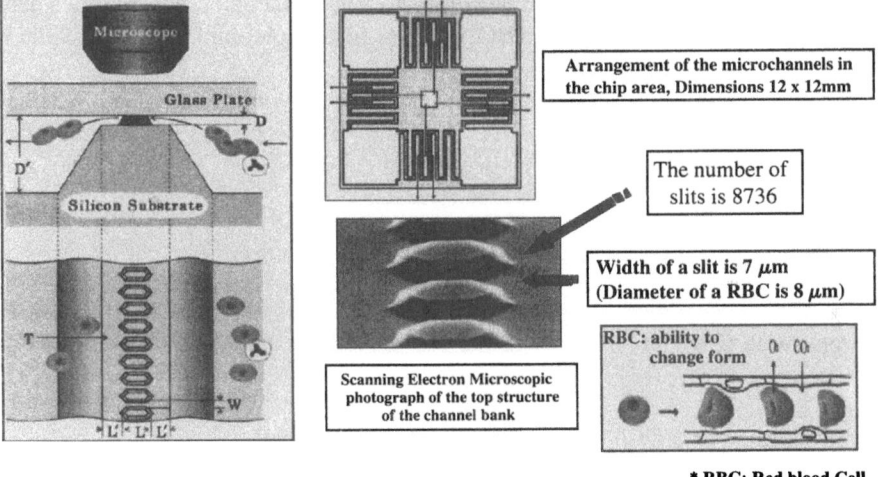

**Figure 1.9.** Outline structure of the microchannels formed between the surface of a Single-crystal Silicon Substrate and a glass plate

Figure 1.9 shows the fine structures of MCFAN. Outline structure of microchannels formed between the surface of a Single-crystal Silicon Substrate and a glass plate. The blood goes through the narrow space between the glass plate and Silicon Substrate. The arrangement of the microchannels in the chip area and the top structure of the channel bank can be seen by the scanning electron microscopic photograph. The number of slits is 8736 and the width of a slit is 7 $\mu$m. Because the average diameter of an RBC is 8 $\mu$m, the size of the red blood cells needs to decrease in order to pass through the slit.

Figure 1.10 is a photograph of normal or abnormal blood flow. In the case of normal blood flow, the blood cells cannot be seen. A total of one hundred micro liters of blood is

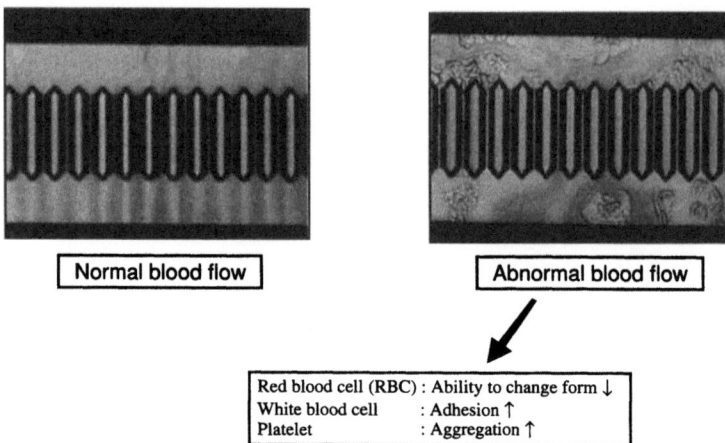

**Figure 1.10.** Flow volume curve of the normal and the abnormal blood flow

injected, and the flow curve is estimated. In the case of abnormal blood flow, blockades of slits by RBC, WBC and platelets can be observed. Abnormal blood flow means a decrease in the ability to change the form of RBC, increase in the adhesion activity of white blood cells and aggregation of platelets.

We are screening the effects of natural products with this equipment. Whether this equipment is useful for the prediction of thrombotic disease such as myocardial infarction or cerebral infarction is still under investigation.

### 9.4.5. Frontal Alpha Wave Pulsed Photic Synchronization

Dr. Kamei developed equipment named FAPPS: Frontal Alpha Wave Pulsed Photic Synchronization (Fig. 1.11). By using the red photodiode light FAPPS, we research the

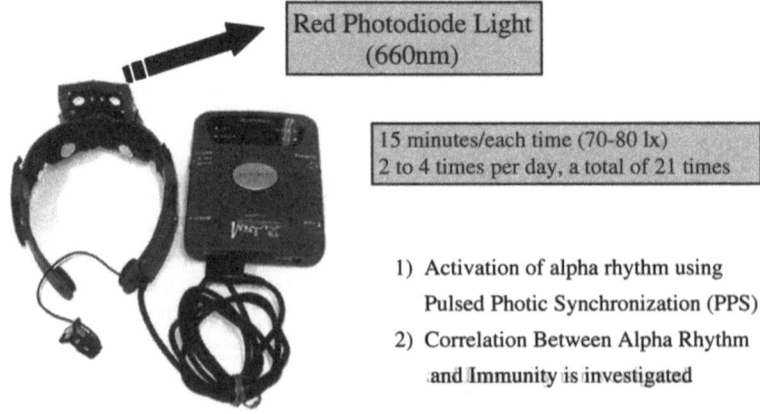

**Figure 1.11.** Equipment of Frontal Alpha Wave Pulsed Photic Synchronization (FAPPS)

activation of Alpha Rhythm and the correlation between Alpha Rhythm and immunity[28-30]. Whether this equipment is useful for a complementary medical therapy for a patient with cancer or immunodeficiency is still under investigation.

### 9.4.6. Exploration of Natural Products that have an Anti-Oxidative Effect on DNA Damage or Preventive Effects against Ultraviolet-Induced DNA Damage

Oxidation of lipids, nucleic acids or protein has been suggested to be involved in the etiology of several chronic diseases including cancer, cardiovascular disease, cataract, and aging. We are investigating the potential role of natural products to prevent or treat these chronic diseases. By using urinary 8-hydroxydeoxyguanosine (8-OHdG) as a marker for oxidative DNA damage, several kinds of open clinical testing are being conducted, such as whether pycnogenol, a strong radical scavenger, can rescue the DNA from damage by cigarette smoke or not, etc.

Exposure to UVB results in formation of cyclobutane pyrimidine dimers (CPDs). By collaborating with the Pharmaceutical Faculty of Kanazawa University; an anti-CPDs monoclonal antibody was completed. Because the stimulation of repair and prevention of immunosuppression have been linked to the prevention of skin cancer, we have started screening natural products which have an anti-UVB effect.

### 9.4.7. Investigation of the Side Effects of CAM

In recent years, private importation has often been the source of troublesome problems. Unfortunately, in Japan, liver dysfunction and death by a Chinese diet food occurred last year. The cause of this accident was a non-permitted chemical agent, N-Nitroso-fenfluramine. In addition, Kava Kava, which is sold in most parts of the world, has also caused fatal liver dysfunction.

Ephedra, which is a food material in the United States, is often added to dietary pills. The main ingredient of ephedra is ephedrine, which can be dangerous if taken excessively because it induces high blood pressure. Ephedra is dealt with as a Kampo medicine in Japan, and is classified strictly as a remedy. In Japan, Kampo medicine requires the prescription of a doctor or pharmacist.

We are making efforts to obtain up to date information to help enlighten the public on the side effects of CAM. International exchange on various levels may also be needed.

## 10. Conclusion

The aim of this chapter was to report the actual status of CAM in Japan and to announce the concrete efforts of a new CAM department.

CAM usage is very common among Japanese citizens. Even though physicians are frequently requested by their patients to advise them about CAM, there is an absence of both evidence of the efficacy and side effects of nearly all specific treatments as well as guidelines that could assist the integration of conventional and CAM therapies. Even so, courageous scientists are taking this up as a challenge.

Just as outer space has been an unknown frontier for human beings from the beginning of time, the span of the domain of CAM is also beyond all expectations. CAM is filled with expectation for new discoveries, and we are sure that it will fulfill dreams as well as increase our welfare.

## 11. Acknowledgment

Preparation of this chapter was made possible through the efforts of Miss Sayara Shitano, the medical secretary of our department.

## 12. References

1. W. B. Jonas, Researching alternative medicine, *Nat. Med.* **3**, 824–82 (1997).
2. J. Couzin, Beefed-up NIH center probes unconventional therapies, *Science.* **282**, 2175–2176 (1998).
3. JAMA patient page, Alternative medicine, *JAMA.* **280**, 1640 (1998).
4. D. M. Eisenberg, R. C. Kessler, C. Foster, F. E. Norlock, D. R. Calkins, and T. L. Delbanco, Unconventional medicine in the United States. Prevalence, costs, and patterns of use, *N. Eng. J. Med.* **328**, 246–252 (1993).
5. D. M. Eisenberg, R. B. Davis, S. J. Ettner, S. Wilkey, S. Van, M. Rompay, and R. C. Kessler, Trends in alternative medicine use in the United States, 1990–1997: results of a follow-up national survey, *JAMA.* **280**, 1569–1575 (1998).
6. P. Fisher and A. Ward, Complementary medicine in Europe, *BMJ.* **309**, 107–111 (1994).
7. S. J. Fulder and R. E. Munro, Complementary medicine in the United Kingdom: patients, practitioners, and consultations, *Lancet.* **2**, 542–545 (1985).
8. A.H. Maclennan, D. H. Wilson, and A. W. Taylor, Prevalence and costs of alternative medicine in Australia, *Lancet.* **347**, 569–573 (1996).
9. S. Kamohara, in: *Alternative Medicine* (Tyuou Kouron Shinsha, Tokyo, 2002), pp. 30–35.
10. H. Yamashita, H. Tsukayama, and C. Sugishita, Popularity of complementary and alternative medicine in Japan: a telephone survey, *Complement. Ther. Med.* **10**, 84–93 (2002).
11. M. S. Wetzel, D. M. Eisenberg, and T. J. Kaptchuk, Courses involving complementary and alternative medicine at US medical schools, *JAMA.* **280**, 784–787 (1998).
12. J. Ruedy, D. M. Kaufman, and H. MacLeod, Alternative and complementary medicine in Canadian medical schools, *Can. Med. Assoc. J.* **160**, 816–817 (1999).
13. H. Rampes, F. Sharples, S. Maragh, and P. Fisher, Introducing complementary medicine into the medical curriculum, *J. R. Soc. Med.* **90**, 19–22 (1997).
14. K. Tsuruoka, Y. Tsuruoka, and E. Kajii, Complementary medicine education in Japanese medical schools: a survey, *Complement. Ther. Med.* **9**, 28–33 (2001).
15. S. Hosseini, S. Pishnamazi, S. M. Sadrzadef, F. Farid, R. Farid, and R. R. Watson, Pycnogenol in the management of asthma, *J. Med. Food.* **4**, 201–209 (2001).
16. F. Schonlau and P. Rhodewald, Pycnogenol for diabetic retinopathy. A review, *Int. Opthalmol.* **24**, 161–171 (2001).
17. S. J. Roseff, Improvement in sperm quality and function with French maritime pine tree bark extract, *J. reprod. Med.* **47**, 821–824 (2002).
18. Z. Ni, Y. Mu, and O. Gulati, Treatment of melasma with pycnogenol, *Phytother. Res.* **16**, 567–571 (2002).
19. T. Kohama and N. Suzuki, The treatment of gynaecological disorders with pycnogenol, *Eur. Bull. Drug. Res.* **7**, 30–32 (1999).
20. Y. Yoshiki, K. Okubo, and T. Kanazawa, Effects of cigarette smoking on the photon emission from plasma, *ITF. Letters.* **1**, 65–68 (2000).
21. Y. Yoshiki, T. Iida, K. Okubo, and T. Kanazawa, Chemiluminescence of hemoglobin and identification of related compounds with the hemoglobin chemiluminescence in plasma, *Photochem. Photobiol.* **73**, 545–550 (2001).

22. Y. Yoshiki, T. Iida, Y. Akiyama, K. Okubo, H. Matsumoto, and M. Sato, Imaging of hydroperoxide and hydrogen peroxide scavenging substances by photon emission, *Luminescence.* **16**, 1–9 (2001).
23. H. Matsumoto, Y. Nakamura, M. Hirayama, Y. Yoshiki, and K. Okubo, Antioxidant activity of black currant anthocyanin aglycons and thieir glycosides measured by chemiluminescnece in neutral pH region and in human plasma, *J. Agric. Food Chem.* **50**, 5034–5037 (2002).
24. Y. Kikuchi, K. Sato, H. Ohki, and T. Kaneko, Optically accessible microchannels formed in a single-crystal silicon substrate for studies of blood rheology, *Microvasc. Res.* **44**, 226–240 (1992).
25. Y. Kikuchi, K. Sato, and Y. Mizuguchi, Modified cell-flow microchannels in a single-crystal silicon substrate and flow behavior of blood cells, *Microvasc. Res.* **47**, 126–139 (1994).
26. Y. Kikuchi and H. E. Kikuchi, Measurement of oxidative stress by leukocytes using microchannel array flow analyzer and effects of food substances, *Microcirc. Ann.* **15**, 153–154 (1999).
27. Y. Kikuchi, Relation between whole blood fluidity through capillaries and hematocrit, *Microcirc. Ann.* **16**, 37–38 (2000).
28. T. Kamei, H. Kumano, K. Iwata, and M. Yasushi, Influences of long-and short-distance driving on alpha waves and natural killer cell activity, *Percept. Mot. Skills.* **87**, 1419–1423 (1998).
29. T. Kamei, Y. Toriumi, S. Nagura, H. Kumano, H. Otani, M. Yasushi, and S. Jimbo, Influence of forehead exposure to red photodiode light on frontal alpha wave amplitude and CD57-CD16+ level, *Photomed. Photobiol.* **21**, 115–121 (1999).
30. T. Kamei, Y. Toriumi, H. Kumano, M. Fukada, and T. Matsumoto, Use of photic feedback as an adjunct tratment in a case of Miller Fisher syndrome, *Perceptual. Motor. Skills.* **90**, 262–264 (2000).

# 2

# New Scientific Approach for Natural Medicine

## EXAMPLES OF KAMPO MEDICINE

**Haruki Yamada**

## 1. Introduction

Kampo medicine originated from traditional Chinese medicine and has a history of over 2,000 years. It is not considered to be an alternative medicine for people in China, Japan and Korea. Kampo medicine has often been used for the treatment of hepatitis, menopausal disorders such as autonomic-nervous and hormonal manifestation, autonomic imbalance, bronchial asthma, cold syndrome, digestive disorder, atopic dermatitis, eczema, cold sensitivity, allergic rhinitis, and complaints of general malaise etc. These diseases are still difficult to treat with modern medicine, but Kampo is relatively effective in treatment through restoring whole body balance. Each type of Kampo medicine consists of several types of herbs as the formula and each formula has a particular name. For example, a kind of Kampo formulation, Syo-Saiko-To (Xiao-Chai-Hu-Tang in Chinese) consists of 7 kinds of different herbs. Folk medicine and other types of natural medicine generally use a single herb for the treatment. Kampo medicines are generally taken as oral decoctions. In 1979, only 28% of the medical doctors in Japan used Kampo medicines, but that number has increased year by year. In 1993, about 77% and in 2000, 86% of the medical doctors in Japan had used Kampo medicines.

In China and Korea, medical practitioners are divided into those who use western medicine and those who use Kampo medicine. In Japan, doctors use both types of treatment. Kampo medicine is especially effective for patients who have problems with physical functions, can not adapt well to operations, have a low response to modern medical treatment because of side effects, show improvement upon clinical examination but still have complaints, hope for improvement of their constitution, tend to have psychosomatic disorders and for those who are aged or whose physical strength has declined. In order to adapt Kampo medicine into modern-day therapy, several Kampo extract preparations which have the same

---

**Haruki Yamada** • KitasatoInstitute for Life Sciences & Graduate School of Infection Control Sciences, Kitasato University and Oriental Medicine Research Center, The Kitasato Institute.

*Complementary and Alternative Approaches to Biomedicine*, edited by Edwin L. Cooper and Nobuo Yamaguchi. Kluwer Academic/Plenum Publishers, 2004.

efficacy as the decoction were developed in Japan in 1976. They were manufactured as granules or fine granule forms by freeze-drying the decoction. Since 1976, the Ministry of Health and Welfare of Japan approved several Kampo extract preparations, and now 148 kinds of Kampo formulations are allowed to be covered by the National Health Insurance System. Kampo extract preparations in granule form have been quality controlled by good manufacturing practice (GMP) guidelines since 1988. Because natural plant and mineral materials have been used in Kampo extract preparations, quality control of the herbs is required. This control includes the selection of high quality herbs with no contamination of pesticides, which are evaluated by morphological and physicochemical methods. Preservation of herbs is also very important for the quality control. Process control and quality control for manufacturing Kampo extract preparation have also been carried out in Japan. Although Kampo medicines have a long clinical experience to prove their efficacy and safety, their efficacy must be estimated by scientifically clarifying the mechanism of their action and active ingredients. Other natural medicines should also be clarified on a scientific level.

## 2. Nonclinical Basic Studies of Kampo Medicines

Basic and clinical studies of Kampo medicines both need to be made. The former needs to elucidate the active ingredients in Kampo formulas and component herbs and what effect will be had from a combination of formulas. Answers for these questions will be available for quality control of the drugs and will also open up the possibility of finding new drugs. The latter study needs to clarify which Kampo formula is responsible for a specific pathophysiological condition, so-called Sho in Japanese, which means the symptoms of the patient, and how to evaluate the clinical efficacy on a scientific level. Answers for these questions may lead to new applications of Kampo medicines. Both studies should be linked to each other by pharmacological studies to resolve the mode of actions. The efficacy of Kampo medicines can not be explained by the pharmacological activity of just one active ingredient, and several active ingredients may affect the immune, endocrine and neural systems of the whole body by several combination effects such as synergistic and/or antagonistic effects. Some of these active ingredients also work through structural modification to the actual active compounds by endogenous factors such as intestinal bacteria or gastric juice. Because standardization of natural medicine is very important, it should be controlled by indicator compounds, hopefully active ingredients, and finger printed using three dimensional HPLC. Analysis of pharmacological activity related to clinical effects by *in vitro* and *in vivo* biological methods may also be very important to obtain reproducible effects of herbs. Fig. 2.1 shows a three dimensional HPLC pattern of one Kampo formula, Juzen-taiho-to (Si-Quan-Da-Bu-Tang in Chinese). Juzen-taiho-to has been used for the treatment of patients recovering from surgery or suffering from diseases by promoting the improvement of their debilitated general condition. Juzen-taiho-to also has been administered to patients with anemia or anorexia. The clinical effects observed suggest that the formula enhances immune responses and improves the functioning of the hematopoietic systems. Since it is possible to prepare HPLC finger printing patterns based on the composition of the constituents in each formula, this method is useful for the standardization of not only Kampo medicines but also natural medicines.

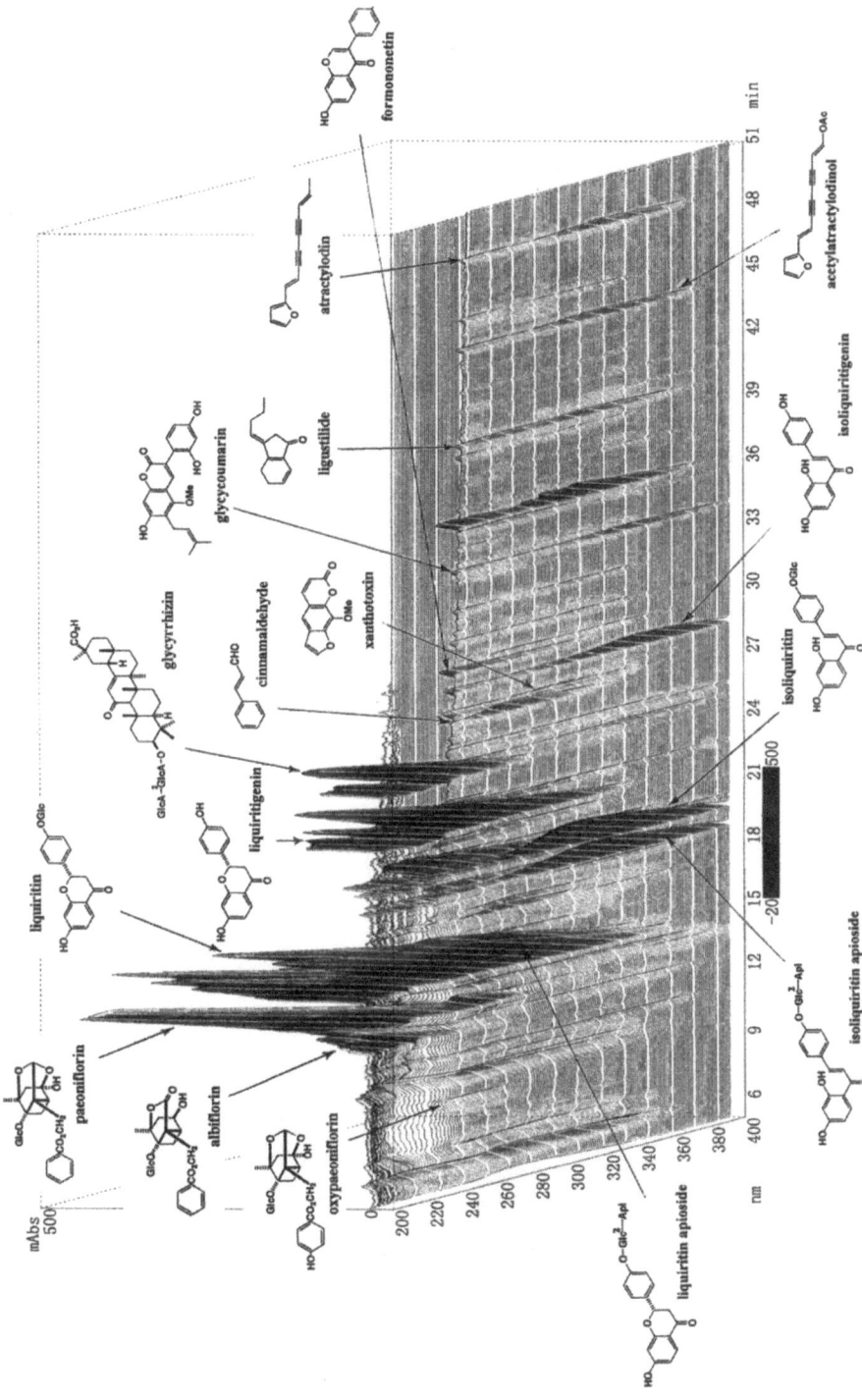

**Figure 2.1.** Three dimensional HPLC profile of Juzen-taiho-to extract.

## 3. Intestinal Immune System Modulating Activity of Kampo Medicine

Because Kampo medicines are taken orally, some of the active ingredients may be absorbed from the intestine, but some other active ingredients like polysaccharides and proteinous molecules may affect intestinal immunity even if they can not be absorbed directly. We have developed an assay method of Peyer's patch cells mediated proliferative response of bone marrow cells as the evaluation of intestinal immunity[1]. After oral administration of one of the formulas, Juzen-taiho-to extract preparation, TJ-48 for 7 consecutive days, C3H/HeJ mice were sacrificed and Peyer's patches were carefully dissected out from the small intestine. Then Peyer's patch cells were cultured and the resulting culture medium was tested for the proliferation of bone marrow cells. When Juzen-taiho-to extract preparation, TJ-48 was administrated to C3H/HeJ mice orally at doses of 500mg/kg to 2g/kg, the conditioned medium of Peyer's patch cells stimulated the production of the hematopoietic factors such as IL-6 and GM-CSF, and enhanced the proliferation of bone marrow cells as shown in Fig. 2.2 Active ingredients were clarified to be polysaccharide and lignin-carbohydrate complex[2–4]. When Peyer's patch cells were incubated in the presence of Juzen-taiho-to extract preparation, the resulting culture supernatant also enhanced the proliferative response of bone marrow cells in comparison with the controls. Because Juzen-taiho-to consists of ten component herbs, the intestinal immune system modulation activities of the active fractions, which were prepared from each ten component herbs, were compared. None of

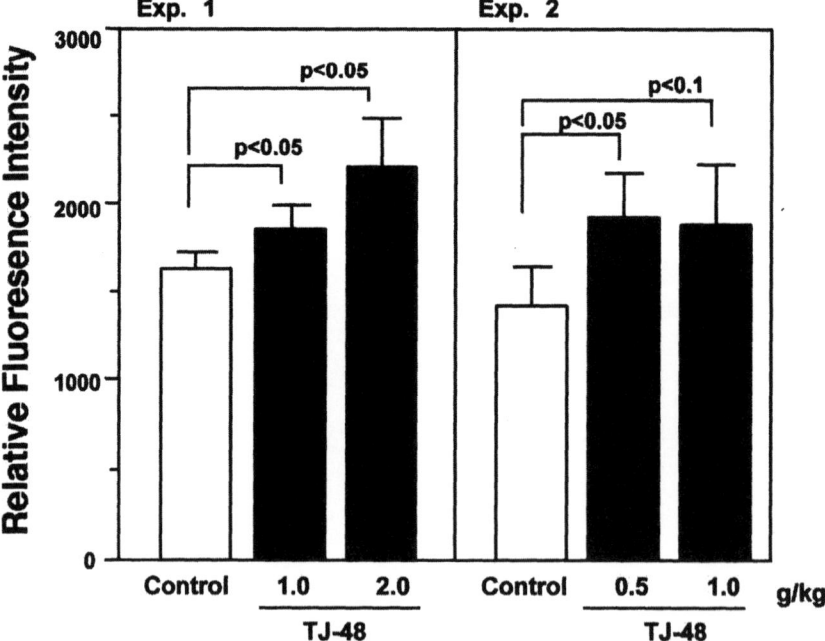

**Figure 2.2.** Effect of orally administered Juzen-taiho-to (TJ-48) on Peyer's patch cells mediated hematopoietic response of bone marrow cells.

the active fractions showed any activity, however, when five component herbs among the formula were decocted together, the resulting active fraction showed potent activity[5]. This result indicates that some combination effects may be involved in the expression of the activity. Although biological active substances with low molecular weight in medicinal herbs have been well studied and characterized, they can not account for all of the clinical effects achieved. Among the high-molecular-weight components of medicinal herbs, polysaccharides have been shown to possess a variety of pharmacological activities. Most especially, pectic polysaccharides containing arabinogalactan have been shown to have immunestimulating, anti-ulcer, anti-metastasis, hypoglycemic or other activities[6,7,8]. These observations suggest that pectic polysaccharides are involved in the efficacy of traditional herbal medicines[6].

## 4. Basic Approach to New Clinical Application of Kampo Medicine

Basic studies also lead to the discovery of new effects of Kampo medicines, and the results can be fed back to new clinical applications such as the following example. Alzheimer disease patients are known to have a decreased level of acetylcholine content and acetylcholine synthesizing enzyme, choline acetyltransferase (ChAT) activity, and these levels correlate well with the severity of dementia. Because Kampo medicines have a mild effect with relatively low adverse effects, they are suitable for treating eldery people. We have screened neurotrophic activity for cholinergic neurons of Kampo formulas, which have been used traditionally but have not been known for the efficacy of dementia by analyzing ChAT activity of rat primary cultured septal cells which are rich cholinergic neurons. Among the Kampo formulas tested, the most potent ChAT activity was observed in Kami-untan-to (Ji-wei-wen-dan-tang in Chinese; KUT), which consists of 13 kinds of component herbs[9]. KUT has been used clinically for the treatment of insomnia and neurosis. When rat embryo septal cells were cultured in the presence of KUT at 50–500 µg/ml for 3days, a significant increase of ChAT activity was observed like nerve growth factor, NGF[10](Fig. 2.3). Ibotenic acid induced basal forebrain lesioned rats or two year old (aged) rats showed decreased memory and learning activity on passive avoidance tests. But if the lesioned rats or aged rats took KUT orally, the mean latency significantly increased similar to that of the mature rats (Fig. 2.4)[10,11]. Two weeks after the end of the passive avoidance test, the aged rats significantly decreased their ChAT activity in the frontoparietal cortex compared with the mature control rats. However, when KUT was administered to the aged rats orally, enhanced ChAT activity was observed in the frontoparietal cortex of the rats. The ChAT mRNA expression in the basal forebrain of the KUT administered aged rats had a tendency to increase compared with the aged rats using the RT-PCR method[8]. Oral administration of KUT also apparently increased the NGF mRNA expression in the frontoparietal cortex[11]. The treatment with KUT increased levels of ChAT mRNA in the basal forebrain cells and NGF mRNA in the cerebral cortex cells, respectively[12]. Among the component herbs of KUT, Polygalae radix extract enhanced ChAT activity in the cerebral cortex and NGF secretion in astroglial cells[13], and onjisaponins were identified as active ingredients in Polygalae radix[14]. Clinical effects of KUT for the treatment of Alzheimer's disease were tested by a collaboration with Arai and Hanawa. The rate of cognitive decline was assessed by changes in the MMSE score per year. Although untreated Alzheimer's disease patients had decreased scores, the KUT

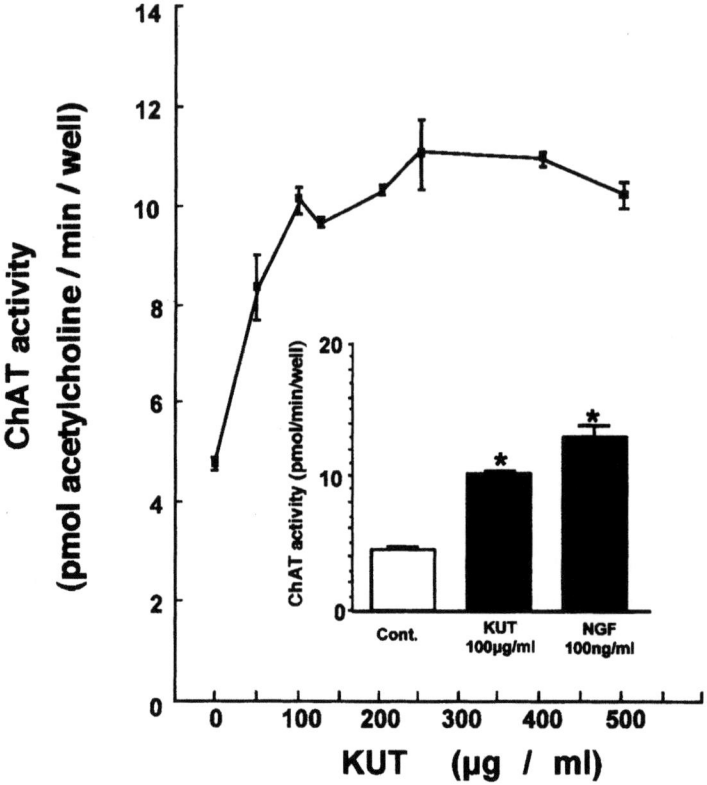

**Figure 2.3.** Effect of KUT on ChAT activity in basal forebrain cultured cells.

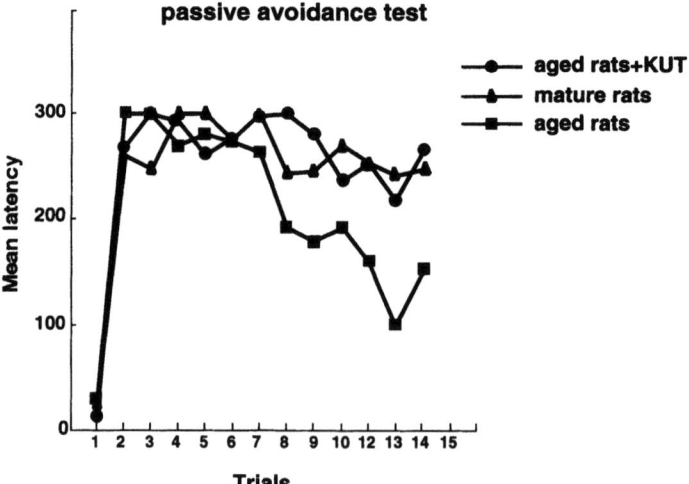

**Figure 2.4.** Effect of KUT on the passive avoidance test in aged rats. Mean latency periods before entering into the dark room during retention testing for mature rats (3 months old), aged rats (2 years old) and KUT administered aged rats.

treated group and KUT combination group which is KUT + estrogen + new anti-dementia drug, donepesil (Aricept®) treated group showed increased scores during around the first 2 months, and then gradually declined. Both groups showed much higher scores than the donepesil or estrogen alone treated groups[15]. This observation suggests that KUT is useful as a potential therapeutic agent for Alzheimer's disease.

## 5. Conclusion: Union of Kampo and Western Strategies

Because Kampo medicines have already been used together with western medicine in modern day-therapy in Japan, and have clinical theories for their use, they are very different from some alternative medicines which are apart from traditional medicine. Although alternative medicines are growing globally, these medicines need to establish evaluation methods for their efficacy and safety. I believe that the immunological and neurobiological approaches for Kampo medicines presented here may also be available for this purpose.

## 6. References

1. T. Hong, T. Matsumoto, H. Kiyohara and H. Yamada, Enhanced production of hematopoietic growth factors through T cell activation in Peyer's patches by oral administration of Kampo (Japanese herbas) medicines, 'Juzen-taiho-to', *Phytomedicine*, **5**, 353–360 (1998)
2. H. Kiyohara, T. Matsumoto and H. Yamada, Lignin-carbohydrate complexes: Intestinal immune system modulating ingredients in Kampo (Japanese herbal) medicine, Juzen-taiho-to, *Planta Med.*, **66**, 20–24 (2000)
3. K.-W. Yu, H. Kiyohara, T. Matsumoto, H.-C. Yang and H. Yamada, Characterization of pectic polysaccharides having intestinal immune system modulating activity from rhizomes of *Atractylodes lancea* DC., *Carbohydr. Polymer*, **46**, 125–134 (2001)
4. K.-W. Yu, H. Kiyohara, T. Matsumoto, H.-C. Yang and H. Yamada, Structural characterization of intestinal immune system modulating new arabino-3, 6-galactan from rhizomes of *Atractylodes lancea* DC., *Carbohydr. Polymer*, **46**, 147–156 (2001)
5. H. Kiyohara, T. Matsumoto and H. Yamada, Combination effect of component herbs of a Japanese herbal (Kampo) medicine, Juzen-taiho-to on expression of intestinal immune system modulating activity, *Phytomedicine, in press*
6. H. Yamada, Pectic polysaccharides from Chinese herbs: Structure and biological activity, *Carbohydr. Polymer*, 25, 269–276 (1994)
7. H. Yamada and H. Kiyohara, Complement-activating polysaccharides from medicinal herbs, In Immunomodulatory Agents from Plants (H. Wagner, ed.), Birkhäuser Verlag, Basel, pp. 161–202 (1999)
8. H. Yamada, Bioactive arabinogalactan-proteins and related polysaccharides in Sino-Japanese herbal medicines, In Cell and Developmental Biology of Arabinogalactan-Proteins (A. Nothnagel, A. Basic and A. E. Clarke eds.), Kluwer Academic Plenum Publishers, New York, pp. 221–229 (2000)
9. T. Yabe, K. Toriizuka and H. Yamada, Effects of Kampo medicines on choline acetyltransferase activity in rat embryo septal cultures, *Journal of Traditional Medicine*, **12**, 54–60 (1995)
10. T. Yabe, K. Toriizuka and H. Yamada, Choline acetyltransferase activity enhancing effects of Kami-untan-to (KUT) on basal forebrain cultured neurons and lesioned rats, *Phytomedicine*, **2**, 41–46 (1995)
11. T. Yabe, K. Toriizuka and H. Yamada, Kami-untan-to (KUT) improves cholinergic deficits in aged rats, *Phytomedicine*, **2**, 253–258 (1996)
12. T. Yabe and H. Yamada, Kami-untan-to enhances choline acetyltransferase and nerve growth factor mRNA levels in brain cultured cells, *Phytomedicine*, **3**, 361–367 (1996/97)
13. T. Yabe, S. Iizuka, Y. Komatsu and H. Yamada, Enhancements of choline acetyltransferase activity and nerve growth factor secretion by Polygalae Radix-extract containing active ingredients in Kami-untan-to, *Phytomedicine*, **4**, 199–205 (1997)

14. T. Yabe, H. Tuchida, H. Kiyohara, T. Takeda and H. Yamada, Induction of NGF synthesis in astrocytes by onjisaponins of *Polygala tenuifolia*, constituents of Kampo (Japanese herbal) medicine, Ninjin-Yoei-To, *Phytomedicine*, **10**, 106–114 (2003)
15. T. Suzuki, H. Arai, K. Iwasaki, H. Tanji, M. Higuchi, N. Okamura, T. Matsui, M. Maruyama, T. Yabe, K. Toriizuka, H. Yamada, T. Hanawa, Y. Ikarashi and H. Sasaki, A Japanese herbal medicine (Kami-untan-to) in the treatment of Alzheimer's disease: A pilot study, *Alzheimer's Reports*, **4**, 177–182 (2001)

# 3

# The Modernization of Traditional Chinese Medicine in Taiwan—Past, Present and Future

Chieh Fu Chen, Yau Chik Shum and Sze Piao Yang

## 1. Introduction

Honed by clinical wisdom accumulated over thousands of years, a time-tested system encompassing theory, system, structure, education, examination, and practice has evolved for traditional Chinese Medicine. Although unassuming, as characteristic of Oriental tradition and heritage, Chinese Medicine nonetheless has faithfully and steadfastly served to protect the health of the Chinese and neighboring populace over thousands of years. In Mainland China, Chinese Medicine has never been relegated to the status of alternative medicine. Instead, the government has spearheaded the modernization of Chinese Medicine since 1950. Instead of pitting the two major health protection systems, viz. Chinese Medicine and the Western system against each other in confrontations and competitions, recent efforts have been directed at integrating the complementary aspects of the two systems into a comprehensive system, integrative medicine. Young doctors, known as East-West Integrative Medical doctors ("New Generation" Chinese Medicine medical doctors in Taiwan) are trained in both disciplines to assume primary and secondary medical care in the community. The results are increasingly recognized and accepted by Western nations.

In view of such progress, what directions should Taiwan, separated by just a narrow strait, take? We would like to report on what has been done in Taiwan and appeal in earnest to members in all walks of life to be concerned and work together to turn the confrontation between Eastern and Western medicines into cooperation, to elevate the status of TCM and realize that East-West Integrative medicine is of benefit to the people of Taiwan.

---

**Chieh Fu Chen** • National Research Institute of Chinese Medicine, No. 155-1, Set 2, Li-Nung ST., Peitou, Taipei, (112), Taiwan   **Yau Chik Shum** • Institute of Pharmacology, National Yang-Ming University, School of Medicine, No. 155, Set. 2, Li-Nung ST., Peitou, Taipei, (112) Taiwan   **Sze Piao Yang** • Society for Integration of Chinese, and Western Medicine, R.O.C., No. 32, Wen Sin St., Changhua, Taiwan

*Complementary and Alternative Approaches to Biomedicine*, edited by Edwin L. Cooper and Nobuo Yamaguchi. Kluwer Academic/Plenum Publishers, 2004.

## 2. Status of Chinese Medicine in Taiwan in the Japanese Occupation Era

Chinese Medicine was brought from Mainland China to Taiwan during the Ming Dynasty.[1] In the waning years of the Ming Dynasty at around mid-seventeenth century, the more prominent practitioners were:
Chuen-Chi Shen (1608–1682)
Kwong-Wen Shen (1612–1685)
Rapid development of Chinese Medicine occurred during the Ching Dynasty. Statistics show that by 1897 there were 1046 practitioners, among whom 29 enjoyed good reputations. Ninety-one were also well known scholars. Ninety-seven were from well known medical families while the rest were common practitioners.

During the Ching Dynasty, Chinese Medicine was the only available health care, although there were no locally trained practitioners. All processed Chinese medicines were imported from China. Taiwan produced only herbs.

The Japanese occupation extended from (1895–1945). Only one board examination was held in 1901 for Chinese Medicine practitioners by the occupying Japanese government. 1,097 passed the examination and the 156 who failed were nonetheless issued practicing permits out of sympathy. In addition, 650 were issued licenses despite never attempting the exam.

There were 1,903 practitioners in Taiwan. By 1945 only 97 remained.

Nevertheless, the dawn of the modernization of Chinese Medicine had begun. Scholars in the predecessor of the National Taiwan University, such as the internationally famed Drs Tsung-Ming Tu and Chen-Yuan Lee, had begun to re-examine, employing modern scientific techniques, the biomedical properties of opium, Chinese medicines and snake venom.[2]

## 3. The Current Status of the Modernization of Chinese Medicine in Taiwan

The modernization of Chinese medicine in Taiwan has been moderately fruitful. The traditional ways of preparation of medically usable Chinese Medicines include the processing of raw or partially processed materials into 'tablets', 'formulations', 'ointments', 'pills', alcoholic extraction or through indirect heating in a boiling water bath. Such processes are still being practiced but are not covered by the National Health Care System. Users are responsible for the expenses.

There have been concerted efforts in recent years to extract, isolate, identify and develop the medicinally active principles, even from compound formulations. Preparations of scientific Chinese Medicine was started in the 1960s by Professor Hong-Yen Hsu of the Sun-Ten Drug Company.

Following active promotion by the advocates of Chinese Medicine, the National Research Institute of Chinese Medicine was established in 1963 in Shin-Dian. In the early years, however, the Institute was virtually a token institution with little real achievement. With the erection of new research facilities, the Institute has become a leading research institution with annual publications of 60 scientific reports, most of which appear in international acclaimed scientific journals.[3]

Supported by the National Science Council, the 'herbal medicines' group was established in the National Defense Medical Center in the early 1970s to focus on research of Chinese Medicines for their anti-neoplastic, antihypertensive and analgesic properties. In the meantime, pharmacologists at the National Taiwan University devoted their attention to snake venom research.[2]

During the last decade, various research units including pharmaceutical and chemistry departments at various universities in Taiwan have actively participated in research of Chinese Medicines. Since the 1960s Taiwanese scientists have published over 1700 research articles concerning Chinese medicines. The areas covered spanned the identifications of the origins, isolation of bioactive principles, purification, structural identification, pharmacokinetics, metabolism and drug interactions, pharmacodynamics, mechanisms of action and therapeutic effectiveness.[4]

Currently used biotechnological techniques include the genomic sequencing of Chinese Medicines, robot-assisted automated high output efficacy analyses and the use of bioinformatics for the analysis of genomic and protein chips generated experimental results in the development of potential drugs for treatment of hepatitis, cancer, cardiovascular diseases, neurodegenerative diseases, diabetes and inflammation.[2]

While successful development of drugs is serendipitous, certain established compound formulations have been in use clinically for over 3,000 years, the equivalent of billions of clinical trials. Their therapeutic indices/profiles are largely known and success rates are consequently relatively high.

Currently, the more controversial issue is the quality regulation of Chinese Medicines as Taiwan does not produce the raw materials. Fortunately, Mainland China has begun to pay attention to this area. Good agriculturing practice, (GAP) and specialized Chinese Medicine culturing are being actively promoted.

## 4. A History of the Modernization of Chinese Medicine in Taiwan[2]

1946 Qualifying Chinese Medicine licensing examination was held. Two candidates applied in the greater Taipei area.
1947 Twelve candidates in the greater Taipei area qualified through that licensing examination.
1948 The Association of Pharmacists and the Association of Chinese herbal drugs Practitioners were established.
1949 Mainland China fell to the Communists. As the Nationalist Government moved its seat to Taiwan and many reputable Chinese Medicine Practitioners also moved.
1950 The College of Chinese Medicine Practitioners was established in Taiwan. Licensing examinations were held periodically. By 2001, 32 examinations had been held, producing 3,047 qualified practitioners.
1956 The legislature passed the Chinese Medicine Education Act, formally endorsing the establishment of Chinese Medicine education and research institutions.[5,6]
1958 The China Medical College was established. Integration of East-West medicines was actively promoted. Two medical and one pharmacy class were established. The curricula included Western medicine supplemented by Chinese Medicine courses. Owing to the lack of a teaching hospital, internship of Western Medicine was carried

out in other largely Western medicine hospitals. Internship of Chinese Medicine was carried out in various Chinese Medicine clinics during the summer recess. Clinical experience was inadequate.

1963 The Research Institute of Chinese Medicine was established. Although comprehensive promotion of Chinese Medicine, including training of practitioners and research workers, was its mandate, the Institute has focused mainly on research. Due to difficulties in the establishment of facilities education has unfortunately not materialized.[5,6]

1966 The inaugural class at the China Medical College graduated. However, when class members were unable to qualify for practicing licenses, a 7-year Chinese Medicine program was established. Those who successfully passed the Chinese Medicine qualifying examination could then sit in for the Western Medicine Practitioner Board Examination, becoming dually qualified in the process if successful. However, even with such dual qualification, the licensee had to choose only one of the areas for practice.[2] By the end of 2001, 2,188 were dually qualified. Of these 439 or about 20% chose to practice Chinese Medicine.

1975 The teaching hospital of the China Medical College opened its doors and provided facilities for internships for students in the Chinese Medicine program but not residency training. A Master's program was established in the Graduate Institute of Chinese Medicine. At last count, that program has produced 230 graduates, mainly focusing on research in herbal medicines, clinical efficacy of formulations and acupuncture.

1984 A five-year post baccalaureate program in Chinese Medicine in which the graduates could only qualify as practitioners of Chinese Medicine was established at the China Medical College. By the end of 2000, there had been 857 graduates, practicing Chinese Medicine was their only choice.

1991 Master's program was established in the Graduate Institute of Traditional Medicine at the Yang-Ming University. By 2000, there had been 81 graduates, most of whom have been engaged in basic Chinese Medicine research. A doctoral program was established in the Institute of Chinese Medicine at the China Medical College. By 2000, thirty-four doctoral graduates in Chinese Medicine had been produced.[2]

1993 The Society for Integration of Chinese and Western Medicine, R.O.C. was established. Legislator Min-Huo Huang was appointed the founding president.

1994 The China Medical College established supplementary Chinese Medicine courses for graduates in Western Medicine. Successful completion of 45 credits entitled the student to sit in for the Chinese Medicine qualifying examination. By the end of 2001 those who qualified through this route numbered 75.

1995 The Committee for Chinese Medicines was established in the Department of Health, Executive Yuan.

1996 The 7 year Chinese Medicine program at the China Medical College was restructured to become an 8-year Western and Chinese Medicines double major program. The expectation was for the graduates to qualify for and practice both medicines.

1997 An 8-year Chinese Medicine program was established at the Chang-Gung Medical College.

1997 Professor Sze-Piao Yang took over as President of the Society for Integration of Chinese and Western Medicine, R.O.C. Professor Yang advocates, as a means to achieve effective modernization of Chinese Medicine: educational reforms including restructuring of the curriculum or instigation of dual programs of basic Chinese and Western Medicines, simplification of classical Chinese Medical Treatises and pharmacopeia, integrative East-West clinical instructions and meticulous documentation of medical histories to provide a solid foundation of integrative East-West medical training in the development of the new generation of Integrative Medical Practitioners.

1998 The Society for Integration of Chinese and Western Medicine, R.O.C. administered the first East-West Integrative Medicine qualifying examination. By the year 2000, five hundred and nine had passed the examination. However, such qualification has remained only nominal.

1998 A doctoral program was established in the Institute of Traditional Medicine at Yang-Ming University.[2]

1999 The Institute of East-West Integrative Medicine was established at the China Medical College and began recruiting students at the Master's level. By 2000 there had been 9 Master's graduates in Chinese Medicine. However, they have been unable to practice East-West Integrative Medicine.[2]

2002 Following an in depth study, Professor Yang proposed a new 7-year as well as a 5-year post-baccalaureate Chinese Medicine program. Following the consensual endorsement by the council of deans of medical schools in Taiwan, the proposal was submitted to the Ministry of Education for recognition as government sanctioned educational and internship programs for Chinese Medicine for the development of the new generation of medical doctors (East-West Integrative Medical doctors).

2002 The Ministry of Education and the Department of Health arrived at the consensus that the aim of medical programs for Chinese Medicine would be the development of a new generation of medical doctors, effectively replacing the 8-year double major programs.

2005 Discontinuation of the Chinese Medicine Qualifying Examination. The supplementary examination will also be discontinued by 2008.

2011 The Senior Level Qualifying Examination will be discontinued.

In summary, East-West Integrative Medicine up to the present has unfortunately not had a legal footing. The place of its followers has remained that of alternative medical practitioners. Hospitalization has not been integrated into the National Medicare System.

## 5. Current Status of the Practice of Chinese Medicine in Taiwan

By the end of 2001, there were 44 Chinese Medicine hospitals (group clinics) and 2,439 clinics. Sixty-two larger hospitals or medical centers had Chinese Medicine departments in the Taiwanese area. Of the 3,979 licensed practitioners, 3,064 or 77% qualified through the Qualifying examination, 915, or 23% had received formal education (including university programs, post-baccalaureate programs and supplementary programs in Chinese Medicine). Among those qualified for both practices, only 439 chose Chinese Medicine.

Warranting concern is the fact that among the 2,263 dually qualified practitioners (2,188 graduated from Chinese Medicine programs while 75 from Western Medicine programs) only 439 or 19% chose to practice Chinese Medicine. The remaining 1,824 or 81% chose Western Medicine, underscoring that the fact the path to modernization of Chinese Medicine in Taiwan is still laden with obstacles as there are still unsolved problems associated with education, evaluation and practice.

For Chinese Medicine, fees payable under the National Medicare Programs are:

1. Diagnostic fees: NT 220 (1 US$ = 35 NT$ per case for daily caseloads of 30–50, coming down to NT 220 for daily caseloads of 50–70 and progressively reduced as the daily caseloads increase. The maximum is 320 while the minimum is 50.
2. Drugs: NT 30 per day's prescription with supplementary charges if a prescription exceeds NT. 200.
3. Acupuncture and therapeutic massage: 6 visits at NT 200 per visit for each treatment course.

In Taiwan, there are about 28,170,000 Chinese Medicine clinical visits, accounting (not including fees collectible from the patients) for about 5.72% of Medicare payments. The National Program does not pay for hospital stays for Chinese Medicine treatments.

**Table 3.1.** Statistics of Payment from the National Medicare Program to Contracted Chinese Medicine Institutions (to 2002.09.31)

| Branch | Chinese Medicine hospitals & Clinics* | | |
|---|---|---|---|
| | Hospitals** | Clinics | Subtotals |
| Taipei | 13 | 614 | 627 |
| Northern | 5 | 267 | 272 |
| Central | 6 | 718 | 724 |
| Southern | 3 | 325 | 628 |
| Kaoshiung-Ping Tung | 9 | 334 | 343 |
| Eastern | 2 | 40 | 42 |
| Total | 38 | 2298 | 2336 |

\* The are still 79 Chinese Medicine departments in general hospitals
\*\* Chinese Medicine Hospitals are largely group practices

**Table 3.2.** Payments from The National Medicare Program (in units of millions of dollars)

| Year | Total Medical Payments | Western Medicine | Chinese Medicine | Dentistry | Pharmacy |
|---|---|---|---|---|---|
| 1998 | 268,669 | 234,526 | 11,297 | 21,721 | 1,125 |
| 1999 | 391,319 | 254,208 | 11,792 | 23,284 | 2,035 |
| 2000 | 296,840 | 258,063 | 11,303 | 25,121 | 2,353 |
| 2001 | 311,547 | 270,288 | 11,832 | 26,712 | 2,715 |

\* Under 5.72% and not including fees collectible from patients does not pay for hospital stays

## 6. The Roles of the New Generation of Chinese Medicine Practitioners

The new generation of Chinese Medicine practitioners should be able to take advantage of the strength of East-West Integrative Medicine. Apart from detailed diagnosis, meticulous physical examination, diagnostically he or she is expected to be able to take advantage of commonly available Western diagnostic instruments such as X-ray, electrocardiogram, ultrasonic and other clinical examination aids. Diagnostic results should be expressed in universally acknowledged terms. Treatment should be aimed at the roots of the diseases. Recommended treatments should principally be Chinese medicines, acupuncture and Chi-Kung supplemented by Western medicines, innovative integrated East-West treatments including inoculation, comprehensive and complementary East-West treatments for cancer and hypertension.

New generation Chinese Medicine practitioners should refrain from invasion, inhumane therapeutic procedures such as surgery on internal organs, intubation, artificial respiration, X-ray irradiation, etc.

New generation Chinese Medicine practitioners should refrain from ethically controversial hi-tech medical procedures such as organ transplantation, artificial insemination, therapeutic abortion, gene manipulation, etc.

New generation Chinese Medicine practitioners should refrain from specializing in obstetrics, ophthalmology, ENT (eyes-nose-throat), dermatology, anesthesia, pathology and radiology.

Therefore new generation Chinese Medicine practitioners are best suited for demands of the National Medicare Program, handling common community ailments with careful and loving medical care, utilization of commonly available medical instruments and equipment in providing readily available, fast, accurate and yet comprehensive, continuing and humane medial services. Such East-West Integrative services, embodying 'truth' 'goodness' 'thrift', along with family medicine or general practitioners, should best fulfil the needs of communities, particularly the aged and the chronically ill.

New generation Chinese Medicine practitioners should pursue advanced studies in Chinese Medicine, scientific exposition of the working of Chinese Medicines, help in the education or be otherwise engaged in basic research or development of Chinese Medicines.

## 7. Conclusions

The modernization, scientific re-examination and internationalization of Chinese Medicine is the current trend. Foreseeing the opportunity, the dander and the need, Mainland China pioneered the development of "East-West Integrative Medical Practitioners ("new generation" Chinese Medicine Practitioners in Taiwan due to regulatory measures) to provide integrative medical services for the general citizenry in hospitals at the county level.

In embodying the virtues of 'truth' 'kindness' and 'thrift', the 'new generation' Chinese Medicine practitioners assume the roles of providers of medical services at the grass root and community levels and are best suited to meet the needs of the communities.

The inclusion of both modern Western and traditional Chinese Medicines in the current 8-year double major programs creates problems as instruction and training of the two vastly different systems are run separately. Even if legally dually qualified, such practitioners would be ill equipped to practice integrative medicine and must choose between one or the other. Those who are qualified only for Chinese Medicine could only practice Chinese Medicine. Such programs could not produce qualified integrative practitioners.

The 5-year post-baccalaureate program, as it is presently run, exacts a high cost as the graduates would have had a combined total of 9 years of post-secondary education and yet could not become capable, able and responsible Chinese Medicine practitioners under such a program. Such a program should either be discontinued or restructured with the 7-year program as a model and aimed at developing 'new generation' integrative practitioners.

East-West Integrative Medicine is obviously the Westernization of Chinese Medicine but also at the same time the Easternization of Western Medicine. In some sense it also represents the integration of Western and Eastern cultures. We have reasons to believe that such a movement could become the mainstream of grass-root community medical services.

The Ministry of Education and the Department of Health have arrived at the consensus of setting the development of "new generation Chinese Medicine Practitioners" as the goal. Based on such a goal as the guiding principle, Tzu-Chi University has developed a "7-year Chinese Medicine" program for both prospective students and faculty members. Since commitment to such a program represents costly investments both on the part of the university and the prospective students, Tzu-Chi would like to have more definitive policies regarding "evaluation" and "practice" to ensure that the efforts will not be in vain. Thus it is calling for meetings between the Department of Examination/Qualification, Department of Health and related departments to discuss and form policies on such related matters.

The next step that warrants serious examination is the degree of concern by the mainstream medical profession, in particular those of the Western School. To protect the public and to ensure the maintenance of high professional standards, regulatory measures regarding training, evaluation and conducts of practice must be established and be sustainable. The deans of medical schools in private and public universities within the country have jointly resolved to provide one or two delegates to establish a "study group' for such purposes. It is hoped that the collective wisdom will soon lead to the establishment of definitive policies regarding such matters.

## 8. References

1. *The Medical History of Taiwan*, edited by Y.M. Jaung (Yuan-Lion Pub. Co. Ltd., Taipei, 1998).
2. C.F. Chen, Y.C. Shum, and M.T. Hsieh, in: *The Way Forward for Chinese Medicine*, edited by K. Chan and H. Lee (Laylor & Francis, London, 2002) p. 303–314.
3. *Annual Publications of National Research Institute of Chinese Medicine*, edited by C.F. Chen (National Research Institute of Chinese Medicine, Taipei, 1997, 1998, 1999, 2000, 2001).
4. *Abstracts of publications on Chinese medicine from Taiwan (1991–1996)*, edited by C.F. Chen (National Research Institute of Chinese Medicine, Taipei, 1998).
5. *Past, present, and future of National Research Institute of Chinese Medicine*, edited by C.F. Chen (National Research Institute of Chinese Medicine Taipei, 1998).
6. *Brief Introduction of the National Research Institute of Chinese Medicine*, edited by C.F. Chen (National Research Institute of Chinese Medicine, Taipei, 1998).

# 4

# Quality Requirements for Herbal Medicinal Products in Europe

G. Franz

## 1. Introduction

The actual discussion on herbal remedies has a long history within the EU with, however, different clinical and economical impact in the individual member states.

The world market on herbal medicinal products (HMP's) clearly reflects the actual situation (Fig. 4.1) with Germany being the leading country in production and utilisation of these medicinal products (Fig. 4.2). It is interesting to note that herbal remedies in this country are an important factor for both prescription and OTC-medicines (Fig. 4.3). Because of their legal status as medicinal products, finished drugs of herbal origin are required to be authorised according to article 4 of CD 65/65 of the EU. Applicants must always document quality, safety and efficacy of their products. The term 'herbal remedies' or 'herbal medicinal products' applies to medicinal products whose active ingredients consist exclusively of plant material or herbal drug preparations as for example powdered herbal drugs, plant secretions, essential oils or herbal extracts. Homeopathic preparations are excluded as are chemically defined natural compounds such as menthol, eugenol, digitoxin, reserpin or others.

The exact definition of herbal drugs is given in the latest issue of the European Pharmacopoeia (Ph Eur) (Fig. 4.4) where a clear specification for herbal drug preparations can be found as well (Fig. 4.5). The most frequently utilised herbal medicinal products, as both OTC and prescribed medicines are listed in Fig. 4.6. However, in the actual Ph Eur more than 140 herbal drugs are described and specified in their respective quality. This number of herbal drug monographs is still increasing and clearly reflects the actual importance of HMP's in the EU.

The change in social attitudes towards the so-called 'natural medicine' ensures continued growth in the utilisation of HMP's. In general, consumers and patients accept these products as therapeutic agents for treatment and cure of diseases and pathological conditions, as prophylactic agents to prevent diseases over the long term and in some cases as proactive agents to maintain health and wellness. Finally, HMP's might be used in adjunct

---

G. Franz • University of Regensburg, Germany
*Complementary and Alternative Approaches to Biomedicine*, edited by Edwin L. Cooper and Nobuo Yamaguchi. Kluwer Academic/Plenum Publishers, 2004.

therapy in order to support conventional therapies. Many herbal remedies are traditionally used to treat minor conditions and illnesses such as coughs, colds, upset stomach and others. The utilisation of herbal drugs as dietary supplements for preventive measures, at the moment is not very common in the EU but markets for these semi-medical products will certainly be opened in the future.

## 2. Quality of HMP's as the Basis of Safety and Efficacy

Since herbal remedies represent complex biological mixtures, reproducible pharmaceutical quality could appear to be a frequent problem. It is obvious that a consistent quality for any HMP can only be assured if the starting materials are defined rigorously and in great detail, whereby the botanical identification of the plant material, the geographical source and the growth conditions must be indicated. Collection of wild grown species often has the problem of an apparent inhomogeneity of the herbal material. Cultivation of defined plant materials under controlled conditions can better guarantee the homogeneity and avoid a possible contamination. Since an European harmonisation of multisource products was officially wanted, uniform definitions of herbal drugs of all EU member states are an essential prerequisite. The Ph Eur as a consequence, has to reflect the practical needs of the EU member states and to provide a broad collection of official, high standard monographs on herbal drugs. One great advantage of the Ph Eur is the multilingual publication, which can be relevant even for countries outside the EU with the consequence that third countries producing herbal raw materials for export to EU-member states have a common basis for all quality requirements.

Quality assessment of HMP's does not pose fundamental problems in practice due to the high standards and long scientific tradition in the fields of pharmacognosy and phytochemistry in the scientific community of the EU member states. More controversy is found when safety and efficacy of herbal remedies are officially addressed. As to the quality requirements, the identical standards are required for the starting biological material i.e. the herbal drug, further for the industrially prepared and specified herbal extract and for the final herbal remedy utilised for both, OTC- and prescription purpose.

According to the official CPMP declaration, 'The herbal drug or herbal drug preparation in its entirety is regarded as the active substance'. Looking on the complex composition of herbal drugs, this might not be easy to understand (Fig. 4.7). The effectors, i.e. constituents with known therapeutic activity, are specified as chemically defined substances or groups of substances which are generally accepted to contribute substantially to the therapeutic activity of a specific HMP (Fig. 4.8). The so-called 'markers' are chemically defined constituents of a specific HMP which are of interest mainly for control purposes independent of whether they have any therapeutic activity or not (Fig. 4.9). With the aid of modern analytical techniques, these marker substances should be easy to detect and to quantify in the respective preparations.

The quality requirements which are specified for each individual herbal drug or the respective HMP are documented in the Ph Eur in a rigorous way, systematically following an identical scheme (Fig. 4.10). Firstly a clear-cut 'Definition' is most important (Fig. 4.11) for the subsequent 'Identification section' (Fig. 4.12), which is based upon botanical characteristics of the starting material, macroscopic and microscopic specifications are provided. In addition, a thin-layer chromatography (TLC)-fingerprint of a specified extract is always required as an additional identification-method.

In the 'Test-section' the purity requirements have to be documented by a series of different analytical techniques including HPLC- and GLC-methods. Emphasis is given today on the detection of potential contaminants such as heavy metals, pesticides, fumigants, radioactive compounds, microbes and mycotoxins. For all these impurities the limiting values are specified and the respective mandatory experimental methods are laid down in separate sections.

The determination and quantification of constituents with known therapeutic activity (effectors) or specified markers (coeffectors) are listed in the subsequent 'Assay section'. A global assay procedure by spectro-photometric methods is more and more often replaced by highly sophisticated HPLC- and GLC-techniques which should allow a more precise determination of individual chemically defined compounds.

A recent issue is the stability of natural compounds in herbal drugs and herbal drug preparations, since it is known that many of these effectors, coeffectors or marker/compounds are being degraded during storage. For the moment, the Ph Eur does not require a detailed stability testing of HMP's but the respective National Authorities request intensive stability testing according to the actual ICH-guidelines (Fig. 4.13).

## 3. Herbal Drug Preparations: Actual Specifications in the Ph Eur

Approximately 75% of the HMP's commercialised in the EU contain herbal extracts mainly in the form of dry extracts which are manufactured for tablets, pills and capsules. This important industrial reality is not reflected yet in the actual Ph Eur. The regulatory authorities such as the European Agency for the Evaluation of Medicinal Products (EMEA) and the respective National Authorities have a great necessity for the specification of defined extracts on the EU market which should be the basis for a subsequent registration of these products. These requirements will be based upon a generally accepted frame monograph on extracts including all forms of extracts such as liquid-, dry- and semisolid preparations including tinctures as well. However, there are many parameters influencing the quality of any extract which should be considered (Fig. 4.14).

Furthermore a clear cut classification-scheme was needed to distinguish between the different grades of proven- or non proven clinical efficacy of the respective HMP's. As a consequence, three different extract-types were defined (Fig. 4.15) to reflect the actual knowledge about therapeutically active principles, markers, coeffectors and further preparations which are therapeutically effective but whose responsible constituents are unknown.

The highest ranking category of the proposed extract type i.e. *standardised extract* should contain chemically defined constituents which, when isolated have effects similar to those of the total extract. There should be a close relation between the daily dose taken and the therapeutic response. These extracts must be standardised to a defined content of the active constituents; the given range should be as narrow as possible due to the dose-response relation.

The second category (extract type) contains the *quantified extracts* having constituents with known therapeutic or pharmacological activity, which are not solely responsible for the clinical efficacy. In most cases there is no clear-cut dose-response relationship for these constituents as the overall efficacy of the extract may be due to other not known compounds.

The last category (extract type) contains the so-called *other extracts* where no constituents, according to the actual scientific knowledge are known to be responsible for the therapeutic effect of the total extract. The quality of these extracts largely depends on the production process. The quality influencing parameters derived from the extract production demonstrated in Fig. 4.16 must be specified to insure batch-to-batch consistency of the final product. At any moment, when scientific data are provided, the defined constituents are responsible for the overall clinical efficacy, the type 'other extract' becomes a 'quantified' or 'standardised' extract.

A special extract-type contains the so-called 'refined' extracts, where the level of characterised constituents is enhanced by industrial purification processes above the levels which normally would be expected when the herbal drug is extracted. The above mentioned classification into the three categories requires an assignment of all the extract types for the actual EU market.

As to the production, special emphasis is given to the quality of the solvents and a possible pre-treatment of the herbal drug (Fig. 4.17). Finally, the labelling of any batch of an extract should follow strict rules (Fig. 4.18). This should provide the appropriate information for any user of the product in question.

## Conclusion

A strong revival in the interest and use of herbal drugs as therapeutic systems can actually be observed. An increasing number of patients appear to have a tendency to limit the use of synthetic therapeutic systems and as a consequence, often prefer to use herbal drugs and herbal drug preparations for treating their complaints. HMP's have a well defined position in the overall market of medicinal products. Recent results in the research of natural products from herbal drugs has considerably contributed to this renewed interest. Various methods have been established which should enable the improved handling of the complexity of HMP's. Highly sophisticated HPLC- and GLC techniques provide selective relative and specific analytical tools for the investigation of the starting biological material and the final products. This rapid increase in scientific knowledge was the basis for the establishment of monographs on herbal drugs and preparations thereof which are being continuously published in the European Pharmacopoeia. These monographs reflect the high quality standards that benefit the patient and play an important role in general health care.

## World Market of HMP'S (2000)

|  | % |
|---|---|
| Europe | 6.9 |
| Asia | 5.1 |
| North America | 3.9 |
| Japan | 2.3 |
| South America | 0.6 |
| Others | 0.8 |
| World | 19.6 |

*This is 39% of the global dietary supplements market of USD 50.6 billion*

Grünwald & Herzberg, ICMAP News (9) 11-16 (2002)

Gerhard.Franz@chemie.uni-regensburg.de

**Figure 4.1.**

## Total Market of HMP'S in the European Union (year 2000)

|  | % |
|---|---|
| Germany | 47 |
| France | 27 |
| UK | 6 |
| Italy | 6 |
| Poland | 4 |
| Spain | 3 |
| all others | 7 |

Gerhard.Franz@chemie.uni-regensburg.de

**Figure 4.2.**

| Market for Phytomedicines in Germany (2000) | |
|---|---|
| | **%** |
| **Phytomedicines upon prescription** | 13 |
| **Phytomedicines (OTC) Pharmacy** | 17 |
| **Total of Phytomedicines** | 30 |
| **Total of synthetic Products (OTC +prescription)** | 70 |

Gerhard.Franz@chemie.uni-regensburg.de

Figure 4.3.

## HERBAL DRUGS
*Plantae medicinales*

**DEFINITION**

**Herbal drugs are mainly whole, fragmented or cut plants, parts of plants, algae, fungi, lichen in an unprocessed state, usually in dried form but sometimes fresh. Certain exudates that have not been subjected to a specific treatment are also considered to be herbal drugs. Herbal drugs are precisely defined by the botanical scientific name according to the binominal system (genus, species, variety and author).**

Gerhard.Franz@chemie.uni-regensburg.de

Figure 4.4.

> **HERBAL DRUG PREPARATIONS**
> *Plantae medicinales praeparatore*

**DEFINITION**
Herbal drug preparations are obtained by subjecting herbal drugs to treatments such as extraction, distillation, expression, fractionation, purification, concentration or fermentation. These include comminuted or powdered herbal drugs, tinctures, extracts, essential oils, expressed juices and processed exudates.

Gerhard.Franz@chemie.uni-regensburg.de

Figure 4.5.


## Most frequently prescribed monopreparation phytomedicines in Germany (year 2002)

| HMP | Therapeutic category |
|---|---|
| Ginkgo biloba leaf extract | circulatory preparations |
| St. John's Wort | antidepressant |
| Horse chestnut seed | vein preparations |
| Yeast | antidiarrheal, acne |
| Hawthorn flower and leaf | cardiac preparations |
| Myrtle | cough remedy |
| Saw palmetto | urological |
| Stinging nettle root | urological |
| Ivy | cough remedy |
| Mistletoe | cancer treatment |
| Milk thistle | hepatoprotective |
| Bromelain – pineapple | enzymeantiinflammatory |
| Echinacea | immunostimulant |
| Chamomile | dermatological |
| Artichoke | hypocholesteremic |
| Kava Kava | tranquilizer |

Gerhard.Franz@chemie.uni-regensburg.de

Figure 4.6.

## Complex composition of Herbal Drugs:

1. Constituents with known therapeutic activity
   (EFFECTORS)
2. Drug-specific compounds
   (MARKERS)
3. Ingredients contributing to efficacy such as increasing absorption
   (COEFFECTORS)
4. Accompanying ingredients such as low amounts of inorganic salts, sugars or amino acids
   (INERT SUBSTANCES)
5. Ingredients with potential negative impact
   (ALLERGENS, TOXINS)
6. Matrix-substances usually not soluble
   (CELLULOSE, LIGNIN)

Gerhard.Franz@chemie.uni-regensburg.de

Figure 4.7.

## Constituents with known therapeutic Activity

> Senna (Cassia angustifolia / - acutifolia, folium, fructus)
  **Sennosides**
> Belladonna extract (Atropa belladonna, radix, folium)
  **L-Hyoscyamine**
> Kava-Kava *(Piper methysticum, rhizoma)*
  **Kava-pyrones**
> Milk thistle extract (*Silybum marianum*, fructus)
  **Silymarin complex**
> Horse chestnut (*Aesculus hippocastanum*, semen)
  **Escin(s)**

Gerhard.Franz@chemie.uni-regensburg.de

Figure 4.8.

# MARKERS

*Any appropriate marker can be selected.*
*The choice has to be justified.*

**Valerian** *(Valeriana officinalis, radix)*
    Valerenic acid *and/or*
    Hydroxy-valerenic acid *and/or*
    Acetoxy-valerenic acid

**Ginkgo extract** *(Ginkgo biloba, folium)*
    Ginkgolides *and/or*
    Flavonoid(s) *and/or*
    Bilobalides

Gerhard.Franz@chemie.uni-regensburg.de

**Figure 4.9.**

---

| European Pharmacopoeia |
|---|

*Each monograph contains the following sections:*

→ **Definition**
→ **Characters**
→ **Identification**
→ **Tests**
→ **Assay Storage**

*For products like herbal extracts*

→ **Production**
→ **Title**
→ **Labelling**

*may be added*

Gerhard.Franz@chemie.uni-regensburg.de

**Figure 4.10.**

## DEFINITION

The definition of a herbal drug in an individual monograph must comply with the definition of the General Monograph herbal drugs (1433).
The following can be included in the definition:

- The state of the drug: whole, fragmented, cut, fresh or dry.

- The complete scientific Latin name of the plant (genus, species, subspecies, variety, author).

- The part of the plant used.

- Where appropriate, the time of harvesting.

- Where appropriate, the minimum content of quantified constituents (constituents with known therapeutic activity or markers).

The statements dried herbal drug prescribes a test or a determination of water.

Gerhard.Franz@chemie.uni-regensburg.de

Figure 4.11.

## IDENTIFICATION

Mandatory tests have to be performed to identify the drug. All the identifications are not necessarily included, some may be absent when they are not feasible or are not fit for the purpose of identification.

### A. MACROSCOPIC BOTANICAL CHARACTERS

The macroscopic botanical characters of the herbal drug are specified to permit a clear identification.

### B. MICROSCOPIC BOTANICAL CHARACTERS

The microscopic examination of the herbal drug reduced to a powder describes the dominant or the most specific characters, including, if necessary, examination of the stomata and stomatal index.

The colour of the powder, the sieve number and the reagent used for the microscopic examination are specified.

Gerhard.Franz@chemie.uni-regensburg.de

Figure 4.12.

## ICH-Guideline: Stability Program

| No. of batches to be examined | 3 |
|---|---|
| Storage conditions | 25°C 60 % rel. humidity<br>30°C 60 % rel. humidity<br>40°C 75 % rel. humidity |
| Time schedule | 0, 3, 6, 9, 12, 18, 24, 36, 48, 60 months ( 25 and 30°C)<br>0, 3, 6 months (40°C) |

Gerhard.Franz@chemie.uni-regensburg

Figure 4.13.

## Parameters Determining the DER Quality of any extract

**1. Herbal drug**

content of extractable matter
content of water
cut size

**2. Solvent utilised**

polarity
concentration
volume

**3. Manufacturing process**

time
pressure
temperature
type of extraction
batch size

Gerhard.Franz@chemie.uni-regensburg

Figure 4.14.

## Types of Herbal Drug Preparations in the European Pharmacopoeia according to their Therapeutically Active Compounds

| Type A | Type B | Type C |
|---|---|---|
| *Examples* | *Examples* | *Examples* |
| Senna leaf dry extract<br>Frangula bark dry extract | Ginkgo leaf dry extract<br>Horse chestnut dry extract<br>St. John's wort dry extract | Valerian root dry extract<br>Passion herb dry extract |
| Compound(s) solely responsible for clinical efficacy are known: <u>active compounds</u> | Compound(s) contributing to clinical efficacy are known but no proof that they are solely responsible for it: <u>active markers</u> | Clinically active compounds not known |
| Adjustment by adding excipients or mixing different lots to a defined content is appropriate | Standardisation by blending before extraction or mixing of different lots is appropriate | <u>selected Markers to:</u><br>Monitor GMP determine content in herbal medicinal products<br>Exact definition of extraction solvent and drug to extract ratio.<br>Minimum content of markers |

Gerhard.Franz@chemie.uni-regensburg.de

Figure 4.15.

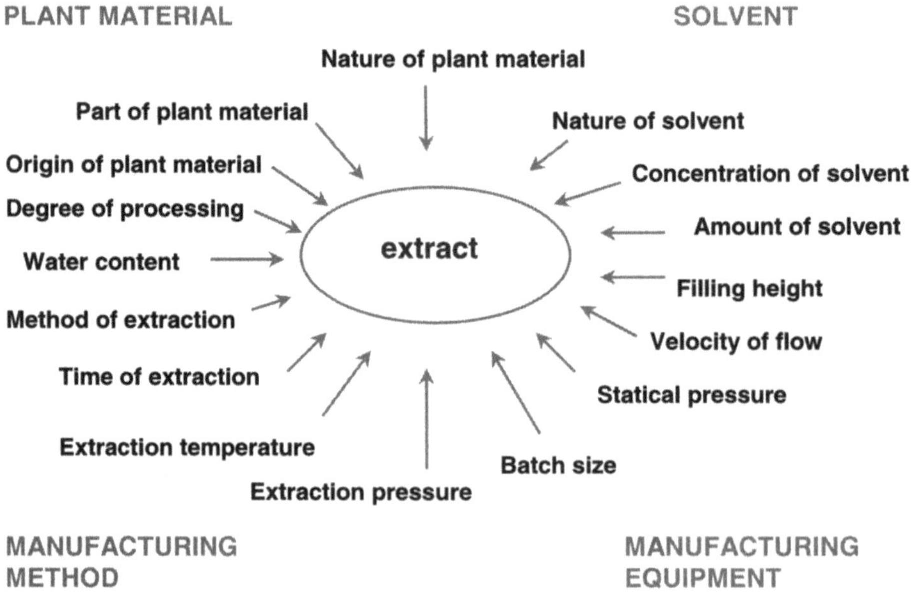

**PLANT MATERIAL**

- Part of plant material
- Origin of plant material
- Degree of processing
- Water content
- Method of extraction
- Time of extraction
- Extraction temperature

Nature of plant material

Extraction pressure

Batch size

**SOLVENT**

- Nature of solvent
- Concentration of solvent
- Amount of solvent
- Filling height
- Velocity of flow
- Statical pressure

extract

**MANUFACTURING METHOD**

**MANUFACTURING EQUIPMENT**

Modified after Gaedcke 1999)

Gerhard.Franz@chemie.uni-regensburg

Figure 4.16.

> **Extracts - *Extracta***
> **Ph Eur 2002**

Production:
Extracts are prepared by suitable methods using ethanol or other suitable solvents.
Different batches of the herbal drug or animal matter may be blended prior to extraction. The herbal drug or animal matter to be extracted may undergo a preliminary treatment, for example, inactivation of enzymes, grinding or defatting. In addition, unwanted matter may be removed after extraction.
Water used for the preparation of extracts is of suitable quality. Except for the test for bacterial endotoxins, water complying with the section on Purified water in bulk of the monograph on *Purified water (0008)* is suitable. Potable water may be suitable if it complies with a defined specification that allows the consistent production of a suitable extract.

Gerhard.Franz@chemie.uni-regensburg

Figure 4.17.


> **Extracts-Extraxcta**

Ph Eur 1997   Ph Eur 2002
Labeling:     Labeling

**The label states:**
- the herbal drug or animal matter used,
- whether the extract is liquid, soft or dry, or whether it is a tincture,
- for standardised extracts, the content of constituents with known therapeutic activity,
- for quantified extracts, the content of constituents (markers) used for quantification,
- the ratio of the starting material to the genuine extract (DER),
- the solvent or solvents used for extraction,
- where applicable, that a fresh herbal drug or fresh animal matter has been used,
- where applicable, that the extract is refined,
- the name and amount of any excipient used including stabilisers and antimicrobial preservatives,
- where applicable, the percentage of dry residue.

Gerhard.Franz@chemie.uni-regensburg.de

Figure 4.18.

## Selected and Pertinent References

Gaedcke, F. The example: St. John's wort. Manufacturing and quality aspects of plant extracts. Pharm Unserer Zit. 2003;43(3):192–201.

Vierling W, Brand N, Gaedcke, F, Sensch KH, Schneider E, Scholz M. Investigation of the pharmaceutical and pharmacological equivalence of different Hawthorn extracts. Phytomedicine 10:8–16.

# 5

# Integrating Medical Knowledge: Old Roots and a Modern Science-Based Approach at the University of Milan

E. Minelli, F. Marotta, U. Solimene

## 1. Traditional Phytotherapy and Traditional Medicine

Traditional phytotherapy is a complex system of skills and practice that belong to the wider system of traditional medicine. Traditional medicine includes great and ancient systems like traditional Chinese medicine, Ayurvedic medicine, Unani medicine and other ethnic systems such as African phytotherapy, Amazonic phytotherapy and so on.

Traditional medicine was born in the period of human history when scientific methods had not yet been invented. For this reason, the criteria for the practice of traditional phytotherapy and the use of herbal remedies may often be esoteric, mysterious, related sometimes to natural or metaphysical forces that are conceived as present in man and the nature that surrounds him.

Nevertheless, these practices may be very effective and often present a great deal of specific therapeutic indications derived from thousands of years of experience. Modern clinical trials are now proving the efficacy of many traditional medical treatments.

In many countries where the primary health care system is based on allopathic medicine, phytotherapy belongs to the kingdom of medicine called Complementary/Alternative Medicine (CAM), Non-Conventional-Medicine (NCM) or Integrative Medicine (IM). The relationship between allopathic medicine and CAM may be stated in terms of:

1. integration
2. inclusion
3. tolerance

---

**E. Minelli** • WHO-Center for Traditional Medical Research in Biotechnology & Natural Medicine, University of Milan, Italy, **F. Marotta** • Hepato-GI Dept., S. Giuseppe Hospital, Milan, Italy, **U. Solimene** • WHO-Center for Traditional Medical Research in Biotechnology & Natural Medicine, University of Milan, Italy

*Complementary and Alternative Approaches to Biomedicine*, edited by Edwin L. Cooper and Nobuo Yamaguchi. Kluwer Academic/Plenum Publishers, 2004.

1) In an *integrative system*, phytotherapy is officially recognized and incorporated into all aspects of primary health care. There is a pharmaceutical national policy; herbs and formulas are officially registered and regulated; phytotherapy is available in hospitals and clinics; phytotherapy care is reimbursed through the public assistance system; there is public and national research and there are formal courses in phytotherapy. China, Korea and Vietnam are the only countries in the world with systems that integrate phytotherapy into allopathic medicine.

2) In an *inclusive system*, phytotherapy is officially recognized but not fully integrated into primary care, educational programs or health policies and regulations. Nevertheless, in these systems the integration of phytotherapy into allopathic medicine may be very advanced. Many countries in the East and West such as Great Britain, Canada, Niger, Mali and Guinea are in this phase of advancement.

3) In a *tolerant system*, the public health is all in the hands of allopathic medicine, but some alternative practices are just tolerated or left aside.

## 2. Diffusion of Traditional Phytotherapy in Western Countries

In ancient times, phytotherapy was widely spread throughout the western world. It had a sudden decline in the first half of the XX century in relation to the rapid growth of chemical drug production. In the last thirty years, traditional phytotherapy has experienced a resurgence of interest in industrialized countries. It's very hard to evaluate the diffusion of traditional phytotherapy with all of its different aspects, but if we look at the wide diffusion of CAM, where phytotherapy plays a big role, we must stress that CAM and traditional phytotherapy are areas of modern medicine that are rapidly growing in the industrialized world. In some recent studies we can see that the percentage of people who use traditional medicine is 48% in Australia, 70% in Canada, 42% in the USA, 38% in Belgium and 75% in France.

In accordance with an old evaluation (1974) of WTO we can realize that the use of natural herbs for the preparation of phytotherapeutic drugs is almost incredible. The following are examples:

3.000 tons per year of aloe
10.000 tons per year of artichoke
5.000 tons per year of chinabark
1000 tons per year of belladonna, giusquiamo and stramonium
1000 tons per year of digitalis
5000 tons per year of senna

There are myriad reasons for this phenomenon, including great criticism about adverse reactions to chemical drugs, their high cost, and the increase in chronic diseases that necessitate the long-term drug use to improve the QOL of patients who often have no hope of recovery from the disease. In this case, many patients and doctors prefer to use natural drugs or natural remedies that have a gentler impact on the patient. Diseases such as AIDS or cancer often use therapies that greatly worsen the QOL of the patients. Traditional phytotherapy combined with chemotherapy can improve the QOL of the patient.

## 3. Pros and Cons

Many providers of traditional phytotherapy keep looking for continuous acknowledgment and support. At the same time, many providers of allopathic medicine, who are often from countries characterized by a strong history in traditional phytotherapy, express great criticism, and often mistrust the claimed benefits. Regulators fight against the problems connected to safety and efficacy of traditional herbal medicines, while consumers do not want their access to traditional phytotherapy limited in any way. Therefore, together with an increasing demand for traditional phytotherapy and CAM, a demand for evidence about safety, efficacy and quality of the products and the practice has also been increasing.

However, generally speaking, the increasing use of traditional phytotherapy and CAM has not been accompanied by an increase in the quantity, quality and accessibility of clinical evidence that could support its indications.

In order to understand and evaluate the diffusion and framework of the use of herbal medicine in industrialized countries, as well as the connected issues, it is important to consider two different aspects:

a) The importance of the drug and the consequences of its use.
b) The need for raw materials of vegetal origin.

a) Industrialized countries are more and more characterized by a progressive and constant increase in the use of pharmaceutical products. This increase is the consequence of multiple features. For example, the fact that many chronic diseases (diabetes, hypertension, etc.) are cured in a more tailored way is one of the reasons. However, besides the use of essential medicines, one can observe an increasing use of medicines which have been defined as "medicines of civilization" and which tend to be widely diffused. These medicines include analgesics, tonics, and above all, psychotropic drugs, as well as anorectic and hormonal products for contraceptive or substitutive therapy, etc. Looking for milder and natural therapies is also probably due – often for questionable reasons – to iatrogenic chemical and environmental pollution. This has led to a new therapeutic approach that is characterized by the association of herbal substances to pharmacological therapy, and sometimes by the total substitution of pharmacological therapy with herbal products.

Thus, in the industrialized world, there is a tendency to emphasize the paradigm 'one problem, one pill' and to include traditional phytotherapy and CAM in this axiomatic formula as a possible answer to the problems of civilization. Thus, traditional phytotherapy and CAM are often abstracted from their natural context, which consists of a holistic approach to life, the importance of an equilibrium between the mind, the body and the environment, and an emphasis on health rather than on ill-health. In fact, in traditional phytotherapy and CAM, in general, the practitioner focuses more on the global condition of the patient than on a single symptom or the ascertained organic disease of the patient.

Therefore, it often happens that instead of considering answers or behaviour that could bring to review the whole lifestyle, the paradigm 'one problem, one herb' also takes place in traditional medicine. However, almost none of the traditional herbal medicines are usually regarded as exclusive therapeutic modalities, and manual, hygienic, dietary and spiritual techniques participate at the same level in determining the therapeutic action of a traditional medical act.

A remedy for hypercholesteremia, for example, would be determined by studying and modifying the patient's lifestyle, diet, and mental and spiritual attitude towards the world. At the same time, arthrosic pain could be treated with advice concerning lifestyle, movement, relaxation, breathing, diet and manual techniques to deal with the pain. The use of herbs should be only one step of the cure.

The reiteration of the model 'one problem, one pill' when applied to traditional medicine raises several issues. The principals are:

- The increase in the quantity of drugs used may cause problems relative to the sustainability of such an increase, in respect to the ethno-botanic environment where the plants grow.
- The chronic and continuous use of herbal medicines may cause problems relative to its stock and adverse effects linked to chronic toxicity often unknown in the traditional use of herbs.
- The impact of traditional drugs on pharmacological polytherapy may develop interactions among various drugs that are often unknown.

b) The great use of vegetal raw materials in industrialized societies, as previously shown, is strictly connected to the first issue. Beyond this, we also have to underline the way increasing diffusion of chronic diseases and constant attempts to ameliorate QOL actually contribute to this increased use. The use of herbal medicines to overcome the adverse effects of major allopathic therapies has brought a parallel increase in the use of herbal medicines. As a consequence, there are strong concerns about how to sustain such a development on the one hand and protect the traditional ethno-botanical resources, on the other. This may lead to the chemical production of remedies which are normally collected in nature.

## 4. The Need of Scientific Research

The great spread of herbal medicine in industrialized countries has to deal with the prevalent medical-epistemological or scientific model. Therefore, an immense amount of traditional knowledge has to be tested through scientific research and validation procedures. The aim is to prove the safety and the efficacy either of a single drug or several drugs in combinations that have been used for thousands of years. In this respect, the WHO has more than once underlined not only the need to prove the safety of utilized drugs, but also the need to prove the efficacy, especially when – this could happen only in Western health care systems - these substances are used as substitutes for medicines whose efficacy has already been verified. In these cases, it is evident that the medical doctor should evaluate the relationship of efficacy/safety between the herbal drug and the chemical drug that will be substituted.

For these several reasons, it seems clear that the spread of herbal medicines in Western countries requires deep study in order to establish the safety and efficacy of commonly used herbal medicines. With this aim, the joint activity of different specialists is needed in order to introduce the safe use of traditional prescriptions into industrialized societies.

There is no question that herbal medicines, which have not been used for a long time, and which have not been previously analyzed, need testing procedures on a par with those used for chemical drugs.

Concerning herbal medicines with a documented traditional use, the following procedures can be followed in order to carry out studies and research about safety and efficacy.

## 5. Review of Data in the Literature

In order to evaluate the safety and/or the efficacy of herbal medicines, either from a single plant or from a combination of plants, the first step consists of carefully analyzing the information already available in the literature.

Bibliographic research must include reference texts, article reviews, systematic control of primary sources and/or databases. Of course, such sources could contain inexact information but they could also quote important references useful for a deeper analysis.

The profile of research should be described and each quoted reference should be reported. Publications by independent researchers of similar data concerning safety and efficacy are highly important.

In a holistic approach of traditional medicine, the patient is treated from physical, emotional, mental and spiritual points of view, also taking into account the surrounding environment. As a consequence, the majority of traditional medical systems provide the use of herbal medicines with certain behavioral rules, diet regimes and healthy habits.

When one reviews the bibliographic material concerning traditional medicine (including both herbal medicines and therapies based on traditional procedures), one must take into account the theories and concepts that characterized the single operator, and the cultural backgrounds of the people involved.

Biochemical or cellular in vitro data about safety are studied as indicators of a potential toxicity. The data derived from studies on animals in vivo are major indicators of toxicity. The pharmacological effects, which can be observed in vitro and are useful to identify action mechanisms in the animal, are not necessarily applicable to human beings, in terms of both safety and efficacy.

Finally, such data have to be confirmed by clinical studies.

**Safety.** When one wants to proceed with new pharmacological and toxicological studies, one must also consider the documented adverse effects (recorded according to well defined principles of pharmaco-supervision) of herbs, mixtures of herbs, similar species, constituents of herbs, their preparations and finished products.

The absence of any documented adverse effect does not totally guarantee the safety of herbal products. However, it may not be necessary to have a complete series of toxicological tests.

Yet, there are certain effects that are either difficult or impossible to be determined from a clinical perspective. For this reason, tests to evaluate immunotoxicity (e.g. tests for allergic reactions), genotoxicity, risk of oncogenesis and toxicity on the reproductive system, should definitely be carried out.

**Efficacy** According to many documents approved by the WHO, it is important, both for simple herbal products and for the mixtures derived from them, that the requirements to prove their efficacy, as well as the documentation required to support the indications for use, are correlated to the nature and level of the indications.

In order to determine the efficacy of a product for the treatment of minor disorders, for non specific indications and for preventive uses, less strict requirements can be considered sufficient (e.g. observational studies), especially if the product has been used for a long time, if there is certain experience about the use of herbal medicines, or if there are data supported by pharmacological studies.

If the drug is indicated for a major pathology or for specific indications and if there is no documentation of long term use, the indications must be supported by a normal set of clinical trials.

## 6. Clinical Trials

The scope and the project of these studies should be based on information concerning the traditional use obtained through national official documents, specific literature, as well as through traditional medicine practitioners.

Ethnopharmacognosy research is fundamental to all of the non-codified medical systems. These systems require knowledge of the utilized vegetal species, parts of the plant containing the drug, collection period and conditions, preliminary treatments to make the drug safe (e.g. aconitum), the modality of administration to the patient (e.g. difference between methods for the natural or chemical extraction), possible additives to stabilize the products when preserved (e.g. the difference between fresh and preserved product), the extraction method for the products (decoction, infusion) and the proper dosage.

In the case of a new herbal medicine or a new indication of a known one, a change in the times of administration, the general principles and the requirements for a clinical trial should be the same as those for conventional medicine.

However, in certain cases, the profile of these studies has to be adapted to the specific peculiarities of herbal medicines.

Well-structured trials, through randomized clinical controls, guarantee the highest level of evidence concerning the efficacy. This kind of control promotes the acceptance of herbal medicines in different regions and in different populations characterized by different traditional cultures. Randomization methods and the placebo cannot always be adopted, since they may imply ethical or technical problems. For instance, a control by placebo cannot be utilized when an herbal medicine has a very strong smell or taste, such as those products that contain some essential oils. Furthermore, the patients, to whom herbal medicines have previously been given cannot be inserted in a control group during the evaluation phase of an herbal medicine with a particular smell or taste.

As an alternative, it is possible to use a positive control, as, for example, a well-defined treatment. Since observational studies involve a relevant number of patients, they can be considered a valid evaluation tool for herbal medicines.

**Molecular research.** The scientific approach aims at isolating and identifying pure molecules, starting from the drugs utilized in traditional medicine, in order to arrive at classical pharmacological screenings on animals and evaluate their therapeutic potency. This methodology is very complex and expensive when weighed against the expected results. Thus, such an approach should be a second step in phytotherapy studies. Even so, it is a priority in many specialized institutes and in major universities of developing countries.

**The rational use: guaranteeing proper use**
Various aspects characterize the rational use of traditional medicine:

- Education and qualification of providers
- The proper use of good quality products
- Communication between traditional phytotherapy providers and allopathic medical doctors and patients
- Preparation of scientific information and guiding documents for the public

The education faces two challenges:

1. To ensure adequate knowledge, education and qualification of the providers
2. To use the education to ensure that traditional phytotherapy providers and health care professionals understand and appreciate the integration between the two kinds of care that they can offer.

The proper use of good quality products can reduce most of the risk linked to traditional phytotherapy. Intensive work is also needed to increase the data on safety and the proper use of traditional phytotherapy. For example, adverse effects may arise from the interaction of herbal medicines with chemical drugs. There are still many patients who don't reveal to their allopathic doctors that they are taking herbal medicines. Increased information, education and communication can overcome these problems.

The WHO Collaborating Center for Traditional Medicine of the University of Milan faces all these aspects of traditional phytotherapy with a strategy that includes research, educational training, and legal advice to several public institutions.

Indeed, the Research Center in Bioclimatology, Biotechnology and Natural Medicine is one of the oldest centers within the University of Milan. It was founded in 1969, and is connected to the Departments of Normal Human Anatomy, Biochemistry, Microbiology, Medical Chemistry and Thermal Medicine chaired by Professor Umberto Solimene.

Since its foundation, the center has dealt with issues concerning the QOL and developed research and training in this field. Particular attention has been given to natural and complementary medicines. The approach has especially focused on research and training, both from scientific and multidisciplinary points of view.

In 1997, the Research Center in Bioclimatology, Biotechnology and Natural Medicine became a collaborating center of the WHO for traditional medicine. Now, the center is the focal point for the European Region of the WHO, and it collaborates with the WHO headquarters for projects that study or experiment with CAM (Complementary and Alternative Medicine), relative to the evaluation of their use in primary health care, in particular.

Professor Emilio Minelli, member of the Evaluation Commission for the Control of the Observational Projects on CAM of the General-Direction for Health of the Regional Council of the Lombardia Region, deals with the methodology of the territorial application of CAM.

During thirty years of activity, in addition to research, the center has organized several meetings and international symposia for studying, as well as deepening and spreading CAM. Some examples are:

- 1995-9 International Meeting on Integrated Sciences
- 1997 International Symposium on Hypertension: one Medicine, two Cultures
- 1997 Nutrition & Climate
- 1998 Research Methodology in CAM
- 1998 The Unconventional Medicine Market in Italy
- 1998 "Bush Doctors": a Traditional Medicine Experience in Jamaica
- 1999 National Report on Medical Doctors Practising CAM
- 1997-2000 WHO: Research in Methodology in Traditional Medicine

Moreover, the center organizes and runs post-graduate specialization courses for medical doctors in acupuncture, non-conventional medicine and complementary techniques. According to the statute, the courses are under the Faculty of Medicine and their main features are:

♦ **Quadriennial Course of Non-conventional Medicine and Complementary Techniques**
- Admission: National Board Certified-MDs (a few openings for: biologists, pharmacists and psychologists)
- Availability: 30 post-graduate students
- Workload: ≥ 1000h
- Schedule: Saturday/Sunday
- Cost: € 1250/year
- Advancement through the course: only those attending ≥ 80% of lessons
- Credits: regularly certified educational credits
- Final certification: After the successful completion and defence of a thesis
- Certification: "Expert in Nonconventional Medicine and Complementary Techniques"

♦ **Quadriennial Course of Acupuncture**
- The teaching programme is based on the Guidelines for Basic Training on Acupuncture of the WHO. It covers the study of basic traditional and modern acupuncture, of its application and selected problem-solving clinical cases.
- Admission: only National Board Certified-MDs
- Availability: 100 post-graduate students
- Workload: ≥ 1000h
- Cost: € 955/year
- Advancement through the course: only those attending ≥ 80% of lessons
- Credits: regularly certified educational credits
- Final Certification: Thesis and practical examination
- Certification: *"Expert in Acupuncture"*

The center has started several international co-operations with various institutions such as: Kitasato Institute (Japan), Russian Ministry of Health, the Beijing University (China), Sidney University (Australia), Pharmacology Department of Cape-Town (South Africa), Traditional Medicine University of Mongolia, Indian Ministry of Health, Brasilia University (Brazil), S. Marco University in Lima (Perù) and the Ministry of Health of Vietnam.

In particular, to study QOL in extreme conditions, it has started a collaboration with the Aero Space Center 'Y. Garin' in Moscow, Russia. This relationship is aimed at studying the QOL of astronauts during the preparation phases, the flight, as well as during the psycho-physical rehabilitation after returning to earth.

Furthermore, the center is also Secretary of the World Federation of Thermalism and Climatology. Connected with its interest concerning the QOL, the center has also started a collaboration with Professor Francesco Marotta of the Fatebenefratelli Foundation of Milan. Its aim is to evaluate the impact of certain natural products on the QOL as well as on clinical management of patients with chronic gastrointestinal diseases, using a strict science-based conventional approach.

## 7. A Lesson from Integration of Natural Medicine and Science-Based Studies

To give an example of our activities, we report on a natural fruit-derived compound that has been clinically analyzed under strict methodological parameters. Indeed, for the past five years or so, we have conducted a number of clinical studies on a fermented papaya preparation (Immune-Age, Osato Research Institute, Gifu, Japan, composition shown in Table 1) which has been shown to be endowed with effective acid- and heat-resistant antioxidant properties.

Preliminary studies further showed that this compound (originally named FPP, fermented papaya preparation) when administered to healthy individuals at midnight and 3h prior to the ingestion of 80g of ethanol, could significantly prevent gastric damage, as assessed either by microscopic or macroscopic evaluation (Fig 1). A more detailed analysis has also proven a significant decrease in ethanol-induced free radical generation at the gastric mucosal level (Fig. 2).

Such a potent antioxidant effect was also confirmed in a set of ex vivo studies in alcoholics. Indeed, the gastric mucosa of ongoing alcoholics, when incubated in vitro with labelled cyanocobalamin and intrinsic factor, appeared to markedly impair a proper binding. On the other hand, when the same patients took the FPP supplement, the previously observed inhibition of binding, significantly improved (Fig. 3).

Table 5.1. Composition of Fermented Papaya Preparation as ascertained by a certified laboratory

| Fermented Papaya Preparation (100g) | | | |
| --- | --- | --- | --- |
| Japan Food Research Lab., report n. 397100396-007 Tokyo, December 16, 1997 | | | |
| CHD | 90.7 g | Phenylalanine | 11 mg |
| Moisture | 8.9 g | Tyrosine | 9 mg |
| Protein | 0.3 g | Leucine | 18 mg |
| Fat | none | Isoleucine | 9 mg |
| Folic acid | 2 µg | Methionine | 5 mg |
| Niacin | 0.24 mg | Valine | 13 mg |
| Lysine | 6 mg | Glycine | 11 mg |
| Histidine | 5 mg | Proline | 8 mg |
| Aspartic acid | 27 mg | Tryptophan | 2 mg |
| Serine | 11 mg | Threonine | 8 mg |
| Arginine | 16 mg | Glutamic acid | 37 mg |

**EFFECT OF FERMENTED PAPAYA PREPARATION (FPP) ON ALCOHOL-INDUCED GASTRIC MUCOSAL DAMAGE IN HEALTHY INDIVIDUALS**

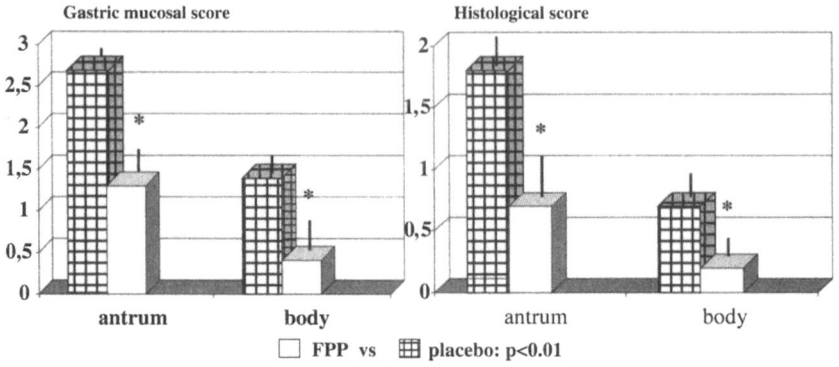

Figure 5.1. Gastric mucosal and histology score after challenge with alcohol: effect of FPP

**EFFECT OF FERMENTED PAPAYA PREPARATION (FPP) ON GASTRIC MUCOSAL CHEMILUMINESCENCE AFTER ETHANOL CHALLENGE IN HEALTHY INDIVIDUALS**

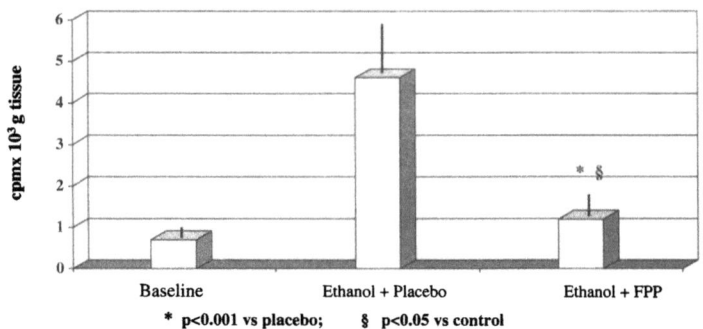

Figure 5.2. Chemiluminescence assay after alcohol challenge: effect of FPP

Clinically, even though alcoholics generally showed a Dual Isotope Shilling Test (DIST) comparable to healthy controls, a subgroup of patients with overtly abnormal vitamin B12 absorption was identified (DIST <10%). This subgroup significantly improved their DIST when regularly supplemented with FPP (Fig. 4).

All of the above findings are of interest when considering the increasing evidence of the involvement of active oxygen species in the promotion stage of carcinogenesis (1). Indeed, one of the radical-modified DNA adducts, i.e. 8-hydroxydeoxyguanosine (8-OhdG) has recently received a great deal of attention as a potential biomarker (2). 8-hydroxyguanine is one of the major products of base damage when DNA is exposed to physiologically relevant systems producing OH and $^1O_2$. The purpose of the present investigation was to test the effect of oral supplementation of a certified papaya-fermented product, (i.e. Immun-Age), (3), on enzymatic abnormalities and free radical-damaged DNA parameters associated with pre-malignant changes in the upper gastrointestinal mucosa.

## VITAMIN $B_{12}$ BINDING BY INTRINSIC FACTOR
## IN VITRO INCUBATION OF IF WITH $^{57}$Co-LABELLED COBALAMIN

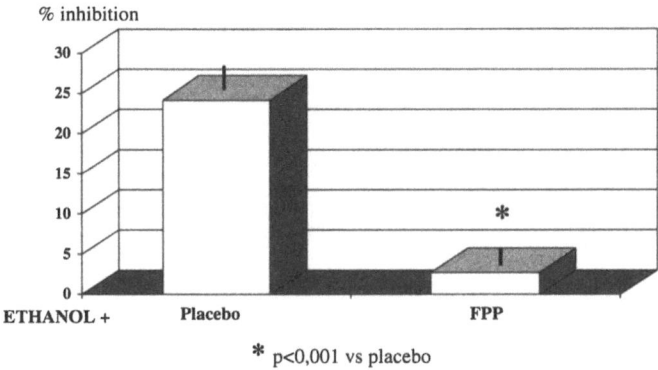

\* p<0,001 vs placebo

**Figure 5.3.** Effect of FPP administration in improving in vitro binding between IF and cyanocobalamin

## DUAL ISOTOPE SCHILLING TEST

*Free $^{58}$Co-Vitamin $B_{12}$ Excretion*

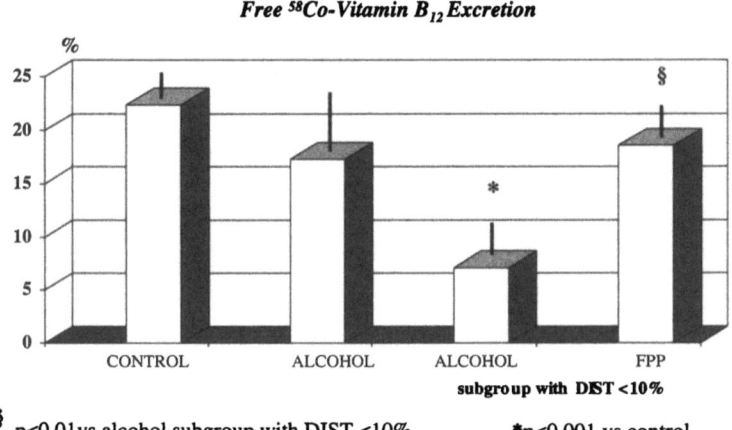

§ p<0.01 vs alcohol subgroup with DIST <10%    \*p<0.001 vs control

**Figure 5.4.** Effect of FPP administration in improving cyanocobalamin absorption test

## 7.1. Materials and Methods

Sixty patients with known atrophic gastritis and intestinal metaplasia and with recent negative results from a urea breath test were selected as our population study group. Each of these patients was carefully interviewed about dietary habits with particular attention to estimated intake of vitamins, alcohol, smoking and drugs or nutraceuticals. In particular, patients were instructed not to consume any vitamin or "health supplement" during the study period. All subjects underwent a routine upper GI endoscopy during which multiple biopsy samples were taken in the antrum. Biopsy samples were processed for routine

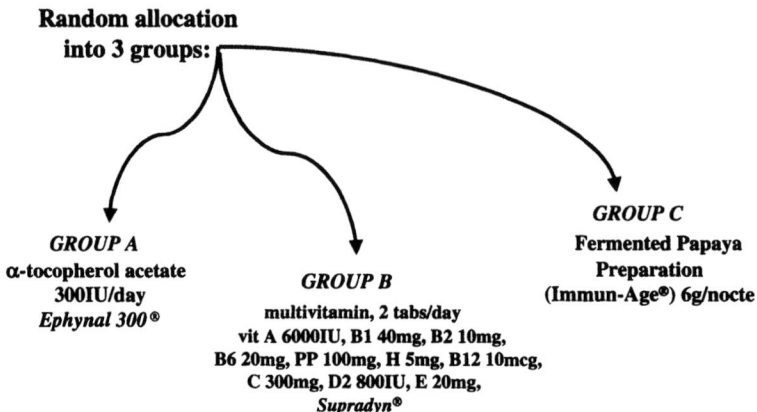

**Figure 5.5.** Study design of oral antioxidant supplementation in patients with chronic atrophic gastritis

histology with particular attention to intestinal metaplasia and H.P. presence. The other samples were processed to test: alpha-tocopherol, malonyldialdehyde, xanthine oxidase, ODC activity and 8OHdG. Overall plasma antioxidant status was assessed as well at entry. Patients were divided into three groups according to age, gender, drinking and smoking habits. After overnight fasting, all patients underwent baseline blood chemical evaluation as described below. All subjects were randomly allocated into one of the following 6-month supplementation trials: A) Immune-Age 3g/day (obtained from biofermentation of carica papaya, pennisetum purpureum and sechium edule, Osato Research Institute, Gifu, Japan); B) Vitamin E 300mg/day; C) Multivitamin preparation ( Supradyn®, Roche, Switzerland ) 2 tablets/day (Fig. 5).

All the above histological and biochemical parameters were repeated at 3 and 6 months. Histological assessment was carried out in a blind fashion and reviewed by one experienced investigator.

## 7.2. Results

There was no statistical difference among the groups as far as dietary composition during the study period. Routine blood chemistry was within a normal range at entry and did not change during the study period, irrespective of the treatment employed.

**Plasma oxidant/antioxidant status.** The baseline oxidant/antioxidant assessment at entry, as measured by the serum levels of $\alpha$-tocopherol, malonyldialdehyde, superoxide dismutase, hydroperoxide and glutathione peroxidase, was within normal limits and did not change irrespective of the antioxidant treatments employed.

**Gastric mucosal oxidant/antioxidant assessment.** As compared to controls, the mucosal concentrations of MDA and XO in patients with atrophic/metaplastic changes were significantly higher ($p < 0.05$). Each of the three antioxidant treatments employed brought about a normalization of these parameters ($p < 0.05$) without any appreciable difference among them (Fig. 6).

**8-OhdG concentration in gastric mucosa.** Patients with CAG showed a significant increase of 8-0hdG ($p < 0.05$) which was unrelated to other tested parameters. This

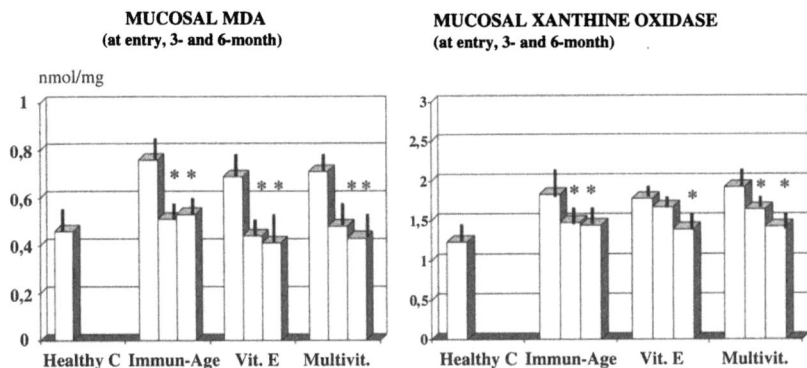

**Figure 5.6.** Effect of antioxidants supplementation of mucosal markers of oxidative stress

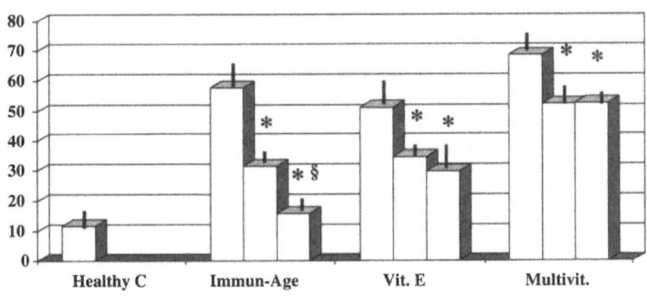

**Figure 5.7.** Effect of antioxidants supplementation on ornithine decarboxylase activity

abnormality was not affected by vitamin E or by multivitamin supplement throughout the study period. Immune-Age, however, brought about a significant, although partial, decrease at the end of the study period ($p < 0.05$) (Fig. 8).

## 7.3. Conclusion

Although the incidence of gastric cancer is declining it remains a common cause of death from malignancy worldwide. There is substantial evidence to support a sequence of histological changes in the mucosa prior to the development of intestinal type gastric carcinoma. The development of inflammatory gastritis progresses to gastric atrophy, to intestinal metaplasia, to dysplasia and finally to intestinal type carcinoma. On the other hand, increased concentrations of lipid peroxidation products have been found in the serum of gastric cancer patients. Oxidative damage to DNA may result in base modification, sugar damage, stand break and DNA-protein cross-links. Of these, modification of guanine by hydroxyl radicals at the C-8 site, frequently estimated as 8-OhdG, is the most commonly studied lesion. However, there have been few reports tackling the issue of vitamin/antioxidant supplementation in patients with pre-malignant gastrointestinal lesions in view of reducing ODC activity (3)

## MUCOSAL 8-OHdG LEVEL IN CAG: EFFECT OF ANTIOXIDANTS SUPPLEMENTATION (at entry, 3- and 6-month)

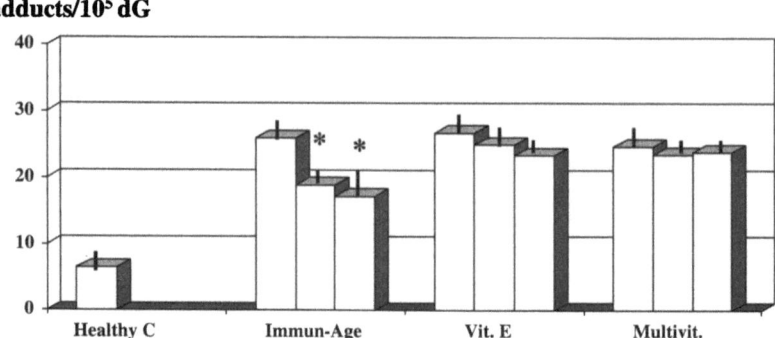

**Figure 5.8.** Effect of antioxidants supplementation of mucosal level of 8-OhdG

### A Multistep Carcinogenesis Model

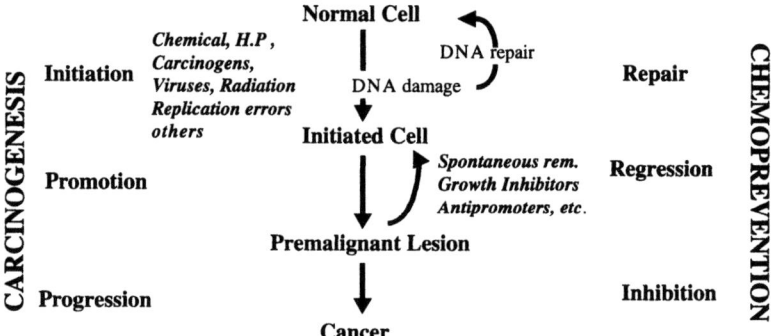

**Figure 5.9.** An overview on the multistep theory of carcinogenesis

and there is a need for a more comprehensive study of the issue. From the present study it appears that patients with CAG, although maintaining a normal plasmatic redox status, show a significant impairment of it at a mucosal level. Virtually any antioxidant supplement (4) seems to improve this abnormality to a certain extent. However, this is the first study showing that nutritional intervention with a product, endowed with immuno- and NO-modulator (6) properties associated with lipid- and protein- (7) antioxidant effects, can significantly improve an oncological biomarker in the gastric mucosa. These positive preliminary data are encouraging when considering the rather inconclusive results coming from survey trials (5) which are likely to have been biased by the intrinsic limitations involved in any large scale epidemiological study. Indeed, besides the biochemical data *per se* such findings have the potential to hold practical importance since gastric cancer meets many of the criteria of the multi-step carcinogenesis model where intermediate phases of the process can be theoretically targeted by therapeutic interventions (Fig. 9).

# 8. References

1. Kensler TW and Taffe BG. Free radicals in tumor promotion. Advances in Free Radical Biology and Medicine. 2:347-387 (1986).
2. Shigenaga MK et al. Urinary 8-hydroxy-2'-deoxyguanosine as a biomarker of in vivo oxidative DNA damage. Proc Natl Acad Sci USA 86:9697-9701 (1989).
3. Bukin YV, Draudin-Krylenko VA, Orlov EN, Kuvshinov YP, Poddubniy BK, Vorobyeva PV, Shabanov MA. Effect of prolonged beta-carotene or DL-alpha-tocopheryl acetate supplementation on ornithine decarboxylase activity in human atrophic stomach mucosa. Cancer Epidemiol Biomarkers Prev 4:865-870 (1995).
4. Dawsey SM, Wang GQ, Tailor PR, Li JY, Blot WJ, Li B, Lewin KJ, Liu FS et al. Effects of vitamin/mineral supplementation on the prevalence of histological dysplasia and early cancer of the esophagus and stomach: results from the Dysplasia Trial in Linxian, China. Cancer Epidemiol Biomarkers Prev 3:167-172 (1994).
5. Xu GP, Song PJ, Reed PI. Effects of fruit juices, processed vegetable juice, orange peel and green tea on endogenous formation of N-nitrosoproline in subjects from a high-risk area for gastric cancer in Moping County, China. Eur J Cancer Prev 2:327-335 (1993).
6. Rimbach G, Park YC, Guo Q, Moini H, Qureshi N, Saliou C, Takayama K, Virgili F, Packer L. Nitric oxide synthesis and TNF-$\alpha$ secretion in RAW 264.7 macrophages. Mode of action of a fermented papaya preparation. Life Sci 2000; 67:679-694.
7. Rimbach G, Guo Q, Akiyama T, Matsugo S, Moini H, Virgili F, Packer L. Ferric nitrilotriacetate-induced DNA and protein damage: inhibitory effect of a fermented papaya preparation. Anticancer Res 2000; 20:2907-2914.

# CAM Approaches to Specific Diseases and Biomedical Conditions

# 6

# Identification of Diseases that may be Targets for Complementary and Alternative Medicine (CAM)

**Aristo Vojdani and Edwin L. Cooper**

## 1. Introduction

### 1.1. What is the Place of Complementary and Alternative Medicine (CAM)?

Complementary and alternative medicine has an enormous challenge to determine how environmental factors such as infection and xenobiotics interact. Effective strategies for early detection, diagnosis, prevention and therapy should then emerge. Despite significant advancements in medicine and discovery of human genome, little attention has been paid to early diagnosis and prevention of complex diseases. Diseases such as Crohn's, ulcerative colitis, cardiovascular, autoimmune, cancer, diabetes and others are multifactorial, meaning that they cannot be ascribed to mutation in a single gene or to a single environmental factor. Rather they arise from the combined action of many genes, environmental factors such as infection, xenobiotics and risk-conferring behaviors (Kimberstis and Roberts, 2002). Complex diseases have reached epidemic proportions in the United States and in a particularly disturbing trend, striking at a young age. With this explosion, we should be committed to expanding the approach to healing. This opens classical approaches to complementary and alternative therapies.

The clear physiological link between diabetes and obesity, or environmental toxicants and lupus or multiple sclerosis (MS) makes a strong case for the causal role of diet, exercise,

---

**Aristo Vojdani, Ph.D., M.T.** • Laboratory of Comparative Neuroimmunology, Department of Neurobiology, David Geffen School of Medicine at UCLA, University of California, Los Angeles, Los Angeles, California 90095, USA • Section of Neuroimmunology, Immunosciences Lab., Inc. 8693 Wilshire Blvd., Ste. 200, Beverly Hills, California 90211, USA    **Edwin L. Cooper, Ph.D., D.Sc.** • Laboratory of Comparative Neuroimmunology, Department of Neurobiology, David Geffen School of Medicine at UCLA, University of California, Los Angeles, Los Angeles, California 90095, USA

Correspondence: Aristo Vojdani, Ph.D., M.T. 8693 Wilshire Blvd., Ste. 200, Beverly Hills, CA 90211; Tel: (310) 657-1077; Fax: (310) 657-1053

*Complementary and Alternative Approaches to Biomedicine*, edited by Edwin L. Cooper and Nobuo Yamaguchi. Kluwer Academic/Plenum Publishers, 2004.

infections and environmental pollutants (Marx, 2002). In patients suffering from autoimmune disorder such as lupus or MS, the immune system recognizes host tissue and cells as foreign and produces serious damage to organs such as the kidney or brain (Marshall, 2002). The working model for disease etiology suggests that people with these autoimmune diseases are genetically predisposed to have a hyperactive immune system. But manifestation of symptoms is also influenced by environmental factors, including bacterial toxins, viruses, drugs and xenobiotics such as mercury (Kimberstis, 2002; Griem et al., 1998). One of the challenges facing complementary medicine today is sorting out how these environmental factors interact with different body organs, so effective strategies for disease diagnosis, prevention and therapies can be developed.

Despite this involvement of environmental factors and their contribution to complex diseases, for the last 25 years, medical research has been dominated by a genocentric view of discovery. Only now are we beginning to make inroads into a consideration of complementary and alternative approaches to treatment. At the same time, clinical discovery, patient-oriented research and implementation have become less common (Strohman, 2002; Rees, 2002). As was emphasized by Jonathan Rees, now that we have solved the easy problem, which is "identification of the gene that causes a myriad of Mendelian disorders," we need a new postgenomic approach to solve complex diseases and move from bench to clinic. No longer are we to be satisfied with discovering the cause of complex diseases, but we must develop proteomics information. Simultaneous with proteomic development, we must convert the existing information into a form of biomarkers and diagnose diseases at the earliest stage possible (Rees, 2002; Willet, 2002). Once biomarkers and diagnoses are identified, we would then be ready for complementary and alternative approaches to healing.

## 1.2. The Use of Biomarkers for Early Detection of Disease Signs and Signals as a Component of Complementary and Alternative Medicine

Biomarkers are physiological manifestations of changes that may occur on the pathway to disease. If an intervention reduces the incidence of these signs and signals in a population, chances are that this strategy will lower the incidence of disease as well. These markers aid clinicians in evaluating a patient's risk of acquiring complex diseases, much as blood lipid levels are used in standard medical practice to monitor heart disease. Like heart disease, complex diseases are the culmination of many years of subtle pathology. It is never too soon to intervene – but it is often too late (Greenwald, 1996). Based on the success in cardiovascular intervention strategies in reducing death from heart disease, we should revise our cancer research and treatment techniques to emphasize the disease's beginnings rather than its terminal stages. That is the role of complementary and alternative medicine. Realizing the importance of complex disease prevention, we have developed several biomarkers capable of signaling a greater risk of illness. These and many other laboratory biomarkers as they relate to the gastrointestinal, cardiovascular and autoimmune diseases will be discussed in this chapter. This consideration should lay the foundation for opening up treatment to complementary and alternative approaches to medicine.

## 2. Gastrointestinal Health in Relation to Bacterial Toxins and Xenobiotics

Healthy intestinal function is critical to overall health and can respond to effector cells capable of protecting the body from potentially harmful organisms or local antigens. This protection against harmful effects of bacterial toxins and xenobiotics is done by the mucosal immune system. The mucosal immune system is the largest and most complex part of the immune system. This system not only encounters more antigens than any other part of the body, but it must also discriminate between harmless antigens such as food proteins, commensal bacteria and invasive microorganisms (Iijima et al., 2001). Most human pathogens enter the body through a mucosal surface such as the lungs and intestine and strong immune responses against these pathogens are required to protect this physiologically important tissue. In addition, this immune response in a form of local mucosal IgA antibody is essential to prevent further dissemination of such infections (Bengmark, et al.). By contrast, active immune response against dietary antigens and normal bacterial flora may result in hypersensitivity and inflammatory disorders, which are seen in coeliac and Crohn's disease and ulcerative colitis (Jackson et al., 1981; Phillip and Mackowiak, 1982; Marshall et al., 1988). Overall, a complex interplay of regulatory mechanisms in the intestinal immune system discriminate between harmful and harmless antigens to ensure maintenance of homeostasis in the gut. Although strong protective immunity is essential to prevent invasion by pathogens, in the absence of homeostasis equivalent responses against dietary proteins or commensal bacteria could lead to chronic diseases (Marshall et al., 1988; Dupont et al., 1989; Jalonen et al., 1991; Majama and Isolauri, 1996). These may in fact be the targets for the CAM approach.

### 2.1. Features that Influence the Mucosal Immune Regulation

The intestinal immune system is characterized by a distinct profile of cells, adhesion molecules, cytokines and chemokines. In addition, it has a predisposition to the induction of tolerance and bias towards productive or protective immunity that are dominated by production of IgA antibodies against food and commensal antigens. However, it is not clear why intestinal microenvironment results in polarized immune function (Marshall et al., 1988).

It is believed that food proteins and antigens of commensal bacteria are taken up by immunoregulatory dendritic cells (DCs). In the absence of inflammation, prostaglandin $E_2$ ($PGE_2$) is produced by mesenchymal cells and macrophages. Transforming growth factor-$\beta$ (TGF-$\beta$) and IL-10 is produced by epithelial cells, resulting in the maturation of DCs in the Peyer's patch or *Lamina properia*. These food and bacterial antigens are then presented to the naive CD4+ T-cells in mesenteric lymph nodes or Peyer's patch. These T-cells differentiate into regulatory T-cells, which produce interferon-$\gamma$ and Interleukin-10 or differentiate into T-helper-3 cells, which produce TGF-$\beta$. The immunological consequences are local IgA production, local immune homeostasis and systemic tolerance. However, when the body encounters pathogens, xenobiotics or some dietary peptides in the presence of inflammation, mesenchymal cells and macrophages not only fail to produce $PGE_2$ but they also express toll-like receptors. As a result, DCs in the Peyer's patch or *Lamina properia* become completely mature, taking up the antigen(s) and producing significant amounts of IL-12. After migrating to mesenteric lymph nodes, these DCs prime the gut T-helper-1 cells, which produce IFN-γ and cause further inflammation and mucosal destruction, which is referred to as Inflammatory Bowel Disease (Mowat, 2003).

From this introduction to the mucosal immune system, we learned that as with the systemic system, the gastrointestinal immune system has enormous potential for harm if it turns against itself. Therefore, it is important to understand the pathology of various gastrointestinal diseases induced by infection, xenobiotics and dietary peptides.

## 2.2. The Importance of IgA in Mucosal Immunity

Secretory IgA is an important factor in mucosal immunity and in intraluminal microbial defense. The mucosal immune system is a comprehensive local systemic mechanism in which IgA plays a major role by disposing of microbial and dietary antigens locally and preventing them from entering the blood. IgA blocks complement-mediated immune effector mechanisms and functions as an anti-inflammatory immunoglobulin. Mucosal IgA neutralizes viruses. IgA blocks the attachment of pathogens to the relevant mucosal tissue and cells. Mucosal antibodies bind to and inhibit absorption of soluble macromolecules. For this reason, IgA deficient individuals have high levels of serum antibodies to food antigens, particularly bovine milk proteins. These individuals experience chronic hyperabsorption of macromolecules and have a tendency to develop autoantibodies and even autoimmune disease (Majama, 1996; Shi et al., 2000). The importance of measuring secretory IgA in saliva is shown below:

| Secretory IgA Level | Indication |
|---|---|
| • Normal (10–28 μg/ml) | 1. Good Mucosal Immunity |
| • Low (< 10 μg/ml) ↓ | 1. Mucosal Immune Deficiency<br>2. Serum Antibody to Dietary Proteins and Peptides<br>3. Possible Autoimmune Disease |
| • High (>28 μg/ml) ↑ | 1. Bacterial Overgrowth<br>2. Enterotoxins<br>3. Viral Infection |

## 2.3. The Role of Intestinal Flora in Gastrointestinal Function

After IgA, intestinal microflora and intestinal permeability play an important role in diseases related to the gastrointestinal tract. Intestinal permeability plays a very important role in the pathogenesis and pathophysiology of a variety of intestinal disorders. Bacterial flora and their colonization are greatly influenced by eating habits and chemical contamination of the food, which plays a significant role in the integrity of the intestinal flora. From

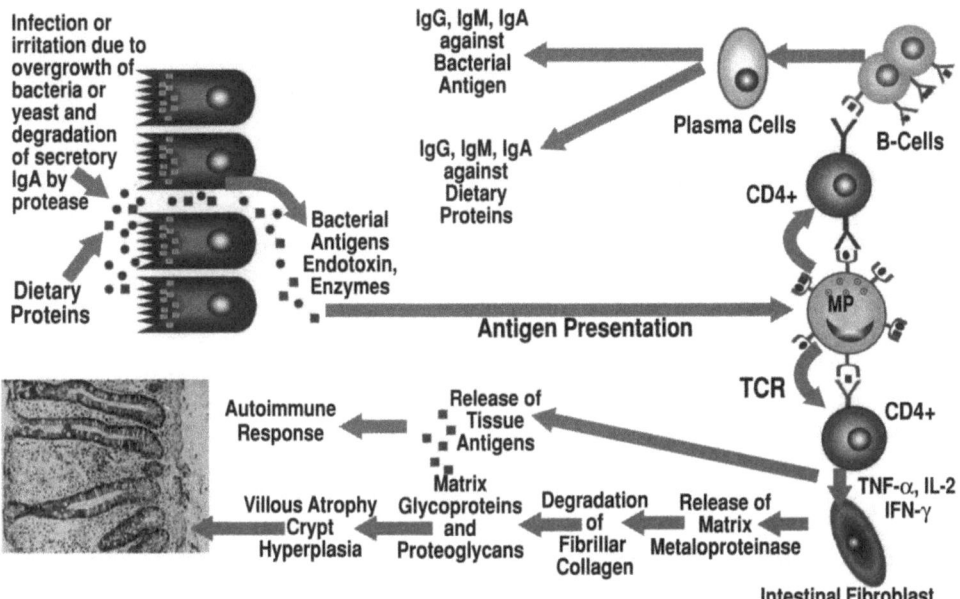

**Figure 6.1.** Immune responses and tissue destruction in the pathogenesis of intestinal barrier dysfunction which may end with food allergy and autoimmunities.

colonized sites, pathogenic bacteria invade deeper tissues by secreting antigens and/or toxins, which result in abnormal systemic immunity. Virulence factors (i.e., fimbriae) enable bacteria to attach to and multiply on mucosal surfaces and to evade the defense mechanisms of the host. This results in bacterial product, proteolytic and hydrolytic enzymes and food antigens being absorbed into the circulation (Webster et al., 1974; Walker and Isselbacher, 1977; Pollack et al., 1997; Shi et al., 2000). Excessive uptake of bacterial, fungal, viral and food antigens into the circulation induces immune response first in the form of IgM and, thereafter in the form of IgG and IgA resulting in clinical conditions. This is the immunological basis for The Intestinal Barrier Function Test (see Fig. 6.1).

The diseases of the gastrointestinal tract cannot be fully understood in their diagnosis and therapeutic implications without coordination of all components of the intestinal flora (Iijima et al., 2001; Marshall et al., 1988; Walker and Isselbacher, 1977). These include: yeast, aerobic bacteria, anaerobic bacteria, dietary proteins and xenobiotics. By implementing these antigens and an ELISA assay, we developed "a single blood test for detection of candidiasis, microflora imbalance, intestinal barrier dysfunction, humoral immune deficiency, food allergy and autoimmunity". This test was developed since in our experience these diseases cannot be fully understood in their diagnostic and therapeutic implications without coordination of all the components of the intestinal flora (yeast, aerobic bacteria, anaerobic bacteria, and dietary antigens).

This systemic translocation of enteric bacteria and yeast may result in excessive uptake of bacterial, fungal, viral and food antigens into the circulation and induction of immune response first in the form of IgM and, thereafter, in the form of IgG and IgA antibodies.

For this reason, measurement of circulating IgM, IgG and IgA antibodies against specific antigens of intestinal bacterial and fungal flora is of considerable importance in the pathogenesis of immunologically mediated diseases including food allergies and autoimmunities. This is the basis for our "Intestinal Barrier Function Test."

In this test, we utilize a highly sensitive and accurate ELISA test method that measures the serum IgG, IgM and IgA specific antibody titers to the purified antigens from five different dietary proteins, three aerobic, two anaerobic microbes and a mixture of three different Candida species (*Candida albicans, Candida tropicalis* and *Candida krusei*) (Vojdani, 1999; Webster et al., 1974). Such quantitative and comparative test results may allow the determination of primary clinical conditions such as:

- Food Allergy
- Intestinal Imbalance
- Gut Barrier Dysfunction
- Bacterial Translocation
- Immunodeficiencies
- Candidiasis
- Autoimmunities

Intestinal barrier dysfunction may result in coeliac disease, Crohn's disease, ulcerative colitis and possibly systemic autoimmune diseases. Therefore, in complementary and alternative medicine, biomarkers for early detection and differentiation of diseases are needed.

## 3. Laboratory Detection of Coeliac and Inflammatory Bowel Disease

### 3.1. Coeliac Disease

*Definition*: intestinal enteropathy caused by immunological hypersensitivity to the gliadin component of wheat and other cereal grains:

1. Occurs as gluten-sensitive enteropathy (GSE) limited to GI manifestation
2. Occurs as dermatitis herpetiformis (skin disease)

*Pathophysiology*: villous atrophy caused by T cell-mediated immune damage to the small intestine mucosa.
*Diagnosis*: jejunal biopsy, response to gluten-free diet, high titers of IgA antigliadin and IgA anti-endomysial antibodies (Shan et al. 2002).
*Clinical effects*: malabsorption, fatty diarrhea, tumors of the GI tract in long-term disease.
*Immunological basis*: immune response gene (HLA) mediated hyper-responsiveness to gliadin.

IgA deficiency occurs far more often in patients with GSE than in normal individuals (Cataldo et al., 1998). Mucosal IgA and IgM production are increased and have been found to closely follow gluten intake, which indicates that gluten ingestion has an almost immediate effect on the immune system. High titers of IgA antigliadin antibodies are quite

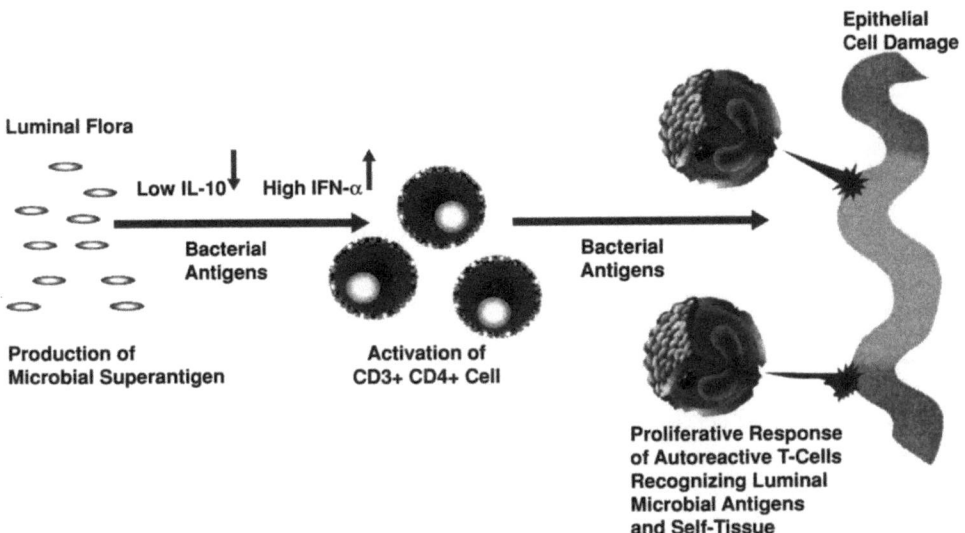

**Figure 6.2.** Immune-mediated disease of the intestine induced by bacterial antigens, showing how staphylococcal enterotoxin B stimulates the expansion of autoreactive T-cells that induce apoptosis in intestinal epithelial cells.

specific for GSE and occur in the vast majority of patients. Recently, a tissue enzyme (transglutaminase) was identified as the autoantigen of coeliac disease. Interestingly enough, gliadin is a preferred substrate for this enzyme, giving rise to novel antigenic epitopes (Dietrich et al., 1997; Sblattro et al., 2000). Immunological detection of IgA and IgG antibodies to tissue transglutaminase (specific peptide), the newly discovered autoantigen to coeliac disease, is a useful tool in the diagnosis and follow-up of the disease.

## 3.2. Detection of Inflammatory Bowel Disease (IBD) and its Subdivisions

Chronic inflammation of the GI tract that is not due to infections is called IBD. Genetic and environmental factors (i.e., bacterial, dietary, drugs, chemicals, cigarettes) play an important role in IBD. This immune-mediated disease induced by *Staphylococcus* superantigens (enterotoxin B) is shown in Fig. 6.2. IBDs are subdivided into ulcerative colitis and Crohn's disease (Ito et al., 2000; Crotty, 1994). While several lines of evidence suggest that ulcerative colitis and Crohn's disease are different diseases, many patients cannot be easily classified into either category. Therefore, a final diagnosis of indeterminate colitis is made. Making a more accurate diagnosis of IBD is important, as the management of Crohn's disease and ulcerative colitis are different, especially when surgery is planned. A search for serological markers to differentiate Crohn's disease from ulcerative colitis has been underway for quite some time. Recently, perinuclear anti-neutrophil cytoplasmic autoantibodies (PANCA) were recognized as a marker for ulcerative colitis. The prevalence of PANCA varies from 40–80% in ulcerative colitis and from 0–20% in Crohn's disease (Quinton et al., 1998). On the other hand, antibodies to oligomannosidic epitopes of the

**Table 6.1.** Laboratory differentiation between coeliac, Crohn's disease and ulcerative colitis

| TEST | COELIAC DISEASE | CROHN'S DISEASE | ULCERATIVE COLITIS |
|---|---|---|---|
| IgA | LOW | NORMAL | NORMAL |
| Gliadin 9IgG, IgA) | HIGH | NORMAL | NORMAL |
| Transglutaminase (IgG, IgA) | HIGH | NORMAL | NORMAL |
| *Saccharomyces cervisiae* (IgG, IgM) | NORMAL | HIGH | NORMAL |
| PANCA | NORMAL | NORMAL | HIGH |
| Tropomyosin and MUC-1 Antibody | NORMAL | NORMAL | HIGH |

yeast *Saccharomyces cervisiae* (ASCA) are a new marker associated with Crohn's disease, which is 64% sensitive and 77% specific for discriminating Crohn's disease from ulcerative colitis. In addition, 90% of patients with ulcerative colitis produce antibodies against colon cytoskeletal proteins called tropomyosin.

As a complementary approach to GI disorders, a combination of these tests could help the diagnosis of IBD and differentiate between Crohn's and ulcerative colitis. Finally a combination of antibody testing against gliadin peptides, transglutaminase, ASCA, PANCA and tropomyosin (shown in Table 6.1) can separate coeliac, Crohn's and ulcerative colitis from each other.

## 4. Relationship Between Commensal and Pathogenic Bacterium with Mucosal Barrier Dysfunction, Cardiovascular and Systemic Immunity, and Autoimmunity

Humans harbor an incredibly complex and abundant ensemble of microbes. These resident bacteria shape our physiology in many ways. To emphasize the importance of commensal bacteria in overall health, germ-free mice were colonized with bacteroides and intestinal transcriptional responses measured using DNA microarrays. Colonized bacteria modulated expression of genes involved in important intestinal function including:

1. Nutrient absorption
2. Lipid absorption capacity
3. Mucosal barrier fortification
4. Xenobiotic metabolism
5. Angiogenesis
6. Postnatal intestinal maturation

This portion of the chapter emphasizes the importance of host-microbial relationships in the GI tract and how gut bacteria and their products play a role in induction and expression of normal immune responses, suggesting that changes in this flora may mediate abnormalities of systemic immunity (Hopper et al., 2001). For example, peritonitis in rats is accompanied by massive small-bowel overgrowth with *E. coli* and this overgrowth contributes to suppression of systemic immunity. This suppression occurred in the absence of significant numbers of translocating bacteria, indicating that endotoxins may be the mediators of immune suppression (Ito et al., 2000). Indeed, endotoxins are the best-characterized microbial products affecting systemic immunity. Other microbial products, however, are capable of altering immune responses (Ito et al., 2000; Marshall et al., 1988). For example, lymphocyte proliferation by mitogens *in vitro* is suppressed by *S. epidermidis* and by cell wall antigens of *Candida albicans*; heat-killed *Pseudomonas aeruginosa* induces *in vivo* suppression of delayed-type hypersensitivity (DTH) by a macrophage-dependant mechanism. Furthermore, to investigate the role of microbial products from the gastrointestinal tract in the pathogenesis of immunosuppression in critical illness, rats have been gavaged daily for three weeks with killed *S. epidermidis*, *Candida* and *Pseudomonas* organisms. These microorganisms are frequently isolated from the upper GI tract of critically ill patients (Marshall et al., 1988). This introduction of bacteria into the GI of rats was followed by delayed-type hypersensitivity response. Interestingly enough, gavaging with either *Pseudomonas* or *Candida* resulted in significant suppression of DTH response, while gavaging with *S. epidermidis* produced modest enhancement of DTH. Rats given *Pseudomonas* demonstrated impaired ability to localize a subcutaneous *S. aureus* challenge and decreased IgM but enhanced anti-*Pseudomonas* IgA and IgG responses after intraperioneal immunization. These findings suggest that gut colonization with *Candida* or *Pseudomonas* may contribute to impairment of cell-mediated immunity in individuals with GI disorders or in critically ill patients. At the point of impaired cell-mediated immunity, one may predict a locale for CAM intervention.

## 4.1. Xenobiotic and Drug Metabolism in GI Disorders

Although several decades of research have not discovered the cause of ulcerative colitis, a diverse range of observations about this severe disease of the GI tract have been made. In one of these observations, it was proposed that ulcerative colitis is caused by toxic metabolites of exogenous agents such as environmental chemicals not usually present in the body (xenobiotics), which are excreted in bile and activated during passage through the colon. Intermittent exposure to the parent compound or their metabolites is the major contributing factor to GI disorders. Genetic influence described in this disease could be explained by inherited differences in the capacity of hepatic enzymes responsible for its metabolism, resulting in decreased elimination of the parent compound and increased transformation into reactive metabolites. The most likely candidate, enzymes are members of the cytochrome P-450 superfamily of mixed-function oxidases. Reactive metabolites produced by this system are commonly coupled to an endogenous conjugate such as glucuronic acid before excretion into bile (Crotty, 1994).

Bacteria in the gut have enzymes, which can act on luminal substrates. In particular, bacterial β-glucuronidase and sulphatase are capable of hydrolyzing the products of hepatic conjugation. If the xenobiotic metabolite were to be slowly reactivated by intestinal bacteria, its luminal concentration would rise with passage down the colon. If one of the concentrations became toxic, the colonic epithelial barrier would be breached, allowing the mucosal immune system to react against luminal contents distal to that point. In susceptible individuals, the biliary epithelium could also be damaged by toxic metabolites, allowing presentation of biliary antigens to surrounding lymphocytes by antigen-presenting cells carrying specific HLA molecules, thereby initiating an inflammatory response in the biliary tree (see Fig. 6.3).

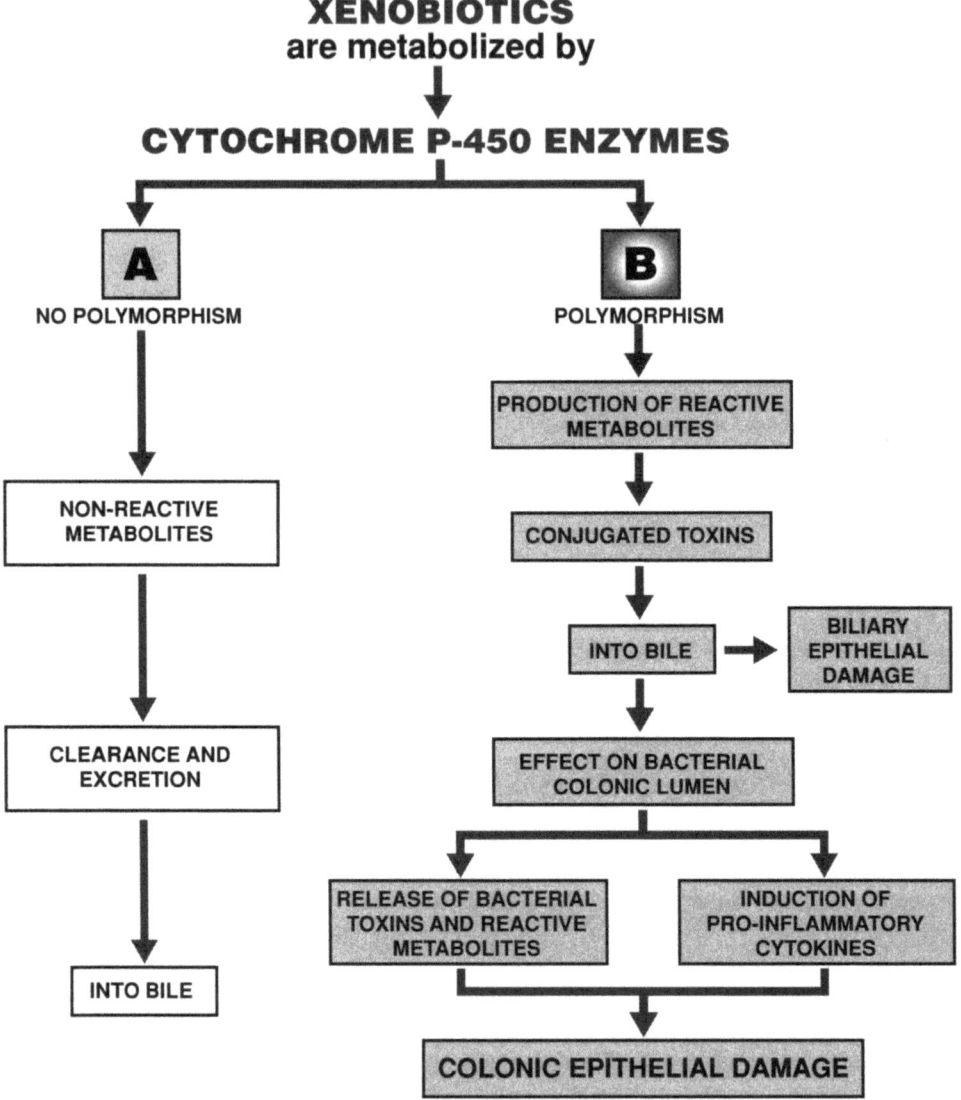

**Figure 6.3.** Ulcerative Colitis Induction by Xenobiotics. In countries with high incidence of ulcerative colitis, xenobiotics serve as substrate for a range of hepatic cytochrome P-450 enzymes. The majority of these individuals with no polymorphism (A) metabolize these chemicals without making significant levels of toxic metabolites. But individuals who inherit a defective enzyme (B) metabolize a greater proportion of xenobiotics and produce reactive metabolites which are conjugated before excretion into bile. Bacterial deconjugation in the colonic lumen release the reactive metabolites, bacterial toxins and proinflammatory cytokines, damaging the colonic epithelial barrier and exposing the mucosal immune system to these products. Biliary and colonic epithelial damage results in systemic immune response, inflammation and autoimmunity.

## 4.2. Xenobiotics as Major Instigators of Autoimmunity

Many toxic chemicals and their metabolites have been shown to be responsible for autoimmune reactions and possibly autoimmune diseases. For a chemical compound to lead to an autoimmune response, it is generally thought that the compound must first become covalently bound to a carrier protein (Griem et al., 1998; Ware et al., 1998). Immune reactions to drugs or their metabolites can develop when a hapten carrier complex interacts with gut-associated lymphoid tissues (GALT) and with peripheral blood lymphocytes (Pohl et al., 1988). If covalent adducts of drugs or other chemical compounds are formed in GALT or in blood, it seems reasonable that they may lead to immune responses and chemically-induced Type I through Type IV-allergic reactions (Salma, 1989). In addition to allergic reactions, xenobiotics such as mercury can induce autoimmunity as well.

Exposure to large doses of mercury results in acute renal tubular lesions and immuno-suppression, whereas chronic administration of smaller doses can lead to development of systemic autoimmunity (Bariety et al., 1971; Hirsch et al., 1982; Robinson et al., 1984; Roman-Franco, 1978). The characteristic features of mercury-induced autoimmunity are similar to manifestations of SLE. These include:

- increased levels of Class II MHC
- antinuclear antibody production hypergammaglobulinemia
- polyclonal antibody to self-antigens
- formation of immune complexes
- lymphocyte proliferation
- necrotizing vasculitis

Given the complexity of metal interaction with cellular and subcellular components of the immune system and numerous molecules that may be affected, genetic studies have been initiated to define the genes responsible for sensitivity of resistance to mercury-induced autoimmunity. A single major quantitative trait locus on chromosome 1, designated as Hm$\gamma$1 has been linked to glomerular immune complex deposits (Kono et al., 2001). This definition of genetic alteration and variation among populations in their response to xenobiotics provide new insights into the relationship of environmental and genetic susceptibility in autoimmune and other complex diseases and may help in preventing allergies and autoimmune diseases (Kono et al., 2001).

## 4.3. Infections as Major Instigators of Autoimmunity

Many infectious agents including measles, rubella virus and cytomegalovirus have long been suspected as etiologic factors in autism. In fact, by reviewing the literature, we found more than 10 species of bacteria or viruses that are involved in autoimmune thyroiditis and rheumatoid arthritis (RA) or multiple sclerosis (MS). Over 60 different microbial peptides have been reported to cross-react with human tissue antigens including brain tissue and MBP. Furthermore, these peptides not only have the ability to cross-react with tissue antigens such as MBP and to induce T-cell responses but can also induce experimental autoimmune disease (Bronze and Dale, 1993; Grogan et al., 1999; Ivarsson, 1990; Lenz et al., 2001; Shi et al., Wakefield et al., 2003). Using the observation that maternal infection increases risk for schizophrenia and autism in offspring, recently, respiratory infection of pregnant mice (both BALB/c and C57BL/6 strains) with the human influenza virus have resulted in offspring that displayed highly abnormal behavioral responses as adults. As in schizophrenia and autism, these offspring displayed deficits in prepulse inhibition (PPI) in the acoustic startle response. Thus, abnormal levels of cytokine production that interfere with neuroimmuno-communications may be responsible for abnormal development of the brain (Fatemi et al., 2002; Shi et al., 2003).

In a different study, antigens from infectious agents may react or interact with lymphocyte receptors that have digestive functions in the gastrointestinal tract. Such a receptor protein is dipeptidylpeptidase IV (DPP IV) or CD26, a closely related family of glycoproteins that is expressed on surfaces and through plasma membranes (Hildebrandt et al., 2000; Misumi et al., 1992). Due to a key role that the membrane-bound DPP IV plays in T-cell mediated immune responses and cytokine production, this enzyme has been analyzed in several autoimmune diseases, such as rheumatoid arthritis (RA) and systemic lupus erythematosus (SLE) (Kameoka et al., 1993; Muscat et al., 1994). In patients with autoimmune disease, the bacterial protein (SK) could bind to DPP IV and induce significant levels of anti-SK and anti-DPP IV antibody production.

CD69 is an additional lymphocyte surface marker involved in autoimmune disease (Hamann et al., 1993). CD69 contributes to deletion of autoreactive lymphocytes by inducing apoptosis; thus, abnormal expression of this molecule could be involved in the pathogenesis of autoimmune diseases. In patients with RA, CD69 is expressed on surfaces of T-cells in synovial membranes but not on surfaces of circulating peripheral blood lymphocytes (PBLs). Levels of CD69 expression on the synovial T-cells in RA are correlated with disease activity (Fernandez et al., 1995; Iannone et al., 1996).

In a different study, autoantibodies to CD69 have been reported in the sera of 38.3% of RA patients, 14.5% of SLE, and 4% of Behcet's disease. In addition to *Streptokinase*, other infectious antigens can bind to different human tissue cell antigens or receptors and modulate their function. For example, infectious agents produce molecules called superantigens or heat shock proteins (HSPs) that can modulate immune function within host organisms. Bacterial superantigens can trans-activate endogenous superantigens by stressful stimuli, including exposure to oxidative radicals, heavy metals, anoxia and infection (Kol et al., 1991; Murray et al., 1995). Induction of endogenous superantigens by infections and xenobiotics might exert a profound impact on the nature of immune responses and nudge them into autoimmune reactions or disease (Kono et al., 2001; Takeuchi et al., 1995).

## 5. The Role of Infections in Cardiovascular and Autoimmune Disease

Atherosclerotic cardiovascular disease (CVD), already the leading cause of death in the United States, is predicted to be the most common cause of death in all regions of the world by the year 2020. Established risk factors for CVD include elevated lipids, hypertension, tobacco abuse, and positive family history, although the exact mechanism by which these may contribute to the development of atherosclerosis is not clear (Muhlestein, 1998; Ridker, 1999, 2000). Moreover, half of patients with heart disease lack any of these established risk factors.

Historically, it has been assumed that infectious agents induce CVD by direct tissue damage via release of toxins and damage to heart tissue, particularly myosin. These toxins may directly or indirectly induce tissue damage and cause release of tissue antigens. An infectious agent can also be ingested by macrophages and thus transferred into the bloodstream and arteries. When a macrophage burrows into the wall of a blood vessel to engulf irritants such as LDL and oxidized LDL, it transfers the infectious agent into neighboring arterial cells. Infected arterial cells then attract more macrophages and cause a cascade of inflammatory mediators, finally succumbing to the immune attack. If this vicious cycle of inflammation continues, it can result in fibrotic lesions or plaque formation. When pieces of the plaque break loose, they can induce clot formation, leading to ischemic events (Vojdani, 2003, 1,2,3; Stolberger and Finster, 2002).

Many viruses and bacteria have been identified as affecting atherosclerosis plaque deposition, among them, dental and other oral pathogens. *Chlamydia pneumoniae* probably has the strongest association with atherosclerosis. Other major atherosclerosis-associated pathogens are *Helicobacter pylori* and CMV, which accelerate atherosclerosis formation, *H. pylori* by secreting vacuolating cytotoxin and CMV by inducing proliferation of smooth muscle cells (Bachmaier, 2000). Infectious agents can also contribute to the acceleration of atherosclerosis development, either by nonspecific mechanisms, such as hypercoagulation, increased production of adhesion molecules, and elevated C-reactive protein (CRP) levels (Ridker, 2000) or by more specific mechanisms, such as induction of HSP-60 expression and eventually pathogenic anti-HSP-60 antibody production (Tabeta, 2000; Xu, 1993). The proposed mechanism by which oral, gastrointestinal and pulmonary pathogens promote atherosclerosis is shown in Fig. 6.4.

## 5.1. Superantigens in Cardiovascular Disease (CVD)

Many pathogenic bacteria and viruses have been shown to produce molecules that interfere with or modulate immune functions of their host organism. Superantigens are potent immunostimulatory molecules that activate large numbers of T-cells. Substantial evidence suggests that certain human bacteria and viruses encode superantigens, which are essential for the virus or bacteria in order to infect cells and tissues (Kol, 1991; Vayssier, 1994; Yamazaki, 2002).

For example human retroviral superantigens can be trans-activated by interferon-$\alpha$ or by Epstein-Barr virus infection. This induction or expression of endogenous superantigens by infection might have a profound impact on the course of a disease or infection. This bacterial or viral upregulation of expression of host superantigen-encoding genes has the potential to dramatically affect the nature of subsequent immune responses, possibly nudging them into autoimmune reactions or disease. The superantigens, which are induced by stressful stimuli, including exposure to oxidative radicals, heavy metals, anoxia and infection are called heat shock proteins (HSPs) (Woodland, 2002).

HSPs are present in all species analyzed so far and exhibit a remarkable sequence homology among various counterparts in bacteria, plants, and mammals. More than half of the residues are identical between bacterial and mammalian HSP-60. The ubiquitous occurrence and remarkable evolutionary conservation suggests that HSP-60 may play an essential role in the cell. Bacterial HSP-60 proteins are major targets of immune responses during infection, and the highly conserved nature of bacterial and mammalian HSP-60 has

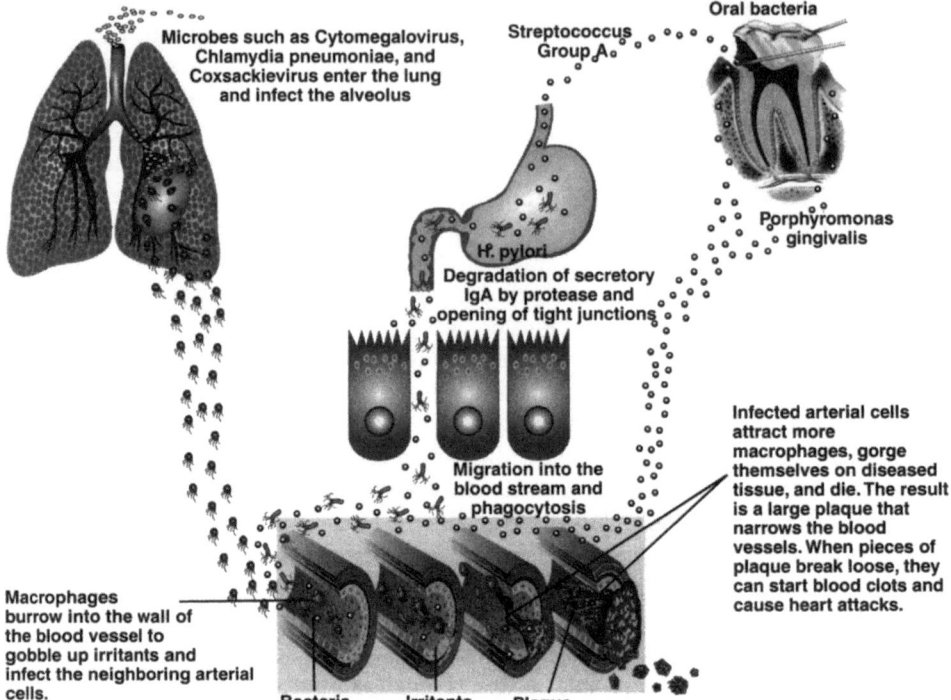

**Figure 6.4.** The role of oral, gastrointestinal and pulmonary pathogens in the induction of atherosclerosis, showing how a bacteria or virus could set the stage for a heart attack. From Vojdani, A., in Laboratory Medicine, Vol. 34, March 2003, reprinted by permission.

led to speculation that immune reactivity to these stress proteins may be a component of certain autoimmune diseases and atherosclerosis (Yamazaki, 2002).

Following is a summary of the role of superantigens in CVD and immune activation:

- Heat shock protein 60, a remarkably immunogenic superantigen, has been detected in a variety of infectious agents, including *Porphyromonas gingivalis*, *Chlamydia*, *Staphylococcus*, *Streptococcus*, and *Escherichia coli*.
- Both T-cell and antibody responses to HSP-60 and HSP-65 have been reported in various inflammatory conditions including atherosclerosis.
- HSP antibody titers correlate with the degree of atherosclerosis in carotid ultrasound studies.
- Patients with periodontal disease not only have an accumulation of HSP-60-reactive T-cells in the gingival tissue, but also have human HSP-60-reactive T-cells with the Type 1 cytokine profile in their peripheral blood T-cell pools.
- Immunization of rabbits with HSP-65 induced arteriosclerosis in normocholesterolemic rabbits.
- Expression of HSP-60 and T-lymphocytes bearing $\gamma/\delta$ receptors were demonstrated in human atherosclerotic lesions.
- Despite the HSP-60 family of related proteins being highly homologous between prokaryotic and eukaryotic cells, HSP-60s are strongly immunogenic, and immune

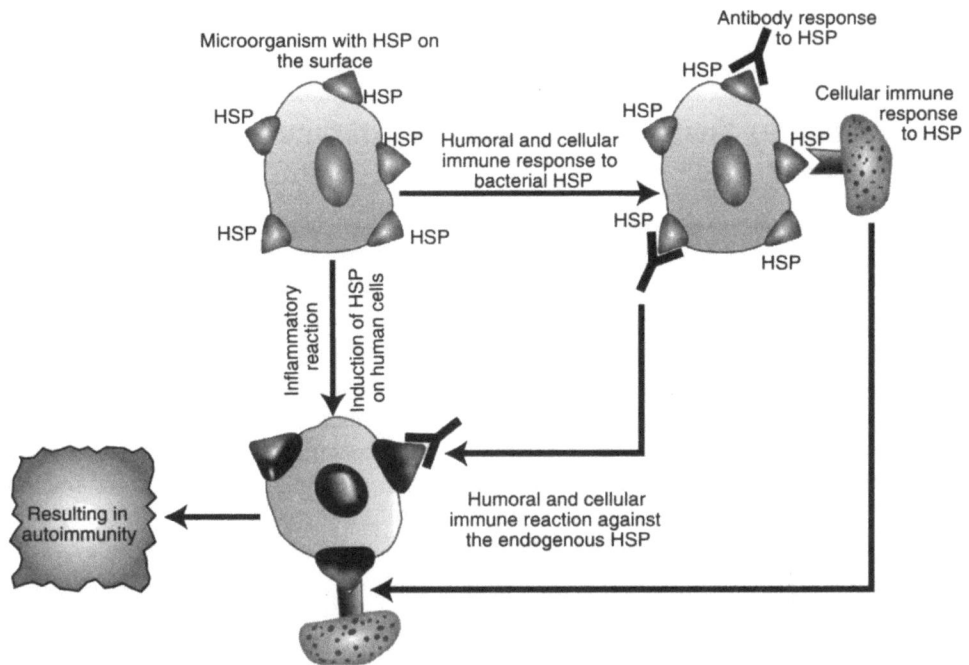

**Figure 6.5.** Mechanism of induction of humoral and cellular immune reaction by HSPs on infecting organisms. Antibodies and T-cells induced by HSPs on the infecting microbes can cross-react with self-HSPs and induce autoimmune reaction. From Vojdani, A., in Laboratory Medicine, Vol. 34, May 2003, reprinted by permission.

responses to microbial HSP-60s are speculated to initiate chronic inflammatory diseases in which autoimmune responses to human HSP-60 may be central to the pathogenesis of cardiovascular diseases.

The proposed mechanism of induction of autoimmunity by HSPs on an infecting microorganism is shown in Fig. 6.5. Based on this mechanism of action many biologists believe that HSP-60 of infectious origin is responsible for a certain percentage of cardiovascular diseases. This conclusion is based on the observation that human and bacterial anti-HSP-60 is detected in a high percentage of patients with cardiovascular disease. Moreover, the anti-HSP antibody titers correlate with the degree of atherosclerosis as revealed in carotid ultrasound studies. An increase in anti-HSP-60 antibody levels could result from direct turbulence damage to bifurcated arteries or could be caused by infectious agents (i.e., C. pneumoniae) by releasing HSP-60, which becomes immunogenic. T-cell lines cultured from atherosclerosis plaque proliferate when exposed to HSP-60 and both autoantibodies and autoantigens can be found in the plaque (Zhou, 2000; Zhu, 2000). Finally, active immunization of rabbits with apolipoprotein-E or low-density lipoprotein (LDL)-receptor knockout of mice with HSP-60, leads to accelerated formation of atherosclerosis plaque (Wick, 1995). Therefore, the presence of HSP-60 antibodies in the blood is indicative of infection-induced cardiovascular disease, hence may justify the use of antibiotic therapy in this subgroup of patients. This treatment with antibiotics may reverse the true underlying causes of disease and improve health and well-being of the patient, which is the major goal of complementary and alternative medicine.

## 5.2. Laboratory Tests for Early Detection of Cardiovascular Diseases

Our laboratory is actively investigating the mechanisms by which infectious agents can cause autoimmune disease. This centers on molecular mimicry or antigenic cross-reactivity between components of infectious agents and cardiac tissue. Immunological cross-reactivity is found in proteins from a multitude of viruses, bacteria, fungi, protozoa and different human tissues. Immune responses against an infectious agent such as *Chlamydia* may result not only in antibodies against the infectious agents but also against cross-reactive tissue antigens such as heart myosin or endothelial cell antigens. Therefore, simultaneous detection of antibodies against viral and bacterial antigens and human tissue may indicate infection-induced autoimmune reaction, possibly resulting in autoimmune disease. Detection of such antibodies in blood can guide clinicians concerning the involvement of infections, suggesting treatment with antibiotics and hopefully, prevention of CVD in patients who lack established risk factors for the disease. These antibodies are measured against different infectious agents, human myosin, oxidized LDL, beta-2 glycoprotein-1 and HSP-60. When these parameters are measured simultaneously, it is possible to differentiate between protective versus pathogenic antibodies.

Based on our observations, we believe that if antibodies are produced in the blood against any of the infectious antigens, but not against any of the target antigens, then these antibodies are protective and not pathogenic. On the other hand, if antibodies are produced in the blood against any of the infectious agents, and simultaneously produced against one of or combinations of the target antigens, then these antibodies are pathogenic and not protective (Table 6.2).

**Table 6.2.** Summary and interpretation of laboratory test results: the role of infections in cardiovascular and autoimmune diseases differentiation between protective versus pathogenic antibodies

Reactivity of serum antibody against:

| Infectious Agents | Myosin Antibody | Oxidized LDL Antibody | Heat Shock Protein-60 Antibody | β-2 Glycoprotein-1 Antibody | Immune Complex | Lupus Peptides Antibody | Arthritic Peptides Antibody | MEDICAL CONDITION |
|---|---|---|---|---|---|---|---|---|
| − | − | − | − | − | − | − | − | Optimal |
| + | − | − | − | − | − | − | − | Unknown Pathology |
| + | + | + | + | + | + | − | − | Possible Atherosclerosis |
| + | − | − | − | − | + | + | + | Possible Autoimmunity |
| + | + | + | + | + | + | + | + | Possible Cardiovascular and Autoimmune Diseases |

We firmly believe that this laboratory detection of high IgG levels against antigens of infectious agents as well as tissue antigens, will aid in differentiation between protective and pathogenic antibodies, and would help in early detection and prevention of cardiovascular and autoimmune diseases. Again, points of CAM intervention may be clearly indicated.

## 6. Autism and Multiple Sclerosis as Models of Complex Diseases, Which are Influenced by Infections, Xenobiotics and Dietary Proteins and Peptides

### 6.1. Autism

Since so little is known about the range of intestinal immune functions that are shaped by dietary proteins, xenobiotics and infectious agents in autism, we decided to test the hypothesis that infectious agents such as antigens, dietary peptides and haptenic chemicals may bind to DPP IV and other tissue antigens or receptors, resulting in autoantibody production, modulation and expression of immune and inflammatory diseases in autism. Similar to many complex autoimmune diseases, genetic and environmental factors including diet, infection and xenobiotics play a critical role in the development of autism. In a very recent study, we postulated that infectious agent antigens such as *Streptokinase* (SK), dietary peptides (gliadin and casein) and ethyl mercury (xenobiotic) bind to different lymphocyte receptors and tissue antigens. We assessed this hypothesis first by measuring IgG, IgM and IgA antibodies against CD26, CD69, SK, gliadin and casein peptides and against ethyl mercury bound to human serum albumin in patients with autism. A significant percentage of children with autism developed anti-SK, anti-gliadin and casein peptides and anti-ethyl mercury antibodies concomitant with the appearance of anti-CD26 and anti-CD69 autoantibodies. These antibodies are synthesized as a result of SK, gliadin, casein and ethyl mercury binding to CD26 and CD69. Immune absorption demonstrated that only specific antigens, like CD26, were capable of significantly reducing serum anti-CD26 levels, indicating that they are specific. However, for direct demonstration of SK, gliadin, casein and ethyl mercury to CD26 or CD69, microtiter wells were coated with CD26 or CD69 alone or in combination with SK, gliadin, and casein or ethyl mercury and then reacted with enzyme labeled rabbit anti-CD26 or anti-CD69. Adding these molecules to CD26 or CD69 resulted in 28–86% inhibition of CD26 or anti-CD69 binding to anti-CD26 or anti-CD69 antibodies. The highest % binding of these antigens or peptides to CD26 or CD69 was attributed to SK and the lowest to casein peptides. We, therefore, proposed that bacterial antigens (SK), dietary peptides (gliadin, casein) and thimerosal (ethyl mercury) in individuals with predisposing HLA molecules, bind to CD26 or CD69 and induce antibodies against these molecules as well as to lymphocyte receptors and tissue antigens. The summary of these findings is depicted in Fig. 6.6 (Vojdani et al., 2003).

### 6.2. Mechanisms by which Infectious Agents and Xenobiotics Induce MS

Autoimmune neurological disorders occur when immunologic tolerance to myelin and other neurologic antigens of Schwann cells, axons and motor neurons are lost. The

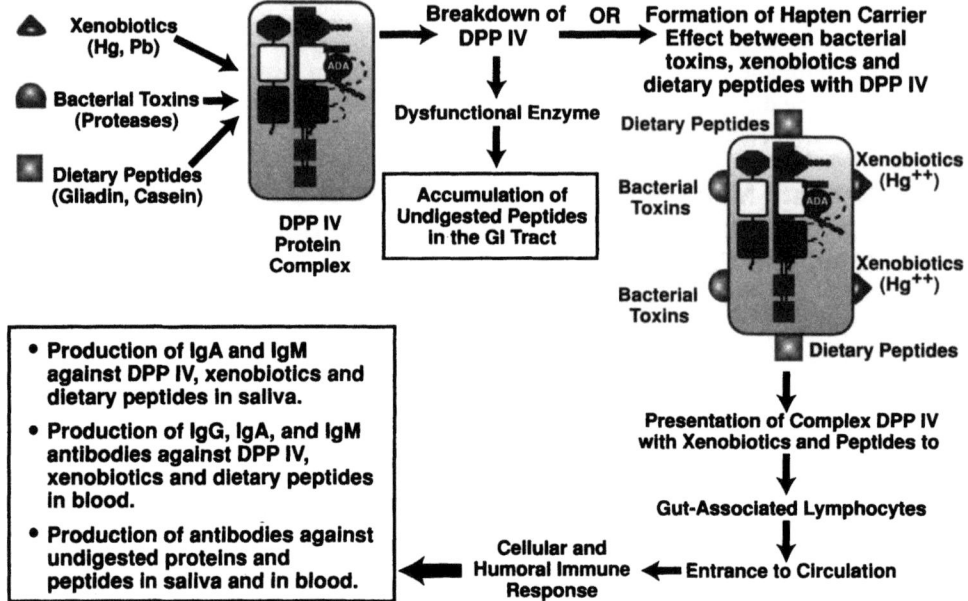

**Figure 6.6.** Xenobiotics, bacterial toxins and dietary peptides binding to DPP IV, formation of hapten carrier effect, and production of antibodies against DPP IV, xenobiotics, peptides and bacterial toxins. This may result in dysfunction of DPP IV molecule and accumulation of peptides in the GI tract and in circulation.

resulting demyelinating diseases share pathologic features characterized by destruction of myelin and accompanied by an inflammatory infiltration into the brain, spinal cord, or optic nerve (Noronha, 1995; Steinman, 1996). MS is the most common demyelinating disease of the myelin central nervous system (CNS), characterized clinically by episodes of neurologic dysfunction separated by time and space.

In recent decades, investigators have been relentless in their search for understanding the causes of MS and the development of new treatment protocols aimed at throwing a stumbling block in its path. They have succeeded in ferreting out a great deal of information about the role of infections and xenobiotics in the initiation of MS. Viruses have been implicated by various researchers as being the initiating event in MS. The evidence supporting a viral etiology of MS includes geographic clustering of MS cases, animal models in which a viral infection mimics the effects of MS, and high levels of antibodies to certain viral infections in MS. Moreover, specific peptides from *Chlamydia pneumoniae*, *Streptococcus* group A and milk butyrophilin were capable of inducing encephalomyelitis in mice and rats (Challoner, 1995; Grogan, 1999, Lenz, 2001, Olson, 2002; Stefferl, 2000; Vojdani et al., 2002; Wucherpfenning, 2002).

Recently, a human herpesvirus (human herpesvirus 6 or HHV-6) has been reported to be found at high levels in the plaque of MS patients (Challoner, 1995). Furthermore, an elevated antibody response to HHV-6 has also been detected in patients with relapsing-remitting types of MS. Recently, a specific antigen of HHV-6 has been used in a serological assay for the detection of antibodies specifically directed against U94/REP protein. Different populations were analyzed by enzyme-linked immunosorbent assay (ELISA), including healthy controls, MS patients and subjects with diseases unrelated to HHV-6 infection, including other neurological diseases. There were statistically significant differences (p > 0.01) between MS patients and control groups, both in antibody prevalence (87 and 43.9%) and in geometric mean titer (1:515 and 1:190). Detection of antibodies specific for HHV-6 U94/REP shows that the immune system is exposed to this antigen during natural infection. The higher prevalence and higher titers of antibodies to U94/REP suggest that MS patients and control groups might experience different exposures to HHV-6 (Caselli, 2002).

When antibodies are produced against bacteria, viruses, dietary proteins or peptides, the immune system becomes confused and may either ignore the invader completely or go on the offensive against self components. This "molecular mimicry" is especially relevant to MS. Although the mechanism is still not completely understood, scientists now believe that previous exposure to one of these mimicking viruses or dietary proteins causes the immune system to create antibodies to that virus. When the virus is no longer around, something prompts the immune system to create antibodies to the body's own brain myelin – which has molecular components bearing an uncanny resemblance to that virus.

In MS patients, all of the immune system's components are involved in one way or another in this inadvertent attack on the brain's myelin. These components include B-cells, T-cells, macrophages, all of the associated cytokines, the chemical mediators serving in this biological warfare. Disruption of myelin or demyelination impairs the ability of nerve fibers to transmit impulses and is responsible for many symptoms of MS, including weakness, sensory loss and visual disturbances. How severely MS affects patients may be directly related to profusion and locations of scarring or plaque on the brain and spinal cord (Olson, 2002; Wucherpfenning, 2002).

## 6.3. MS: A Cascade of Myelin Damage by Cellular Immune Response

In this cascade of events, the body is exposed early to a virus, which mimics myelin basic protein, such as HHV-6, EBV, adenovirus, or measles. Exposure to xenobiotics and bacterial toxins increases levels of adhesion molecules on brain endothelial cells and leukocyte function-associated antigen on activated T-cells. Increased adhesion molecules on endothelial cells may allow transmigration of auto-reactive T-cells across the blood brain barrier, resulting in cellular and humoral immune responses against nerve cells (Raine, 1999; Noseworthy, 2000).

The result of cellular and humoral immune responses is cytokine production (tumor necrosis factor-$\alpha$ or TNF-$\alpha$ and interferon-$\gamma$ or INF-$\gamma$) release of myelin basic protein (MBP) neurofilaments, myelin-associated glycoprotein (MAG), myelin oligodendrocyte glycoprotein (MOG), $\alpha$-$\beta$-crystallin and other neural antigens and IgG, IgM and IgA production against neurologic antigens (Vojdani, 2003). Continuation of this vicious cycle of cellular and humoral attacks on brain cells results in demyelinating oligodendrocyte damage and plaque, which is detected by MR I (see Fig. 6.7).

## 6.4. Diagnosing MS

Until recently, there has been no specific laboratory test for the diagnosis of MS. Usually, diagnoses are based on clinical indications that may vary from patient to patient.

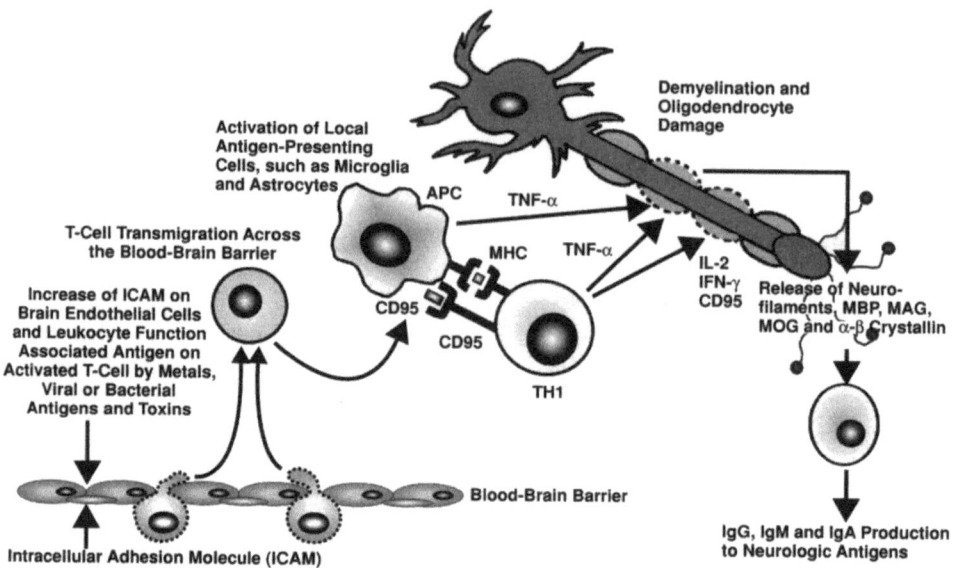

**Figure 6.7.** Cellular and humoral immune mechanisms in infections and xenobiotics-induced neurotoxicity, which includes neuronal degeneration, secondary demyelination, and possibly reactive astrogliosis. Under pathological conditions, pre-existing autoreactive T-cells are generated by molecular mimicry as a result of sequence homologies or matched motifs between autoantigen and viral, bacterial or parasitic proteins. Increased ICAM on endothelial cells by xenobiotics and bacterial toxins may allow transmigration of these auto-reactive T-cells across the blood-brain barrier resulting in cellular and humoral immune responses against nerve cells.

Supportive evidence is derived from MRI of the brain, analysis of cerebrospinal fluid (CSF), and evoked responses (Cole, 1998; Gronseth, 2000; Miller, 1998). MRI is usually the procedure of choice for corroborating a clinical diagnosis of MS. MRI is not specific for MS since several diseases of white matter including ischemic, infectious or neoplastic, as well as mercury exposure and gluten ataxia present a similar picture (Hadjivassiliou, 2002). CSF examination is an additional supportive technique for diagnosing MS. CSF total protein is usually normal but CSF IgG levels may be increased and the ratio of CSF IgG to CSF albumin is often elevated. The presence of a discrete IgG oligoclonal band revealed by immunofixation electrophoresis is more characteristic, but not specific for MS (Cole, 1998). This oligoclonal band may be found in many conditions including subacute sclerosing panencephalitis, neurosyphilis, Lyme disease, HTLV-1 associated myelopathy, Sjögren syndrome, sarcoidosis, meningeal carcinomatosis and HIV infection (Esposito, 1999; Levin, 2002; Martin, 2001). The third technique for supporting diagnosis of MS is revealed by pattern-sensitive visual-evoked potential and brainstem auditory-evoked potential (Gronseth, 2000).

Overall, a combination of MRI, CSF examination and evoked responses support a clinical diagnosis of MS in most instances. However, these three determinants are not always positive in the same patient. Based on immunological mechanisms underlying injury and repair in MS, we developed laboratory markers for early detection and confirmation of MS diagnosis measuring antibodies, activated T-cells and proinflammatory cytokine involvement in the destruction of myelin sheath.

MS patients showed the highest levels of IgG, IgM or IgA antibodies against one or all three tested antigens. Moreover, in the presence of MBP, MOG peptides or α-β-crystallin, a significant percentage of lymphocytes from MS patients underwent blast transformation, which resulted in high levels of IFN-γ, tumor necrosis factor-alpha (TNF-α) and tumor necrosis factor-beta (TNF-β) production.

Sensitivity of these assays was 60–80% and specificity, 65–70%. We concluded that detection of antibodies against MBP, MOG peptides, α-β-crystallin, lymphocyte stimulation and production of proinflammatory cytokines in response to these antigens could be used as surrogate markers for the confirmation of MS diagnosis. A combination of antibodies, lymphocyte activation or cytokine production with abnormal MRI may significantly increase sensitivity and specificity of MS diagnosis (Vojdani et al., 2003). Moreover, these abnormalities are detected in the blood a few years before the appearance of symptomatologies and abnormal MRI of the brain. Therefore, these humoral and cellular immune measurements will help in early detection and intervention of MS and may actually open up new approaches using CAM intervention.

## 7. Antioxidants and MS As Well As Other Complex Diseases: Protecting Cells Subjected to Oxidative Stress

Recall that one of the major mechanisms of injury in cells is oxidative stress, and that it is increased when the immune system metabolizes more oxygen and spins off greater numbers of oxygen free radicals desperately searching for electrons. While these free radicals accomplish their goal of destroying invaders, they may also damage healthy cells in the process, tearing apart cell membranes and dissolving valuable genetic material. Cell membrane destruction may result in cellular death. This oxidative stress death warrant may have a greater impact on brain cells because of their constant high rate of oxygen consumption and high numbers of mitochondria, the cellular energy factories our bodies depend upon to convert glucose into energy. Patients with MS and other complex diseases may suffer from increased lipid oxidation that is potentially damaging to the myelin sheath so essential to conducting messages along nerve cells. Basically, free radicals amplify injury to myelin by disrupting its fatty acid components.

Evidence of direct free radical damage in MS patients has been reported by different researchers (Giulian, 1993; Hu et al., 1996). MS patients have significantly increased levels of lipid peroxidation during exacerbation of attacks, as well as lower vitamin E levels. The free radical nitric oxide, in particular, appears to wreak havoc on the myelin of patients with MS. Nitric oxide is released by macrophages during immune system attacks. Nitric oxide is a profound free radical that performs its activities as a gas. Antioxidants that inhibit nitric oxide production and therefore its attack on myelin may be of benefit in MS patients. Certain antioxidants, including proanthocyanidins, are known to inhibit nitric oxide. Nitric oxide is also inhibited by cholesterol-lowering drugs that explains why some MS patients benefit from the use of statins.

While no clinical studies have demonstrated the beneficial effects of antioxidants in MS patients, based on testimony from patients with MS, it seems that there are certain antioxidants that are indeed effective. These include coenzyme-Q10, epigallocathecin from green tea, pycnogenol extract from pine bark, resveratrol extract from grape seeds, ginkgo biloba, phosphatidyl choline, phophatidyl serine, lecithin, omega-3 fish oil, glutathione, N-acetylcysteine, L-carnitine, lipoic acid, vitamin $B_6$, $B_{12}$, folic acid, vitamin A, vitamin C, vitamin E with selenium, lycopene and trace elements showed a significant improvement in their clinical conditions and therefore these antioxidants are recommended.

## 7.1. Antioxidants and Other Biological Response Modifiers for Prevention of Complex Diseases: Precursor of CAM Intervention

A majority of people over 60 years of age has vitamin deficiencies. The appreciation of the role of diet in determining the level or state of health continues to grow. A substantial amount of research has now solidly established that certain dietary practices cause, as well as prevent, a wide range of diseases. In addition, the research has also shown that certain diets and foods can provide immediate therapeutic benefits (Gann, 1999; Garewall, 1995; Vojdani, 1997). Diet and nutrition are both essential for sustenance, health and overall individual well-being. But it is also true that dietary factors contribute substantially to the burden of preventable illness and premature death in the United States and elsewhere in the world. Dietary factors are associated with 5 of the 10 leading causes of death. They are:

- Coronary heart disease
- Cancer
- Stroke
- Non-insulin-dependent diabetes
- Atherosclerosis

Diet can also play a major role in the prevention of these diseases. Based on a recent research study, the international differences in cancer incidence are largely accounted for by lifestyle practices that include nutrition, exercise, as well as alcohol and tobacco use. About 50% of cancer incidence and 35% of cancer mortality in the U.S (represented by cancers of the breast and colon) are associated with Western dietary habits (Caygill, 1998; Greenwald, 1995).

Many dietary components are involved in diet and health relationships. Chief among them is the disproportionate consumption of foods high in complex carbohydrates and dietary fiber that may be more conducive to health. *The Dietary Guidelines for Americans* recommends that to stay healthy, one should eat a variety of foods; maintain healthy weight; choose a diet low in fat, saturated fat, and cholesterol; and choose a diet with abundant fruits and vegetables, which contain high levels of fiber and antioxidants (Zhang et al., 1999).

It is well known that both synthetic and natural antioxidants inhibit carcinogenesis and mutagenesis. Natural antioxidants include ascorbic acid (vitamin C), selenium, oxidized glutathione (GSSG) and reduced glutathione (GSH), which are water-soluble, and the fat-soluble antioxidants $\alpha$-carotene, $\beta$-carotene (precursor of vitamin A), $\alpha$-tocopherol (vitamin E), $\gamma$-tocopherol, lycopene and coenzyme $Q_{10}$. These antioxidants may be obtained from various food sources, or can be taken as nutritional supplements. Vitamin deficiencies (hypovitaminosis) underlie many human diseases or, conversely, many diseases lead to vitamin deficiencies. Therefore, measurement of endogenous antioxidant levels is of great importance because the resultant values may be used as indicators of future health (Vojdani et al., 1993, 1997, 2000; Williams et al., 1999).

Through a variety of mechanisms, including improvement in natural killer cell cytotoxic activity, antioxidants can block oxygen radicals from damaging DNA, induction of cell mutation and development of cancer.

## 7.2. Enhancement of Immune Function by Using Antioxidants

Clearly patients with a variety of solid malignancies and large tumor burdens have decreased NK activity in the circulation and this low NK activity may be significantly associated with development of distant metastases. Furthermore, in patients treated for metastatic disease, the survival time without metastases correlates positively with levels of NK activity. In patients with hematologic malignancies, there appears to be a correlation between NK activity and the status of disease: the more advanced the disease, the lower the NK activity. Decreased NK activity may also be an important risk factor for development of malignancy in humans. The prognostic significance of low NK activity in patients with cancer has been recently emphasized. Thus, low NK activity may have prognostic value in predicting relapses, poor response to treatment, and especially decreased survival time without metastases (Pross and Lotzova, 1993; Whiteside and Herberman, 1990).

NK cells are sensitive indicators of activation by biologic response modifiers. Their monitoring has been used to document alterations in the activity of circulating immune cells during therapy with these agents. In cancer patients treated with radiation or chemotherapy, NK activity becomes depressed as a result of therapy. Given a role for natural immunity in tumor control, it may be clinically important to monitor the extent and/or duration of cyto-reductive therapy. In human bone marrow transplants, NK cells may influence the outcome through help in engraftment and control of viral infections. They may also mediate anti-leukemic effects important for eliminating residual tumor cells. In contrast, there is evidence for the ability of NK cells to suppress hematopoietic development. Furthermore, recent evidence indicates that there is a relationship between NK activity and an individual's reaction to emotional stress. Attempts are being made to define the mechanism responsible for low NK activity in individuals who have difficulty in managing stress and in those suffering from behavioral disorders.

The role of NK cells in viral disease has been known for a long time. The correlation between low NK activity and serious viral infections in immunocompromised hosts – e.g., in AIDS, after transplantation, and in certain congenital immunodeficiencies – is well documented. Abnormalities in NK function have been described in several autoimmune diseases, since these diseases are frequently associated with serious viral infections and malignancy, low levels of NK activity may be biologically important in individuals with autoimmune disorders. Finally, chronic fatigue immune dysfunction (CFIDS) is characterized by a number of immunologic abnormalities, the most consistent being a significant depression of NK activity (Aoki et al., 1987; McIntyre and Welsh, 1986). A similar phenomenon (low NK cytotoxic activity) in patients with a history of toxic chemical exposure or silicone breast implants was recently reported by our laboratory (Vojdani et al., 1992).

With the usage of biologic response modifiers, it is possible to enhance NK cell activity (Greenwald, 1996). However, the best modifiers of NK activity are antioxidants, especially ascorbic acid. In fact, in our three different studies, when ascorbic acid was given to patients with low NK cell activity, significant enhancement was observed after oral usage of 60 mg/Kg body weight ultra-potent-C or buffered vitamin-C (Vojdani et al., 1993, 1997, 2000). In the first two studies we used a buffered preparation of vitamin C or ultra-potent-C concentration of 60 mg/Kg. This preparation was well tolerated at high doses. So far, many patients have been followed by physical examination, hematology, and blood chemistry, including liver enzyme and complete urinalysis for a period of three years post ultra potent C usage; no signs of liver toxicity (liver or kidney) or other abnormalities have been detected. While vitamin C was capable of enhancing the NK, T, and B cell functions, non-significant improvement in symptomatologies was observed. However, enhancement in functions of NK, T and B cells by vitamin C may prevent or delay infections or other health problems in patients who may suffer from complex diseases (Vojdani et al, 1993, 1997).

Our third study was initiated because a study by Podmore et al (1998) claimed that vitamin C exhibits pro-oxidant properties. This debate was based on the assessment of oxidative damage to peripheral blood leukocytes (PBLs) after administering 500 mg/day of ascorbic acid to healthy volunteers for 6 weeks. The level of 8-oxoguanine or 8-hydroxyguanosine decreased, whereas the level of 8-oxoadenine increased. This increase in a potentially mutagenic lesion, 8-oxoadenine, after a typical ascorbic acid supplementation raised questions about the antioxidant nature of ascorbic acid (Podmore et al 1998). The recommended dietary allowance for ascorbic acid is 60 mg daily, based on threshold urinary excretion of the vitamin and preventing the ascorbic acid deficiency disease scurvy with a margin of safety (Sub-Committee of the RDA 1989). As other factors, such as steady-state plasma concentration, bioavailability, and cell concentration, play an important role, data from a more recent study and Institute of Medicine criteria prompted the recommendation that the current recommended dietary allowance of 60 mg should be increased to 200 mg daily. According to conclusions safe doses of ascorbic acid are less than 1,000 mg daily and ascorbic acid doses above 400 mg have no evident value (Levine 1996).

We decided to analyze the effects of 500 mg and much higher doses (up to 5,000 mg) of ascorbic acid on DNA adducts, NK cell activity, programmed cell death, and cell cycle phases that involve signal transduction pathways, along with markers of DNA damage relating to cellular absorption of ascorbic acid that will better clarify the pro-oxidant and antioxidant properties of ascorbic acid. We confirmed the antioxidant properties of ascorbic acid not only by measuring DNA adducts and demonstrating no change in the number of DNA lesions in white blood cells, but also by measuring as well the numbers and functionality of some of these cells that did not decrease.

Several possible mechanisms of action for ascorbic acid in cancer prevention have been described. Ascorbic acid plays a major role in free radical scavenging and protection against lipid peroxidation. Moreover, ascorbic acid and other antioxidants exert different effects on cancer and normal cells (Frei et al, 1998, 1999, Prasad et al, 1993). For example, high doses of different vitamins can induce direct or indirect apoptosis in cancer cells whereas they can protect normal cells against apoptosis (Prasad et al, 1993). This protection of normal cells against apoptosis by vitamins is in complete agreement with our findings.

We propose that measurement of biologic markers such as DNA and protein adducts, DNA damage, programmed cell death, DNA repair system, levels of antioxidants, efficient immune functions and other tests described in this chapter should be measured. These measurements are biological clues indicating that the body has been assaulted by infections or cancer-causing agents. This early identification of biomarkers for special vulnerability to infections and xenobiotics and development of complex diseases may result in preventive measures and save millions of lives (Vojdani et al, 1997). In the meantime, establishing an identification of biomarkers provides key points appropriate for complementary and alternative interventions.

## 8. Opening up to Complementary and Alternative Approaches

During the last 50 years, medical research and development have been dominated by drugs for treatment of symptomatologies of disease. We have been brainwashed by the manufacturers of these medications that it is acceptable to obtain temporary relief from the symptoms of a variety of complex diseases and forgo the search for root causes and ideally eliminate the word "cure" from the dictionary. We have popular brands for allergies, for acid reflux, anti-acids and $H_2$ blocker for peptic ulcers, anti-inflammatory medications for autoimmune diseases, sugar-lowering medications for diabetes, statins for lowering cholesterol, anti-depressants for psychiatric disorders, others with a combination of anti-depressants for autism spectrum and attention deficit disorders, and betaserones (interferon-$\beta$) for MS. This culture of pill-popping is so deep-rooted in us that not only do we dismiss the role of environmental factors in development of complex diseases, but clinicians have even begun to prescribe medications such as anti-depressants for treatment of depression in dogs and cats!

When the completion of the human genome project was announced in early 2002, every human being on the planet had something (albeit unrealistic) to be optimistic about - the potential cure for ALL diseases of the 21st century. Since then we have read numerous articles that enthusiastically promote the genome-based therapeutic revolution and have promised to transform medicine. However, despite an explosion of knowledge in basic biology and new discoveries about the root causes of disease, none dealt with curing illnesses. This is the point where the new CAM could be operative.

Clearly acceptable genetic and environmental factors, including diet, lifestyle, infection, smoking and pollution all contribute to complex diseases that are major causes of mortality. Indeed, various lines of evidence, which are presented in this chapter, indicate that environmental factors are most important in development of diseases. Overly enthusiastic expectations regarding genetic research and drug development for disease treatment only have the potential to distort research priorities and spending for health. However, integration of new genetic information, laboratory biomarkers and CAM approaches into epidemiologic results can only help to clarify the causal relationships among infections, xenobiotics, diet, genetic factors and risk of disease. Thus consideration of root causes of disease using new lab markers for their detection should provide the best data to make informed choices about the most effective means to prevent and treat complex diseases. If infectious agents are involved, then patients should be treated with proper antibiotics. If toxic environmental chemicals are involved, avoidance and detoxification techniques may be applied. If patients are synthesizing antibodies against dietary proteins and peptides, then those molecules should be removed from the diet.

A complementary and alternative approach to easing/curing complex diseases rests first with early detection of disease signals by using biomarkers. These biomarkers are physiological manifestations of changes that may occur on the pathway to diseases. As described in this chapter, biomarkers can aid clinicians in evaluating a patient's risk of acquiring possible complex disease, much as blood lipid levels are used in standard medical practice to monitor heart disease. If an intervention such as treating for infectious agents or reducing chemical load or changing diet can reduce the incidence of complex disease signs and signals in the population, chances are that this strategy will also lower the incidence of complex diseases as well.

In conclusion we believe that a complementary and alternative approach to medicine is defined in two ways: 1) practitioners working on the cutting edge of healthcare using the best evidence-based assessments and interventions available, regardless of source and method, to reverse the true underlying causes (infections, xenobiotics and others) of disease to improve health and well-being; 2) practitioners working collaboratively to achieve the best outcomes for their patients.

## 9. References

Aoki T, Usuda Y, Miyakoshi H, Tamura K and Herberman RB, 1987, Low natural killer syndrome: clinical and immunological features. *Nat. Immun. Cell Growth Regul.* **6**:116–128.
Bariety J, Druet P, Laliberte F and Sapin C, 1971, Glomerulonephritis with γ- and 1C-globulin deposits induced in rats by mercuric chloride. *Am. J. Pathol.* **65**:293–297.
Bengmark S, 1996, Econutrition and health maintenance – a new concept to prevent GI inflammation, ulceration and sepsis. *Clin. Nutr.* **15**:1–10.
Bronze MS and Dale JB, 1993, Epitope of streptococcal M. proteins that evoke antibodies that cross-react with the human brain. *J. Immunol.* **151**:2820–2828.
Caselli E, Boni M, Bracci A, Rotola A, Cermelli C, Castellazi M, Di Luca D and Cassai E, 2002, Detection of antibodies directed against human herpesvirus-6 U 94/REP in sera of patients affected by multiple sclerosis. *J. Clin. Microbiol.* **40**:4131–4137.
Cataldo F, Marwo V, Ventura A, Bottaro G and Corazza GR, 1998, Prevalence and clinical features of selective immunoglobulin A deficiency in coeliac disease: an Italian multicenter study. *Gut* **42**:362–365.
Caygill CPJ, Charlett A and Hill MJ, 1998, Relationship between the intake of high-fibre foods and energy and the risk of cancer of the large bowel and breast. *Eur. J. Cancer Prev.* **7**:S11–17.
Challoner PB, Smith KT and Parker JD, 1995, Plaque-associated expression of human herpesvirus 6 in multiple sclerosis. *Proc. Natl. Acad. Sci. USA* **92**:7440–7444.
Cole SR, Beck RW, Moke PS, Kaufman DI and Tourtellotte WW, 1998, The predictive value of CSF oligoclonal banding for MS 5 years after optic neuritis. *Neurology* **51**:885–887.
Crotty B, 1994, Ulcerative colitis and xenobiotic metabolism. *Lancet* **343**, 35–38.
Das KM, Dasgupta A, Mandal A and Geg X, 1993, Autoimmunity to cytoskeletal protein Tropomyosin. *J. Immunology* **150**:2487–2495.
Dieterich W, Ehnis T, Bauer M, Donner P, Volta U, Riecken E and Schuppan D, 1997, Identification of tissue transglutaminase as the autoantigen of celiac disease. *Nature Medicine* **3**:797–801.
Dupont C, Barau E, Molkhou P, Raynaud F, Barbet JP and Dehennin L, 1989, Food-induced alterations of intestinal permeability in children with cows' milk-sensitive enteropathy and atopic dermatitis. *J. Ped. Gastroentero Nutr.* **8**:459–465.
Esposito M, Venkatesh V, Otvos L, Weng Z, Vajda S, Banki K and Perl A, 1999, Human transaldolase and cross-reactive viral epitopes identified by autoantibodies of multiple sclerosis patients. *J. Immunology* **163**:4027–4032.
Fatemi SH, Earle J, Kamodia R, Kist D, Emamian ES and Patterson PH, 2002, Prenatal viral infection leads to pyramidal cell atrophy and macrocephaly in adulthood: implications for genesis of autism and schizophrenia. *Cell. Mol. Neurobiol.* **22**:25–33.
Fernandez-Gutierrez B, Hernandez-Garcia C, Banares AA and Jover JA, 1995, Characterization and regulation of CD69 expression on rheumatoid arthritis synovial fluid T-cells. *J. Rheumatol.* **22**:413.

Frei B, Stocker R and Ames BN, 1988, Anti-oxidant defenses and lipid peroxidation in human blood plasma. *Prod. Natl. Acad. Sci. USA* **86**:6377–6381.

Gann PH, Ma J, Giovannucci E, Willet T, Sacks FM, Hennekens CH and Stampfer MJ, 1999, Lower prostate cancer risk in men with elevated plasma lycopene levels: results of a prospective analysis. *Cancer Res.* **59**:1225–1230.

Garewal H, 1995, Antioxidants in oral cancer prevention. *Am. J. Clin. Nutr.* **62**:1410–1416S.

Giulian D and Corpuz M, 1993, Microglial secretion products and their impact on the nervous system. *Adv. Neurol.* **59**:315–321.

Greenwald P, 1996, Chemoprevention of cancer. *Scientific American* **275**:96–99.

Greenwald P, Kellof G, Burch-Whitman C and Kramer BS, 1995, Chemoprevention. *CA Cancer J. Clin.* **45**:31–49.

Griem P, Wulferink M, Sachs, Gonzalez JB and Gleichmann E, 1998, Allergic and autoimmune reactions to xenobiotics: how do they arise? *Immunology Today* **19**:133–142.

Grogan JL, Kramer A, Nogai A, Dong L, Ohde M, Schneider-Mergener J, et al., 1999, Cross-reactivity of myelin basic protein-specific T-cells with multiple microbial peptides: experimental autoimmune encephalomyelitis induction in TCR transgenic mice. *J. Immunol.* **163**:3764–3770.

Gronseth GS and Ashman EJ, 2000, Practice parameter: the usefulness of evoked potentials in identifying clinically silent lesions in patients with suspected multiple sclerosis (an evidence-based review): report of the Quality Standards Subcommittee of the American Academy of Neurology. *Neurology* **54**:1720–1725.

Hadjivassiliou M, Boscolos S, Davies-Jones GAB, et al., 2002, The humoral response in the pathogenesis of gluten ataxias. *Neurology* **58**:1221–1226.

Hamann J, Fiebig H and Strauss M, 1993, Expression cloning of the early activation antigen CD69, a type II integral membrane protein with a C-type lectin domain. *J. Immunol.* **150**:4920.

Heyman M, Grasset E, Ducroc R and Desjeux JF, 1988, At absorption by the jejunal epithelium of children with mild allergy. *Pediatr Res.* **24**:197–202.

Hildrebrandt M, Reutter W, Archk P, Rose M and Klapp B, 2000, A guardian angel: the involvement of Dipeptidylpeptidase IV in psychoneuroendocrine function, nutrition and immune defense. *Clin. Sci.* **99**:93–98.

Hirsch F, Couderc J, Sapin C, Fournie G and Druet P, 1982, Polyclonal effect of $HgCL_2$ in the rat, its possible role in an experimental autoimmune disease. *Eur. J. Immunol.* **12**:620–626.

Hopper LV, Wong MH, Thelin A, Hansson L, Falk PG and Gordon JI, 2001, Molecular analysis of commensal host-microbial relationship in the intestine. *Science* **291**:881–884.

Hu S, Chao CC, Khanna KV, Gekker G, Peterson PK and Molitor TW, 1996, Cytokine and free radical production by prcine microglia. *Clin. Immunol. Immunopathol.* **78**:93–98.

Iannone F, Corrigal VM and Panayi GS, 1996, CD69 on synovial T-cells in rheumatoid arthritis correlates with disease activity. *Br. J. Rheumatol.* **35**:397–401.

Iijima H, Takahashi I, Kiyono H, 2001, Mucosal immune network in the gut for control of infectious diseases. *Rev. Med. Virol.* **2**:117–133.

Ito K, Takaishi H, Jin Y, Song F, Denning TL and Ernst PB, 2001, *Staphylococcal enterotoxins* B stimulates expansion of autoreactive T-cells that induce apoptosis in intestinal epithelial cells: regulation of autoreactive responses by IL-10. *J. Immunl.* **164**:2994–3001.

Ivarsson SA, Bjerre L, Vegfors P and Ahlfors K, 1990, Autism as one of several abnormalities in two children with congenital cytomegalovirus infection. *Neuropediatrics* **21**:102–103.

Jackson PG, Baker RW, Lessof MH, Ferret J and MacDonald DM, 1981, Intestinal permeability in patients with eczema and food allergy. *Lancet* **1**:1295–1286.

Jalonen T, 1991, Identical intestinal permeability change children with different clinical manifestation of cow's milk allergy. *J. All. Clin. Immunol.* **88**:37–742.

Kameoka J, Tanaka T, Nojima Y, Schlossman SF and Morimoto C, 1993, Direct association of adenosine deaminase with a T-cell activation antigen CD26. *Science* **261**:466–470.

Kiberstis P and Roberts L, 2002, It's not just the genes. *Science* **296**:685–686.

Kohl A, Bourcier T, Lichtman AH, et al., 1991, Chlamydia and human heat shock protein 60s activate human vascular endothelium, smooth muscle cells and macrophages. *J. Clin. Invest.* **103**:571–577.

Kono DH, Park MS, Szydlik A, Haraldsson KM, Duan JD and Pearson DL, 2001, Resistance to xenobiotic-induced autoimmunity maps to chromosome 1. *Immunol.* **167**:2396–2403.

Lenz DC, Lu L, Conant SB, Wolf NA, Gerard HC and Whittum-Hudson JA, 2001, A *Chlamydia pneumoniae*-specific peptide induces experimental autoimmune encephalomyelitis in rats. *J. Immunol.* **167**:1803–1808.

Levin MC, Lee SM, Klaume F, et al., 2002, Autoimmunity due to molecular mimicry as a cause of neurological disease. *Nature Medicine* **8**:509–513.

Levine M, Conry-Cantilena C, Wang Y, et al., 1996, Vitamin C pharmacokinetics in healthy volunteers: evidence for a recommended dietary allowance. *Proc. Natl. Acad. Sci. USA* **93**:3704–3709.

Majama H and Isolauri E, 1990, Evaluation of the gut mucosal barrier for increased antigen transfer in children with atopic eczema. *J. All. Clin. Immunol.* **97**:985–990.

Marshal E, 2002, Lupus: mysterious disease holds it secrets tight. *Science* **29**:689–691.

Marshall JC, Christou NV and Meakins JL, 1988, Immunomodulation by altered gastrointestinal tract flora. *Arch Surg.* **123**:1465–1469.

Martin R, Gran B, Zhae Y, et al., 2001, Molecular mimicry and antigen-specific T-cell. *J. Autoimmunity* **16**: 187–92.

Marx J, 2002, Unraveling the causes of diabetes. *Science* **296**:686–689.

McIntyre KW and Welsh RM, 1986, Accumulation of natural killer and cytotoxic T large granular lymphocytes in the liver during virus infection. *J. Exp. Med.* **164**:1667–1681.

Miller, DH, Grossman RI, Reingold SC, 1998, The role of magnetic resonance techniques in understanding and managing multiple sclerosis. *Brain* **121**, 3–24.

Misumi Y, Hayashi Y, Arakawa F and Ikehara Y, 1992, Molecular cloning and sequence analysis of human dipeptidylpeptidase IV, a serine proteinases on the cell surface. *Biochem. Biophys. Acta.* **1131**:333–336.

Mowat A, 2003, Anatomical basis of intolerance and immunity to intestinal antigens. *Nature Reviews Immunology* **3**:331–340.

Muhlestein JB, 1998, Chronic infection and coronary artery disease. *Sci. Med.* Nov/Dec, 16–25.

Muscat C, Bertotto A, Agea E, Bistoni O, Ercolani R, Tognelli R, et al., 1994, Expression and functional role of 1F7 (CD26) antigen on peripheral blood and synovial fluid cells in rheumatoid arthritis patients. *Clin. Exp. Immunol.* **98**:252–256.

Noronha A and Arnanson B, 1995, Demyelinating diseases. In: Rich, Fleisher, Schwartz, Shearer and Strober, eds. *Clinical Immunology* pp. 364–1375. Mosby.

Noseworthy JH, Lucchinetti C, Rodriquez M and Weinshenker BJ, 2000, Multiple sclerosis, a review. *New England Journal of Medicine* **343**:938–952.

Olson JK, Eagar TN and Miller SD, 2002, Functional activation of myelin-specific T-cells by virus-induced mimicry. *J. Immunol.* **169**:2719–2726.

Phillip A and Mackowiak MD, 1982, The normal microbial flora, medical progress section. *New Engl. J. Med.* **307**:83–93.

Phillips DIM and Matthews N, 1987, The measurement of antibodies to C. albicans as a screening test for humoral deficiencies. *J. Immunol. Methods* **105**:127–131.

Pohl LR, Satoh H, Christ DD and Kenna JG, 1988, The immunological and metabolic basis of drug hypersensitivities. *Annu. Rev. Pharmacol. Toxicol.* **28**:367–387.

Pollack M, Ohl CA, Goldenbock DT, 1997, Dual effects of LPS antibodies on cellular uptake of LPS and LPS-induced pro-inflammatory functions. *J. Immuno.* **159**:3519–3530.

Podmore ID, Griffiths HR and Herbert KE, 1998, Vitamin C exhibits pro-oxidant properties, *Nature* **392**: 559–560.

Prasad KN, Edwards-Prasad J, Kumar S, et al., 1993, Vitamins regulate gene expression and induce differentiation and growth inhibition in cancer cells: their relevance in cancer prevention. *Arch. Otolaryngol. Head Neck Surg.* **119**:1133–1140.

Pross HF and Lotzova E, 1993, Role of natural killer cells in cancer. *Nat. Immunol.* **12**:279–292.

Quinton JF, Charrier G, Colombel JF, Sendid B, Poulain D, Grandbastien B, Reumaux D, Duthilleul P and Targan SR, 1998, Anti-*Saccharomyces cerevisiae* mannan antibodies combined with antineutrophil cytoplasmic autoantibodies in inflammatory bowel disease: prevalence and diagnostic role. *Gut* **42**:788–791.

Raine CS, Cannella B, Hauser SL and Genain CP, 1999, Demyelination in primate autoimmune encephalomyelitis and acute multiple sclerosis lesions: a case for antigen-specific antibody mediation. *Ann. Neurol.* **46**: 144–160.

Ridker PM, 1999, Evaluating novel cardiovascular risk factors: can we better predict heart attack? *Ann. Intern. Med.* **130**:933–937.

Ridker PM, Hennekens CH, Buring JE et al., 2000, C-reactive protein and other markers of inflammation in the prediction of cardiovascular disease in women. *N. Eng. J. Med.* **342**:836–843.

Rees J, 2002, Complex disease and the new clinical sciences. *Science* **296**:698–701.

Robinson CJ, Abraham AA and Balazs T, 1984, Induction of anti-nuclear antibodies by mercuric chloride in mice. *Clin. Exp. Immunol.* **58**:300–307.

Roman-Franco AA, Turiello M, Albini B, Ossi E, Milgrom F and Andres GA, 1978, Anti-basement membrane antibodies and antigen-antibody complexes in rabbits injected with mercuric chloride. *Clin. Immunol. Immunopathol.* **9**:464–470.

Salama A, Schutz B, Kietel V, Breithaupt H and Mueller-Eckhardt C, 1989, Immune-mediated agranulocytosis related to drugs and their metabolites:mode of sensitization and heterogeneity of antibodies. *Br. J. Haematol.* **72**:127–132.

Sblattero D, Berti I and Trevisiol C, 2000, Human recombinant tissue transglutaminase ELISA: an innovative diagnostic assay for celiac disease. *Am. J. Gastroenterol.* **95**:1253–1257.

Shan L, Molberg O, Parrot I, Hausch F, Filiz F, Gray GM, Sollid LM and Khosla C, 2002, Structural basis for gluten intolerance in Celiac Sprue. *Science* **297**:2275–2279.

Shi HN, Liu HY and Nagler-Anderson, 2000, Enteric infection as an adjuvant for the response to a model food antigen. *J. Immunol.* **166**:6174–6182.

Shi L, Fatemi SH, Sidwell RW and Patterson PH, 2003, Maternal influenza infection causes marked behavioral and pharmacological changes in the offspring. *J. Neurosciences* **23**:297–302.

Stefferi A, Schubart A, Storch M, Amina A, mather I, Lassman H and Linington C, 2000, Butyrophilin, a milk protein, modulates the encephalitogenic T-cell response to myelin oligodendrocyte glycoprotein in experimental autoimmune encephalomyelitis. *J. Immunol.* **165**:2859–2865.

Steinman L, 1996, Multiple sclerosis: a coordinated myelin attack against myelin in the central nervous system. *Cell* **85**:299–302.

Stollberger C and Finsterer J, 2002, Role of infections and immune factors in coronary and cerebrovascular arteriosclerosis. *Clin. Diag. Lab. Immunol.* **9**:207–215.

Strohman R, 2002, Maneuvering in the complex path from genotype to phenotype. *Science* **296**:701–704.

Sub-Committee on The 10th Edition of the RDA's Food and Nutrition Board, commission of Life Sciences, National Research council. Recommended dietary allowances, ed 10. Washington, DC, National Academy Press, 1989:115–124.

Tabeta K, Yamazaki K, Hotokezaka H et al., 2000, elevated humoral immune response to the heat shock protein 60 (hsp60) family in periodontitis patients. *Clin. Exp. Immunol.* **120**:285–293.

Vayssier C, Mayrand D and Grenier D, 1994, Detection of stress proteins in *Porphyromonas gingivalis* and other oral bacteria by Western immunoblotting analysis. *FEMS Microbiol. Lett.* **121**:303–307.

Vojdani A, 1999, A single blood test for detection food allergy, candidiasis, microflora imbalance, intestinal barrier dysfunction and humoral immune deficiencies. *Biomedical Therapy* **17**:129–134.

Vojdani A, 2003, A look at infectious agents as a possible causative factor in cardiovascular disease: part I. *Lab. Medicine* **34** (3):7–11.

Vojdani A, 2003, A look at infectious agents as a possible causative factor in cardiovascular disease: part II. *Lab. Medicine* **34** (4):5–9.

Vojdani A, 2003, A look at infectious agents as a possible causative factor in cardiovascular disease: part III. *Lab. Medicine* **34** (5):24–31.

Vojdani A, Barzagan M, Vojdani E and Wright J, 2000, New evidence for antioxidant properties of Vitamin C. *Cancer Detec. And Prev.* **24**:508–523.

Vojdani A, Campbell AW, Anyanwu E, Kashanian A, Bock K and Vojdani E, 2002, Antibodies to neuron-specific antigens in children with autism: possible cross-reaction with encephalitogenic proteins from milk, Chlamydia pneumoniae and Streptococcus group A. *J. Neuroimmunol.* **129**:168–177.

Vojdani A, Ghoneum M and Brautbar N, 1992, Immune alteration associated with exposure to toxic chemicals. *Toxicol. Ind. Health* **8**:231–246.

Vojdani A and Ghoneum M, 1993, In vivo effect of ascorbic acid on enhancement of natural killer cell activity. *Nutr. Res.* **13**:753–764.

Vojdani A and Namatalla G, 1997, Enhancement of human natural killer cell cytotoxic activity by Vitamin C in pure and augmented formula. *J. Nutr. Environ. Med.* **7**:187–195.

Vojdani A, Pangborn JB, Vojdani E and Cooper EL (in press) Infections, toxic chemicals and dietary peptides binding to lymphocyte receptors and tissue enzymes are responsible for autoimmunity in autism. *Int. J. Immunopath. Pharmacol.* **16**:189–200.

Vojdani, A., Vojdani, E. and Cooper, E., 2003, Antibodies to myelin basic protein, myelin oligodendrocyte peptides, $\alpha$-$\beta$-crystallin, lymphocyte activation, and cytokine production in patients with multiple sclerosis. *J. Internal Medicine* **254**:1–12.

Wakefield AJ, Murch SH, Anthony A, Linnell J, Casson DM, Malik M, et al., 1998, Ileal-lymphoid-noduclar hyperplasia, non-specific colitis, a pervasive developmental disorder in children. *Lancet* **351**:637–641.

Walker WA and Isselbacher KJ, 1977, Intestinal antibodies. *New Engl. J. Med.* **297**:767–773.

Ware JA, Graf MLM, Bartin BM, Lustberg LR and Pohl LR, 1998, Immunochemical detection and identification of protein adducts of diclofenac in the small intestine of rats: possible role in allergic reactions. *Chem. Res. Toxicol.* **11**:164–171.

Webster AD, Efter T and Asherson GL, 1974, *Escherichia coli* antibody: a screening test for immunodeficiency. *BMJ* **3**:16–18.

Whiteside TL, Bryant J, Day R, et al., 1990, Natural killer cytotoxicity in the diagnosis of immune dysfunction: criteria for a reproducible assay. *J. Clin. Lab. Anal.* **4**:102–110.

Wick G, 1995, Is atherosclerosis an immunologically mediated disease? *Immunol. Today* **16**:27–33.

Willett WC, 2002, Balancing life-style and genomics research for disease prevention. *Science* **296**:695–698.

Williams GM, Williams CL and Weisburger JH, 1999, Diet and cancer prevention: the fiber first diet. *Toxicological Sciences* **52**:72–86.

Woodland DL, 2002, Immunity and retroviral superantigens in humans trends *Immunology* **23**:57–58.

Wucherpfenning KW, 2002, Infectious triggers for inflammatory neurological diseases. *Nature Medicine* **8**:455–457.

Xu Q, Willeit J, Marosi M, et al., 1993 Association of serum antibodies to heat-shock protein 65 with carotid atherosclerosis. *Lancet* **341**:255–259.

Yamazaki K, Ohsawa Y, Tabeta K, et al., 2002, Accumulation of heat shock protein 60-reactive T-cells in the gingival tissues of periodontis patients. *Infect. Immun.* **70**:2492–2501.

Yoshida S and Gershwin ME, 1993, Autoimmunity and selected environmental factors of disease induction. *Semin Arthritis Rheum.* **22**:399–405.

Yu X, Matsui T, Otsuka M, Senine T, Yamamoto K, Nishioka K, et al., 2001, AntiCD69 autoantibodies cross-react with low density lipoprotein receptor-related protein 2 in systemic autoimmune diseases. *J. Immunol.* **166**:1360–1369.

Zhang S, Hunter DJ, Forman MR, Rosner BA, Speizer FE, Colditz GA, Manson JE, Hankinson SE and Willet WC, 1999, Dietary carotenoids and Vitamins A, C, and E and the risk of breast cancer. *J. Natl. Cancer Inst.* **91**:547–556.

Zhou YF, Shou M, Harrell RF et al., 2000, Chronic non-vascular cytomegalovirus infection: effects on the neointimal response to experimental vascular injury. *Cardiovascular Research* **45**:1019–1025.

Zhu J, Shearer GTM, Norman JE, et al., 2000, Host response to cytomegalovirus infection as a determinant of susceptibility to coronary artery disease: gender-based differences in inflammation and type of immune response. *Circulation* **102**:2491–2496.

# 7

# Complementary and Alternative Medicine During HIV Infection

Milena Nasi, Marcello Pinti, Leonarda Troiano, Andrea Cossarizza

## 1. The Relevance of HIV/AIDS Epidemics

According to the Joint United Nations Program of HIV/AIDS (UNAIDS) and the World Health Organization (WHO) (Joint United Nations Program of HIV/AIDS, 2001), as of the end of 2001, there were about 40 million adults and children living with human immunodeficiency virus (HIV) infection. This total does not include the 20 million people around the world who already died of AIDS. Of the 40 million currently alive, 37.2 are adults, 17.6 are women, and more than 2.7 are children. In 2001, there were 5 million new cases of HIV infection in the world, and 3 million AIDS related deaths. The large majority (almost three quarters) live in Sub-Saharan Africa where the prevalence rate of the infection among adults is 8.4%; more than 55% of infected individuals are women. The second major pocket of HIV infection is in South and Southeast Asia, with more than 6 million people infected. In North America where the epidemic was first described, there are 940,000 individuals who are HIV+, and in Western Europe, 540,000. Furthermore, South America, China, and Eastern Europe are characterized by a rapid increase in infection rates. These dramatic numbers clearly indicate that the fight against HIV/AIDS is an absolute health, social, economical and political priority in all parts of the world.

## 2. Treatment of HIV/AIDS and Side Effects of the Therapies

In 1995, the introduction of highly active antiretroviral therapy (HAART) for HIV-infected patients had a terrific impact and changed the story of the infection[2]. In a consistent percentage of cases HAART can efficiently suppress HIV production, by using the strategy of inhibiting different steps of viral cycle and replication with the combination of drugs such as nucleosidic (zidovudine, didanosine, stavudine, lamivudine, abacavir) or non-nucleosidic (efavirenz, nevirapine, delavirdine) reverse transcriptase inhibitors, nucleotidic reverse transcriptase inhibitors (tenofovir), and viral protease inhibitors (indinavir, saquinavir, ritonavir,

---

**Andrea Cossarizza** • MD PhD, University of Modena and Reggio Emilia, Dept. of Biomedical Sciences, Section of General Pathology, Via Campi 287, 41100 Modena, Italy. E-mail: cossarizza.andrea@unimore.it

*Complementary and Alternative Approaches to Biomedicine*, edited by Edwin L. Cooper and Nobuo Yamaguchi. Kluwer Academic/Plenum Publishers, 2004.

lopinavir, amprenavir, nelfinavir). Such an antiviral activity can be evidenced in a relatively simple manner either by measuring plasma viral load, that is often undetectable (depending on the assay, lower than 20 or 50 copies of the virion/ml plasma)[3], or analysing the recovery of peripheral CD4+ T cell count[4]. Since its introduction, HAART caused an immediate and consistent decrease of mortality rates in HIV+ patients[5], along with a consistent reduction of the incidence of systemic opportunistic infections in AIDS patients, who do not require any prophylaxis in a consistent number of cases[6,7]. However, paradoxically, the decreased mortality and morbidity due to HAART has provoked a dangerous relaxation of appropriate public and private health measures, which in Western Countries threaten a recrudescence of the epidemics.

After the first successful years, the attention of clinicians is currently devoted to several difficulties linked to the use of potent regimens. Indeed, in a significant number of patients they are unable to reduce HIV production. One of the main causes for this phenomenon is the onset of viral resistances, caused by the high rate of the mutation of HIV genome, but that can also be strictly linked to the adherence to therapy, and indirectly to the long-term complications. Adherence is a crucial problem that has an enormous impact on therapy effectiveness: when adherence is >95% (meaning that a patient is able to take almost all of the prescribed drugs with an extreme precision—which is not easy because of the high daily number of pills), virological failure can occur in 22% of cases, whereas an adherence <80% leads to a 80% virological failure[8]. HAART is also associated with serious side effects that significantly decrease adherence, the most relevant of which are redistribution of fat (lipodystrophy) and hyperlactatemia/lactic acidosis, among others that are more specific for each drug and can involve almost all tissues and organs[9].

People living with HIV/AIDS are struggling with a chronic, life-threatening disease, and have to be treated for long periods, if not for all life. Acceptance that a definitive cure (clearly, vaccine is the Holy Grail) for HIV will likely not be discovered in the near future, and that the pharmacological treatment is an absolute necessity have justifiably shifted the attention to the quality of life concerns for people living with HIV/AIDS. Consequently, the attention has been redirected to improving the management of the most common symptoms related to the infection and its treatment, such as dermatological problems (occurring in 45% of the patients), nausea (44%), depression (36%), insomnia (36%) and weakness (34%), peripheral neuropathy, diarrhoea and muscle pain, among others. Effective drugs that can successfully cope with such constellation of side effects and symptoms are lacking, and, as a result, in Western Countries about half of adults with HIV/AIDS are using complementary and alternative medicine (CAM) with the aim of improving general health, treating symptoms, reducing side effects caused by drugs, and in some cases hoping to prevent opportunistic infections without the need of primary or secondary prophylaxis.

## 3. The Use of CAM in HIV/AIDS

Various historical and socio-political aspects have contributed to the current popularity of CAM. On the one side, potent antiretroviral treatments are often accompanied by severe side effects, on the other the access to HAART is dramatically limited by the cost of the drugs. In this regard, the fact that most HIV+ people live in poor countries and have as little as 1 US $ a day for living – considering that the cost of HAART can reach several thousand dollars a year – is so dramatic that no comments are possible.

A study among physicians in places where there is access to antiretroviral therapy (Washington, New Mexico, and Israel) found that, at least once, more than 60% of them recommended CAM to their patients in the past year[10]. A recent report indicates that 84% of the patients participating in the National Institute of Health (NIH, Bethesda, MD) clinical trials were taking at least one CAM therapy[11]. Other studies showed that on average one-third to one-half of HIV/AIDS patients use some type of CAM[12–15]: 29–58% of subjects enrolled in clinical trials outside NIH were indeed using CAM therapies. Subjective assessment of the most helpful effects of these therapies entailed collapsing all therapies into five generally recognized categories (developed by the former Office of Alternative Medicine of the NIH in 1994): (1) mind-body, (2) structural/energetic, (3) lifestyle/diet/nutritional, (4) traditional/ethnomedicine and (5) pharmacologic/biologic.

According to recent data, 71% of patients were using CAM therapies before the diagnosis of HIV infection[16]. After diagnosis, such percentage increased of another 13%. Moreover, 91% of HIV patients had used at least one CAM therapy sometime in their lives[11]. Therapies that became significantly more popular after the diagnosis were high-dose vitamins, weight gain, massage, relaxation, herbals, spiritual healing and acupuncture[17]. Patients who benefit from CAM use report: feeling better, increased ability to cope, feeling in control, and stated that CAM therapy not only enhanced treatment outcome but was as or more effective than conventional antiretroviral treatment.

Despite the popularity and the massive diffusion of CAM among HIV/AIDS patients, it is poorly known that its use can affect the success of HAART. Indeed, whether and how CAM therapies interact with antiretrovirals is not well known, nor which is effective inducing negative side effects. In fact, outcome studies examining the efficacy of CAM among people with HIV/AIDS are often conducted on a small number of patients, and with short follow-up. Most studies just report trends, and are unable to provide any statistical evidence. The main problem is that patients may invest energy and resources in expensive, demanding, ineffective and perhaps even hazardous treatments, sometimes abandoning conventional and effective treatments[18,19]. Furthermore, certain treatments especially with herbal medicines have direct, objective risks for the HIV+ patient. Herbal medicine, the most popular CAM therapy in the US, has grown faster than any other alternative treatment method, and physicians regularly see patients who self prescribe such products but do not volunteer this information. The public perception that whatever is "natural" is safe often results in almost complete ignorance of eventual serious complications through hepatotoxic or nephrotoxic effects, or through interactions with drugs used for "official" therapy. For example, relevant interactions exist between antiretroviral drugs and *Hypericum perforatum* (known as St. John's Wort, SJV), a natural antidepressant with more than 2.7 million prescriptions every year. Recent data have shown that SJV may influence the metabolism of coadministered drugs[20]: administration of SJV with the viral protease inhibitor indinavir produces an 81% reduction in drug plasma concentration, either altering the expression of the multidrug resistance P-glycoprotein 1, or inducing expression of the hepatic enzyme CYP3A4. As a result, the efficacy of antiretroviral therapy is seriously compromised, and the production of HIV is no longer inhibited.

Other complementary therapies (e.g., homeopathy, acupressure, touch therapy and spiritual healing, among others) might be judged as entirely free of direct risk. However, the indirect risk is that these CAM could fully replace conventional treatment in spite of the absence of any scientific demonstrations of their efficacy. Homeopathy is one of the most diffuse types of CAM therapy in US and Western Europe. Nevertheless, homeopathy cannot be considered an evidence-based and scientifically recognized form of therapy. The hypothesis that any given homeopathic remedy leads to clinical effects that are relevantly different from placebo unfortunately is not supported by evidence from systematic reviews, and the risk exists that homeopatic remedies could be substituted for conventional therapy (particularly in the case of HIV infection treatment), leading to serious consequences for the health of the patient[21–23].

Acupuncture is becoming a common adjunctive therapy for the management of peripheral neuropathy, a pain syndrome that affects up to 1/3 of drug-treated HIV+ patients. However, there has been very little research on patient responses and outcomes, with controversial results. In fact, all findings are severely limited by the small number of patients enrolled in clinical trials, high drop-out rate and by the frequent omission of a control group[24].

Touch therapy involves manual manipulation of soft tissues which is reported to have an effect on pain reduction and sense of well-being, with reduction of muscular tightness and tension. A possible explanation of these effects is that touch modalities can stimulate the release of endorphins, but, again, very little empirical data exist in regard to the effect of these modalities. The molecular interactions between the nervous and immune system have been under investigation for several years[25], and it is well known that the mind can heavily influence the immune response. However, there is convincing and scientific data that go beyond the description of the sensation of well being and demonstrate which molecules and biochemical mechanisms are triggered by manipulation of tissues, and how they could influence the immune response, and thus a disease such as HIV infection have still to be produced.

The use of alternative therapies that could interact with conventional treatments has often been underestimated, and a bias could exist in several studies as far as effectiveness of antiretrovirals is concerned. Many physicians do not ask information about alternative treatments taken by their patients and indeed, as reported above, a large percentage of subjects enrolled in clinical trials are using CAM therapies. Only one half of them are specifically asked by physicians whether they were using adjunct therapies. This could render false the results of several studies. Thus, physicians' need for reliable information on effects of alternative medicine is considerable, and one of the most relevant problems linked to the use and the effectiveness of CAM therapy in HIV treatment is the relative small number of clinical trials aimed at evaluating the effects of such treatments.

## 4. Conclusions

Many people who practise CAM are quite resistant to using scientific protocols such as those commonly employed in clinical trials to estimate and understand the effects of their therapies. It is often stated that a thousand years of experience provide more evidence than modern clinical trials, that patients are single individuals while trials just consider "populations", that there are no placebo for some CAM therapies, and that alternative medicine has a long term perspective, whereas trials have a short one.

Even if these considerations seem plausible for people who usually choose this kind of therapy, they are not acceptable for the scientific community. Years of experience clearly could represent a precious source of information, but do not demonstrate anything *per se*. The history of medicine is full of examples of practices that have been employed for centuries on the basis of empirical experience or traditions, but which are useless if not deleterious. The advent of evidence based medicine (EBM) as the golden rule to perform clinical studies today represents a unique chance for therapies which require and need validation, as for many CAM therapies. EBM can help to demonstrate in a rigorous manner that they are objectively useful, as such a modern approach asks whether or not a therapy works, not why it works. Thus, there is an urgent need for trials performed with scientific criteria for the evaluation of the real benefits of such therapies.

The objection that CAM therapies are individualized is not a real obstacle for the design of clinical trials: many conventional treatments are indeed individualised, and this fact did not thwart the final result. One of the most used arguments for justifying the lack of clinical trials for CAM is the absence of an adequate placebo, for certain therapies such as acupuncture, aromatherapy, or massage. However, concerning acupuncture, a simple design could be the application of needles in the "wrong" places, where they are supposed to have no effect. Many practices of official medicine cannot be studied utilizing a placebo group. EBM indeed does not insist on this, since they are not available for many areas of therapeutics (for example, surgery): sometimes the gold standard is not a placebo trial, but a randomized study design.

In conclusion, understanding the contraindications of CAM therapies is necessary to prevent deleterious outcomes and to facilitate the safe and efficacious use of CAM in the management of HIV infection and treatment. Even if certain CAM interventions seem to improve the quality of life of HIV+ patients (and psychological benefits should not be underestimated), rigorous studies are needed using longitudinal, controlled designs to accurately assess the safety and utility of such interventions.

## 5. References

1. M. S. Gottlieb, R. Schroff, H. M. Schanker, J. D. Weisman, P. T. Fan, R. A. Wolf and A. Saxon, Pneumocystis carinii pneumonia and mucosal candidiasis in previously healthy homosexual men: evidence of a new acquired cellular immunodeficiency. *N Engl J Med* **305**(24), 1425–31 (1981).
2. M. Louie and M. Markowitz, Goals and milestones during treatment of HIV-1 infection with antiretroviral therapy: a pathogenesis-based perspective. *Antiviral Res* **55**(1), 15–25 (2002).
3. J. W. Mellors, C. R. Rinaldo, Jr., P. Gupta, R. M. White, J. A. Todd and L. A. Kingsley, Prognosis in HIV-1 infection predicted by the quantity of virus in plasma. *Science* **272**(5265), 1167–70 (1996).
4. T. S. Li, R. Tubiana, C. Katlama, V. Calvez, H. Ait Mohand and B. Autran, Long-lasting recovery in CD4 T-cell function and viral-load reduction after highly active antiretroviral therapy in advanced HIV-1 disease. *Lancet* **351**(9117), 1682–6 (1998).
5. G. Ippolito, V. Galati, D. Serraino and E. Girardi, The changing picture of the HIV/AIDS epidemic. *Ann N Y Acad Sci* **946**, 1–12 (2001).
6. C. Mussini, P. Pezzotti, A. Govoni, V. Borghi, A. Antinori, A. d'Arminio Monforte, A. De Luca, N. Mongiardo, M. C. Cerri, F. Chiodo, E. Concia, L. Bonazzi, M. Moroni, L. Ortona, R. Esposito, A. Cossarizza and B. De Rienzo, Discontinuation of primary prophylaxis for Pneumocystis carinii pneumonia and toxoplasmic encephalitis in human immunodeficiency virus type I-infected patients: the changes in opportunistic prophylaxis study. *J Infect Dis* **181**(5), 1635–42 (2000).
7. C. Mussini, P. Pezzotti, A. Antinori, V. Borghi, A. Monforte, A. Govoni, A. De Luca, A. Ammassari, N. Mongiardo, M. C. Cerri, A. Bedini, C. Beltrami, M. A. Ursitti, T. Bini, A. Cossarizza and R. Esposito,

Discontinuation of secondary prophylaxis for Pneumocystis carinii pneumonia in human immunodeficiency virus-infected patients: a randomized trial by the CIOP Study Group. *Clin Infect Dis* **36**(5), 645–51 (2003).
8. R. P. van Heeswijk, A. Veldkamp, J. W. Mulder, P. L. Meenhorst, J. M. Lange, J. H. Beijnen and R. M. Hoetelmans, Combination of protease inhibitors for the treatment of HIV-1-infected patients: a review of pharmacokinetics and clinical experience. *Antivir Ther* **6**(4), 201–29 (2001).
9. M. John, D. Nolan and S. Mallal, Antiretroviral therapy and the lipodystrophy syndrome. *Antivir Ther* **6**(1), 9–20 (2001).
10. J. Borkan, J. O. Neher, O. Anson and B. Smoker, Referrals for alternative therapies. *J Fam Pract* **39**(6), 545–50 (1994).
11. A. Sparber, J. C. Wootton, L. Bauer, G. Curt, D. Eisenberg, T. Levin and S. M. Steinberg, Use of complementary medicine by adult patients participating in HIV/AIDS clinical trials. *J Altern Complement Med* **6**(5), 415–22 (2000).
12. B. R. Bates, P. Kissinger and R. E. Bessinger, Complementary therapy use among HIV-infected patients. *AIDS Patient Care STDS* **10**(1), 32–6 (1996).
13. J. T. Dwyer, A. M. Salvato-Schille, A. Coulston, V. A. Casey, W. C. Cooper and W. D. Selles, The use of unconventional remedies among HIV-positive men living in California. *J Assoc Nurses AIDS Care* **6**(1), 17–28 (1995).
14. R. M. Greenblatt, H. Hollander, J. R. McMaster and C. J. Henke, Polypharmacy among patients attending an AIDS clinic: utilization of prescribed, unorthodox, and investigational treatments. *J Acquir Immune Defic Syndr* **4**(2), 136–43 (1991).
15. C. Rowlands and W. G. Powderly, The use of alternative therapies by HIV-positive patients attending the St. Louis AIDS Clinical Trials Unit. *Mol Med* **88**(12), 807–10 (1991).
16. D. M. Eisenberg, R. B. Davis, S. L. Ettner, S. Appel, S. Wilkey, M. Van Rompay and R. C. Kessler, Trends in alternative medicine use in the United States, 1990–1997: results of a follow-up national survey. *JAMA* **280**(18), 1569–75 (1998).
17. L. J. Standish, K. B. Greene, S. Bain, C. Reeves, F. Sanders, R. C. Wines, P. Turet, J. G. Kim and C. Calabrese, Alternative medicine use in HIV-positive men and women: demographics, utilization patterns and health status. *AIDS Care* **13**(2), 197–208 (2001).
18. E. Ernst, The dark side of complementary and alternative medicine. *Int J STD AIDS* **13**(12), 797–800 (2002).
19. E. Ernst, "Alternative" medicine and science–like fire and water? *Wien Klin Wochenschr* **114**(15-16), 655–6 (2002).
20. M. Hennessy, D. Kelleher, J. P. Spiers, M. Barry, P. Kavanagh, D. Back, F. Mulcahy and J. Feely, St Johns wort increases expression of P-glycoprotein: implications for drug interactions. *Br J Clin Pharmacol* **53**(1), 75–82 (2002).
21. E. Ernst and J. Barnes, Meta-analysis of homoeopathy trials. *Lancet* **351**(9099), 366; author reply 367–8 (1998).
22. E. Ernst, A systematic review of systematic reviews of homeopathy. *Br J Clin Pharmacol* **54**(6), 577–82 (2002).
23. E. Ernst and M. H. Pittler, Re-analysis of previous meta-analysis of clinical trials of homeopathy. *J Clin Epidemiol* **53**(11), 1188. (2000).
24. R. Power, C. Gore-Felton, M. Vosvick, D. M. Israelski and D. Spiegel, HIV: effectiveness of complementary and alternative medicine. *Prim Care* **29**(2), 361–78 (2002).
25. E. Ottaviani, E. Caselgrandi, M. Bondi, A. Cossarizza, D. Monti & C. Franceschi. The "immuno-mobile" brain: evolutionary evidence. *Adv. Neuroimmunol.* **1**(1), 27–39 (1991).

# 8

# A Polyherbal Formulation for a Wide Spectrum of Reproductive Tract and Sexually Transmitted Infections

## G.P. Talwar

### 1. Introduction & Rationale

Sexually transmitted infections (STI) are on an increase worldwide. According to WHO[1], 333 million new cases of STIs occur globally every year *excluding* HIV, and HIV is assuming pandemic proportions in many countries. Besides this, reproductive tract infections (RTIs) caused by aerobic, anaerobic bacteria, protozoa, yeast (Candida) etc, manifested in the syndrome of abnormal vaginal discharge (AVD) are widespread. Its prevalence is around 30% in women visiting public hospitals in cities in India[2] and is over 50% in rural areas[3]. One billion events of bacterial vaginosis are believed to occur globally in a year[4]. Although antibiotics and oral ± local vaginal use therapies are available for most of the pathogens, their rational use requires diagnosis of the causative organism, but facilities for their diagnosis are not available universally, specially in economically developing countries. The objective of our work was to develop a vaginal formulation, whose use be totally under the control of the woman, and which could protect her as much as possible from sexually transmitted infections. In addition, the formulation as a single locally applied "drug" was aimed to act across the spectrum of micro-organisms responsible for the syndrome of AVD. This charter demanded not one but a combination of ingredients as there is no single compound known which acts on all micro-organisms responsible for AVD. Furthermore, as a frequently used preparation, it has to be safe and free from causing inflammation or mucosal damage, which the presently marketed vaginal suppositories, creams, sponges do. These preparations are based on powerful surfactants such as Nonoxynol-9 (N-9), benzalkonium chloride etc. Although N-9 has *in vitro* high anti-HIV activity, clinical trials with preparations containing N-9 failed to prevent transmission of HIV through hetero-sexual intercourse[5], and instead facilitated the uptake of the virus, presumably because of the damage it caused to the vaginal epithelium. Our strategy was therefore to avoid synthetic chemical compounds. We

---

G.P. Talwar • Talwar Research Foundation, E-8 Neb Valley New Delhi-110068, India.
Talwar37@rediffmail.com

*Complementary and Alternative Approaches to Biomedicine*, edited by Edwin L. Cooper and Nobuo Yamaguchi.
Kluwer Academic/Plenum Publishers, 2004.

searched for plants whose extracts may have antibacterial, antiviral and antifungal properties and whose safety for humans has been amply tested by long years of use. Preference for selection was given to plants, whose active principles are chemically defined. It is in this context and with these considerations that our choice fell on 3 plants: Neem (Azadirachta indica), Sapindus mukerosi and Cinchona. For fragrance, an essential oil from Mentha citrata was included.

## 2. Ingredients

### 2.1. Neem (Azadirechta indica)

This hardy tree grows in abundance in India and in many countries of Africa. Its plantations in Australia are doing well. It is an evergreen shady tree. All parts of the tree, leaves, seed, and bark have traditionally been used by people in India for various purposes. Twigs are chewed in the morning and teeth cleaned by the brush so formed at the extremity of the twig chewed upon. The extracts of the Neem twig bathing the mouth are rinsed out and these have apparently antiseptic action against the diverse micro-organisms inhabiting the oral cavity. Good strong teeth and oral hygiene of the common people are maintained without recourse to any other toothpaste or mouth wash. In case of injuries, the wounds are washed by the villagers with an infusion or extract in hot water of Neem leaves to prevent sepsis. Dry Neem leaves are put in canisters of rice and wheat to protect from insects. Neem leaf compost or crushed Neem seeds are used by the farmers as fertilizer and as an antidote to infections. Neem leaves are burnt to repel mosquitos and Neem leaves are taken orally to ward off malaria and other infectious diseases. The age-old traditional uses of this remarkable tree point to the presence in this tree (leaves, seed, bark) of compounds which have a wide spectrum of action against bacteria, viruses, insects and nematodes with apparently no ill effects on humans.

Many of the attributes of traditional empirical use have been tested scientifically. Rao et al[6] reported action of Margosa (a variety of Neem) leaves on vaccinia and variola viruses. Pant et al[7] observed antifungal action of Neem seed oil. Various fractions of Neem seed extracts were found to kill Plasmodium falciparum, the malarial parasite at both the trophozoites and schizont stages[8]. In addition, all the maturation stages of the gametocytes were killed by Neem, suggesting the antiparasitic effect on stages causing clinical infection and those responsible for transmission. The antiplasmodium effect of Neem components was exercised also on parasites resistant to chloroquine and pyrimethamine[8].

Neem has volatiles that on short exposure, suppress the laying of eggs (oviposition), by the females of both Anopheles stephensi and Anopheles culcifaciens[9]. Long term exposure causes impairment of the gonotrophic cycle resulting in their inability to reproduce and multiply. Thus the same tree has compounds acting on both the vector (mosquito) and the parasite.

Another elegant demonstration of the effect of various compounds in Neem at more than one level is provided by studies of Rembold[10] on Azadirachtin. The chemical structure of this triterpenoid is known[10], though its total synthesis has not yet been achieved. This compound blocks the development of larvae into insects. Neem also contains compounds with an antifeedant effect, discouraging insects to feed on plants sprayed by Neem extracts. In fact the attention to this plant of the German scientist Schmutterer who did pioneering work on the chemistry of Neem was drawn by the observation of a Neem tree standing untouched in a field devastated by locusts. Thus action is exercised at 2 levels: do not feed or if you are already fed, do not multiply and stay blocked as larvae. The USFDA approved the use of Neem extracts for agricultural & horticultural purposes following the success achieved by Neem extracts in saving a crop of oranges in Florida attacked by pests which were resistant to all available pesticides. The National Academy Press brought out a monograph on Neem[11] in 1992, which describes uses of Neem and gives key references.

### 2.1.1. Immunomodulatory properties of Neem

Besides the presence of about 25 tri, tetra and penta terpenoids and their derivative compounds, which have action on a variety of bacteria, fungi, viruses, parasites and insects, there are also compounds which stimulate immune response, in particular of $Th_1$ type, the cell mediated immune response and macrophage function. An intraperitoneal injection of emulsified Neem oil in mice caused leukocytic infiltration. The phagocytic activity of macrophages was enhanced along with induced expression of MHC-class II antigens[12]. Splenocytes from the Neem treated animals had higher lymphocyte proliferative response with mitogens such as Con A and antigens such as tetanus toxoid[12]. Oral administration of Neem oil caused elevation of $Th_1$ cytokines, TNF-$\alpha$ and $\gamma$-interferon[13].

## 2.2. Sapindus Mukerosi

It is a tree growing wild in India, Nepal and neighbouring countries. Its high yield of seeds and pericarp have saponins, which are easily extractable in hot water. These have been used for centuries by villagers to wash hair. Fine woollens, delicate Kashmere and precious carpets are also washed with aqueous extracts of this soap nut. The saponins contained in Sapindus have a mild detergent action. Saponins have been purified, their chemical structure is known.

## 2.3. Quinine Hydrochloride

It is the pharmacologically active compound from Cinchona. Its action on malarial parasites is well known. As a component of TONIC WATER, it is used worldwide and no toxic effects of its ingestion in indicated amounts have been noted. With Roger Le Grand, we tested its action on HIV-1, and observed that the compound has a potent virucidal effect on this virus.

## 3. Polyherbal Formulation

The active principles employed were (a) Praneem: purified deodourized, colourless extracts of Neem leaves (b) Purified saponins from Sapindus mukerosi (c) Quinine hydrochloride and Mentha citrata oil. Each component was purified to the extent that its stability at room temperature (25° C) for 1 year was not compromised. It may be mentioned that ultrapurification to single compounds makes those from Neem highly unstable. Physico-chemical characteristics and finger print profiles were developed on HPLC, HPTLC and gas-liquid chromatography for quality control to assure reproducibility from batch to batch. The components in defined amounts were dispensed in the form of cream, pessary or tablet with pharmacopially approved excipients, antioxidants and preservatives. The shelf life of the products was determined by accelerated stability studies at three temperatures and studied for various physico-chemical parameters and bioactivity at different time points.

## 4. Spermicidal Action (and Contraceptive Efficacy)

An easily performed bioactivity was the spermicidal action of the Praneem polyherbal formulation on human sperm. Every component of the formulation has spermicidal properties, however their combination is synergistic[14], and potentiates the action 8 fold, as determined by the Sander Cramer slide test on human sperm from a panel of donors[15]. Spermicidal action of Praneem is manifest not only *in vitro* but also *in vivo* as determined by post-coital tests conducted as per WHO guidelines in 9 volunteers[15]. This property of Praneem polyherbal (PPH) indicates its potential for preventing unwanted pregnancies. The contraceptive efficacy of PPH has however been tested so far only in rabbits. None of the 15 rabbits mated after vaginal application of Praneem became pregnant, whereas 13/15 rabbits receiving physiological saline conceived and delivered pups.

## 5. Anti-Microbial Action

### 5.1. Neisseria Gonorrhoea

Tests were performed on standard strains and clinical isolates from the STD clinic of the All India Institute of Medical Sciences (AIIMS). The specimens were plated on Thayer-Martin agar plates, cultured at 37° C for 48 hrs in a candle extinction jar. Identification was confirmed on the basis of colony morphology, gram staining and cytochrome C oxidase. By a calibrated dichotomous sensitivity test, Praneem inhibited the growth of N. gonorrhoea including the strains resistant to penicillin[16].

### 5.2. Herpes Simplex Virus-2 (HSV-2)

The tests for HSV and Chlamydia trachomatis were carried out at Johns Hopkins University by Kevin Whaley, Larry Zeitlin and Sharon Achilles[16]. These were *in vivo* investigations performed in progestin primed mice. 10 µl of Praneem pessary dispersed in phosphate buffered saline was delivered to the vagina followed about 20 seconds later with 10 µl of HSV-2 containing 10 $ID_{50}$. Coitus was simulated by moving the pipette in

Table 8.1. Prevention by Praneem polyherbal of vaginal transmission of herpes simplex virus-2

| Treatment | No. of mice shedding the virus/no. of mice inoculated | No. of mice with lesions/ no. of mice inoculated |
| --- | --- | --- |
| 1. Praneem polyherbal pessary | 0/8 (0%) | 0/8 (0%) |
| 2. Phosphate buffered saline | 14/14 (100%) | 12/14 (85.7%) |

**Table 8.2.** Prevention of Chlamydia trachomatis serovar D infection in CF-1 mice by the vaginal route

| Treatment | No. of positive/ | No. of inoculated |
|---|---|---|
| 1. Praneem polyherbal cream | 1/24* | (4.17%) |
| 2. Praneem polyherbal pessary in PBS | 5/36* | (13.9%) |
| 3. Phosphate buffered saline (PBS) | 31/36 | (86%) |

*$p<0.001$

and out 4 times. Infection was assessed by culturing vaginal lavages taken 3 days after the viral challenge. The animals were also observed for lesions. Table 8.1 gives the results of a representative experiment.

It will be observed that Praneem was highly effective in preventing the uptake of Herpes by vaginal route

### 5.3. Chlamydia Trachomatis

Every year about 89 million new cases of C. trachomatis infection occur in adults according to WHO. These invariably lead to serious long term sequale such as PID, ectopic pregnancy and infertility. The ability of Praneem to protect against vaginal transmission of this infection was studied at Johns Hopkins in progestin primed mice[16]. Table 8.2 summarizes the results.

### 5.4. Action on Aids Virus HIV-1

Although a large number of microbicides are under development, the one product, which has undergone Phase 3 trials under the UN Aids Initiative, failed to demonstrate proper protection against transmission of this infection. Efforts thus continue to provide women with self controlled methods for safe sex. Praneem polyherbal pessary and cream were tested at Institut Pasteur at 10% or 50% concentrations after exposure for one or ten minutes and then the viable virus was titrated by culture in activated lymphocytes. These studies showed potent virucidal action of Praneem cream and pessary on HIV-1[16]. Gustavo Doncel at Conrad Norfolk obtained similar results following cytopathic effect assay[16]. In the course of clinical trials in women with RTIs with the cream, pessary and tablet forms of Praneem, the the tablet was the most acceptable to the women. Ritu Malhotra with Roger Le Grand at CEA Paris recently examined the effect of Praneem tablets on HIV-1. Table 8.3 gives their observations.

### 5.5. Action on Candida, and Multidrug Resistant Urinary Tract E.coli

While Praneem cream inhibited the growth of all the clinical isolates of Candida albicans, C. tropicalis and C. krusei, the pessary acted on only some and not all isolates of Candida.[16] On the other hand, both cream and pessary inhibited the growth of E.coli A, B, C strains resistant to Ciprofloxacin, Cephalexin, Gentamycin, Cefatoxime, Amikacin and Augmentin.

## 6. Toxicity and Safety Studies

Preclinical toxicity studies were carried out for Praneem in 2 species, rats and monkeys. Both acute (3 weeks) and subacute toxicity testing (3 months) were done at the PGI centre designated by the Indian Council of Medical Research. The data was submitted to the Drugs controller of India to seek permission for Phase I clinical trials to evaluate the safety of the formulations in women. With due approval and clearance of the Ethics Committees, Phase I clinical trials on Praneem pessary (PP) were conducted in 23 women in 3 major institutions. Healthy, protocol eligible, women of reproductive age were asked to use PP once a night for seven consecutive days. They were asked to record irritation or any other side effects. Clinical examination was conducted before enrolment and after using Praneem for 7 days. Laboratory analysis included vaginal cytology, urine and blood chemistry and hematology. These studies showed that the Praneem pessary could be used safely with no local or systemic side effects.

Phase I studies on Praneem cream, conducted in Brazil by Ladipo and Spinola,[17] showed no side effects.

Phase I studies on Praneem Tablet were conducted in 20 women, 10 each at the Postgraduate Institute of Medical Education, Chandigarh and at the Institute for Research in Reproductive Health, Mumbai. In these studies, colposcopy was added to the other tests. These studies confirmed that the Praneem tablet could be safely used for 7 consecutive days, with out any local or systemic side effects.

## 7. Phase II Efficacy Studies

### 7.1. Praneem Polyherbal Cream (PPC)

Three pilot studies were conducted on PPC under South to South Co-operation in Reproductive Health. In Santa Domingo, Drs. Vivian Brache and Frank Alvarez tried PPC in 6 women with candidiasis. After 7 days use of 5ml of PPC every night, all of them were relieved clinically and 5/6 were negative for candida microbiologically.

In Salzburg, Austria, Prof. Julian Frick tried PPC in 3 men with genital herpes lesions. Application of the cream thrice every day for 3 days led to the disappearance of the lesions. Even more note-worthy was the relief given to 2 patients of drugs resistant balanoposthitis.

Table 8.3. HIV titres on day 14 after exposure for 10 min to Praneem or Placebo tablets

| Treatment | $TCID_{50/ml}$ Ultracentrifuged | Non ultracentrifuged |
|---|---|---|
| 1. None (virus only) | 34,939 | 102,161 |
| 2. Tablet (1/20) | 19 | 19 |
| 3. Tablet (1/100) | 11 | 56 |
| 4. Placebo (1/20) | 11,949 | 11,949 |
| 5. Placebo (1/100) | 59,744 | 59,744 |

In Assiut, Egypt, Prof. Shaaban used PPC in 20 women with abnormal vaginal discharge. 5 women were kept as controls to assess normal regression. After 7 days of treatment with Praneem 95% (19/20) of the women were relieved clinically. The viable count of micro-organisms in the vagina diminished from a mean of $1.01 \times 10^{14} \pm 3.07 \times 10^4$ at the time of enrolment to $177,000 \pm 357,000$.

## 7.2. Praneem Polyherbal Pessary (PPC)

Trials were conducted in 28 women with abnormal vaginal discharge due to RTIs at PGI, Chandigarh. After 7 days of treatment with PPC, 23 women completely recovered from their symptoms and 5 had partial response. Six women who had Gardnerella vaginalis and 2 who had Trichomonas to begin with, were negative for these organisms after 7 days treatment with Praneem. 2/3 were microbiologically negative for Candida. The remaining had uncharacterised aerobic and anaerobic flora, which appeared to be reduced with clinical improvement.

## 7.3. Praneem Polyherbal Tablet (PPT)

PPT is the form preferred by women. Two series of Phase II trials have been conducted. The first was on 45 women at the Postgraduate Institute of Medical Education and Research Chandigarh. 43/45 women were relieved of abnormal vaginal discharge after 7 days of use of the tablet every night. The remaining 2 who had partial response asked for continuation of the treatment by another week, at the end of which they were cured. Microbiological examination showed that 10 of these women had Gardenerlla and 9, Candida at the time of enrolment. All of them responded and were negative to these organisms at the end of the treatment.

The Indian Council of Medical Research has extended Phase II efficacy trials in 6 other centres located in Chandigarh, New Delhi, Kolkata, Mumbai and Pune. At present, 128 women suffering from abnormal vaginal discharge due to pathogens have been investigated. 70% of the patients enrolled had the syndrome for longer than 3 months duration and 38.5% had a chronic problem for over a year. 123/128 patients experienced full or partial relief after 7 days use of PPT (subject statement). 4 subjects had no relief and one had increased symptoms. On speculum examination, 68.7% had no or normal discharge after treatment. Regarding microbiological examination, 100% of those with Trichomonas vaginalis were cured, 92% with Candida and 72–75% of patients with Bacterial vaginosis as per Nugent score, Wet Smear and Amine test.

### 7.3.1. Other on Going Trials

We have at present data on about 200 women suffering from abnormal vaginal discharge due to Reproductive tract infections (RTIS). 95% of the subjects get partially or fully cured with 7 daily intakes of Praneem Polyherbal formulation. Experimental data indicates also the virucidal properties of Praneem on Herpes simplex-2 and HIV-1. A Phase II trials have recently been approved by NACO (National Aids Control Organization) which will be conducted in 2 cohorts of seronegative women: monogamous married women and professional sex workers at NARI (National Aids Research Institute) Pune. Another pilot trial has started at the Institute of Cytology and Preventive Oncology, New Delhi with Praneem/ Placebo Tablets in women with cervical erosions and dysplasia who by molecular probes are positive for the human papilloma virus of the type associated with carcinoma of the cervix, HPV 16 and HPV 18. Early data on 4 patients completing 30 days use of PPT is encouraging.

## 8. Overview and Summary

In addressing a major problem of female Reproductive Health, we have drawn on traditional knowledge and age old practices of using some plants to cope with a wide variety of infections. Reproductive tract infections are caused by a plethora of microorganisms ranging from aerobic, anaerobic bacteria, viruses, Mycoplasma, Candida and protozoa. No single compound can be expected to act against all these organisms, hence a combination has been formulated with purified extracts from Neem (Azadirachta indica), Sapindus mukerosi, Cinchona and Mentha citrata oil. Praneem polyherbal is dispensed as cream (for use by men or on skin), pessary and tablets. The tablet is the preferred form and is the most highly acceptable by women. Each ingredient is standardized and quality controls have been established to ensure consistency of properties from batch to batch. Large scale requirement of Praneem Tablets for ongoing clinical trials are being provided by a company manufacturing the product under GMP conditions.

Praneem polyherbal inhibits the growth of N. gonorrheae (including penicillin resistant strains), multidrugs resistant urinary tract E.coli, Candida albicans, C. krusei, and C. tropicalis. It has potent virucidal action on HIV-1. It prevents transmission of Herpes simplex-2 and Chlamydia trachomatis through the vaginal route in progestin primed mice.

Praneem polyherbal is in clincial trials with the approval of the Drugs Regulatory Authorities and Ethics Committees. Three phase I trials with cream, pessary and tablet have shown the formulation's safety, with no local or systemic side effects observed.

Phase II efficacy trials are in progress and results are available in 201 women investigated in the course of 2 series of trials. 95% of the subjects were fully or partially relieved of the syndrome of abnormal vaginal discharge due to reproductive tract infections (RTIs) by taking Praneem 7 times a day. Microbiological examination showed that Praneem was highly effective against Gardenerella vaginalis, Trichomonas, Candida and flora associated with Bacterial vaginosis.

## 9. Acknowledgement

Thanks are due to Ms. Sayantani Basak for her competent typing of the manuscript.

## 10. References

1. World Health Organization: Report on Fighting Disease Fostering Development, Geneva, (1996).
2. RA Bhujwala, K Buckshee, Sriniwas. Gardnerella vaginitis and associated aerobic bacterials in nonspecific vaginitis. Ind J Med Res, 81; 251–256, (1985).

3. RA Bang, AT Bang, M Baitula, Y Chaudhry, S Sarmukaddam, O Tale. High prevalence of gynaecological diseases in rural Indian Women. Lancet 1:85–86, (1989).
4. M Morris, A Nicoll, I Simms, Jeffrey M Spieler. Bacteial vaginosis: a public health review. British J Ob & Gy; 108:439–450, (2001).
5. J Kreiss, E Ngugi, K Holmes, J Ndinya-Achola, P Waiyaki, PL Roberts et al. Efficacy of nonoxynol-9 contraceptive sponge use in preventing heterosexual acquisition of HIV in Nairobi prostitutes. JAMA 268; 477–82, (1992).
6. AR Rao, SSU Kumar, TB Paramsivan, S Kamalakshi, AR Parashuraman, M Shantha. Study of antiviral activity of tender leaves of margosa tree (Melia azadirachta) on vaccinia and variola virus, a preliminary report. Ind J Med Res 57; 495–502, (1969).
7. N Pant, HS Garg, KP Madhusudanan, DS Bhakuni. Sulfurous compounds from Azadirachta indica leaves. Fitoterapia, 57; 302–304, (1986).
8. Ravi Dhar, Zhang Kuniyan, GP Talwar, S Garg, N Kumar. Inhibition of the growth and development of asexual and sexual stages of the drug sensitive and resistant strains of the human malarial parasite plasmodium falciparum by Neem (Azadirachta indica) fractions. J Ethnopharmacology 61:31–39, (1998).
9. R Dhar, H Dawar, S Garg, SF Basir, GP Talwar. Effect of volatiles from neem and other natural products on gonotrophic cycle and oviposition of Anopheles stephensi and An culcifacies. J Med Entomol. 33:195–201, (1996).
10. H Rembold. Azadirachtin—a botanical growth inhibitor and its relation to Biosemiotics In Recent Trends in Life Sciences. (Indian National Science Academy, New Delhi) pp. 471, (1994).
11. National Research Council. Neem—A tree for solving Global Problems. Washington DC: National Academy Press. (1992).
12. SN Upadhyay, S Dhawan, S Garg, GP Talwar. Immunomodulatory effects of Neem (Azadirachta indica) oil. Int J Immunopharmacol 14:1187–1193, (1992).
13. GP Talwar, S Shah, S Mukherjee, R Chabra. Induced termination of pregnancy by purified extracts of Azadirachta indica (Neem): Mechanisms involved. Am J Reprod Immunol 37:485–491, (1997).
14. S Garg, G Doncel, S Chhabra, SN Upadhyay, GP Talwar. Synergistic spermicidal activity of neem seed extract, reetha saponins and quinine hydrochloride. Contraception. 50:185–190, (1994).
15. P Rahuvanshi, R Bagga, D Malhotra, S Gopalan, GP Talwar. Spermicidal and contraceptive properties of Praneem polyherbal pesary. Ind J med Res 113:135–141, (2001).
16. GP Talwar, P Raghuvanshi, R Mishra, U Bannerjee, A Rattan, KJ Whaley, L Zeitlin, SL Achilles, FB Sinoussi, A David, GF Doncel. Polyherbal formulations with wide spectrum antimicrobial activity against reproductive tract infections and sexually transmitted pathogens. Am J Reprod Immunol 43:144–1581, (2000).
17. OA Ladipo. South to South experience with polyherbal cream in the treatment of vaginal infections. In Barrier contraceptives Eds. Christine K Mauck, Milton Cardero, Henry L Gabelnick, Jeffrey M Spieler, Roberto Rivera. New York. Wiley-Liss pp. 277–282, (1994).

# 9

# Anticancer Therapeutic Potential of Soy Isoflavone, Genistein

Mepur H. Ravindranath, Sakunthala Muthugounder, Naftali Presser, and Subramanian Viswanathan

## Abstract

Genistein (4'5, 7-trihydroxyisoflavone) occurs as a glycoside (genistin) in the plant family Leguminosae, which includes the soybean (*Glycine max*). A significant correlation between the serum/plasma level of genistein and the incidence of gender-based cancers in Asian, European and American populations suggests that genistein may reduce the risk of tumor formation. Other evidence includes the mechanism of action of genistein in normal and cancer cells. Genistein inhibits protein tyrosine kinase (PTK), which is involved in phosphorylation of tyrosyl residues of membrane-bound receptors leading to signal transduction, and it inhibits topoisomerase II, which participates in DNA replication, transcription and repair. By blocking the activities of PTK, topoisomerase II and matrix metalloprotein (MMP9) and by down-regulating the expression of about 11 genes, including that of vascular endothelial growth factor (VEGF), genistein can arrest cell growth and proliferation, cell cycle at G2/M, invasion and angiogenesis. Furthermore, genistein can alter the expression of gangliosides and other carbohydrate antigens to facilitate their immune recognition. Genistein acts synergistically with drugs such as tamoxifen, cisplatin, 1,3-bis 2-chloroethyl-1-nitrosourea (BCNU), dexamethasone, daunorubicin and tiazofurin, and with bioflavonoid food supplements such as quercetin, green-tea catechins and black-tea thearubigins. Genistein can augment the efficacy of radiation for breast and prostate carcinomas. Because it increases melanin production and tyrosinase activity, genistein can protect melanocytes of the skin of Caucasians from UV-B radiation-induced melanoma. Genistein-induced antigenic alteration has the potential for improving active specific immunotherapy of melanoma and carcinomas. When conjugated to B43 monoclonal antibody, genistein becomes a tool for passive immunotherapy to target B-lineage leukemias that overexpress the target antigen CD19. Genistein is also conjugated to recombinant EGF to target cancers overexpressing the EGF receptor. Although genistein has many potentially therapeutic actions against cancer, its biphasic bioactivity (inhibitory at high concentrations and activating

---

**Mepur H. Ravindranath** • Laboratory of Glycoimmunotherapy, John Wayne Cancer Institute, 2200 Santa Monica Blvd. Santa Monica, CA 90404-2302, USA. *ravi@jwci.org (310) 449-5263, fax 310 449-5259.

*Complementary and Alternative Approaches to Biomedicine*, edited by Edwin L. Cooper and Nobuo Yamaguchi. Kluwer Academic/Plenum Publishers, 2004.

**Table 9.1.** Genistein and daidzein rich plant foods. (Values are expressed as µg/100 g dry weight)

| Plant | Genistein | Daidzein | Source |
|---|---|---|---|
| Soybean, *Glycine max* | 26,800–102,500 | 10,500–85,000 | Mazur, 1998 |
| Kudzu root, *Pueraria lobata* | 126,000 | 185,000 | Mazur, 1998 |
| Kudzu leaf, *Pueraria lobata* | 2520 | 375 | Mazur, 1998 |
| Black gram, *Vigna mungo* | 1900 | 745 | Adlercruetz & Mazur, 1997 |
| Kidney bean, *Phaseolus vulgaris* | 18–518 | 7–40 | Mazur, 1998 |
| Chick Pea, *Cicer arietinum* | 69–214 | 11–192 | Mazur, 1998 |
| Peanut, *Arachis hypogaea* | 64 | 58 | Mazur & Adlercruetz, 2000 |

at low concentrations) requires caution in determining therapeutic doses of genistein alone or in combination with chemotherapy, radiation therapy, and/or immunotherapies. Of the more than 4500 genistein studies in peer-reviewed primary publications, almost one fifth pertain to its antitumor capabilities and more than 400 describe its mechanism of action in normal and malignant human and animal cells, animal models, *in vitro* experiments, or phase I/II clinical trials. Several biotechnological firms in Japan, Australia and in the United States (e.g., Nutrilite) manufacture genistein as a natural supplement under quality controlled and assured conditions.

## 1. Introduction

Phytoestrogens are plant compounds with weak estrogenic activity. They include lignans, flavones and isoflavones. Genistein (4'5, 7-trihydroxyisoflavone), genistin (glycoside of genistein), daidzein, daidzin (glycoside of daidzein) are isoflavones from plants primarily belonging to the family Leguminosae. Both genistein and daidzein are found in several legumes (Table 9.1).

Their glycosides are present in plants, although food processing such as fermentation may cause deglycosylation. When the dietary isoflavones are ingested, their sugar moieties are hydrolyzed by gastric hydrochloric acid, β-glycosidases in gut-associated bacteria, thereby increasing the free isoflavone concentration (Kelly et al., 1993). Chemical structures of genistein, daidzein and its excretory metabolite equol (Cruz et al., 1994) are shown in Fig. 9.1. Infants fed soy formulas during the first four postnatal months, when gut microflora are underdeveloped, do not excrete appreciable amounts of equol (Cruz et al., 1994). Genistein and daidzein have diphenolic structures and differ by only one hydroxy group.

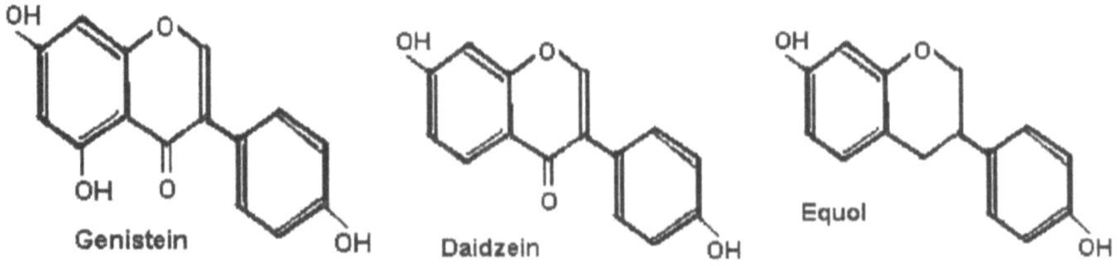

**Figure 9.1.** Diphenolic structure of Genistein, Daidzein and Equol

**Table 9.2.** Number of published studies on Genistein (based on a search of the National Library of Medicine Medline database)

| Subject | <2000 | 2000 | 2001 | 2002 | 2003 | Total |
|---|---|---|---|---|---|---|
| All aspects | 2196 | 733 | 796 | 766 | 261 | 4752 |
| All Cancers | 403 | 161 | 165 | 160 | 48 | 937 |
| Breast | 69 | 47 | 50 | 48 | 13 | 227 |
| Prostate | 10 | 25 | 28 | 41 | 13 | 117 |
| Liver | 28 | 9 | 7 | 4 | 1 | 49 |
| Melanoma | 19 | 11 | 7 | 6 | 1 | 44 |
| Colon | 25 | 7 | 4 | 2 | 2 | 40 |
| Pancreas | 10 | 0 | 4 | 5 | 2 | 21 |
| Brain | 12 | 2 | 2 | 1 | 0 | 17 |
| Bladder | 5 | 1 | 1 | 1 | 0 | 8 |
| Uterus | 6 | 0 | 6 | 0 | 0 | 12 |
| Ovary | 2 | 1 | 4 | 4 | 0 | 11 |
| Kidney | 4 | 1 | 0 | 1 | 0 | 6 |
| Bone and others | 201 | 109 | 116 | 117 | 32 | 575 |

A review of the literature shows that genistein has been studied extensively (Table 9.2). Two clinical trials conducted under the auspices of the National Cancer Institute (NCI-98-C-0099B and NCI-G00-1788) are examining genistein as an oral chemotherapeutic for prostate cancer. A company based in the United States (Nutrilite) manufactures two forms of genistein under stringent good manufacturing practices (GMP) and good laboratory practices (GLP): genistein in combination with black cohosh as a food supplement to avoid the risk of cancer, and genistein in combination with phaseolamin as an agent to block enzymatic digestion of complex carbohydrates. An Australian company (Novogen) used the isoflavone genistein as the basis for synthesis of an isoflavonoid called phenoxodiol.

Our focus is the anticancer activity and therapeutic potential of genistein. Dietary analyses suggest that genistein may lower the risk of cancer formation. The individuals on Western diet typically have low levels of blood isoflavones. Mean plasma/serum genistein and daidzein concentrations were 4.9 nmol/L and 4.2 nmol/L, respectively, in a Finnish study and 5.7 nmol/L and 2.1 nmol/L, respectively, in an American study. By comparison, concentrations were 248 nmol/L (range, 90 to 1204 nmol/L) and 163 nmol/L (range, 58 to 924 nmol/L) in a Japanese study (Pumford et al., 2002). The estimated plasma half-life for genistein and daidzein is about 8 hours in adults, with peak plasma concentrations 6 to 8 hours after ingestion of the pure compounds (Setchell, 1998) and soy powder (Watanabe et al., 1998). The amount of daidzein excreted in the urine also differs between US and Japanese populations (Table 9.3).

With soy supplementation, excretion half-lives for genistein, daidzein and equol are 7, 4 and 9 hours, respectively (Shelnutt et al., 2000; Lu et al., 1996). Excretion half-lives were longer in women than in men after an initial exposure to soy milk, but were shorter in women and longer in men after chronic ingestion (Lu and Anderson, 1998). Improved assay detection systems together with more efficient extraction from smaller samples of urine, serum and tissue will augment the use of isoflavones as biomarkers in large

**Table 9.3.** Level of urinary daidzein in American and Japanese populations

| Location | Mean (µmol/day) | Range |
|---|---|---|
| California, USA (multiethnic) | 0.28 | 0 to 14 |
| Minnesota, USA (Caucasian) | 0.56 | 0.04 to 29.48 |
| Japan (traditional) | 2.58 | 1.6 to 5.25 |

population-based studies (Lampe, 2003). The clinical trials testing division of Laboratory Corporation of America Holdings (LabCorp) has entered into a relationship with ESA Laboratories, Inc., a unit of ESA, Inc., (Chelmsford, Mass.) for high-performance-liquid-chromatography (HPLC)-based analytical technologies and assays. These include validated procedures using a multi-channel Coularray® system for plasma phytoestrogens (daidzein, genistein, and equol).

## 2. Mechanisms of Action of Genistein

### 2.1. Genistein, a Specific Inhibitor of Protein Tyrosine Kinase

Protein tyrosine kinases (PTKs) play an important role in cell growth. Retroviral oncogenes such as *src, yes, fgr, abl, fps, fes*, and *ras* code for PTKs (Bishop, 1983). PTKs are associated with cell receptors for EGF, platelet-derived growth factor (PDGF), insulin and insulin-like growth factors (IGF), suggesting that tyrosine phosphorylation plays an important role in cell proliferation and transformation. Recent studies show that PTKs are involved in integrin signal transduction pathways (Scholar and Toews, 1994). While testing the hypothesis that a specific inhibitor for tyrosine kinases could be an antitumor agent, Akiyama and co-investigators (1987) demonstrated that genistein inhibited *in vitro* tyrosine-specific PTK activity of the EGF receptor, $pp60^{v-src}$ and $pp110^{gag-fes}$. By contrast, genistein scarcely inhibited the serine- and threonine-specific protein kinases such as cAMP-dependent protein kinase, protein kinase C and phosphorylase kinase. The PTK-inhibitory activity of genistein is hypothesized to be responsible for the lower rate of breast cancer observed in Asian women consuming soy. However, although genistein inhibits the growth of breast cancer cells *in vitro*, it does so without gross inhibition of PTK activity (Paterson and Barnes, 1996).

### 2.2. Genistein Inhibits Topoisomerase II

Topoisomerases introduce transient breaks in DNA. They participate in DNA replication, transcription, recombination, integration and transposition (Sternglanz et al., 1981, Steck and Drlica, 1984, Kikuchi and Nash, 1979). They are also related to transformation by ras-oncogenes. Genistein inhibited the formation of a covalent complex between topoisomerase II and DNA (Markovits et al., 1986) and suppressed the growth of the transformed cells (Okura et al., 1988).

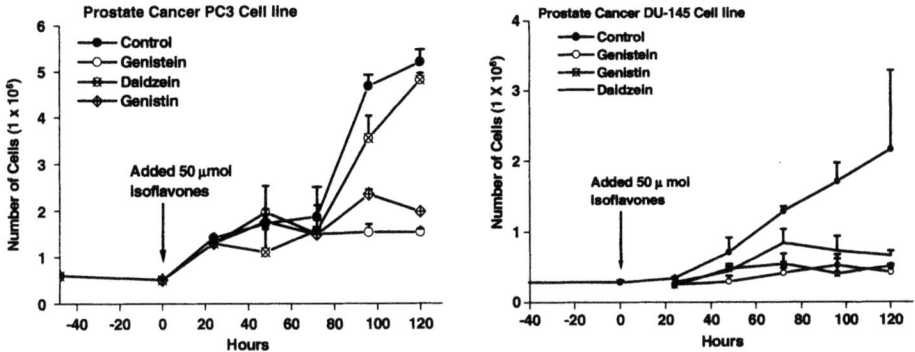

**Figure 9.2.** Specific growth inhibition of prostate cancer cells (PC3 and DU145) by genistein. Genistein but not its glycoside inhibits tumor growth. Forty eight hours before exposing the cells to genistein, flasks [T25 Becton Dickinson 353107] were seeded with 0.5 million PC3 cells and 0.3 million DU145 cells. [Unpublished data from Ravindranath, MH., Presser, N., Muthugounder, S.)

## 2.3. Genistein Arrests Tumor Cell Growth: The Biphasic Effect

Table 9.4 summarizes published evidence that genistein can inhibit the growth of various cancers including melanoma, leukemia, lymphoma, breast, prostate, colon, pancreas and lung carcinoma. The most striking finding is that the bioactivity of genistein is biphasic. Genistein at low concentrations stimulated growth and at high concentrations inhibited growth of estrogen-positive MCF-7 and estrogen-negative MDA-MB-438 breast cancer cells (Miodini et al., 1999; Balabhadrapathruni et al., 2000, Maggiolini et al., 2001) and SCC-25 oral squamous-cell carcinoma cells (Elattar and Virji, 2000). In preclinical studies, genistein at a level corresponding to its serum/plasma concentration inhibits tumor growth and proliferation (Santell et al., 2000).

## 2.4. Genistein Arrests the Cell Cycle Between G2 and M Phases

Genistein arrests the cell cycle at the G2/M phase (Table 9.5). The possible mechanism is indicated in Fig. 9.3. A number of studies (Table 9.6) suggest that genistein-induced cell cycle arrest may involve upregulation of P21WAF1 and consequent downregulation of cyclin B1. Recent studies of Choi and others (2000) on human prostate cancer cells, Casagrande and Darbon (2000) on human melanoma cells, Park and co-investigators (2001) on hepatoma cells, and Frey and Singletary (2003) on breast cancer cells confirm increased expression of Cdk2 inhibitor P21 (WAF/CIP1) after exposure to genistein. Induction of P21, which inhibits threshold kinase activities of Cdks and associated cyclin B, leads to cell-cycle arrest.

## 2.5. Genistein Induces Apoptosis

Studies of human cancer cells show that genistein can induce apoptosis by the following mechanisms (Table 9.4): fragmentation of DNA; activation of caspase-3 (CPP32b); cleavage of poly (ADP-ribose) polymerase (PARP); downregulation of Bcl-2 (Bcl-2

**Table 9.4.** Effects of Genistein *in vivo* and *in vitro* in Human Cancers [G: Genistein; D: Daidzein]

| Authors | Cell Lines | Results & Comments |
|---|---|---|
| **BRAIN CANCER *In vitro*** | | |
| Oude Weernink et al. (1996) | A172 glioblastoma & Hs683 glioma | G inhibited EGF & PDGF stimulated autophosphorylation of the receptors and subsequent DNA synthesis. G appears to be cytotoxic. |
| Cataldi et al. (1996) | GH3 pituitary adenoma | G (20-200 μ) modified Ca+ channel activity. |
| Jones et al. (1997) | pituitary adenomas, prolactinoma, somatotropinoma & mammosomatropinoma | G at 74 and 740 μmol/L inhibited thymidine uptake in tumors by 62% and 93%, respectively. G also inhibited EGF-stimulated thymidine uptake. Growth factors released by tumors may promote tumor growth by stimulating tyrosine kinase activity. |
| Khoshyomn et al. (2000) | medulloblastoma: HTB-186, CRL-8805 & MED-1 | G (6 μM) caused 2.8-fold increase in growth inhibitory effect, 2.6-fold inhibition of colonigenic survival, 1.3-fold increase in antiproliferative effect compared to that of cisplatin in HTB186 cells. G at dietary plasma level enhanced antiproliferative and cytotoxic action of cisplatin and to a lesser extent vincristine. |
| Khoshyomn et al. (2002) | U87 glioblastoma multiforme (GBM) | G at dietary plasma level (4 μM) enhanced antiproliferative and cytotoxic action of 1,3-bis 2 chloroethyl-1-nitrosourea (BCNU). |
| **BREAST CANCER *In vivo*** | | |
| Barnes et al. (1996) | | Based on pharmacokinetic calculations involving daily intake of G, absorption from the gut, distribution to peripheral tissues, and excretion, it is unlikely that blood G concentrations even in high soy consumers could be greater than 1 to 5 μM. G is well absorbed from small intestine, undergoes enterohepatic circulation, and gets excreted in bile as its 7-*O*-β-glucuronide. The plasma G levels are unlikely to be sufficient to inhibit the growth of mature, established breast cancer cells by chemotherapeutic-like mechanisms. However, these levels are enough to regulate proliferation of epithelial cells in breast and thereby may cause a chemopreventive effect. |
| Lu et al. (1996) | | Urinary recovery of G, D and equol was measured in six healthy women (22 to 28 yrs old) after daily ingestion of 35-oz of soymilk. Chronic ingestion of soya in women alters the metabolism and disposition of ingested isoflavones, thus altering their estrogenic potency. |
| McMichael-Phillips et al. (1998) | | Women (n = 48) with benign or malignant breast disease were randomly assigned to receive their normal diet alone or with a 60 g soy supplement (containing 45mg isoflavones) daily for 14 days. The proliferation rate of breast lobular epithelium significantly increased after 14 days of soy supplementation when the day of menstrual cycle and the age were accounted for. G significantly increased progesterone receptor expression, and stimulated breast proliferation, which could be due to estrogen agonist activity. |

| Reference | | Description |
|---|---|---|
| Zheng et al. (1999) | | The urinary excretion of G and D is greater in age-matched controls than in breast cancer patients, suggesting that a high intake of isoflavones may reduce the risk of breast cancer. |
| Lu et al. (2000) | | This study was done on ten healthy, regularly cycling women, who consumed soy diet starting on day 2 of the cycle until day 2 of the next cycle. G and D levels, ovarian hormones and progesterone were measured in blood and urine. No change in the length or luteal phase of the cycle was observed. Systemic bioavailability of G & D occurred within 15 hours after consumption of soymilk. There is a decrease in follicular and luteal phase 17-b-estradiol levels, which correlated with urinary G & H levels. Decrease in progesterone level was inversely related with the increase in G level. Decrease in hormone levels may have implications for reducing breast cancer risk. |
| Murkies et al. (2000) | | Association between G/D and the risk of breast cancer in Australian postmenopausal women and 20 controls was examined. Women with breast cancer had lower 24-h urinary D compared with controls (p = 0.03). A similar but not significant trend was observed with G. |
| den Tonkelaar et al. (2001) | | Individuals with high urinary excretion of G had almost 50% risk of breast cancers. Protective effects of G on BrCA risk may be smaller in postmenopausal Dutch women. |
| Kumar et al. (2002) | | 68 premenopausal women ages 25 to 55 yrs were randomized into an experimental group (supplemented with 40 mg of G/day) or to a control group which did not consume G. Serum-free estradiol and estrone levels decreased moderately in the experimental group. Serum hormone-binding globulin level increased in 41.4% of those compared to 37.5% of women in placebo group. The menstrual cycle is increased by 3.52 days compared with 0.06 days for Placebo group [p = 0.4]. Isoflavone intake affects estrogen metabolism by altering the steroid hormone concentrations and menstrual cycle length, thereby showing a potential to reduce the risk of breast cancer. |
| **BREAST CANCER In vitro** | | |
| Peterson and Barnes (1991) | MCF-7, MCF-7-D-40 (ER+) & MDA-468 (ER-) | G inhibited cell growth independent of ER status. ER is not needed for G to inhibit cell proliferation. The effect of G is not attenuated by overexpression of the multi-drug resistance gene product. |
| Pagliacci et al. (1993) | MCF-7, Human BrCA, Jurkat cells T-cell leukemia & L-929 | G inhibited cell growth at G2/M cell cycle. G treated cells displayed an increase in cell volume and in mitochondrial number and/or activity. |
| Pagliacci et al. (1994) | MCF-7 | G caused dose-dependent inhibition of cell growth [ID50 = 40 μM], by causing reversible G2/M arrest in cell cycle progression. G caused marked fall in S-phase associated with persistent arrest in G2/M phase, and decreased DNA content and nuclear fragmentation. Since the mitogenic action of insulin and IGF-1 in MCF-7 cells is a tyrosine kinase dependent phenomenon, G impact was analyzed after addition of insulin. Insulin addition produced a strong increase in the percentage of S-phase cells, which was almost completely blocked by 100 μM of G. Anti-phosphotyrosine antibody stained intensely MCF-7 cells after addition of insulin but not after addition of G. G arrests cell cycle by inhibiting tyrosine kinase activity and the signal transduction pathway. |

*(Continued)*

Table 9.4. (Continued)

| Authors | Cell Lines | Results & Comments |
|---|---|---|
| Monti & Sinha (1994) | MCF-7/WT, MCF-7/ADR & MDA-231 | G inhibited cell proliferation in all 3 cell lines [IC 50 = between 7 and 37 µM] and acted synergistically with Adriamycin. This combination may have potential in hormone-resistant and multidrug-resistant tumor cells. |
| Hoffman (1995) | MCF-7[ER+], T47D [ER+], & SKBR3 [ER−] | G is about 5-fold more potent than Kievitone as an inhibitor of soluble EGFR kinase activity and EGFR autophosphorylation. DNA synthesis of MCF-7 cells by IGF-1 and IGF-2, FGF or TGF-a and 17 b-estradiol was inhibited by G. |
| Das & Vonderhaar (1996) | T47D [prolactin sensitive] & NOG-8 [prolactin sensitive] | Prolactin [prl] promotes growth and differentiation of normal and malignant breast cells. The two cell lines respond to prl and kinases were activated. Raf-1 was activated within 2–5 min, followed by activation of MEK and MAP kinase. Increased MAP kinase activity was accompanied by tyrosine phosphorylation. G blocked the increase in MAP kinase activity, and prl induced growth of T47D cells. |
| Panno et al. (1996) | MCF-7 | Insulin increased ER content and its binding capacity. G decreased ER content and completely abolished its binding capacity. Insulin increased progesterone [PR] receptor levels; G abolished this effect. Insulin induced phosphorylation of ER and PR, and G abrogated tyrosine phosphorylation. |
| Clark et al. (1996) | SKBR3 & MCF-7 | Protein tyrosine kinases of src proto-oncogene family form membrane associated PTK complex, pivotal for signaling. Inhibiting PTK is a potential approach to treat BrCa. G showed dose-dependent inhibition of cell growth with decreased Shc tyrosine phosphorylation, decreased Shc's association with Grb-2 and MAP kinase activity. |
| Peterson et al. (1996) | HME (normal human breast epithelial) cells & MCF-7 | In HME cells, G caused 5-fold more growth inhibition than Biochanin A, and in MCF-7 cells both are equally potent growth inhibitors. G uptake by cells and the resultant metabolites were quantitated. Once G was metabolized there was a decrease in growth inhibition. |
| Peterson & Barnes (1996) | MCF-7 | Growth inhibition of G was cytostatic and reversible at IC50 concentrations, which ranged from 2.6 to 20 µg/mL. Growth inhibition was without gross inhibition of PTK activity. |
| Wang et al. (1996) | MCF-7 | G stimulated growth of the cells at $10(-8)$–$10(-6)$ M. G inhibited the growth at concentrations $>10(-5)$ M. Inhibition appears to be via ER pathway; at higher concentrations it appears to be independent of ER pathway. Since G is estrogenic it may interfere with the effects of estradiol. Prolonged exposure to G resulted in decrease in ER mRNA level. |
| So et al. (1997) | MCF-7 | IC50 for G was calculated based on thymidine incorporation and found to range from 4.2 to 18.0 µg/mL. The cells were viable at this concentration. When cells were exposed to 17-b-estradiol inhibition of cell proliferation by G was reversed. |
| Verma et al. (1997) | MCF-7 | G and curcumin in the diet acted synergistically to reduce the proliferation of estrogen positive cells in response to pesticides or 17-beta estradiol. |
| Katdare et al. (1998) | 184-B5 & non cancerous HMEC | G in combination with green tea polyphenol EGCG induced apoptosis and enhanced p53 immunoreactivity. |

| | | |
|---|---|---|
| Uckun et al. (1998) | MDA-MB-231[EGFR+] BT-20 [EGFR+] NALM-6 [EGFR-] & HL-60 (EGFR -) | EGF receptors are overexpressed on BrCa cells and associated with protein kinases of ErbB2, ErbB3 and Src proto-oncogene family to form membrane associated PTK complex pivotal for signaling. Human EGF-conjugated G construct directed against EGFR, arrested PTK activity in EGFR+ but not EGFR-negative cells. The construct inhibited PTK activity and triggered rapid apoptosis at nanomolar concentrations, whereas G alone inhibited at >10 μM concentration. GCSF (granulocyte colony stimulating factor) conjugated G did not inhibit or induce apoptosis, suggesting the specificity of EGF-G. EGF-G is 100-fold more potent inhibitor of PTK activity and 100-fold more potent cytotoxic than G alone. |
| Choi et al. (1998) | MCF-7 & MDA-MB-231 | G induced G2/M arrest. G did not change the steady state level of cdks, cyclin-A, D-type cyclins and cyclin E protein but inhibited expression of cyclin B1 protein in a time-dependent manner. Decrease in the level of cyclin B1 protein correlated with a decrease in the level of cyclin B1 mRNA. G induced expression of p21 and increased binding of p21 with cdc2 and cdk2. G mediated arrest of cell cycle could be due to inhibition of kinase activities of cdc2 and cdk2 and decrease in cyclin B1 expression. |
| Shao et al. (1998a) | MDA-MB-231 & MDA-MB-468 (ER+) | G inhibited both ER+ and ER− cell lines by arresting specifically G2/M phase and inducing p21WAF/C1P1 expression (mRNA expression 5 to 6-fold and protein expression 3-to 4-fold). Increased p21WAF1/CIPI was followed by apoptosis. |
| Shao et al. (1998b) | MCF-7 & MDA-MB-231 | G down-regulated MMP-9 (matrix metalloprotein) and up-regulated tissue inhibitor of metalloproteinase-1. G inhibited tumor growth in nude mice, stimulated apoptosis, upregulated p21WAF1/CIP1 expression. G also inhibited angiogenesis in xenografts by decreasing the vessel density and the levels of VEGF & TGF-β1. |
| Satyamoorthy et al. (1998) | HMEC & MCF-7 | G caused dose dependent increase in TGF-1 mRNA expression (causing growth inhibition and apoptosis) in normal human mammary cells (HMEC) but not in tumor derived MCF-7 cells. |
| Willard & Frawley (1998) | MCF-7 | G has agonistic effect on estrogen-dependent gene expression. |
| Hsieh et al. (1998) | MCF-7 | G enhanced proliferation of E dependent BrCa cells *in vitro*, induced estrogen responsive gene pS2 expression. |
| Fioravanti (1998) | MCF-7, MDA-MB-231 & HBL-100 | G inhibited cell growth [IC 50 approximately 10 μM] of the three breast cancer cell lines but not skin-derived fibroblasts which counteracted growth stimulatory effects of estradiol and growth factors. It abolished the paracrine stimulation observed in MCF-7 in co-culture with MDAMB-231 or fibroblasts. G caused accumulation of cells in S and G2M phases of cell cycle and underwent apoptosis. G decreased tyrosine phosphorylation induced upon treatment with TGFα. G bound to ER (Kd = 4 nm), induced pS2, and cathepsin-D transcription and increased nuclear ER levels. |
| Wang & Kurzer (1998) | MCF-7 | MCF-7 is estrogen dependent cell line. G enhanced insulin and EGF induced DNA synthesis at 0.01–1.00 and 0.1 to 10 μM concentration respectively. At higher concentration inhibition was observed. DNA synthesis is suggested to be mediated by tyrosine-kinase. |
| Shao et al. (1998c) | MDA-MB-231(ER), MDA-MB-468(ER−) | G caused downregulation of MMP-9 and upregulation of TIMP-1 and trypsin inhibitors. |

*(Continued)*

Table 9.4. (Continued)

| Authors | Cell Lines | Results & Comments |
|---|---|---|
| Peterson et al. (1998) | ZR-75-1, BT-20, MCF-7 & T47D | ZR-75-1 and BT-20 are more sensitive to G than MCF-7. The cells metabolized G and produced methylated and sulfated by-products. |
| Constantinou et al. (1998a) | MCF-7 | G at 0.15 mM caused apoptosis of MCF-7 accompanied by cell cycle delay in the G2/M phase. 24 hrs after treatment, 47.3% of cells accumulated at G2/M compared to 19.9% in the untreated controls. At 0.015 mM, G caused an increase in the wild-type tumor suppressor p53. p53 mRNA did not increase. Prior to upregulation of p53, bcl-2 phosphorylation was observed. After 24 hrs G inhibited phosphorylation. |
| Constantinou et al. (1998b) | MCF-7 (ER+) & MDA-MB-468 (ER−) | G at 15, 30 and 45 μmol/L, inhibited cell growth and accompanied by expression of maturation markers. The markers were optimally expressed after 9 days of treatment with 30 μmol G/L. Both ER+ and ER− cells responded to G treatment suggesting that antiestrogenic function of G is unrelated to the mechanism of cell differentiation. D did not induce such differentiation in both the cell lines. |
| Miodini et al. (1999) | MCF-7 | G exerted a biphasic effect on growth, stimulating at low concentrations and inhibiting at high concentration. At 5 μM, G counteracted estrogen and TGF promoted cell growth stimulation. G regulated expression of pS2 and cathepsin-D. G competed with estradiol for binding to ER. G downregulated cytoplasmic ER levels. G was able to upregulate progesterone receptor protein level. |
| Li et al. (1999a) | MDA-MB-231 | G caused upregulation of Bax and p21WAF1 expressions and down regulation of Bcl-2 and p53, leading to apoptosis and arrest of cell growth. Upregulation of Bax and p21WAF1 may be the molecular mechanisms by which G induced apoptosis. |
| Li et al. (1999b) | MDA-MB-435 & MDA-MB-435.eB | MDA-MB-435 was established by transfecting c-erb-2 cDNA. Erb-2 influenced the cell line to secrete matrix metalloproteinase (MMP). G inhibited cell growth of both the cell lines and induced apoptosis of both the cell lines. G upregulated Bax and P21 WAF1 and downregulated Bcl-2 and c-erbB-2. |
| El-Zarruk & van den Berg (1999) | ZR-45-1, ZR-75-9a1 (Tmxfn resis) & ZR-PR-LT (ER resist) | Tamoxifen resistant cell line showed increased expression of EGF-receptor and was least sensitive to G. |
| Shen et al. (1999) | MDA-MB-435(ER−) | IC50 of G and tamoxifen in growth inhibition assays of the cell lines were compared. IC50 for tamoxifen and G is 17 ± 0.9 and 27 ± 1.6μM, respectively. In clonogenic assays Tam and G showed an IC50 of 0.9± 0.4 and 12.5 ± 1.1 μM, respectively. Tam and G when added together showed synergism in growth inhibition, cytotoxicity and in the reduction of inositol 1,4,5-triphosphate concentration. |
| Balabhadrapathruni et al. (2000) | MDA-MB-468 | G inhibited cell growth with IC50 values of 8.8 μM. Flow cytometry showed G2/M cell cycle arrest with 22 μM. At 100 μM, G caused 70% accumulation at G2/M by 24 hrs. D is ineffective. At 10 μM, G caused apoptosis as measured by APO-BRDU analysis. G treatment resulted in a biphasic response on cyclin B1. 70% increase in cyclin B1 level at 25 μM, 50 and 70% decrease in cyclin B1 at 50 and 100 μM. |

| Reference | Cell line | Description |
|---|---|---|
| Santell et al. (2000) | MDA-MB-231 | G inhibited the cell growth and induced cell cycle arrest at G2/M in vitro. But in nude mice, G did not inhibit growth even after increasing plasma G concentration to 7 μmol/L. This suggested that it is unlikely that the plasma concentration required to inhibit cell growth in vivo can be achieved from a dietary dosage of G. Established tumors can not be regressed in vivo with dietary G. |
| Shao et al. (2000) | MCF-7 (ER+) T47D (ER+) T 549 (ER+) MDA-MB-231 (ER−) & MDA-MB-438 (ER−) | Antiproliferative effect of G is estrogen dependent in ER+ cell lines. G also inhibited the expression of ER-downstream genes including pS2 and TGF-b in ER+ cell lines. Inhibition is also dependent on estrogen. G decreased steady state ER mRNA only in the presence of estrogen in ER+ lines, therefore suppression through ER pathway. |
| Nakagawa et al. (2000) | DD-762, MDA-MB-231, MCF-7, Sm-MT & HBL-100 | G stimulated ER+ MCF-7 at 3.7 nmol to 37 μM. G induced apoptosis and induced cell cycle arrest at G2/M at higher concentrations. G caused upregulation of Bax protein, downregulation of bcl-XL protein and activation of caspase-3. G acted in synergism with eicosapentaenoic acid (EPA). G in combination with EPA may be beneficial. |
| Cappelletti et al. (2000) | T47D (ER+), BT-20 (ER−) ZR75.1 (ER+) & MDA-MB-231 (ER−) | G arrested cell cycle at G2/M. The percentage of cells arrested with cell cycle varied as follows: Compared to controls (9% of cells with cell cycle arrest), T47D (7 fold) > MDA-MB-231 & ZR75.1 (4-fold) > BT-20 (2.4 fold). Cdc-2 (p34) was unaffected by G in all cell lines, except T47D cell line. Cyclin B (p62) increased after G treatment. |
| Leung & Wang (2000) | MCF-7 | G at 25 μmol/L caused significant apoptosis. At or above this concentration c-jun N-terminal kinase was activated, Bax and Bcl-2 expression was elevated thus G is competitive ER agonist. Elevated Bcl-2 protein might neutralize the proapoptotic effect of Bax. |
| Frey et al. (2001) | MCF-10F | MCF-10F is non-neoplastic. G caused reversible G2/M block in cell cycle progression with IC (50) approximately 19-22 μM. G, at 45 μM, increased phosphorylation of Cdc2 by 3 fold, decreased the activity of Cdc2 by 70% by 8 hr. G induced expression of cell cycle inhibitor p21 (waf/cip1) as well as its interaction with Cdc2. G inhibited expression of phosphatase cdc25c by 80%. |
| Xu & Loo (2001) | MCF-7 (wild type gene) MDA-MB-231 (p53 gene) | G increased the level of p21 expression, independent of p53 gene level. G at 50 μM concentration induced apoptosis in MCF-7 cells. Significant DNA fragmentation was observed in MCF-7. MCF-7 in contrast to MDA-MB-231 cells were sensitive to the induction of apoptosis by genistein. |
| Upadhyay et al. (2001) | MCF10A, MCF12A (normal) MCF10CA1a, MDA-MB-231 (malignant) & HCT116 | G caused greater G2/M arrest and apoptosis in malignant cell lines as compared to normal Br epithelial cells. Sensitivity of normal and malignant BrCA cells to G is determined by p21(WAF1) content. G induced greater degree of cell cycle arrest and apoptosis in P21 (WAF1)−/− HCT116 compared with p21 (WAF1)+/+ HCT116 cells. P21 (WAF1) was downregulated with antisense p21 (WAF1) cDNA. Both normal and malignant p21 (WAF1) antisense expressing clones became more sensitive to G2/M arrest after G treatment. These results suggest that P21(WAF1) may play a role in determining the sensitivity of normal and malignant breast epithelial cells to G. |

(Continued)

Table 9.4. (Continued)

| Authors | Cell Lines | Results & Comments |
| --- | --- | --- |
| Allred et al. (2001a) | MCF-7 | MCF-7, ER+ tumors grew well in athymic mice fed with genistein. G in diet did not inhibit the tumor growth, however when animals were fed with G-free diet, tumor regressed notably. |
| Allred et al. (2001b) | MCF-7 | Soy protein isolates with increasing concentrations of genistein stimulated the growth of estrogen-dependent breast cancer cells in athymic nude mice in a dose-dependent manner. Cell growth was greatest in animals given 150 to 300 ppm G. Expression of pS2 was increased in tumors from animals exposed to G. |
| Maggiolini et al. (2001) | MCF-7 | G induced down regulation of ERα mRNA and protein levels. G is agonist of ERα and ERβ. At low concentration G stimulated proliferation of ERα-dependent cell lines, at high concentration G was cytotoxic and killed ER independent HeLa cells. |
| Dampier et al. (2001) | MCF-7, ZR-75.1, T47-D, MDA-MB 468, MDA-MB 231 & HBL 100 | G, at 1 μM, stimulated growth in MCF-7 cells. At 10 μM, G arrested the growth of all cell lines. G arrested growth at G2/M phase of cell cycle of all cell lines. G induced apoptosis only in MDA-MB 468 cells. G inhibited c-fos expression, transcription factor activator protein [AP-1] activity and extra-cellular signal regulated kinase ERK phosphorylation. G mediated growth inhibiting mechanism cannot be generalized for all cell lines. |
| Diel et al. (2001) | MCF-7 | G decreased ER protein content in BrCA cells. Molecular properties of G are similar to selective estrogen receptor modulators (SERMs) like ICI (faslodex) and Ral (raloxifene). |
| Ju et al. (2001) | MCF-7 in athymic nude mice | Dietary genistein increased MCF-7 cell growth in athymic nude mice in a dose dependent manner. Expression of pS2 mRNA was also significantly increased with increasing G levels. Total plasma concentration was between 0.39 and 3.36 μmol/L. |
| Messina & Loprinzi (2001) | MCF-7 | G exhibited a biphasic effect on the growth of MCF-7 cells in vitro. In ovariectomized athymic mice implanted with MCF-7. G stimulated tumor growth in a dose-dependent manner. In contrast, in intact mice fed with estrogen, G inhibited tumor growth. |
| Hu et al. (2001) | MCF-7 | G down-regulated Mcl-1 and Bcl-2 expression and induced growth inhibition by blocking at G2/M phase of the cell cycle, followed by apoptosis, leading to chromatin condensation and DNA fragmentation |
| Tanos et al. (2002) | MCF-10A(1), MCF-ANeoT, MCF-T(6)3B) [dysplastic] MCF-7, MDA-231 & MDA-435[malignant] | Growth inhibition of dysplastic and malignant epithelial cancer cells by G was dose-dependent [1–10 μg/ml]. Growth inhibition was higher [p < 0.0001] for dysplastic than for malignant cells. Inhibition was augmented by tamoxifen, reflecting a synergistic antiproliferative effect on dysplastic cells [p < 0.0001] and an additive growth inhibition effect [p < 0.0003] on malignant cells. Estradiol stimulated the growth of dysplastic cells while G had a significant antiproliferative effect on growth of the malignant cells. Effect of G is ER independent. |
| Ju et al. (2002) | MCF-7 in athymic nude mice | The experiment was performed in ovariectomized athymic mice. Experiment groups were (1) Control; (2) 0.25 mg of estradiol [E2]; E2+2.5 mg of Tamoxifen (TAM); E2+2.5 mg of Tamoxifen (TAM)+1000 ppm G; E2+2.5 mg of Tamoxifen (TAM); E2+2.5 mg of Tamoxifen (TAM) + 1000 ppm G; G negated the inhibitory effect of TAM on MCF-7, lowered E2 level in plasma, and increased expression of E-responsive genes (pS2, PR, cyclin D1). |

| Reference | Cell line | Findings |
|---|---|---|
| Nomoto et al. (2002) | MDA-MB-231[ER−] | G as well as estradiol induced cell cycle arrest and apoptosis through ERα-independent pathways. |
| Katdare et al. (2002) | 184-B5 / HER [Her2/neu +] (a model for Human ductal carcinoma in situ [DCIS] | G inhibited cell growth in a dose-dependent manner, increased ratio of G0/G1: S+G2/M and enhanced apoptotic population. G down-regulated HER-2/neu-mediated signal transduction, increased expression of cyclin dependent kinase inhibitor p16INK4, and induced Bcl-2 dependent apoptosis. |
| Po et al. (2002) | MCF-7 | G induced apoptosis in MCF-7, involving BaK and Bcl-x without evidence of anti-estrogenic effects. |
| Mitropoulou et al. (2002) | MCF-7 (ER+) & BT-20 (ER−) | G affected synthesis of glycosaminoglycans, proteoglycans, depending on presence or absence of ER and localization of proteoglycans. |
| Frey & Singletary (2003) | MCF-10F (non-malignant) | G arrested growth of MCF-10F in the G2 phase, induced Tyr15 phosphorylation of Cdc2, increased expression of p21 (WAF/CIP1) and decreased expression of the cell cycle phosphatase Cdc25C. G activated p38 mitogen-activated protein kinase, inactivated ERK1/ERK2 and had no effect on SAPK/JNK activity. Therefore p38 is involved in G-induced changes in Cdc2 phosphorylation and downregulation of Cdc25C. G mediated cell growth inhibition may involve p38 pathway and interplay between p38 pathway and G2 cell cycle. |
| **ENDOMETRIAL CANCER** *In vitro* | | |
| Markiewicz et al. (1993) | endometrial cells of Ishikawa-Var1 | The relative estrogenic potencies based on estrogen-specific enhancement of Alk phosphatase activity were similar to estradiol for coumestrol, G and D. These isoflavonoids exert their estrogenic effect by interacting with estrogen receptor. |
| **GASTROINTESTINAL CANCER** *In vitro* | | |
| Yanagihara et al. (1993) | HSC-41E6, HSC-45M2 & SH101-P4 | G is cytotoxic at low concentration (< 10 μg/ml). It induced apoptosis and suppressed tumor growth at high concentration. |
| Milovic et al. (1995) | Caco-2 | G completely inhibited EGF-stimulated polyamine uptake in Caco-2 cells in a dose-dependent manner. The effect of EGF on polyamine uptake by Caco cells could be due to translocation of intracellular protein or direct alteration of polyamine transporter. |
| Heruth et al. (1995) | HCT8 & SW837 | G decreased c-myc RNA and inhibited proliferation. |
| Kuo et al. (1995) | SW620 | TPA (12-O-tetradecanoylphorbol-13 acetate) induced differentiation of SW620 mediated by PTK. G (60 μM) potentiated TPA (12-O-tetradecanoylphorbol-13 acetate)-induced increase in PTK activity. |
| Kuo (1996) | Caco-2 & HT-29 | G is antiproliferative in a dose-dependent manner. Chromatin condensation by G was noticed. |
| Kuo et al. (1997) | Caco-2 & HT-29 | G inhibited accumulation of ascorbic acid during rapid cell division as compared to post confluency. The antiproliferative property of G could be linked to G's ascorbic acid deprivation property. |
| Yu et al. (1999) | HCT | G and D inhibited growth; IC50 concentrations are 15 and 40 μM respectively. Both affected the membrane fluidity, cell surface charges and membrane protein conformation. Membrane fluidity was reduced by G but not by D. G-effect was time and dose-dependent. G and D reduced the density of cell surface charges and increased the order of membrane protein conformation. |

(*Continued*)

Table 9.4. (*Continued*)

| Authors | Cell Lines | Results & Comments |
|---|---|---|
| Fiorelli et al. (1999) | HCT8 & HCT116 | Local synthesis of estrogen was assessed by evaluating gene expression and aromatase p450 activity. G increased aromatase gene expression in HCT8 cells only, whereas quercetin increased expression in both cells. |
| Cafferata et al. (2000) | T84 | G [60 μM], like PKC inhibitors, inhibited the stimulatory effect of IL-1beta on CFTR gene expression, which impairs chloride channel function. |
| Salti et al. (2000) | HT-29, SW-620 & SW-1116 | G induced topoisomerase II mediated DNA breakage within 1 hr of treatment. It induced apoptosis at >60 μM, inhibition of growth at 60-150 μM. Aclarubicin, a topo II antagonist reduced G-induced DNA breaks but did not reduce apoptosis. Topo-II-mediated DNA cleavage is not required for the induction of apoptosis. |
| Arai et al. (2000) | HT-29, Lovo, colo320 SW480 & HCT116 | These cell lines expressed ERβ mRNA but lacked ERα mRNA. Estrogen did not influence these cells. G (10 μM) inhibited growth of HT29, Colo320 and Lovo cells. |
| Plewa et al. (2001) | HT 29 | G, genestin, dadzein, dadzin isolated from soybean fraction PCC70 showed wide range of growth suppression on HT-29 cells. |
| Zhu et al. (2002) | Caco2, SW620 & HT29 | G induced vacuole formation and reduced survival of colon cancer cells. G enhanced small GTP binding proteins involved in membrane traffic of golgi apparatus. |
| **LEUKEMIA *in vivo*** | | |
| Uckun et al. (1999) | | 7 children and 8 adults with CD19+ B lineage acute lymphoblastic leukemia as well as one adult with chronic lymphocytic leukemia were treated with CD43-Genistine immunoconjugate. All patients tolerated the I.V. infusion of the conjugate at varying doses for consecutive days for a total of nine doses. Plasma half-life and systemic clearance of the conjugate was determined. Moderate levels of Human antimouse antibody (HAMA) were detected in the day 29 samples. |
| **LEUKEMIA *in vitro*** | | |
| Constantinou et al. (1990) | promyelocytic HL-205erythroid K-562-J | G suppressed growth and induced differentiation in both cell lines. Induction of DNA damage is evident after G treatment by inhibition of topoisomerase II activity. G-Treatment also decreased reactivity of the cells to anti-phosphotyrosine PY-20 murine monoclonal antibody suggesting inhibition of PTK activity by G. |
| Honma et al. (1990) | chronic myelogenous K-562 | Multidrug resistant (MDR) leukemias have abnormally high activity of oncogene tyrosine kinase. MDR subline K562R was induced to undergo erythroid differentiation by G as effectively as parent K562 cells were. It is because G is a PTK inhibitor. Tyrosine kinase inhibition resulted in erythroid differentiation. |
| Makishima et al. (1991) | myelogenous HL-60, myeloblastic ML-1, | PTK and phosphatidyl inositol turnover is associated with cell growth and differentiation of leukemia cells. G inhibited PTK activity, phosphotidyl inositol turnover and topoisomerase II. G induced myeloblastic ML-1 into promyelocytes and promyelocytic HL-60 into mature granulocytes. G effect was enhanced with 25-di (OH)-vitamin D and retinoic acid. |

| Reference | Cell line | Description |
|---|---|---|
| Kuriu et al. (1991) | myeloid MO7E | This study investigates the expression, degree of phosphorylation and activation of the proto-oncogene c-kit product before and after stimulation with the c-kit ligand in MO7E. Murine 3T3 fibroblasts contain the ligand for murine c-kit product was found to stimulate proliferation of MO7E cells in a dose-dependent manner. G inhibited the proliferation significantly, suggesting a role of tyrosine kinase activity, which was confirmed by staining MO7E cells after activation with c-kit ligand with an antibody for phosphotyrosine. |
| Torti et al. (1992) | erythroleukemic HEL | Erythropoietin induced 5-fold increase in the amount of GTP bound to the endogenous p21ras, an effect that was dose-dependent. Correlated with p21ras activation, erythropoietin also caused tyrosine phophorylation of several proteins in a time-dependent manner. G inhibits erythropoietin-induced accumulation of p21ras-GTP complex. |
| Traganos et al. (1992) | myelogenous HL-60 lymphocytic MOLT-4 normal lymphocytes | G, at 50 µg/ml is cytotoxic to both the leukemia cell lines. In HL-60 cultures, apoptotic cells were observed within 8 hrs after adding G. Normal lymphocytes survived exposure to G at 200 µg/ml; however G suppressed proliferation of leukemia cells at 5 to 20 µg/ml. 40% of MOLT-4 cells were arrested at G2 phase with as little as 5µg/ml of G. Mitogen induced proliferation of normal lymphocytes was sensitive to G, in that they were arrested at G2 phase, with IC50 of 1.6 µg/ml. Therefore G is a valuable chemotherapeutic agent against proliferating leukemia cells. Since G arrested normal lymphocytes induced to proliferate by mitogen, G is considered as a strong immunosuppressant. |
| Bergamschi et al. (1993) | myeloid MO7E myelogenous HL-60 | G induced apoptosis of both the cell lines, as evidenced by nuclear condensation and fragmentation, and DNA degradation. |
| Finlay et al. (1994) | T-leukemia Jurkat | G arrested human Jurkat cell line at G2 phase. G is implicated in inhibition of topoisomerase II activity. |
| Ding et al. (1994) | myelogenous HL-60 neutrophils | G inhibited GM-CSF induced tyrosine phosphorylation of MAP kinase in human neutrophils and HL-60 cells. |
| Hagmann (1994) | basophil leukemia MC/9 mastocytoma RBL2H3 | Proliferating basophil leukemia MC/9 cells produce leukotrienes, when grown and stimulated appropriately. The same is true for RBL2H3. G completely inhibited leukotriene generation in RBL2H3 cells suggesting that leukotriene synthesis appears to involve protein tryosine kinase activity. |
| Takeda et al. (1994) | K562/TPA | Leukemia subline resistant to TPA, called K562/TPA is a non-P-glycoprotein-mediated multidrug resistant cell line. G could overcome drug resistance by enhancing the accumulation of drug into the nuclear fraction of K562/TPA. |
| Spinozzi et al. (1994) | T-leukemia Jurkat | G inhibited dose-dependently the proliferation of the cells and arrested them at G2/M phase of cell cycle at low concentrations [5–10 µg/ml]. At high concentrations [20 to 30 µg/ml] there was a perturbation of S phase progression. Cell cycle disturbances resulted in apoptosis. Immunostaining with anti-phosphotyrosine antibody revealed inhibition of protein tyrosine phosphorylation by G. G is considered to be a promising new agent for therapy. |
| Hunakova et al. (1994) | K562 | G arrested cell cycle at G2/M phase. G induced a increase in the expression of cell surface Lewis X (CD15) antigen. G downregulated CD45 and CD14 antigens. |

(*Continued*)

Table 9.4. (Continued)

| Authors | Cell Lines | Results & Comments |
|---|---|---|
| Tsuijishita et al. (1994) | myelogenous HL-60 and U-937 | Permeabilized cells suspended in alkaline medium released various unsaturated fatty acids most abundantly oleic and arachidonic acids. The release under alkaline conditions was potentiated by vandadate and this potentiation was counteracted by G suggesting a role of tyrosine phosphorylation in this release reaction. |
| Tiisala et al. (1994) | monoblastic THP-1 | G upregulated the surface expression of the beta 2-integrins in these cells. This upregulation lead to an increase in the adherence to THP-1 to ICAM-1. G also modulated the expression of ICAM-1 on endothelial cells in potentiating the upregulating effect of TNF and IFN-gamma. G may enhance intracellular binding by affecting both the endothelium and the circulating cells. |
| Constantinou & Huberman (1995) | promyelocytic HL-60 erythroid K-562 | G inhibited cell multiplication in a dose-dependent manner and induced cell differentiation. The maturing HL-60 cells acquired granulocytic and monocytic markers. The differentiating K-562 cells stained positively with benzidine, indicative of hemoglobin production, an erythroid marker. |
| Belka et al. (1995) | U937 monocytic leukemia | IL-6 signalling leads to phophorylation of the small heat shock protein (Hsp-27) through activation of the MAP kinase and MAPKAP kinase 2 pathway in U937 cells. In the presence of NaF inhibitor of serine/threonine phosphatase IL-6 failed to phosphorylate Hsp-27. 1L-6 mediated phosphorylation of Hsp-27 is also inhibited by G. Inhibition was restored after adding MAP kinase to MAPKAP kinase 2, serine-threonine phosphorylase. |
| Uckun et al. (1995) | B cell precursor leukemia | BCP leukemia is the most common form of childhood cancer, and the second most common leukemia in adults. Immunoconjugate B43, an antibody for CD19 receptor-conjugated with G bound to these cells with high affinity and selectively inhibited CD19 associated tyrosine kinase and triggered rapid apoptotic death. 99.99% of BCP cells were killed. The conjugate might be useful in eliminating BCP in patients who have not responded to conventional therapy. |
| Ponnathpur et al. (1995) | Pre-B 697/neo | Taxol induced mitotic arrest and apoptosis in 697/neo cells. Taxol induced polymerization of tubulin into stable microtubules resulting in cell cylce metaphase arrest. G (10 μM) blocked taxol induced apoptosis but not cell cycle arrest, suggesting that taxol induced apoptosis may involve tryosine kinase activity. |
| Carlo-Stella et al. (1996a) | Normal & leukemic Haemopoietic progenitors | G strongly inhibited the growth of normal and leukemic haemopoietic progenitors. Growth inhibition is dose and time dependent. Leukemic progenitors are more sensitive to G-induced growth inhibition than normal progenitor cells. G exerted a direct toxic effect on haemopoietic cells while sparing substantial proportion of longtime culture initating cells. |
| Carlo-Stella et al. (1996b) | Chronic myelo-genous Leukemia | G is capable of exerting a strong antiproliferative effect on colony forming CML cells. |
| Yamada et al. (1997) | human monocytic THP-1 | Monoclonal antibody YH384 upregulated ICAM-1 in THP-1 cells. The effect of YH384 was inhibited in a dose dependent manner by G, suggesting tyrosine phosphorylation by the antibody may precede expression of ICAM-1. |

| | | |
|---|---|---|
| Deora et al. (1997) | K562 | G induced apoptosis as well as cellular differentiation in K562 cells. |
| Li and Weber (1998b) | K562 | Tiazofurin inhibited inosine 5'-monophosphate dehydrogenase, reduced signal transduction, inositol triphosphate concentration and arrested the cell cycle chiefly in S phase. G an inhibitor of 1-phosphatidyl inositol kinase [EC 2.7.1.68], tyrosine kinase and topoisomerase II induced arrest in G2/M phase. Tiazofurin together with G caused 5.9-fold elevation in inducing differentiation and inhibition of proliferation. |
| Yun et al. (2000) | | G showed inhibitory effects on IL-5 and IL-3 bioactivities and did not inhibit GM-CSF and IL-6 bioactivities. |
| Polkowski et al. (2000) | HL-60 | G-piperazine complex and G effects on HL-60 cells were compared. Both inhibited cell proliferation, viability, apoptosis and cell cycle equally. Complexing amine with G did not alter the potential of G. |
| **LIVER CANCER *In vitro*** | | |
| Mousavi & Adlercreutz (1993) | Hep-G2 | G increased sex hormone-binding globulin production and inhibited cell proliferation. SHBG may reduce clearance of sex hormones and probably risk of cancer. |
| Thompson et al. (1993) | Hep-G2 | G inhibited IL-6 stimulated synthesis of acute phase proteins. Tyrosine kinases may be involved in downregulating IL-6 signal. |
| Wegenka et al. (1994) | | IL-6 activated a latent cytoplasmic transcription factor, acute phase response factor (APRF), by tyrosine phosphorylation. APRF binds to IL-6 target genes. G inhibited protein kinase activity and thus activation of APRF. |
| Loukovaara et al. (1995) | Hep-G2 | G, similar to 17β-estradiol increased SHBG level within cells, whereas D and equol also increased the level extracellularly. Regulation occurs at post-transcriptional level. |
| Yang et al. (1996) | Hep-G2 | G inhibited EGF-induced EGF-receptor internalization dose dependently. Preincubation of cells with/without G showed degraded released EGF ratios of 16 & 24 respectively. EGF-induced internalization depends on tyrosine kinase activity. |
| van Rijin & van den Berg (1997) | Reuber H35 hepatoma | G, apigenin & quercetin appear to be cytotoxic when applied to cells following irradiation. Irradiation alone reduced survival of cells by a factor of 20, whereas continuous presence of G during radiation reduced survival by 10,000. G acts as radiation enhancer by inhibiting topoiosmerase II activity, which is involved in replication, transcription, and probably DNA repair. In addition G acts as potential antitumor agent. |
| Marino et al. (1998) | Hep-G2 | Estrogen induced inositol triphosphate (IP3) production and PKC alpha level in the membrane that modulate ion channel. G stimulated cell proliferation and phosphorylation of cytosolic ER. G, while simulating the structure of estrogen, inhibited the above effects of estrogen when preincubated with HEPG2 cells. G may inhibit activation of IP3-PKC-alpha signal transduction pathway. |
| Arts et al. (1999) | Hep-G2 | G inhibited the induction of plasminogen activator inhibitor-1 (PAI-1) mRNA accumulation. |

(*Continued*)

Table 9.4. (Continued)

| Authors | Cell Lines | Results & Comments |
|---|---|---|
| Kikuchi & Hossain (1999) | Hep-G2 | G inhibited the signal transduction mediated effect of benzimidazole compounds on CYP 1A1 production in cells. Inhibition is caused by its anti-tyrosine kinase. This study proposes the mechanism by which G may protect against carcinogenesis. |
| Nakanishi et al. (1999) | Hep3B, HepG2, PLC & Huh-7 | Hepatocyte growth factor (HGF), a potent mitogen stimulated the mobility of HCC cells, in association with tyrosine phosphorylation and phosphatidyl inositol 3-kinase activation. G like PI3-K inhibitor prevented the migration of HCC cells, whereas as calphostin, an inhibitor of protein kinase C did not have an effect. G appears to be a PI3-K inhibitor. |
| Lamon-Fava (2000) | Hep-G2 | G and D at 10 µM/L increased the secretion of apolipoprotein A-1 by 5 and 1 fold respectively. G modulated apo-A1 gene expression, similar to 17-beta estradiol. G increased transcriptional activity of apo-A1 gene. |
| Park et al. (2001) | Hep-G2, Colo320 & HSR | G strongly increased the expression of Cdk inhibitor p21 protein. G is useful to increase the sensitivity of dexamethasone. Both regulated hepatocyte cholesterol metabolism. |
| Borradaile et al. (2002) | Hep-G2 | G and D at 100 µM concentration decreased apo-B secretion by 63% and 71%; cholesterol synthesis by 41% and 18%; cholesterol esterification by 56% and 29%. Both decreased the activity of microsomal triacylglycerol transfer protein and increased the binding, uptake and degradation of LDL. |
| **LUNG CANCER In vitro** | | |
| Oelmann et al. (1995) | Lung carcinoma, HTB119, HTB120 & CCL 185 | Nerve Growth factor (NGF) stimulated the growth of these cell lines dose-dependently. NGF-induced growth was inhibited by G. |
| Versantvoort et al. (1996) | COR-L23/R & MOR/R | G enhanced efflux of rhodamine 123 and influx of daunorubicin in cell lines overexpressing multidrug resistance protein. |
| Klein & McCarthy (1997) | NCI-H441 | G downregulated surfactant protein A [SPA] and SPA mRNA without affecting viability. This effect was reversible and EGFR independent. |
| Hooijberg et al. (1997) | GLC4/ADR | G stimulated ATPase activity of the multi-drug resistant proteins, probably enhancing uptake of anticancer drugs. Genistein had no effect. |
| Cattaneo et al. (1997) | SCLC | G inhibited the action of nicotine on MAP kinase. |
| Lei et al. (1998) | NCI-H596 | G antagonized growth stimulatory EGF signaling upstream of MAP kinase and may simultaneously stimulate an apoptotic pathway. |
| Lian et al. (1998) | H460 (NSCLC) | G (30 µM) arrested cell cycle at G2/M phase. It upregulated p21WAF1 expression. |
| Lei et al. (1999) | NCI-H596 & NCI-H358 | G inhibited phosphorylation of EGFR in NCI-H596. G in combination with drugs enhanced antiproliferative effects and induced programmed cell death in cells, which strongly express EGFR. |
| Lian et al. (1999) | H460 & H322 | G inhibited cell growth dose dependently. It upregulated p21WAF1, Bax and p53 expression. It induced apoptosis through p53-independent pathway. |

| | | |
|---|---|---|
| Ando et al. (1998) | RERF-LC-AI (SCC) | X-ray irradiation (15 Gy) induced VEGF mRNA (2.5 fold) in squamous cell carcinoma (SCC). G (PKC inhibitor) blocks the induction of VEGF mRNA by irradiation. This suggests that the mechanism of induction might be concerned with the pathway which triggers Scrc tyrosine kinase of the cell surface and the protein kinase C pathway. |
| Yamane & Abe (2000) | HepG2 & A549 | Glucose addition into culture medium caused a dose-dependent increase in leukotriene B (4) omega-hydroxylation activity. G decreased glucose-induced omega-hydroxylation of leukotriene B dose dependently. G's effect was reversible by removing G from the medium. |
| Leyton et al. (2001) | H1299 (NSCLC) | The effects of some oncogenes, growth factors and neuropeptides are mediated by Tyrosine phosphorylation of focal adhesion kinase (p125 FAK) and paxillin cytoskeletal proteins. Bombesin is one such factor. G (50 μM) reduced Tyro-phosphorylation of p125 (FAK) and paxillin stimulated by bombesin. |
| Maeno et al. (2002) | NCI-H322 | G blocked all trans-retinoic acid mediated VEGF expression, similar to the effect of cyclohexamide. This finding suggests that tyrosine kinase is necessary for trans-RA mediated VEGF expression. |
| Fujimoto et al. (2002) | A549 | G/Epigallocatechin gallate inhibited the growth of A549 cells in a dose-dependent manner. G/Epigallocatechin gallate reduced the expression of lung cancer biomarker, hnRNP B1. G and EGCG inhibited the promoter activity of hnRNP A2/B1 gene expression with IC50 values 66 and 29 μM. |
| **LYMPHOMA *In vivo*** | | |
| Chen et al. (1999) | B-lineage lympho-ma (BLL), Acute Lymphoblastic Leukemia (ALL) & non Hodgkin's lymphoma (NHL) | BLL: 4 children/12 adults, ALL: 12; NHL: 5 were treated with anti-CD19 (B43) conjugated to G intravenously. Plasma half-life was 19 ± 4 hours. This is the first clinicopharmacokinetics study of a tyrosine kinase inhibitor containing immunoconjugate. |
| **LYMPHOMA *In vitro*** | | |
| Andrew et al. (1995) | TK1-CD8 lymphoma | G inhibited DATK44 (mAb, that binds to T cell activation B cell subset Ag-TABS)-induced aggregation of TK1 cells. TABS, an adhesion inducer that selectively activates LFA-1 mediated lymphocyte aggregation. TABS is highly expressed in CD4+ and CD8+ T cells more than CD4+ or CD8+ alone. CD4- and CD8- T cells expressed two distinct populations. One set did not express TABS and another highly expressed. On activation T cells overexpressed TABS. |
| Myers et al. (1995) | B lineage lymphoma cells | CD19 receptor is expressed at high levels on human B lineage lymphoid cells and is associated with the Scr-protooncogene family tyrosine kinase Lyn. Immunoconjugate B43-Genestein triggered rapid apoptotic cell death in highly radiation resistant p53 Bax-Ramos-BT B-lineage lymphoma cells expressing high level of CD19 receptor. Treatment of scid mice with B43-Gen (at a dose level <1/10 of maximum tolerated dose) after challenging with fatal number of Ramos-BT cells resulted in 70% long-term event free survival. This study suggests that anti-apoptotic CD19-Lyn complex may be as important as Bcl-2/Bax ratio for survival of lymphoma cells. |

*(Continued)*

**Table 9.4.** Effects of Genistein in vivo and in vitro in Human Cancers

| Authors | Cell Lines | Results & Comments |
|---|---|---|
| Bronte et al. (1996) | | Extracellular ATP induced apoptosis and osmotic lysis in several cell lines. Antiphosphotyrosine immunostaining revealed an increase in tyrosine phosphorylation after ATP treatment and the increase in PTK activity was inhibited by G. Consequently, G blocked extracellular ATP mediated DNA fragmentation in cells without affecting cell lysis. |
| Myers et al. (1998) | Raji-S1 lymphoid cells | B43-Gen (anti-CD19 antibody B43 conjugated to genestein) elicited selective and potent cytotoxicity against CD19 + human leukemia cells. |
| Lebakken et al. (2000) | | Syndecan-1 expressing Raji-S1 cells bind and spread rapidly when attaching to matrix ligands containing heparin sulphate-binding domain. G affected spreading of Syndecan 1 when attached to the matrix ligand suggesting the involvement of intracellular signals with tryosine phosphorylation. |
| Kim et al. (2001) | A20 | G partially inhibited Fas induced phospholipase D activation. However G had no effect on Fas-induced activation of phosphatidyl choline specific phospholipase C and protein kinase C. These results suggest tyrosine phosphorylation in FAS induced phospholipase D activation. |
| **MELANOMA *In vitro*** | | |
| Kiguchi et al. (1990) | Ho & SK-MEL-131 | G inhibited growth and increased melanin content after 6 days at 45 µM. The increase was time- and dose-dependent. G increases tyrosinase activity, dendrite like structure formation. |
| Sjoberg et al. (1995) | Melur melanoma cells | G induced antigenic alteration by inducing the expression of de-N-acetyl ganglioside. Gangliosides GM3 and GD3 were expressed without N-acetyl groups. G arrested the cells at G2M phase. |
| Ralph et al. (1995) | SK-MEL-28, SK-MEL3, MM96, HT144 & Hs 294T | G and herbimycin dose dependently inhibited the antiviral action of IFNs. The induction of antiviral state in melanoma cells by IFN requires activation of tyrosine kinase dependent signaling pathway. |
| Ida et al. (1995) | melanoma cells | G inhibited cell spreading on substrata coated with integrin-binding fibronectin synthetic peptide CS1-OVA. When cells were cultured on plates coated with CS1-OVA or anti-NG2 antibody 9.2.27, two proteins were tyrosine phosphorylated in a G-sensitive fashion. |
| Rauth et al. (1997) | melanoma cells | G increased melanin content and tyrosinase activity and caused cells to form dendrite like structures. Cells lacking p53 responded more than cells with p53. G upregulated tyrosinase mRNAs. It arrested cell cycle at G2/M stage. |
| Hartmann et al. (1997) | melanoma cells | G inhibited cell proliferation, induced morphological changes similar to that of drugs used as tyrosine kinase inhibitors, and increased melanin content. |
| Darbon et al. (2000) | OCM-1 melanoma | G inhibited cell proliferation at G2 phase by impairing Cdc25C dependent Tyr-15 dephosphorylation of Cdk1. G induced only weak response on DNA damage. |
| Casagrande & Darbon (2000) | OCM-1 melanoma | G arrested cell cycle at G2/M and induced CDK inhibitor p21CIPI. CDK2 was not affected. CDK inhibitor p21CIPI is not required for G-induced G2 arrest. |

| Reference | Cell line | Description |
|---|---|---|
| **ORAL CANCER** *In vitro* | | |
| Elattar & Virji (2000) | SCC-25 | G and quercetin (0.1, 1.0, 10 μM) had biphasic effect on cell growth and proliferation, based on dose. Curcumin was more potent than G and Q. |
| **OSTEOSARCOMA** *In vitro* | | |
| Burgener et al. (1995) | UMR106 | Fluoride enhances protein tyrosine phosphorylation in osteoblast-like cells by enhancing tyrosine kinase activity. G inhibited fluoride-induced cell proliferation and Na-coupled Pi transport system in a dose-dependent manner by inhibiting signal transduction pathway. |
| Yamashita et al. (1996) | MG-63 | (3H) Thymidine (TdR) incorporation by human osteosarcoma cell line MG-63 was significantly stimulated by metallo-proteinases (TIMP-1 & TIMP-2). Tyrosine kinase inhibitor G inhibited metalloproteinase TIMP-2 induced (3H) Thymidine (TdR) incorporation in cells. |
| Munoz et al. (1997) | U20SDr1 | G inhibited the transport of acidic fibroblast growth factor into cells. |
| Kawase et al. (1999) | UMR106 | G is more effective in attenuating EGF-induced receptor tyrosine phosphorylation than phorbol myristate acetate in permeabilized osteoblasts. |
| Veldman & Schmid (1998) | SAOS-2/B-10 | G blocked Insulin-like growth factor-1 but not fluoride-stimulated sodium-dependent alanine transport. |
| **OVARY CANCER** *In vitro* | | |
| Liu et al. (1994) | | G interfered with taxol (an antineoplastic agent)-induced DNA fragmentation, bcl-2α expression and clonal cell death. Tyr-phosphorylation mediate apoptosis. |
| Bagnato et al. (1997) | OVCA 433 | G prevented endothelin-1 induced stimulation of (3H) thymidine incorporation and cell proliferation, by blocking both PKC and tyrosine kinase. |
| Shen & Weber (1997) | OVCAR-5 | G and quercetin blocked conversion of phosphotidyl inositol to IP3 by inhibiting PI kinase and PIP kinase. The effect of G was synergistic with quercetin. |
| Li & Weber (1998a) | OVCAR-5 | G and tiazofurin (T), an oncolytic drug, independently inhibited cell growth. IC 50s were 18 and 26 μM for growth inhibition assay, 4 and 17 μM for clonogenic assay. As single agents G and T reduced cell counts to 50% and 60%; in combination the effect was synergistic reducing the count to 8%. |
| Venkatakrishnan et al. (2000) | SK-OV-3 | Treatment of the cells with G indicated that tyrosine kinase is involved in the IL-8 activation of Erk1 and Erk2 (extracellular signal regulated kinases). |
| Weber et al. (2000) | OVCAR-5 | G, quercetin and tiazofurin inhibited PIP kinase, PI kinase and PLC, respectively. Combination of tiazofurin with either of the two flavonoids showed synergistic killing of cancer cells. Signal transduction activity provides novel, sensitive targets for chemotherapy. |
| Chen & Anderson (2001) | OVAR-3 & Caov-3 | G and D inhibited DNA synthesis at dietary concentrations 10(-8) and 10(-10) M and decreased viability of cells to 45–75%. Both inhibited IL-6 production by 20% and enhanced transforming growth factor beta 1 production by 30%. Addition of ICI-182780, an estrogen antagonist, neutralized the effects of isoflavones. G and D reduced cell proliferation via estrogen receptor-dependent pathway. |
| Ahmed et al. (2002) | HOSE, OVCA 429, OVCA 433 & OVHS-1 | G inhibited secretion of high molecular weight urokinase plasminogen activator and metallo-proteinase, which are involved in invasion and metastasis. |

*(Continued)*

**Table 9.4.** (*Continued*)

| Authors | Cell Lines | Results & Comments |
|---|---|---|
| **PANCREATIC CANCER** *In vitro* | | |
| Ohno et al. (1993) | MIN6 | G augmented glucose-induced insulin release dose dependently (upto 20 µg/ml), by elevating cAMP level and inhibiting phosphodiesterase enzyme. Deprivation of extracellular calcium antagonists inhibited these effects. |
| Simeone et al. (1995) | AR42J | Tyrosine kinases are receptors for EGF and FGF. Intracellular signaling by an increase in Ca2+ was observed in pancreas cell line AR42J in response to EGF. G blocked the response to EGF. Growth factors act on receptors, which signal through tyrosine kinase activity. |
| Morisset et al. (1995) | AR4-2J | Pituitary adenylate activating peptide (PACAP) caused dose-dependent and parallel activation of tyrosine kinase and phospholipase D. Preincubation of cells with G/pertussis toxin inhibited the activity of both enzymes. |
| Boros et al. (2001) | MIA | G controlled tumor growth by affecting nucleic acid synthesis and the glucose metabolism, dose dependently at 2, 20 & 200 µmol/l. |
| Ding et al. (2001) | PANC-1 & HPAF | G inhibited 12(S) HETE (the lipoxygenase product)-induced proliferation of cancer cells. G also inhibited tyrosine kinase, which is involved in the mitogenic effects of 12(s) HETE. 12(s) HETE phosphorylates ERK, which was inhibited by G, indicating that tyrosine phosphorylation is essential for ERK activation. |
| **PROSTATE CANCER** *in vivo* | | |
| Adlercreutz et al. (1993) | | The mean plasma concentration of G is 7 to 110 fold higher in the Japanese (n = 14) than in Finnish men (n = 14). High G level may inhibit growth of CaP in Japanese men. |
| Barnes et. al. (1995) | | Oriental populations consume 20 to 80 mg of G (derived from soy)/day, whereas the dietary intake of soy G in US is 1 to 3 mg/day. Genistein is a tyrosine kinase inhibitor (TKI) naturally found in Soy; it is cost effective and more reliable than synthetic TKIs. |
| Lu et al. (1995) | | Men consuming 6 x 12 ounce soy milk/day excreted 15 + 2.74 mg of diadzein & 7.73 ± 1.95 mg of G by 24 hrs after soy milk ingestion, whereas females excreted significantly higher 40.4 ± 7.4 mg of daidzein and 9.11 ± 0.84 mg of G (p = 0.02). A simplified method to quantify isoflavones in diets and urine is established. |
| Strom et al. (1999) | | A correlative study of daidzein [D] and G intake and CaP risk was analyzed in 83 Caucasian cases and 107 isoflavone consumers. D but not G showed protective effect |
| Yamamoto et al. (2001) | | Collected valid food frequency questionnaires (FFQ), dietary records (DR), blood/urine samples from 215 subjects. For D mean intake estimated from FFQ & DR, serum concentration and urine excretion were 18.3 mg/d, 14.5 mg/d, 119.9 nmol/L and 17.0 µM/d and for G 31.4 mg/d, 23.4 mg/d, 475.3 nmol/L and 14.2 µM/d respectively. |

# Anticancer Therapeutic Potential of Soy Isoflavone, Genistein

| Reference | Cell lines | Findings |
|---|---|---|
| de la Taille et al. (2001) | | Optimal 40 μ/day of soybeans may lower serum PSA. |
| Pumford et al. (2002) | | In serum samples of Japanese, concentrations of D [men: 281 nmol/L; women 246 nmol/L] and G [men 493/nmol/L; women: 502 nmol/L] were approximately 15 times higher than British men [D 18.2 nmol/L; G 34.1 nmol/L] and women [D: 13.5 nmol/L; G 30.1 nmol/L]. In pharmacokinetic studies the D and G levels in British subjects elevated to that of Japanese after consumption of 20 mg of soy isoflavones after 6 to 8 hours. |
| Akaza et al. (2002) | | 141 CaP patients and 112 cancer free controls were examined for serum D, G and equol (D-nonmetabolizer). The levels in patients were higher than that in controls. The poorly differentiated cancer patients' group included significantly lower % of equol. The role of equol, nonmetabolizer of D in the biology of CaP, deserves elucidation. |
| Ghafar et al. (2002) | | G combined with a polysaccharide (from Basidiomycetes) reduced serum PSA level from 19.7 to 4.2 ng/ml, after 44 days. |
| Morton et al. (2002) | | G mean serum concentration in Japanese men was 492.7 nmol/L, compared with 33.2 nmol/L in men in UK. 58% of Japanese men and 38% of Japanese women had equol >20 nmol/L compared with none of the UK men and 2.2% of UK women. |
| **PROSTATE CANCER *in vitro*** | | |
| Peterson & Barnes (1993) | LNCaP & DU-145 | G not D inhibited both serum and EGF-stimulated growth of the two cell lines (IC50 values from 8.0 to 27 μg/ml for serum and 4.3 to 15 μg/ml for EGF). G inhibition did not depend on EGF receptor tyrosine autophosphorylation. |
| Naik et al. (1994) | PC-3 | G appears to be cytotoxic to PC-3. |
| Rokhlin & Cohen (1996) | LNCaP, DU-145, PC-3, ND1, ALVA 31 & JCA1 | Genistein inhibited serine/threonine and tyrosine PK in a dose-dependent manner. Endoglin (CD105) expression changed with G-treatment. CD105 may be involved in both protein phosphatase and kinase-mediated phosphoprotein turnover. |
| Bergan et al. (1996) | PC3-M , DU-145, PC-3 & MCF-7 | G caused cell flattening/cell adhesion caused by formation of a complex between Beta-1-integrin and FAK (focal adhesion kinase). G is a novel tool for studying the molecular events of cell adhesion |
| Kyle et al. (1997) | PC3-M | No growth inhibitory effect was seen at G concentrations below the low micromolar range (3 logs above that attained in serum). |
| Santibanez et al. (1997) | PC-3, DU145 & LNCaP | Most invasive CaP cell line is PC-3 and other two cell lines are less invasive. G is more cytotoxic to PC-3 and less cytotoxic to other cell lines. Tyr-phosphoproteins in the cell lines were studied. The molecular target of G appears to be a membrane-bound protein [130 kDa]. |
| Geller et al. (1998) | BPH & CaP cells in 3-Dcollagen gel Histoculture | G at 1.25-10 μg/ml decreased the growth of BPH tissue in cell culture in dose-dependent manner. Inhibition is measured by 3H-thymidine incorporation per μg protein. |
| Yang et al. (1998) | PC3 & LNCaP | Proline-directed PK F activity is increased [55.5 ± 3.8 units/mg] in poorly differentiated PC-3 compared to well differentiated LNCaP [28.1 ± 2.3 units/mg]. G blocked the enhanced tyrosine phosphorylation of the PDPK in PC3 but not in LNCaP. Oncogenic PDPK can be differentially regulated in well and poorly differentiated CaP cell lines. |

*(Continued)*

Table 9.4. (*Continued*)

| Authors | Cell Lines | Results & Comments |
|---|---|---|
| Onozawa et al. (1998) | LNCaP | G inhibited the growth of LNCaP most effectively [IC50:40 µM]. Inhibition of cell growth was followed by suppression of DNA synthesis and apoptosis. Expression of PSA was significantly reduced by G. |
| Sun et al. (1998) | LNCaP | LNCaP accumulates intracellular testosterone as glucuronidated metabolites. UDP-glucuronyltransferase [UDPGT] increased linearly on 6 day exposure of cells to G. |
| Davis et al. (1999) | LNCaP & PC3 | G decreased NF-k B DNA binding and abrogates NK-k B activation by DNA damaging agents (h2o2, TNF-α) in both androgen-sensitive LnCaP and androgen-resistant PC-3. G reduced phosphorylation of protein I kappa B alpha and prevents NF-k B activation. |
| Borsellino et al. (1999) | PC-3 | Both endogenous and exogenous IL-6 and exogenous onco-statin M [OM] up regulated cell growth and enhanced resistance of PC-3 cells to both etoposide and cisplatin. Both IL-6 and OM-mediated effects are inhibited by G. G is useful to increase drug sensitivity of CaP cells. |
| Mitchell et al. (2000) | PC-3 & LNCaP | G and D inhibited growth at < 10 µM, the physiological concentration. G induced DNA damage at the same concentration. D did not do so up to 500 µM. |
| Kumi-Diaka et al. (2000) | DU145 & LNCaP | G inhibited growth via apoptosis and necrosis at high concentration; G induced activation and expression of caspase-3 (CPP32); G induced apoptosis and CPP32 activation can be inhibited by caspase-3 inhibitor, z-VAD-fmk. G can work independently of hormone sensitivity. |
| Choi et al. (2000) | P53-null CaP | G suppressed proliferation of p53-null CaP cells reversibly. G arrested cell cycle at G2/M phase with a marked inhibition of cyclin B1 and an induction of Cdk inhibitor p21 (WAF1/CIP1) in a p53 independent manner. After adding G an increased binding of p21 with Cdk2 and Cdc2 paralleled a significant decrease in Cdc2 and Cdk2 kinase activity with no change in their expression. G induced activation of a p21 promoter construct. Anticarcinogenic effect of G possibly involves induction of p21, which inhibits the threshold kinase activities of Cdks and associated cyclins, leading to G2/M arrest in cell cycle progression. |
| Sakamoto (2000) | PC3 | G combined with thearubigin inhibited growth of PC-3 and induced a G2/M phase cell cycle arrest dose dependently. |
| Davis et al. (2000) | VeCaP & LNCaP | G inhibited secretion of PSA by androgen-dependent LNCaP. G also inhibited PSA secretion by androgen-independent VeCaP cell lines. G at concentration found in sera of men consuming soy-rich diet decreased PSA mRNA, protein expression and secretion by LNCaP. VeCaP required much higher concentrations of G for the same inhibition. G inhibited cell proliferation independent of PSA signaling pathway. |
| Agarwal (2000) | DU-145 | In androgen-dependent DU145 cell lines, G significantly inhibited TGF-α induced activation of membrane receptor erbB1, followed by inhibition of downstream cytoplasmic signaling target Shc activation and a decrease in its binding with erbB1, without any alteration in their protein expression. G also inhibited ERK1/2 activation. Thus G impairs the activation of erbB1-Shc-ERK1/2 signaling of DU145 |

| Reference | Cell line | Findings |
|---|---|---|
| Shen et al. (2000) | LNCaP | Physiological concentrations of G [< 20μM] decreased viability in a dose-dependent manner, induced a G1 cell cycle block, decreased PSA mRNA expression, increased p27(K1P1) and p21 (WAF1) (mRNA & Protein), but had no effect on mRNA expression of apoptosis and cell cycle related markers bcl-2, Bax, Rb and proliferation cell nu-antigen. Higher concentration of G [>20 μM] needed to induce apoptosis. |
| Bhatia & Agarwal (2001) | DU-145 | In DU145 cell lines, G [100–200 μM] inhibited TGF-α induced activation of membrane receptor erbB1, followed by inhibition of downstream cytoplasmic signaling target Shc activation and a decrease in its binding with erbB1, without an alteration in their protein expression. G also inhibited ERK1/2 activation. Thus G impairs the activation of erbB1-Shc-ERK1/2 signaling of DU145. G's activity is similar to that of silymarin of Milk Thistle and EGCG of green tea. |
| Hillman et al. (2001) |  | G [15 μM] caused significant inhibition of DNA synthesis, cell growth, colony formation, and potentiated effect of low-dose (200–300 cGy photon or 100-150 cGy neutron) radiation. This study recommends potential combination of G with radiation for the treatment of CaP. |
| Davis et al. (2001) | cultured lymphocytes peripheral blood lymphocytes. | G protected cells from oxidative stress-inducing agents. G but not D inhibited TNF-α induced NFkappa B activation in cultured lymphocytes. G may decrease DNA adduct levels. |
| Xiang et al. (2002) | LNCaP & DU145, | Genistein blocked G2-M phase of cell cycle and caused apoptosis. |
| Kobayashi et al. (2002) | LNCaP | G suppressed cyclin B expression leading to arrest at G2M phase. It induced cyclin dependent kinase inhibitor p21 level. Like G, apigenin and luteolin also increased p21 levels but quercetin did not. Apigenin performed in a p53-dependent way, whereas luteolin was p53-independent. |
| Davis et al. (2002) | LNCaP | The mechanism by which G modulates PSA expression in CaP cells was studied. G decreased the transcriptional activation of PSA by both androgen-dependent and androgen-independent LnCaP cells. The decrease correlated with decreased androgen receptors and m-RNA levels. G had differential effects on 17 beta-estradiol-mediated PSA expression. |
| Suzuki et al. (2002) | LNCaP & PC3 | G inhibited proliferation of the two cell lines in a dose-dependent manner. G downregulated the expression of apoptosis inhibitor. (Survivin, DNA topoisomerase II, cell division cycle 6 (CDC-6) & mitogen activated protein kinase 6, MAPK 6). |
| Li & Sarkar (2002a) | PC3 | G upregulated glutathion peroxidase-1 gene expression. Real time PCR revealed significant increase in GPx enzyme activities G induced no changes in other antioxidant enzymes, superoxide dismutase and catalase. G inhibited growth and induced apoptosis in PC3 cells but not in nontumorigenic CRL-2221 CaP cells. G specifically inhibited Akt kinase activity & abrogates EGF-induced activation. G abrogates EGF induced activation of NF-kappa B, which mediated via Akt signaling pathway. |
| Li & Sarkar (2002b) | PC3 | G downregulated transcription and translation of genes involved in control of angiogenesis, tumor cell invasion and metastasis. |

(*Continued*)

**Table 9.4.** (*Continued*)

| Authors | Cell Lines | Results & Comments |
|---|---|---|
| Li & Sarkar (2002c) | PC3 | G induced > 2 fold change in 832 genes. G down regulated 11 genes [MMP-9, protease M, uPAR, VEGF, neutropiin, TSP BPGF, LPA, TGF-beta2, TSP-1 and PAR-2]. |
| Guo et al. (2002) | | G increased cellular glutathione (reduced). |
| Rao et al. (2002) | HPEC & CaP cells | G with 1-α-25-dihydroxycholecalciferol (active form of Vit D) synergistically inhibited growth of HPEC and CaP cells. G induced arrest at G2/M cell cycle phase of primary human prostatic epithelial cells. |
| Maggiolini et al. (2002) | LNCaP | At low concentration, G stimulated LNCaP growth and at high concentration it is cytotoxic independent of androgen receptor expression. |
| Farhan et al. (2002) | DU-145 | G caused time-, dose-dependent inhibition of CYP24 & CYP27B1 (cytochrome enzyme), thereby causing antitumorigenic effect. CYP27B1 is the cytochrome enzyme that synthesizes 1, 25-D3. G inhibited transcription of the enzyme. |
| Hedlund et al. (2003) | PC3, DU-145, LNCaP, 22Rv1 & LAPC-4, | Genistein arrested cell growth in G2/M phase. Equol, a biological active metabolite of diadzein, also had potent antiproliferative effects on benign and malignant CaP cells. |
| Yu et al. (2003) | LNCaP, DU-145 & PC-3 | G inhibited dose-dependent DHT (5 alpha dihydro testosterone) induced expression of the PART-1 (prostate androgen regulated transcript) in androgen-dependent LNCaP but not in androgen-independent PC-3 and DU-145. |
| **RENAL CARCINOMA** *In vitro* | | |
| Yabunaka (1995) | SMKT-R3-B1/B4 & SMKT-R3-A1/A6 (transfected) | EGF elevated activity in B1 & B4 (transfected mutant v-ras) but not in A1 & A6 (transfected with v-s). G reduced the activity of glycolipid transferase in B1 and B4 but not in A1 & A6. |
| Yabunaka et al. (1997) | SMKT-R3 | G reduced sulfotransferase activity in control cells, but not in v-H-ras expressing cells. |
| Mukhopadhyay et al. (1998) | U87 & HT1080 | Incubation of cells with G for 3 h following seeding reduced VEGF mRNA level in highly confluent culture, but not in sparse cultures. |
| Honke et al. (1998) | SMKT-R1, R2, R3, R4, TOS-1, TOS-2 & ACHN, | G down-regulated glycolipid transferase gene expression |
| Grandaliano et al. (2000) | | G but not D blocked thrombin-induced DNA synthesis and monocyte-chemotactic peptide [MCP-1] expression. |
| Sasamura et al. (2002) | | G has strong inhibitory effect on the expression of VEGF mRNA. |
| **SARCOMA** *In vitro* | | |
| Kusaka (1996) | Human uterine Leiomyosarcoma cell line | G and D inhibited fast Na+ channels dose dependently; therefore the effect was independent of protein tryosine kinase inhibition. |

| Reference | Cell line | Description |
|---|---|---|
| Veldman & Schmid (1998) | Osteosarcoma SAOS-2/B-10 | G blocked insulin growth factor-1 (IGF-1) but not fluoride-stimulated Na-dependent alanine and phosphate transport. |
| Yan & Han (1999) | HT1080 fibrosarcoma | G decreased invasive potential of cells at low concentrations (20, 40 µ mol/L) but not the attachment of cells on fibronectin, laminin or matrigel. G is valuable to prevent invasion and metastasis of cancer. |
| Ihn & Tamaki (2002) | Dermato-Fibrosarcoma | G inhibited DNA synthesis. |
| **TESTIS CANCER *In vivo*** | | |
| Mitchell et al. (2001) | | Consumption of 40 mg isoflavones daily for 2 months, increased plasma G and D concentration to 1 and 0.5 µM respectively. It has no effect on endocrine measurements, testicular volume or semen parameters. |
| **TESTIS CANCER *In vitro*** | | |
| Jenab & Morris (1997) | sertoli cells | G blocked the induction of c-fos and junB gene expression by inhibiting tyrosine phosphorylation of STAT proteins. |
| Kumi-Diaka et al. (1998) | TM3, TM4 & GC-1 | G inhibited the growth and proliferation of testicular cells in a dose and time-dependent manner. G induced apoptosis. |
| Kumi-Diaka et al. (1999) | TM3, TM4 & GC-1 | Incubation of cells with G/sodium azide/dexamethasone(dxm) resulted in cell death. G and dxm induced apoptosis, while naz induced necrosis. G in combination with dxm or naz demonstrated synergistic effect, suggesting its chemotherapeutic potential. |
| Kumi-diaka & Butler (2000) | TM4 | G exerted dose and time-dependent effects on TM4. G induced apoptosis at low concentration and necrosis at high concentration. G-effects are inhibited by caspase 3 inhibitor. G induced caspase 3 activation. G-induced apoptosis is associated with activation of CPP32 enzyme activity. |

**Table 9.5.** Genistein induces cell-cycle arrest at G2/M phase in human cancer cells in vitro

| Cancer | Cell lines | Reference |
|---|---|---|
| Breast | MCF-7 | Pagliacci et al. (1993, 1994) |
| | | Choi et al. (1998) |
| | | Fioravanti (1998) |
| | | Constantinou et al. (1998) |
| | | Nakagaw et al. (2000) |
| | | Dampier et al. (2001) |
| | | Hu et al. (2001) |
| | | Xiang et al. (2002) |
| | MCF-10F non-malignant | Frey et al. (2001) |
| | | Frey & Singletary (2003) |
| | MCF-10A (non-cancer) | Upadhyay et al. (2001) |
| | | Xiang et al. (2002) |
| | MCF-12A | Upadhyay et al. (2001) |
| | MCF-10CA1a | Upadhyay et al. (2001) |
| | MDA-MB-231 (ER−) | Shao et al. (1998a) |
| | | Choi et al. (1998) |
| | | Fioravanti (1998) |
| | | Santell et al. (2000) |
| | | Nakagawa et al. (2000) |
| | | Cappelletti et al. (2000) |
| | | Upadhyay et al. (2001) |
| | MDA-MB-468 (ER+) | Shao et al. (1998a) |
| | | Balabhadrapathruni et al. (2000) |
| | | Dampier et al. (2001) |
| | | Xiang et al. (2002) |
| | HBL-100 | Fioravanti (1998) |
| | DD-762 | Nakagawa et al. (2000) |
| | Sm-MT | Nakagawa et al. (2000) |
| | T47D (ER+) | Cappelletti et al. (2000) |
| | | Dampier et al. (2001) |
| | BT-20 (ER−) | Cappelletti et al. (2000) |
| | ZR75.1 (ER+) | Cappelletti et al. (2000) |
| | | Dampier et al. (2001) |
| | HCT116 | Upadhyay et al. (2001) |
| | 184-B5/HER (Her2/nu+) | Katdare et al. (2002) |
| | SK-BR-3 | Xiang et al. (2002) |
| | | Lian et al. (1998) |
| Melanoma | | Sjoberg et al. (1995) |
| | | Rauth et al. (1997) |
| | OCM-1 | Darbon et al. (2000) |
| | OCM-1 | Casagrande & Darbon (2000) |
| Prostate | | Choi et al. (2000) |
| | PC3 | Sakamoto (2000) |
| | LnCaP, | Xiang et al. (2002) |
| | | Kobayashi et al. (2002) |
| | | Hedlund et al. (2003) |
| | DU145 | Xiang et al. (2002) |
| | | Hedlund et al. (2003) |
| | HPEC | Rao et al. (2000) |

*(Continued)*

**Table 9.5.** (*Continued*)

|          |                    |                       |
|----------|--------------------|-----------------------|
|          | PC3                | Hedlund et al. (2003) |
|          | LAPC-4             | Hedlund et al. (2003) |
|          | 22Rv1              | Hedlund et al. (2003) |
| Leukemia | Myelogenous HL-60  | Traganos et al. (1992) |
|          | Lymphocytic MOLT-4 | Traganos et al. (1992) |
|          | Jurkat             | Finlay et al. (1994)  |
|          |                    | Spinozzi et al. (1994) |
|          | K562               | Hunakova et al. (1994) |

protects cells from apoptosis); enhancement of Bax protein (Bax protein antagonizes the anti-apoptotic function of Bcl-2); increase of the Bax:Bcl-2 ratio; induction of p21WAF1, which downregulates cyclin B and thereby arrests the cell cycle at the G2/M phase and promotes apoptosis by p53-independent pathway; and inhibition of the activation of NF-kB.

## 2.6. Genistein Inhibits Angiogenesis

Inhibition of angiogenesis by genistein is documented in several *in vitro* studies. Shao and co-investigators (2000) showed that genistein decreased vessel density and the production and release of vascular endothelial growth factor (VEGF) and TGF-$\beta$1. Genistein also reportedly blocked VEGF mRNA induced by X-ray irradiation in RERF-LC-A1 squamous cell carcinoma cell line (Ando et al., 1998) and by trans-retinoic acid in NCI-H322 lung cancer cell line, suggesting that VEGF induction may involve tyrosine phosphorylation. Most importantly, Li and Sarkar (2002a, 2002b) have shown that genistein downregulated 11 genes including VEGF. In U87 and HT1080 renal carcinoma cells, Mukhopadhyay and

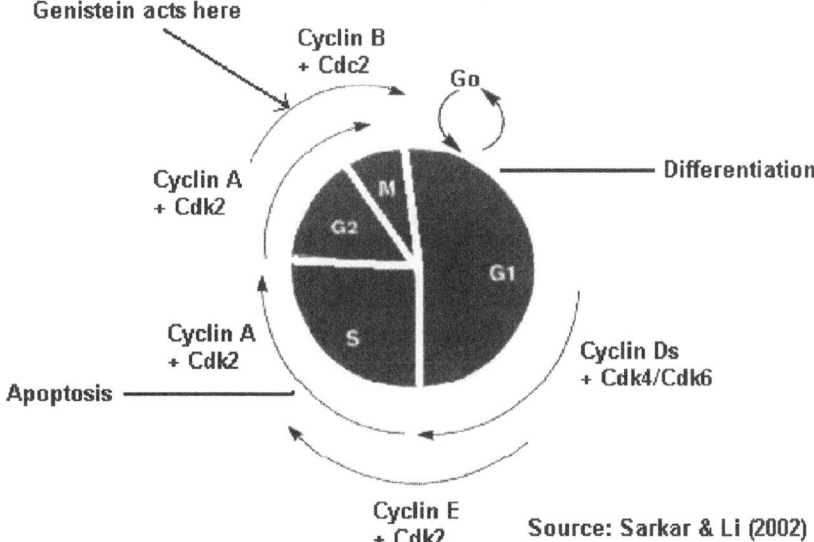

**Figure 9.3.** Genistein arrests cell cycle of most of the human cancer cell lines at G2–M phase

**Table 9.6.** Genistein-induced cell-cycle arrest at G2/M phase involves upregulation of P21WAF1 to downregulate cyclin B1

| Cancer | Cell lines | Reference |
|---|---|---|
| Breast | MCF-7/WT | Shao et al. (1998b, 2000) |
| | | Choi et al. (1998) |
| | | Xu & Loo (2001) |
| | MCF-10A | Upadhyay et al. (2001) |
| | MCF-10CA1a | Upadhyay et al. (2001) |
| | MCF-12A | Upadhyay et al. (2001) |
| | MCF-10F | Frey et al. (2001) |
| | MDA-MB-435; | Li et al. (1999b) |
| | MDA-MB-231 (ER−) | Shao et al. (1998a, 2000) |
| | | Choi et al. (1998a, 2000) |
| | | Li et al. (1999a) |
| | | Xu & Loo (2001) |
| | | Upadhyay et al. (2001) |
| | MDA-MB-435.eB | Li et al. (1999b) |
| | MDA-MB-468 (ER+) | Shao et al. (1998a) |
| | HCT116 | Upadhyay et al. (2001) |
| Hepatoma | Hep G2 | Park et al. (2001) |
| | Colo320 HSR | Park et al. (2001) |
| Lung | H322 | Lian et al. (1999) |
| | H460 (NSCLC) | Lian et al. (1998) |
| | | Lian et al. (1999) |
| Melanoma | OCM-1 | Casagrande & Darbon (2000) |
| Prostate | | Choi et al. (2000) |
| | LnCaP | Shen et al. (2000) |
| | | Kobayashi et al. (2002) |
| | PC3 | Davis et al. (1998) |

co-investigators (1998) and Sasamura (2002) showed that incubation of confluent cultures with genistein for 3 hours reduced the level of VEGF mRNA. Various other bioactivities of genistein are summarized in Table 9.4.

## 3. Anticancer Therapeutic Potential of Genistein

### 3.1. Genistein and Chemotherapeutic Drugs

Genistein enhances antitumor activities of several chemotherapeutic agents. Genistein increases the antiproliferative effect of cisplatin 1.3-fold in HTB-186 medulloblastoma cell line (Khoshyoma et al., 2000). Dietary plasma levels (4 µM) of genistein enhance the antiproliferative and cytotoxic action of 1,3 bis 2-choloroethyl-1- nitrosourea (BCNU) in the U87 glioblastoma multiformed cell line (Khoshyoma et al., 2002). Shen et al. (1999) have shown that tamoxifen and genistein act synergistically with respect to growth inhibition, cytotoxicity and reduction of inositol 1, 4, 5-triphosphate concentration in the estrogen-negative MDA-MB-435 breast cancer cell line. Similarly, Tanos et al. (2002) showed that genistein-mediated inhibition of dysplastic MCF-T6 breast cancer cells was augmented by addition of tamoxifen. A synergistic antiproliferative effect of genistein and doxorubicin has been shown in MCF-7wt, MCF-7-Adr-Resistant, and MDA-MB-231 breast cancer cells; these lines are negative for the estrogen receptor but positive for EGF receptor. Genistein also can increase the sensitivity of dexamethasone in Hep-G2 and Colo 320 HSR hepatoma cells (Park et al., 2001) and promote influx of daunorubicin in COR-L23/R and MOR/R lung cancer cells (Versantvoort et al, 1996). Tiazofurin is an oncolytic drug that inhibits the growth of OVCAR-5 ovarian cancer cells. As single agents, genistein and tiazofurin reduce cell counts to 50% and 60%, respectively. In combination, however, they reduce the count to 8% (Li and Weber, 1998; Weber et al. 2000). In a study involving TM3, TM4 and GC-1 testis cancer cell lines, genistein and dexamethasone had a synergistic effect on the induction of apoptosis. If genistein can augment the efficacy of anticancer chemotherapeutic agents, then its use may allow the dosage of chemotherapeutic drugs to be reduced, thereby reducing toxicity. Clinical phase I and II studies are urgently needed to validate the evidence from *in vitro* studies.

Other *in vitro* studies have reported that the effect of genistein is enhanced by polyphenol food supplements such as quercetin (apple-derived) (Weber et al., 2000; Elatter and Virji, 2000), curcumin (turmeric-derived), epigalloepicatechin, EGCG (green tea-derived) (Katdare, 1998, Fujimoto et al., 2002) and thearubigin (black tea-derived) (Sakamoto, 2000), and by minerals such as vanadium. Genistein in combination with green-tea polyphenol EGCG induced apoptosis and enhanced p53 immunoreactivity in the 184-B5 breast cancer cell line (Katdare, 1998).

## 3.2. Genistein and Radiation Therapy

Genistein is considered to enhance the cytotoxicity of radiation. In Reuber H35 hepatoma cells, survival was reduced by a factor of 20 with irradiation alone and by a factor of 10,000 when radiation was administered in the presence of genistein (van Rijin and van den Berg, 1997). Based on similar findings in prostate cancer, Hillman et al. (2001) recommended a potential combination of genistein with radiation for the treatment of prostate cancer. The radiation enhancement is attributed to inhibition of topoisomerase II activity, which is involved in replication, transcription and probably DNA repair.

## 3.3. Genistein and Passive Immunotherapy

Genistein has been used in passive immunotherapy of B-cell lineage lymphomas. Human B-lineage lymphoid cells overexpress CD19 receptor. It is physically associated with the Src proto-oncogene family protein-tyrosine kinase Lyn. The membrane-associated CD19-Lyn receptor-enzyme complex plays a pivotal role for survival and clonogenicity of immature B-cell precursors from acute lymphoblastic leukemia patients. CD19-associated Lyn kinase can be selectively targeted and inhibited when genistein is conjugated with the B43 murine monoclonal antibody that targets the CD19 receptor. Research investigators from University of Minnesota Health Sciences Center carried out extensive *in vitro*, animal, and phase I/II safety and toxicity studies on the B43-Gen immunoconjugate (Myers et al., 1995, 1998; Ek et al., 1998; Chen et al., 1999; Uckun et al., 1999). Myers et al. (1995) showed that targeting the membrane-associated CD19-Lyn complex *in vitro* with B43-Gen triggers rapid apoptotic cell death in radiation-resistant "p53-Bax-Ramos-Bt" B-lineage lymphoma cells expressing high levels of Bcl-2 protein, without affecting Bcl-2 mRNA expression level. The therapeutic potential of this apoptosis induction strategy was examined in a scid mouse xenograft model for radiation-resistant high-grade human B-lineage lymphoma. *In vivo* treatment of scid mice challenged with a fatal number of Ramos-Bt cells plus B43-Gen at a dose level <1/10, the maximum tolerated dose, resulted in 70% long-term event-free survival. Subsequently, Myers and co-workers (1998) produced highly purified clinical grade B43-Gen. Preclinical murine studies (Ek et al., 1998), phase I/II toxicity studies, and measurements of human antimouse antibody levels were completed (Uckun et al., 1999) before launching clinical studies. Chen et al. (1999) examined the pharmacokinetics of B43-Gen preparation in 17 patients (4 children, 13 adults) with B-lineage lymphoid malignancies, including 12 patients with acute lymphoblastic leukemia (ALL) and 5 patients with non-Hodgkin's lymphoma (NHL). Pharmacokinetic analyses revealed a plasma half-life of $19 \pm 4$ hours, mean residence time of $22 \pm 4$ hours, and a systemic clearance of $18 \pm 2$ mL/h/kg. There was a trend toward faster clearance rates, shorter mean residence time, and shorter systemic clearance time for non-Caucasian patients compared to Caucasian patients. Pediatric patients showed a significantly larger volume of distribution at steady state and a longer elimination of half life than did adults. The pharmacokinetics of B43-Gen was not affected by the patient's gender or by a history of bone marrow transplantation.

Using the same principle, recombinant epidermal growth factor (EGF) has been conjugated to genistein. The receptor for EGF (EGF-R) is expressed at high levels on human breast cancer cells and forms membrane-associated complexes with ErbB2, ErbB3 and Src proto-oncogene family PTKs. The EGF-Gen preparation bound to and entered MDA-MB-231 and BT-20 breast cancer cells, which express EGF-R, but not to NALM-6 or HL-60 leukemia cells, which do not express EGF-R, and it effectively competed with unconjugated EGF in ligand-binding assays. EGF-Gen inhibited EFG-R tyrosine kinase in breast cancer cells at nanomolar concentrations. The IC50 was $>10$ μM for unconjugated genistein and nmol for EGF-Gen. EGF-Gen but not GCSF-Gen induced apoptosis of cells. In essence, EGF-Gen was one hundred times stronger than EGF with respect to inhibition of EFG-R-associated PTK in intact breast cancer cells. Using genistein conjugates to target cell-surface receptors that are closely associated with proto-oncogene family PTKs is a promising area of research that should be further investigated in preclinical or clinical phase I or II studies.

## 3.4. Antigenic Alteration by Genistein and Its Implications in Active Specific Immunotherapy

Numerous investigations document genistein's ability to alter tumor-associated proteins and glycoconjugate antigens. Li and Sarkar (2002) reported that in prostate cancer genistein altered the expression of 832 genes two fold or higher. Other investigators found that genistein downregulated 11 genes including metalloproteinase MMP-9, protease M, uPAR, VEGF, TSP, BPGF, LPA, TGF-β2, Tsp1 and PAR-2; and it upregulated the expression and activity of tyrosinases and melanin biosynthesis in melanocytes and in human melanoma (Kiguchi et al., 1990, Rauth et al., 1997, Hartmann et al., 1997).

In renal carcinoma cell lines (SMKT-R1, R2, R3 (B1/B4), R4, TOS-1, TOS-2 and ACHN), genistein downregulated glycolipid transferase gene expression (Yabunaka 1995, Honke et al., 1998) and sulfotransferase activity (Yabunaka et al., 1997). Mitropoulou and coinvestigators (2002) have shown that in both ER-positive (MCF-7) and ER-negative (BT-20) breast cancer cells, genistein alters synthesis of glycosaminoglycans, proteoglycans and cell surface localization of the proteoglycans. In sarcomas, genistein-mediated antigenic alterations prevented invasion and metastasis (Yan and Han, 1999).

In a study that used murine monoclonal antibody SGR37 specific for De-N-acetyl gangliosides, Sjoberg and co-workers (1995) noted a dose-dependent genistein effect on the expression of De-NAcGD3 in the Melur melanoma cell line. Gangliosides are excellent targets for active specific immunotherapy (Morton et al., 1994). Autologous or allogeneic formulations of purified antigens, cell lysates, or irradiated tumor cells are used as vaccines to induce immune responses against tumor-associated antigens. The success of vaccine therapy depends on immunogenic antigens expressed in the cancer vaccine. Most of the antigens may also be expressed in small amounts in normal cells but are overexpressed in cancer cells. For example, ganglioside GD3 is overexpressed in tumor cells from patients with AJCC stage III or IV melanoma (Portoukalian et al., 1978; Ravindranath et al., 1991). GD3 is poorly expressed in normal melanocytes but highly expressed in human retina (Ravindranath and Morton, 1991). O-acetylation of GD3 induces a strong immunogenic immune response (Ravindranath et al., 1989). O-acetylation is facilitated by activation of an enzyme called O-acetyl transferase. Such alteration in the antigenic structure results in an enhanced immune response. Similarly, genistein-induced de-N-acetylation should enhance the immune response to GD3 or GM3, the ganglioside antigens commonly expressed in human cancers.

## 3.5. Genistein Significantly Lowers Prostate-Specific Antigen (PSA)

Taille et al. (2001) reported that daily intake of 40 grams of soybeans may significantly lower serum levels of prostate-specific antigen (PSA). A recent clinical study by Ghafar and co-investigators (2002) reported that 44 days of treatment with genistein plus a polysaccharide from Basidiomycetes reduced serum PSA levels to 4.2 ng/ml from a pretreatment mean of 19.7 ng/ml. *In vitro* studies on human prostate cancer cell lines document that genistein can decrease PSA mRNA (arresting transcriptional activation) in both androgen-dependent (LnCaP) and androgen-independent (VeCaP) cell lines (Davis et al., 2000, 2002; Shen et al., 2000). Serum PSA appears to be a useful measure of genistein's efficacy alone or in combination with irradiation or chemotherapeutic drugs for prostate cancer.

## 4. Conclusion

Genistein can arrest tumor growth, proliferation, cell cycle, invasion, metastasis and angiogenesis. All these events associated with tumor progression involve PTK activity induced by growth factors such as EGF, IGF, PDGF and VEGF, by hormones such as estrogen, androgen, prolactin and progesterone, and by cytokines such as IL3, IL5 and IL6. Although it is still not clear how genistein orchestrates different cytokines, it appears to block TPK-mediated signal transduction during tumor progression. Genistein alone can be a potential therapeutic agent. However, due to its biphasic nature, the dose required for arresting tumor growth *in situ* should be determined. The pharmacokinetics of genistein alone or in combination with various chemotherapeutic drugs or radiation requires further study.

When conjugated to the B43 murine monoclonal antibody that targets CD19 of B-lineage leukemia, or when conjugated to recombinant EFG, genistein appears to be an effective passive immunotherapy for human cancer. Because of its ability to induce highly immunogenic antigens, it should also be a useful agent for active specific immunotherapy. However, the dose required for antigenic alterations requires further study, and the efficacy of genistein as an anticancer agent can only be determined by carefully controlled phase III multicenter trials. There is tremendous scope for genistein as an anticancer therapeutic agent. However, genistein may interfere with paclitaxel (Taxol).

Genistein also appears to have prophylactic value; studies report a reduced risk of cancer among Japanese and Finnish populations that have a high consumption of genistein. Genistein as a food supplement can be given to women from prepubertal stage of life so that it would be beneficial in arresting tumor initiation. Genistein may avoid the risk of developing cancer in both men and women who have risk factors for gender-based cancers, such as familial expression of BRCA 1 and 2.

## 5. Acknowledgements

We sincerely thank Dr. Donald L. Morton, President, John Wayne Cancer Institute, for all his support and encouragement. We thank Professor Edwin L. Cooper for his long-standing support and encouragement. Thanks are also due to Dr. Meena Verma for her valuable assistance in growth inhibition studies on prostate cancer cell lines and for collecting literature on prostate cancer. We are thankful to Ms. Gwen Berry for editing and to Ms. Posha Green for word-processing. Works related to Prostate Cancer, reported in this study is supported by Department of the US Army Grant No. DAMD17-01-1-0062.

## 6. References

Adlercreutz H., Markkanen H., and Watanabe S., 1993, Plasma concentrations of phyto-oestrogens in Japanese men, *Lancet.* **342**:1209–1210.

Adelcreutz H., and Mazur W., 1997, Phytoestrogens and western diseases, *Ann. Med.* **29**:95–120.

Agarwal R., 2000, Cell Signaling and Regulators of Cell Cycle as Molecular Targets for Prostate Cancer Prevention by Dietary Agents, *J Biochem Pharm.* **60**:1051–1059.

Ahmed N., Pansino F., Baker M., Rice G., and Quinn M., 2002, Association between alphavbeta6 integrin expression, elevated p42/44 kDa MAPK, and plasminogen-dependent matrix degradation in ovarian cancer, *J Cell Biochem.* **84**:675–686.

Akaza H., Miyanaga N., Takashima N., Naito S., Hirao Y., Tsukamoto T., and Mori M., 2002. Is daidzein non-metabolizer a high risk for prostate cancer? A case-controlled study of serum soybean isoflavone concentration, *Jpn J Clin Oncol.* **32**:296–300.

Akiyama T., Ishida J., Nakagawa S., Ogawara H., Wantanabe S., Itoh N., Shibuya M., and Fukami Y., 1987, Genistein, a specific inhibitor of tyrosine-specific protein kinases, *J Biol Chem.* **262**:5592–5595.

Allred C.D., Allred K.F., Ju Y.H., Virant S.M., and Helferich W.G., 2001a, Soy diets containing varying amounts of genistein stimulate growth of estrogen-dependent (MCF-7) tumors in a dose-dependent manner, *Cancer Res.* **61**:5045–5050.

Allred C.D., Ju Y.H., Allred K.F., Chang J., and Helferich W.G., 2001b, Dietary genistein stimulates growth of estrogen-dependent breast cancer tumors similar to that observed with genistein, *Carcinogenesis.* **22**:1667–1673.

Ando S., Nojima K., Majima H., Ishihara H., Suzuki M., Furusawa Y., Yamaguchi H., Koike S., Ando K., Yamauchi M., and Kuriyama T., 1998, Evidence for mRNA expression of vascular endothelial growth factor by X-ray irradiation in a lung squamous carcinoma cell line, *Cancer Lett.* **132**:75–80.

Andrew D.P., Yoshino T., Guh L., Martin–Simonet M.T., and Butcher E.C., 1995, TABS, a T cell activation antigen that induces LFA-1-dependent aggregation, *J Immunol.* **155**:1671–1684.

Arai N., Strom A., Rafter J.J., and Gustafsson J.A., 2000, Estrogen receptor beta mRNA in colon cancer cells: growth effects of estrogen and genistein, *Biochem Biophys Res Commun.* **270**:425–431.

Arts J., Grimbergen J., Toet K., and Kooistra T., 1999, On the role of c-Jun in the induction of PAI-1 gene expression by phorbol ester, serum, and IL-1alpha in HepG2 cells, *Anterioscler Thromb Vasc Biol.* **19**:39–46.

Bagnato A., Tecce R., Di Castro V., and Catt K.J., 1997, Activation of mitogenic signaling by endothelin 1 in ovarian carcinoma cells, *Cancer Res.* **57**:1306–1311.

Balabhadrapathruni S., Thomas T.J., Yurkow E.J., Amenta P.S., and Thomas T., 2000, Effects of genistein and structurally related phytoestrogens on cell cycle kinetics and apoptosis in MDA-MB-468 human breast cancer cells, *Oncol Rep.* **7**:3–12.

Barnes S., Peterson T., and Coward L., 1995, Rationale for the use of Genistein-containing soy matrices in chemoprevention trials for breast and prostate cancer, *J Cell Biochem Suppl.* **22**:181–187.

Barnes S., Sfakianos J., Coward L., and Kirk M., 1996, Soy isoflavonoids and cancer prevention. Underlying biochemical and pharmacological issues, *Adv Exp Med Biol.* **401**:87–100.

Belka C., Ahlers A., Sott C., Gaestel M., Herrmann F., and Brach M.A., 1995, Interleukin (IL)-6 signaling leads to phosphorylation of the small heat shock protein (Hsp) 27 through activation of the MAPKAP kinase 2 pathway in monocytes and monocytic leukemia cells, *Leukemia.* **9**:288–294.

Bergaamaschi G., Rosti V., Danova M., Ponchio L., Lucotti C., and Cazzola M., 1993, Inhibitors of tyrosine phosphorylation induce apoptosis in human leukemic cell lines, *Leukemia.* **7**:2012–2018.

Bergan R., Kyle E., Nguyen P., Trepel J., Ingui C., and Neckers L., 1996, Genistein-stimulated adherence of prostate cancer cells is associated with the binding of focal adhesion kinase to beta-1-integrin, *Clin Exp Metastasis.* **14**:389–398.

Bhatia N., and Agarwal R., 2001, Detrimental effect of cancer preventive phytochemicals silymarin, Genistein and epigallocatechin 3-gallate on epigenetic events in human prostate carcinoma DU145 cells, *Prostate.* **46**:98–107.

Bishop J. M., 1983, Oncogenes and proto-oncogenes, *Hosp Pract.* **18**:67–74.

Boros L.G., B. S., Lim S., and Lee W.N., 2001, Genistein inhibits nonoxidative ribose synthesis in MIA pancreatic adenocarcinoma cells: a new mechanism of controlling tumor growth, *Pancreas.* **22**:1–7.

Borradaile N.M., de Dreu L.E., Wilcox L.J., Edwards J.Y., and Huff M.W., 2002, Soya phytoestrogens, genistein and daidzein, decrease apolipoprotein B secretion from HebG2 cells through multiple mechanisms, *Biochem J.* **366**:531–539.

Borsellino N., Bonavida B., Ciliberto G., Toniatti C., Travali S., and D'Alessandro N., 1999, Blocking signaling through the Gp130 receptor chain by interleukin-6 and oncostatin M inhibits PC-3 cell growth and sensitizes the tumor cells to etoposide and cisplatin-mediated cytotoxicity, *Cancer.* **85**:134–144.

Bronte V., Macino B., Zambon A., Rosato A., Mandruzzato S., Zanovello P., and Collavo D., 1996, Protein tyrosine kinases and phosphatases control apoptosis induced by extracellular adenosine 5' triphosphate, *Biochem Biophys Res Commun.* **218**:344–351.

Burgener D., Bonjour J.P., and Caverzasio J. 1995, Fluoride increases tyrosine kinase activity in osteoblast-like cells: regulatory role for the stimulation of cell proliferation and Pi transport across the plasma membrane, *J Bone Miner Res.* **10**:64–171.

Cafferata E.G., Gonzalez-Guerrico A.M., Giordano L., Pivetta O.H., and Santa-Coloma T.A., 2000, Interleukin-1beta regulates CFTR expression in human intestinal T84 cells, *Biochim Biophys Acta.* **1500**:241–248.

Cappelletti V., Fioravanti L., Miodini P., and Di Fronzo G., 2000, Genistein blocks breast cancer cells in the G (2) M phase of the cell cycle, *J Cell Biochem.* **79**:594–600.

Carlo-Stella C., Dotti G., Mangoni L., Regazzi E., Garau D., Bonati A., Almici C., Sammarelli G., Savoldo B., Rizzo M.T., and Rizzoli V., 1996a, Selection of myeloid progenitors lacking BCR/ABL mRNA in chronic myelogenous leukemia patients after in vitro treatment with the tyrosine kinase inhibitor genistein, *Blood.* **88**:3091–3100.

Carlo-Stella C., Regazzi E., Garau D., Mangoni L., Rizzo M.T., Bonati A., Dotti G., Almici C., and Rizzoli V., 1996b, Effect of the protein tyrosine kinase inhitibor genistein on normal leukaemic haemopoietic progenitor cells, *Br J Haematol.* **93**:551–557.

Casagrande F., Darbon J.M., 2000, P21 CIP1 is dispensable for the G2 arrest caused by genistein in human melanoma cells, *J Exp Cell Res.* **258**:101–108.
Cataldi M., Taglialatela M., Guerriero S., Amoroso S., Lombardi G., di Renzo G., and Annunziato L., 1996, Protein-tyrosine kinases activate while protein-tyrosine phosphatases inhibit L-type calcium channel activity in pituitary GH3 cells, *J Biol Chem.* **19**:9441–9446.
Cattaneo M.G., D'atri F., and Vicentini L.M., 1997, Mechanisms of mitogen-activated protein kinase activation by nicotine in small-cell lung carcinoma cells, *Biochem J.* **328**:499–503.
Chen C.L., Levin A., Rao A., O'Neill K., Mess Inger Y., Myers D.E., Goldman F., Hurvitz C., Casper J.T., and Uckun F.M., 1999, Clinical pharmacokinetics of the CD19 receptors-directed tyrosine kinase inhibitor B43-Genistein in patients with B–lineage lymphoid malignancies, *J Clin Pharmacol.* **39**:1248–1255.
Chen X., Anderson J.J., 2001, Isoflavones inhibit proliferation of ovarian cancer cells in vitro via an estrogen receptor-dependent pathway, *N.utr Cancer.* **41**:165–171.
Choi Y.H., Lee W., Park K.Y., and Zhang L., 2000, p53-independent induction of p21 (WAF1/CIP1), reduction of cyclin B1 and G2/M arrest by the isoflavone Genistein in human prostate carcinoma cells, *Jpn J Cancer Res.* **91**:164–173.
Choi Y.H., Zhang L., Lee W.H., and Park K.Y., 1998, Genistein-induced G2/M arrest is associated with the inhibition of cyclin B1 and the induction of p21 in human breast carcinoma cells, *Int J Oncol* **13**:391–396.
Clark J.W., Santos-Moore A., Stevenson L.E., and Frackelton A.R. Jr., 1996, Effects of tyrosine kinase inhibitors on the proliferation of human breast cancer cell lines and proteins important in the ras signaling pathway, *Int J Cancer* **65**:186–191.
Constantinou A. and Huberman E., 1995, Genistein as an inducer of tumor cell differentiation: possible mechanisms of action. *Proc Soc Exp Biol Med.* **208**:109–15.
Constantinou A., Kiguchi K., and Huberman E., 1990, Induction of differentiation and DNA strand breakage in human HL-60 and K-562 leukemia cells by genistein. *Cancer Res.* **50**:2618–2624.
Constantinou A.I., Kamath N., and Murley J.S., 1998a, Genistein inactivates bcl-2, delays the G2/M phase of the cell cycle, and induces apoptosis of human breast adenocarcinoma MCF-7 cells, *Eur J Cancer.* **34**:1927–1934.
Constantinou A.I., Krygier A.E., and Mehta R.R., 1998b, Genistein induces maturation of cultured human breast cancer cells and prevents tumor growth in nude mice, *Am J Clin Nutr.* **68**:1426S–1430S.
Cruz M.l., Wong W.W., Mimouni F., Hachey D.L., Setchell K.D., Klein P.D., and Tsang R.c., 1994, Effects of infant nutrition on cholesterol synthesis rates, *Pediatr Res.* **35**:135–140.
Dampier K., Hudson E.A., Howells L.M., Manson M.M., Walker R.A., and Gescher A., 2001, Differences between human breast cell lines in susceptibility towards growth inhibition by genistein, *Br J Cancer.* **85**:618–624.
Darbon J.M., Penary M., Escalas N., Casagrande F., Goubin-Gramatica F., Baudouin C., and Ducommun B., 2000, Distinct Chk2 Activation Pathways Are Triggered by Genistein and DNA-damaging Agents in Human Melanoma Cells, *J Biol Chem.* **275**:15363–15369.
Das R., and Vonderhaar B.K., 1996, Activation of raf-1, MEK, and MAP kinase in prolactin responsive mammary cells, *Breast Cancer Res Treat* **40**:141–149.
Davis J.N., Kucuk O., and Sarkar F.H., 2002, Expression of prostate-specific antigen is transcriptionally regulated by Genistein in prostate cancer cells, *Mol Carcinog.* **34**:91–101.
Davis J.N., Kucuk O., Djuric Z., and Sarkar F.H., 2001, Soy isoflavone supplementation in healthy men prevents NF-kappa B activation by TNF-alpha in blood lymphocytes., *Free Radic Biol Med,* **30**:1293–1302.
Davis J.N., Kucuk. O., and Sarkar F.H., 1999, Genistein inhibits NF-kappa B activation in prostate cancer cells, *Nutr Cancer.* **35**:167–174.
Davis J.N., Muqim N., Bhuiyan M., Kucuk O., Pienta KJ, and Sarkar F.H., 2000, Inhibition of prostate specific antigen expression by Genistein in prostate cancer cells, *Int J Oncol.* **16**:1091–1097.
de la Taille A. Katz A., Vacherot F., Saint F., Salomon L., Cicco A., Abbou C.C., and Chopin D.K., 2001, Cancer of the prostate: influence of nutritional factors. A new nutritional approach, *Presse Med.* **30**:561–564.
den Tonkelaar I, Keinan-Boker L, Veer PV, Arts CJ, Adlercreutz H, Thijssen JH, Peeters PH. 2001. Urinary phytoestrogens and postmenopausal breast cancer risk. *Cancer Epidemiol Biomarkers Prev.* **10**:223–228.
Deora A.B., Miranda M.B., and Rao S.G., 1997, Down-modulation of P210bcr/abl induces apoptosis/differentiation in K562 leukemic blast cells, *Tumori.* **83**:756–761.
Diel P., Olff S., Schmidt S., and Michna H., 2001, Molecular identification of potential selective estrogen receptor modulator (SERM) like properties of phytoestrogens in the human breast cancer cell line MCF-7, *Planta Med.* **67**:510–514.

Ding D.X., Rivas C.I., Heaney M.L., Raines M.A., Vera J.C., and Golde D.W., 1994, The alpha subunit of the human granulocyte-macrophage colony-stimulating factor receptor signals for glucose transport via a phosphorylation-independent pathway, *Proc Natl Acad Sci USA.* **91**:2537–2541.

Ding X.Z., Tong W.G., and Adrian T.E., 2001, 12-lipoxygenase metabolite 12(S)-HETE stimulates human pancreatic cancer cell proliferation via protein tyrosine phosphorylation and ERK activation, *Int J Cancer.* **94**:630–636.

Elatter T.M., Virji A.S., 2000, The inhibitory effect of curcumin, Genistein, quercetin and cisplatin on the growth of oral cancer cells in vitro, *Anticancer Res.* **20**:1733–1738.

El-Zarruk A.A., and van den Berg H.W., 1999, The anti-proliferative effects of tyrosine kinase inhibitors towards tamoxifen-sensitive and tamoxifen-resistant human breast cancer cell lines in relation to the expression of epidermal growth factor receptors (EGF-R) and the inhibition of EGF-R tyrosine kinase, *Cancer Lett.* **142**:185–193.

Ek O., Yanishevski Y., Zeren T., Waurzyniak B., Gunther R., Chelstrom L., Chandan-Langlie M., Schneider E., Myers D.E., Evans W., and Uckun F.M., 1998, In vivo toxicity and pharmacokinetic features of B43 (Anti-CD19)– Genistein immunoconjugate, *Leuk Lymphoma.* **30**:389–394.

Farhan H., Wahala K., Adlercreutz H., and Cross H.S., 2002, Isoflavonoids inhibit catabolism of vitamin D in prostate cancer cells., *J Chromatogr B Analyt Technol. Biomed Life Sci.* **777**:261–268.

Finlay G.J., Holaway K.M., and Baguley B.C., 1994, Comparison of the effects of genistein and amsacrine on leukemia cell proliferation, *Oncol Res.* **6**:33–37.

Fioravanti L., Cappelletti V., Miodini P., Ronchi E., Brivio M., and Di Fronzo G., 1998, Genistein in the control of breast cancer cell growth: insights into the mechanism of action in vitro. *Cancer Lett.* **130**:143–152.

Fiorelli G., Picariello L., Martineti V., Tonelli F., and Brandi M.L., 1999, Estrogen synthesis in human colon cancer epithelial cells, *J Steroid Biochem Mol Biol.* **71**:223–230.

Frey R.S., and Singletary K.W., 2003, Genistein activates p38 mitogen-activated protein kinase, inactivates ERK1/ERK2 and decreases Cdc25c expression in immortalized human mammary epithelial cells, *J Nutr.* **133**:226–231.

Frey R.S., Li J., and Singletary K.W., 2001, Effects of genistein on cell proliferation and cell cycle arrest in nonneoplastic human mammary epithelial cells: involvement of Cdc2, p21 (waf/cip1), p27 (kip1), and Cdc25c expression, *Biochen Pharmacol.* **61**:979–989.

Fujimoto N., Sueoka N., Sueoka E., Okabe S., Suganuma M., Harada M., and Fujiki H., 2002, Lung cancer prevention with (-)-epigallocatechin gallate using monitoring by heterogenous nuclear ribonucleoprotein B1, *Int J Oncol.* **20**:1233–1239.

Geller J., Sionit L., Partido C., Li L., Tan X., Youngkin T., Nachtsheim D., and Hoffman R.M., 1998, Genistein inhibits the growth of human-patient BPH and prostate cancer in histoculture, *Prostate.* **34**:75–79.

Ghafar M.A., Golliday E., Bingham J., Mansukhani M.M, Anastasiadis A.G., and Katz A.E., 2002, Regression of prostate cancer following administration of Genistein Combined Polysaccharide (GCP), a nutritional supplement: a case report, *J Altern Complement Med.* **8**:493–497.

Grandaliano G., Monno R., Ranieri E., Gesualdo L., Schena F.P., Martino C., and Ursi M, 2000, Regenerative and proinflammatory effects of thrombin on human proximal tubular cells, *J Am Soc Nephrol.* **11**:1016–1025.

Guo Q., Rimbach G., Moini H., Weber S., and Packer L., 2002, ESR and cell culture studies on free radical-scavenging and antioxidant activities of isoflavonoids, *J Toxicology.* **179**:171–180.

Hagmann W., 1994, Cell proliferation status, cytokine action and protein tyrosine phosphorylation modulate leukotriene biosynthesis in a basophil leukaemia and a mastocytoma cell line, *Biochem J.* **299**:467–472.

Hartman R.R., Rimoldi D., Lejeune F.J., and Carrel S., 1997, Cell differentiation and cell-cycle alterations by tyrosine kinase inhibitors in human melanoma cells, *Melanoma Res.* **7**:S27–S33.

Heruth D.P., Wetmore L.A., Leyva A., and Rothberg P.G., 1995, Influence of protein tyrosine phosphorylation on the expression of the c-myc oncogene in cancer of the large bowel, *J Cell Biochem.* **58**:83–94.

Hedlund T.E., Johannes. W.U., Miller G.J., 2003, Soy isoflavonoid equol modulates the growth of benign and malignant prostatic spithelial cells in vitro, *J Prostate.* **54**:68–78.

Hillman G.G., Forman J., Kucuk O., Yudelev M., Maughan R.L., Rubio J., Layer A., Tekyi-Mensah S., Abrams J., and Sarkar F.H., 2001, Genistein potentiates the radiation effect on prostate carcinoma cells, *Clin Cancer Res.* **7**:382–390.

Hoffman R., 1995, Potent inhibition of breast cancer cell lines by the isoflavonoid kievitone: comparison with genistein, *Biochem Biophys Res Commun.* **211**:600–606.

Honke K., Tsuda M., Hirahara Y., Miyao N., Tsukamoto T., Satoh M., and Wada Y., 1998, Cancer-associated expression of glycolipid sulfotransferase gene in human renal cell carcinoma cells, *Cancer Res.* **58**:3800–3805.

Honma Y., Okabe-Kado J., Kasukabe T., Hozumi M., and Umezawa K., 1990, Inhibition of abl oncogene tyrosine kinase induces differentiation of human myelogenous leukemia K562 cells. *Jpn J Cancer Res.* **81**:1132–113.

Hooijberg J.H., Broxterman H.J., Heijn M., Fles D.L., Lankelma J., and Pinedo H.M., 1997, Modulation by (iso) flavonoids of the ATPase activity of the multidrug resistance protein, *FEBS Lett.* **413**:344–348.

Hsieh C.Y., Santell R.C., Haslam S.Z., and Helferich W.G., 1998, Estrogenic effects of genistein on the growth of estrogen receptor-positive human breast cancer (MCF-7) cells in vitro and in vivo, *Cancer Res.* **58**:3833–3838.

Hu Y., Dragowska W.H., Wallis A., Duronio V., and Mayer L., 2001, Cytotoxicity induced by manipulation of signal transduction pathways is associated with down-regulation of Bcl-2 but not Mcl-1 in MCF-7 human breast cancer, *Breast Cancer Res Treat.* **70**:11–20.

Hunakova L., Sedlak J., Klobusicka M., Duraj J., and Chorvath B., 1994, Tyrosine kinase inhibitor-induced differentiation of D-562 cells: alterations of cell cycle and cell surface phenotype, *Cancer Lett.* **81**:81–87.

Ihn H., and Tamaki K., 2002, Mitogenic activity of dermatofibrosarcoma protuberas is mediated via an extracellular signal related kinase dependent pathway, *J Invest Dermatol.* **119**:954–960.

Iida J., Meijne A.M., Spiro R.C., Roos E., Furct L.T., and McCarthy J.B., 1995, Spreading and focal contact formation of human melanoma cells in response to the stimulation of both melanoma-associated proteoglycan (NG2) and alpha 4 beta 1 integrin, *Cancer Res.* **55**:2177–2185.

Jenab S., Morris P.L., 1997, Transcriptional regulation of Sertoli cell immediate early genes by interleukin-6 and interferon-gamma is mediated through phosphorylation of STAT-3 and STAT-1 proteins, *Endocrinology* **138**:2740–2746.

Jones T.H., Justice S.K., and Price A., 1997, Suppression of tyrosine kinase activity inhibits [3H] thymidine uptake in cultured human pituitary tumor cells, *J Clin Endocrinol Metab.* **82**:2143–2147.

Ju Y.H., Allred C.D., Allred K.F., Karko K.L., Doerge D.R., and Helferich W.G., 2001, Physiological concentrations of dietary genistein dose-dependently stimulate growth of estrogen-dependent human breast cancer (MCF-7) tumors implanted in athymic nude mice, *J Nutr.* **131**:2957–2962.

Ju Y.H., Doerge D.R., Allred K.F., Allred C.D., and Helferich W.G., 2002, Dietary genistein negates the inhibitory effect of tamoxifen on growth of estrogen-dependent human breast cancer (MCF-7) cells implanted in athymic mice, *Cancer Res.* **62**:2474–2477.

Katdare M., Osborne M., Telang N.T., 2002, Soy isoflavone genistein modulates cell cycle progression and induces apoptosis in HER-2/neu oncogene expressing human breast epithelial cells. *Int J Oncol.* **21**:809–815.

Katdare M., Osborne M.P., and Telang N.T., 1998, Inhibition of aberrant proliferation and induction of apoptosis in pre-neoplastic human mammary epithelial cells by natural phytochemicals, *Oncol Rep* **5**:311–315.

Kawase T., Oriskasa M., Oguro A., and Burns DM., 1999, Possible regulation of epidermal growth factor-receptor tyrosine autophosphorylation by calcium and G proteins in chemically permeabilized rat UMR106 cells, *Arch Oral Biol.* **44**:157–171.

Kelly G.E., Nelson C., Waring M.A., Joannou G.E., and Reeder A.Y., 1993, Metabolites of dietary (soya) isoflavones in human urine, *Clin Chim Acta.* **223**:9–22.

Khoshyomn S., Manske G.C., Lew S.M., Wald S.L., and Penar P.L., 2000. Synergistic action of genistein and cisplatin on growth inhibition and cytotoxicity of human medulloblastoma cells, *Pediatr Neurosurg.* **33**:123–131.

Khoshyomn S., Nathan D., Manske G.C., Osler T.M., and Penar P.L., 2002, Synergistic effect of genistein and BCNU on growth inhibition and cytotoxicity of glioblastoma cells, *J Neurooncol.* **57**:193–200.

Kiguchi K., Constantinou A.I., and Huberman E., 1990, Genistein-induced cell differentiation and protein-linked DNA strand breakage in human melanoma cells, *Cancer Commun.* **2**:271–277.

Kikuchi H., and Hossain A., 1999, Signal transduction-mediated CYPIA1 induction by omeprazole in human HepG2 cells., *Exp Toxicol pathol.* **51**:342–346.

Kikuchi Y., and Nash H.A., 1979. Nicking-closing activity associated with bacteriophage lambda int gene product, *Proc Natl Acad Sci USA.* **76**:3760–3764.

Kim J.G., Shin I., Lee K.S., and Han J.S., 2001, D609-sensitive tyrosine phosphorylation is involved in Fas-mediated phospholipase, *Exp Mol Med.* **33**:303–309.

Klein J.M., and McCarthy T.A., 1997, Inhibition of tyrosine kinase activity decreases expression of surfactant protein A in a human lung adenocarcinoma cell line independent of epidermal growth factor receptor, *Biochim Biophys Acta.* **1355**:218–230.

Kobayashi T., Nakata. T., and Kuzumaki T., 2002, Effect of flavonoids on cell cycle progression in prostate cancer cells, *Cancer Lett.* **176**:17–23.

Kumar NB., Cantor A., Allen K., Ricardi D., and Cox C.E., 2002, The specific role of isoflavones on estrogen metabolism in premenopausal women, *Cancer.* **94**:1166–1174.

Kumi-Diaka J., and Butler A., 2000, Caspase-3 protease activation during the process of genistein-induced apoptosis in TM4 testicular cells, *Biol Cell.* **92**:115–124.

Kumi-Diaka J., Nguyen V., and Butler A., 1999, Cytotoxic potential of the phytochemical genistein isoflavone (4', 5', 7-trihydroxyisoflavone) and certain environmental chemical compounds on testicular cells, *Biol Cell.* **91**:515–523.

Kumi-Diaka J., Rodriguez R., and Goudaze G., 1998, Influence of genistein (4', 5' 7-trihydroxyisoflavone) on the growth and proliferation of testicular cell lines, *Biol Cell.* **90**:349–354.

Kumi-Diaka J., Sanderson N., and Hall A., 2000, The mediating role of caspase-3 protease in the intracellular mechanism of Genistein-induced apoptosis in human prostatic carcinoma cell lines, DU145 and LNCaP, *Biol Cell.* **92**:594–604.

Kuo M.L., Huang T.S., and Lin J.K., 1995, Preferential requirement for protein tyrosine phosphatase activity in the 12-O-tetradecanoylphorbol-13-acetate-induced differentiation of human colon cancer cells, *Biochem Pharmacol.* **50**:1217–1222.

Kuo S.M., 1996, Antiproliferative potency of structurally distinct dietary flavonoids on human colon cancer cells, *Cancer Lett.* **110**:41–48.

Kuo S.M., Morehouse H.F., and Lin C.P., 1997, Effect of antiproliferative flavonoids on ascorbic acid accumulation in human colon adenocarcinoma cells, *Cancer Lett.* **116**:131–137.

Kuriu A., Ikeda H., Kanakura Y., Griffin J.D., Druker B., Yagura H., Kitayama H., Ishikawa J., Nichiura T., and Kanayama Y, et al., 1991, Proliferation of human myeloid leukemia cell line associated with the tyrosine-phosphorylation and activation of the proto-oncogene c-kit product. *Blood.* **78**:2834–2840.

Kusaka M., and Sperelakis N., 1996, Genistein inhibition of fast Na+ current in uterine leiomyosarcoma cells is independent of tyrosine kinase inhibition, *Biochim Biophys Acta.* **1278**:1–4.

Kyle E, Neckers L., Takimoto C., Curt G., and Bergan R., 1997, Genistein-induced apoptosis of prostate cancer cells is preceded by a specific decrease in focal adhesion kinase activity, *Mol Pharmacol.* **51**:193–200.

Lamon-Fava S., 2000, Genistein activates apolipoprotein A-I gene expression in the human hepatoma cell line Hep G2, *J. Nutr.* **130**:2489–2492.

Lampe J.W., 2003, Isoflavonoid and lignan phytoestrogens as dietary biomarkers, *J Nutr.* **133**:956S–964S.

Lebakken C.S., McQuade K.J., and Rapraeger A.C., 2000, Syndecan1 signals independently of betal integrin s during Raji cell spreading, *Exp Cell Res.* **259**:315–325.

Lei W., Mayotte J.E., and Levitt M.L., 1998, EGF-dependent and independent programmed cell death pathways in NCI-H596 normal cell lung cancer cells, *Biochem Biophys Res Commun.* **245**:939–945.

Lei W., Mayotte J.E., and Levitt M.L., 1999, Enhancement of chemosensitivity and programmed cell death by tyrosine kinase inhibitors corelates with EGFR expression in non-small cell lung cancer cells, *Anticancer Res.* **19**:221–228.

Leung L.K., and Wang T.T., 2000, Bcl-2 is not reduced in the death of MCF-7 cells at low genistein concentration, *J Nutr.* **130**:2922–2926.

Leyton J., Garcia-Marin L.J., Tapia J.A., Jensen R.T., and Moody T.W., 2001, Bombesin and gastrin releasing peptide increase tyrosine phosphorylation of focal adhesion kinase and paxillin in non-smell cell lung cancer cells, *Cancer Lett.* **162**:87–95.

Li W., and Weber G., 1998a, Synergistic action of tiazofurin and Genistein in human ovarian carcinoma cells, *Oncol Res.* **10**:117–122.

Li W., and Weber G., 1998b, Synergistic action of tiazofurin and genistein on growth inhibition and differentiation of K-562 human leukemic cells, *Life Sci.* **63**:1975–1981.

Li Y., and Sarkar F., 2002a, Inhibition of nuclear factor kappaB activation in PC3 cells by Genistein is mediated via Akt signaling pathway, *J Clin Cancer Res.* **8**:2369–2377.

Li Y., and Sarkar F., 2002c, Gene expression profiles of Genistein-treated PC3 prostate cancer cells, *J Nutr.* **132**:3623–3631.

Li Y., and Sarkar F.H., 2002b, Down-regulation of invasion and angiogenesis-related genes identified by cDNA microarray analysis of PC3 prostate cancer cells treated with Genistein, *J Cancer Lett.* **186**:157–164.

Li Y., Bhuiyan M., Sarkar F.H., 1999b. Induction of apoptosis and inhibition of c-erbB-2 in MDA-MB-435 cells by genistein. *Int J Oncol.* **15**:525–533.

Li Y., Upadhyay S., Bhuiyan M., and Sarkar F.H., 1999a, Induction of apoptosis in breast cancer cells MDAMB-231 by genistein, *Oncogene.* **18**:3166–3172.

Lian F., Bhuiyan M., Li Y.W., Wall N., Kraut M., and Sarkar F.H., 1998, Genistein-induced G2-M arrest, p21WAF1 upregulation, and apoptosis in a non-small-cell lung cancer cell line, *Nutr Cancer.* **31**:184–191.

Lian F., Li Y., Bhuiyan M., and Sarkar F.H., 1999, p53 independent apoptosis induced by genistein in lung cancer cells, *Nutr Cancer.* **33**:125–131.

Liu Y., Bhalla K., Hill C., and Priest D.G., 1994, Evidence for involvement of tyrosine phosphorylation in taxol-induced apoptosis in a human ovarian tumor cell line, *Biochem Pharmacol.* **48**:1265–1272.

Loukovaara M., Carson. M., Palotie A., and Adlercreutz H., 1995, Regulation of sex hormone-binding globulin production by isoflavonoids and patterns of isoflavonoid conjugation in HepG2 cell cultures, *Steroids.* **60**:656–661.

Lu L.J., and Anderson K.E., 1998. Sex and long-term soy diets affect the metabolism and excretion of soy isoflavones in humans, *Am J Clin Nutr.* **68**:1500S–1504S.

Lu L.J., Anderson K.E., Grady J.J., Kohen F., and Nagamani M., 2000, Decreased ovarian hormones during a soya diet: implications for breast cancer prevention, *Cancer Res.* **60**:4112–4121.

Lu L.J., Bromeling L.D., Marshall M.V., and Ramanujam V.M., 1995, A simplified method to quantify isoflavones in commercial soybean diets and human urine after legume consumption, *Cancer Epidemiol Biomarkers Prev.* **4**:497–503.

Lu L.J., Lin S.N., Grady J.J., Nagamani M., and Anderson K.E., 1996, Altered kinetics and extent of urinary daidzein and genistein excretion in women during chronic soya exposure, *Nutr Cancer.* **26**: 289–302.

Maeno T., Tanaka T., Sando Y., Suga T., Maeno Y., Nakagawa J., Hosono T., Sato M., Akiyama H., Kishi S., Nagai R., and Kurabayashi M., 2002, Stimulation of vascular endothelial growth factor gene transcription by all trans retinoic acid through Sp1 and Sp3 sites in human bronchioloalveolar carcinoma cells, *Am J Respir Cell Mol Biol.* **26**:246–253.

Maggiolini M., Bonofiglio D., Marsico S., Panno M.L., Cenni B., Picard D., Ando S., 2001. Estrogen receptor alpha mediates the proliferative but not the cytotoxic dose-dependent effects of two major phytoestrogens on human breast cancer cells. *Mol Pharmacol.* **60**:595:602.

Maggiolini M., Vivacqua A., Carpino A., Bonofiglio D., Fasanella G., Salerno M., Picard D., and Ando S., 2002, The mutant androgen receptor T877A mediates the proliferative but not the cytotoxic dose-dependent effects of Genistein and quercetin on human LNCaP prostate cancer cells, *Mol Pharmacol.* **62**:1027–1035.

Makishima M., Honma Y., Hozumi M., Sampi K., Hattori M., Umezawa K., Motoyoshi K., 1991, Effects of inhibitors of protein tyrosine kinase activity and/or phosphatidylinositol turnover on differentiation of some human myelomonocytic leukemia cells. *Leuk Res.* **5**:701:708.

Marino M., Pallottini V., and Trentalance A., 1998, Estrogens cause rapid activation of IP3-PKC-alpha signal transduction pathway in HEP2 cells., *Biochem Biophys Res Commun.* **245**:254–258.

Markiewicz L., Garey J., Adlercreutz H., and Gurpide E.,1993, In vitro bioassays of non-steroidal phytoestrogens, *J Steroid Biochem Mol Biol.* **45**:399–405.

Markovits J., Pommier Y., Mattern M.R., Esnault C., Roques B.P., Le Pecq JB., and Kohn K.W., 1986. Effects of the bifunctional antitumor intercalator diterealinium on DNA in mouse leukemia L1210 cells and DNA topoisomerase II, *Cancer Res.* **46**:5821–5826.

Mazur W., 1998, Phytoestrogen content in foods, *Baillieres Clin Endocrinol Metab.* **2**:729–742

Mazur W., and Adlecreut H., 2000, Overview of naturally occuring endocrine-active substances in the human diet in relation to human health, *Nutrition .* **16**:654–658.

McMichael-Phillips D.F., Harding C., Morton M., Roberts S.A., Howell A., Potten C.S., and Bundred N.J., 1998, Effects of soy-protein supplementation on epithelial proliferation in the histologically normal human breast, *Am J Clin Nutr.* **68**:1431S–1435S.

Messina M.J., and Loprinzi C.L., 2001, Soy for breast cancer survivors: a critical review of the literature, *J Nutr.* **131**:3095S–3108S.

Milovic V., Deubner C., Zeuzem S., Piiper A., Caspary W.F., and Stein J., 1995, EGF stimulates polyamine uptake in Caco-2 cells, *Biochem Biophys Res Commun.* **206**:962–968.

Miodini P., Fioravanti L., Di Fronzo G., and Cappelletti V., 1999. The two phyto-oestrogens genistein and quercetin exert different effects on estrogen receptor function, *Br J Cancer.* **80**:1150–1155.

Mitchell J.H., Cawood E., Kinniburgh D., Provan A., Collins A.R., and Irvine D.S., 2001, Effect of phytoestrogen food supplement on reproductive health in normal males, *Clin Sci.* **100**:613–618.

Mitchell J.H., Duthie S., and Collins A.R., 2000, Effects of phytoestrogens on growth and DNA integrity in human prostate tumor cell lines: PC-3 and LNCaP, *J Nutr Cancer.* **38**:223–228.

Mitropoulou T.N., Tzanakakis G.N., Nikitovic D., Tsatsakis A., and Karamanos N.K., 2002, In vitro effects of genistein on the synthesis and distribution of glycosaminoglycans/proteoglycans by estrogen receptor-positive and -negative human breast cancer epithelial cells, *Anticancer Res.* **22**:2841–2846.

Monti E., and Sinha B.K., 1994, Antiproliferative effect of genistein and adriamycin against estrogen-dependent and –independent human breast carcinoma cell lines, *Anticancer Res.* **14**:1221–1226.

Morisset J., Douziech N., Rydzewska G., Buscail L., and Rivard N., 1995, Cell signalling pathway involved in PACAP-induced AR4-2J cell proliferation, *Cell Signal.* **7**:195–205.

Morton D.L., Ravindranath M.H., Irie R.F., 1994. Tumor gangliosides as targets for active specific immunotherapy of melanoma in man. *Prog Brain Res.* **101**:251–275.

Morton M.S., Arisaka O., Miyake N., Morgan L.D., and Evans B.A., 2002, Phytoestrogen concentrations in serum from Japanese men and women over forty years of age, *J Nutr.* **132**:3168–3171.

Mousavi Y., Adlercreutz. H., 1993, Genistein is an effective stimulator of sex hormone-binding globulin production in hepatocarcinoma human liver cancer cells and suppresses proliferation of these cells in culture, *Steroids.* **58**:301–304.

Mukhopadhyay D., Tsiokas L., and Sukhatme V.P., 1998, High cell density induces vascular endothelial growth factor expression via protein tyrosine phosphorylation, *Gene Expr.* **7**:53–60.

Munoz R., Klingenberg O., Wiedlocha A., Rapak A., Falnes PO., and Olsnes S.,1997, Effect of mutation of cytoplasmic receptor domain and of Genistein on transport of acidic fibroblast growth factor into cells, *Oncogene.* **15**:525–536.

Murkies A., Dalais F.S., Briganti E.M., Burger H.G., Healy D.L., Wahlqvist M.L., and Davis S.R., 2000. Phytoestrogens and breast cancer in postmenopausal women: a case control study, *Menapause.* **7**:289–296.

Myers D.E., Jun X., Waddick K.G., Forsyth C., C helstrom L.M., Gunther R.L., Tumer N.E., Bolen J., and Uckun F.M., 1995. Membrane-associated CD19 -LYN complex is an endogenous p53-independent and Bcl-2-independent regulator of apoptosis in human B-lineage lymphoma cells, *Proc Natl Acad Sci USA.* **92**:9575–9579.

Myers D.E., Sicheneder A., Clementson D., Dvo rak N., Venkatachalam T., Sev A.R., Chandan-Langlie M., and Uckun F.M., 1998, Large scale manufacturing of B43 (anti–CD19)–genistein for clinical trials in leukemia lymphoma, *Leuk Lymphoma.* **29**:329–338.

Naik H.R., Lehr J., and Pienta K.J., 1994, An in vitro and in vivo study of antitumor effects of Genistein on hormone refractory prostate cancer, *Anticancer Res.* **14**:2617–2619.

Nakagawa H., Yamamoto D., Kiyozuka Y., Tsuta K., Uemura Y., Hioki K., Tsutsui Y., and Tsubura A., 2000. Effects of genistein and synergistic action in combination with eicosapentaenoic acid on the growth of breast cancer cell lines, *J Cancer Res Clin Oncol.* **126**:448–454.

Nakanishi K., Fujimoto. J., Ueki T., Kishimoto K., Hashimoto-Tamaoki T., Furuyama J., Itoh T., Saski Y., and Okamoto E., 1999, Hepatocyte growth factor promotes migration of human hepatocellular carcinoma via phosphatidylinositol 3-kinase, *Clin Exp. Metastasis.* **17**:507–514.

Nomoto S., Arao Y., Horiguchi H., Ikeda K., Kayama F., 2002. Oestrogen causes G2/M arrest and apoptosis in breast cancer cells MDA-MB-231. *Oncol Rep.* **9**:773–776.

Oelmann E., Sreter L., Schuller I., Serve H., Koenigsmann M., Wiedenmann B., Oberberg D., Reufi B., Thiel E., and Berdel W.E., 1995, Nerve growth factor stimulates clonal growth of human lung cancer cell lines and a human glioblastoma cell line expressing high-affinity nerve growth factor binding sites involving tyrosine kinase signaling, *Cancer Res.* **55**:2212–2219.

Ohno T., Kato N., Ishii E., Shimizu M., Ito Y., Tomono S., and Kawazu S., 1993, Genistein augments cyclic adenosine 3'5'-monophosphate (cAMP) accumulation and insulin release in MIN6 cells, *Endocr Res.* **19**:273–285.

Okura A., Arakawa H., Oka H., Yoshinari T., and Monden Y., 1988. Effect of genistein on topoisomerase activity and on the growth of [Val 12] Ha-rastransformed NIH 3T3 cells, *Biochem Biophys Res Commun.* **157**:183–189.

Onozawa M., Fukuda K., Ohtani M., Akaza H., Sugimura T., and Wakabayashi K., 1998, Effects of soybean isoflavones on cell growth and apoptosis of the human prostatic cancer cell line LNCaP, *Jpn J Clin Oncol.* **28**:360–363.

Oude Weernink P.A., Verheul E., Kerkhof E., van Veelen C.W., and Rijksen G., 1996, Inhibitors of protein tyrosine phosphorylation reduce the proliferation of two human glioma cell lines, *Neurosurgery.* **38**:108–113.

Pagliacci M.C., Smacchia M., Migliorati G., Grignani F., Riccardi C., and Nicoletti I., 1994, Growth-inhibitory effects of the natural phyto-oestrogen genistein in MCF-7 human breast cancer cells, *Eur J Cancer* **30A**:1675–1682.
Pagliacci M.C., Spinozzi F., Migliorati G., Fumi G., Smacchia M., Grignani F., Riccardi C., and Nicoletti I., 1993, Genistein inhibits tumor cell growth in vitro but enhances mitochrondrial reduction of tetrazolium salst: a further pitfall in the use of the MTT assay for evaluating cell growth and survival, *Eur J Cancer* **29A**:1573–1577.
Panno M.L., Salerno M., Pezzi V., Sisci D., Maggiolini M., Mauro L., Morrone E.G. and Ando S., 1996, Effect of oestradiol and insulin on the proliferative pattern and on oestrogen and progesterone receptor contents in MCF-7 cells, *J Cancer Res Clin Oncol.* **122**:745–749.
Park J.H., Oh E. J., Choi Y.H., Kang C.D., Kang H.S., Kim D.K., Kang K.I., and Yoo M.A., 2001, Synergistic effects of dexamethasone and Genistein on the expressions of Cdk inhibitor p21WAF1/CIP1 in human hepatocellular and colorectal carcinoma cells, *Int J Oncol.* **18**:997–1002.
Peterson G., and Barnes S., 1991, Genistein inhibition of the growth of human breast cancer cells: independence from estrogen receptors and the multi-drug resistance gene, *Biochem Biophys Res Commun.* **179**:661–667.
Peterson G., and Barnes S., 1993, Genistein and biochanin A inhibit the growth of human prostate cancer cells but not epidermal growth factor receptor tyrosine autophosphorylation, *Prostate.* **22**:335–345.
Peterson G., and Barnes S., 1996, Genistein inhibits both estrogen and growth factor-stimulated proliferation of human breast cancer cells, *Cell Growth Differ.* **7**:1345–1351.
Peterson T.G., Coward L., Kirk M., Falany C.N., and Barnes S., 1996, The role of metabolism in mammary epithelial cell growth inhibition by the isoflavones genistein and biochanin A, *Carcinogenesis* **17**:1861–1869.
Peterson T.G., Ji G.P., Kirk M., Coward L., Falany C.N., and Barnes S., 1998, Metabolism of the isoflavones genistein and biochanin A in human breast cancer cell lines, *Am J Clin Nutr.* **68**:1505S–1511S.
Plewa M.J., Berhow M.A., Vaughn S.F., Woods E.J., Rundell M., Naschansky K., Bartolini S., and Wagner E.D., 2001, Isolating antigenotoxic components and cancer cell growth suppressors from agricultural by-products, *Mutat Res.* **480**:109–120.
Po L.S., Chen Z.Y., Tsang D.S., and Leung L.D., 2002, Baicalein and genistein display differential actions on estrogen receptor (ER) transactivation and apoptosis in MCF-7 cells, *Cancer Lett.* **187**:33–40.
Polkowski K., Skierski J.S., and Mazurek A.P., 2000, Anticancer activity of genistein-piperazine complex. In vitro study with HL-60 cells, *Acta Pol Pharm.* **57**:223–232.
Ponnathpur V., Ibrado A.M., Reed J.C., Ray S., Huang Y., Self S., Bullock G., Nawabi A., and Bhalla K., 1995, Effects of modulators of protein kinases on taxol-induced apoptosis of human leukemic cells possessing disparate levels of p26BCL-2 protein, *Clin Cancer Res.* **1**:1399–1406.
Portoukalian J., Zwingelstein G., Abdul-Malak N., and Dore J.F., 1978, Alteration of gangliosides in plasma and red cells of humans bearing melanoma tumors, *Biochem Biophys Res Commun.* **85**:916–920
Pumford S.L., Morton M., Turkes A, and Griffiths K., 2002, Determination of the isoflavonoids Genistein and daidzein in biological samples by gas chromatography-mass spectrometry, *Ann Clin Biochem.* **39**:281–292.
Ralph S.J., Wines B.D., Payne M.J., Grubb D., Hatzinisiriou I., Linnane A.W., and Devenish R.J., 1995, Resistance of melanoma cell lines to interferons correlates with reduction of IFN-induced tyrosine phorylation. Induction of the anti-viral state by IFN is prevented by tyrosine kinase inhibitors, *J Immunol.* **154**:2248–2256.
Rao A., Woodruff. R., Wade W.N., Kute T.E., and Cramer SD., 2002, Genistein and Vitamin D synergistically inhibit prostatic epithelial cell growth, *J Nutr.* **132**:3191–3194.
Rauth S., Kichina J., and Green A., 1997, Inhibition of growth and induction of differentiation of metastatic melanoma cell sin vitro by Genistein: chemosensitivity is regulated by cellular p53, *Br J Cancer.* **75**:1559–1566.
Ravindranath M.H., Morton D.L., 1991. Role of gangliosides in active immunotherapy with melanoma vaccine. *Int Rev Immunol.* **7**:303–329.
Ravindranath M.H., Morton D.L., Irie R.F., 1989. An epitope common to gangliosides O-acetyl-GD3 and GD3 recognized by antibodies in melanoma patients after active specific immunotherapy. *Cancer Res.* **49**:3891–3897.
Ravindranath M.H., Tsuchida T., Morton D.L., Irie R.F., 1991. Ganglioside GM3: GD3 ratio as an index for the management of melanoma. *Cancer.* **67**:3029–3035.
Rokhlin O.W., and Cohen M.B., 1995, Differential sensitivity of human prostatic cancer cell lines to the effects of protein kinase and phosphatase inhibitors., *Cancer Lett.* **98**:103–110.

Sakamoto K., 2000, Synergistic effects of thearubitin and genistein on human prostate tumor cell (PC-3) growth via cell cycle arrest, *Cancer Lett.* **151**:103–109.
Salti G.I., Grewai S., Mehta R.R., Das Gupta T.K., Boddie A.W., J.R. 2000, Genistein induces apoptosis and topoisomerase II-mediated DNA breakage in colon cancer cells, *Eur J Cancer.* **36**:796–802.
Santell R.C., Kieu N., and Helferich W.G., 2000, Genistein inhibits growth of estrogen-independent human breast cancer cells in culture but not in athymic mice, *J Nutr.* **130**:1665–1669.
Santibanez J.F., Navarro A., and Martinez J., 1997, Genistein inhibits proliferation and in vitro invasive potential of human prostatic cancer cell lines, *Aniticancer Res.* **17**:1199–1204.
Sasamura H., Takahasi A., Miyao N., Yanase M., Masumori N., Kitamura H., Itoh N., and Tsukamoto T, 2002, Inhibitory effect on expression of angiogenic factors by antiangiogenic agents in renal cell carcinoma, *Br J Cancer.* **86**:768-773.
Sathyamoorthy N., Gilsdorf J.S., and Wang T.T., 1998, Differential effect of genistein on transforming growth factor beta 1 expression in normal and malignant mammary epithelial cells, *Anticancer Res,* **18**:2449–2453.
Savickieni J., Gineitis A., Shanbhag V.P., and Stigbrand T., 1995, Protein kinase inhibitors exert stage specific and inducer dependent effects on HL-60 cell differentiation, *Anticancer Res.* **15**:687–692.
Scholar E.M., and Toews M.L., 1994. Inhibition of invasion of murine mammary carcinoma cells by the tyrosine kinase inhibitor genistein, *Cancer Lett.* **87**:159–162.
Setchell K.D., Zimmer-Nechemias L., Cai J., and Heubi J.E., 1998, Isoflavone content of infant formulas and the metabolic fate of these phytoestrogens in early life, *Am J Clin Nutr.* **68**:1453S–1461S.
Shao Z.M., Alpaugh M.L., Fontana J.A., and Barsky S.H., 1998a, Genistein inhibits proliferation similarly in estrogen receptor-positive and negative human breast carcinoma cell lines characterized by P21WAF1/CIP1 induction, G2/M arrest, and apoptosis, *J Cell Biochem.* **69**:44–54.
Shao Z.M., Shen Z.Z., Fontana J.A., and Barsky S.H., 2000, Genistein's "ER-dependent and independent" actions are mediated through ER pathways in ER-positive breast carcinoma cell lines, *Anticancer Res.* **20**: 2409–2416.
Shao Z.M., Wu J., Shen Z.Z., and Barsky S.H., 1998c, Genistein inhibits both constitutive and EGF-stimulated invasion in ER-negative human breast carcinoma cell lines, *Anticancer Res.* **18**:1435–1439.
Shao Z.M., Wu J., Shen Z.Z., and Barsky S.H., 1998b, Genistein exerts multiple suppressive effects on human breast carcinoma cells, *Cancer Res.* **58**:4851-4857.
Shelnutt S.R., Cimino C.O., Wiggins P.A., and Badger T.M., 2000, Urinary pharmacokinetics of the glucuronide and sulfate conjugates of genistein and daidzein, *Cancer Epidemiol Biomarkers Prev.* **9**:413–419.
Shen F., and Weber G., 1997, Synergistic action of quercetin and Genistein in human ovarian carcinoma cells, *Oncol. Res.* **9**:597–602.
Shen F., Xue X., and Weber G., 1999, Tamoxifen and genistein synergistically down-regulate signal transduction and proliferation in estrogen receptor-negative human breast carcinoma MDA-MB-435 cells, *Anticancer Res.* **19**:1657–1662.
Shen J.C., Klein R., Wei Q., Guan Y., Contois J.H., Wang T.T., Chang S., and Hursting S.D., 2000, Low-dose Genistein induces cyclin-dependent kinase inhibitors and G (1) cell-cycle arrest in human prostate cancer cells, *Mol Carcinog.* **29**:92–102.
Simeone D.M., Yule D.I., Logsdon C.D., ane Williams J.A., 1995, Ca2+ signaling through secretagogue and growth factor receptors on pancreatic AR42J cells, *Regul Pept.* **55**:197–206.
Sjoberg E.R., Chammas R., Ozawa H., Kawashima I., Kho K.H., Morris HR Dell., A., Tai T and Varki A.,1995, Expression of De-N-actetyl-gangliosides in Human melanoma Cells Is Induced by Genistein or Nocodazole, *J Bio Chem.* **270**:2921–2930.
So F.V., Guthrie N., Chambers A.F., and Carroll K.K., 1997, Inhibition of proliferation of estrogen receptor-positive MCF-7 human breast cancer cells by flavonoids in the presence and absence of excess estrogen, *Cancer Lett.* **112**:127–133.
Spinozzi F., Pagliacci M.C., Migliorati G., Moraca R., Grignani F., Riccardi C., and Nicoletti I., 1994, The natural tyrosine kianse inhibitor genistein produces cellcycle arrest and apoptosis in Jurkat T-leukemia cells, *Leuk Res.* **18**:431–439.
Steck T.R., and Drlica K., 1984, Bacterial chromosome segregation: evidence for DNA gyrase involvement in decatenation, *Cell.* **36**:1081–1088.
Sternglanz R., DiNardo S., Voelkel K.A., Nishimura Y., Hirota Y., Becherer K., Zumstein L., and Wang J.C., 1981. Mutations in the gene coding for Escherichia coli DNA topoisomerase I affect transcription and transposition, *Proc Natl Acad Sci USA.* **78**:2747–2751.

Strom S.S, Yamamura Y., Duphorne C.M., Spitz M.R., Babaian R.J., Pillow P.C., and Hursting S.D., 1999, Phytoestrogen intake and prostate cancer: a case-control study using a new database, *Nutr Cancer.* **33**:20–25.

Sun X.Y., Plouzek C., Henry J.P., Wang T.T., and Phang J.M., 1998, Increase d UDP-glucuronosyltransferase activity and decreased prostate specific antigen production by biochanin A in prostate cancer cells, *Cancer Res.* **58**:2379–2384.

Suzuki K., Koike H., Matsui H., Ono Y., Hasumi M., Nakazato H., Okugi H., Sekine Y., Oki K., Ito K., Yamamoto T., Fukabori Y., Kurokawa K., and Yamanaka H., 2002, Genistein, a soy isoflavone, induces glutathione peroxidase in the human prostate cancer cell lines LNCaP and PC-3, *Int J Cancer.* **99**:846–852.

Takeda Y., Nishio K., Niitani H., and Saijo N., 1994, Reversal of multidrug resistance by tyrosine-kinase inhibitors in a non-P-glycoprotein-mediated multidrug-resistant cell line, *Int J Cancer.* **57**:229–239.

Tanos V., Brzezinski A., Drize O., Strauss N., Peretz T., 2002. Synergistic inhibitory effects of genistein and tamoxifen on human dysplastic and malignant epithelial breast cells in vitro. *Eur J Obstet Gynecol Reprod Biol.* **102**:188–194.

Thompson D., Whicher J.T., and Evans S.W., 1993, Interleukin 6 signal transduction in a human hepatoma cell line (Hep G2), *Immunopharmacol Immunotoxicol.* **15**:371–386.

Tiisala S., Majuri M.L., Carpen O., Renkonen R., 1994, Genistein enhances the ICAM-mediated adhesion by inducing the expression of ICAM-1 and its counter-receptors, *Biochem Biophys Res Commun.* **203**:443–449.

Torti M., Marti K.B., Altschuler D., Yamamoto K., and Lapetina E.G., 1992, Erythropoietin induces p21ras activation and p120GAP tyrosine phosphorylation in human erythroleukemia cells, *J Biol Chem.* **267**:8293–8298.

Traganos F., Ardelt B., Halko N., Bruno S., and Darzynkiewicz Z., 1992, Effects of genistein on the growth and cell cycle progression of normal human lymphocytes and human leukemic MOLT-4 and HL-60 cells. *Cancer Res.* **52**:6200–6208.

Tsujishita Y., Asaoka Y., and Nishizuka Y., 1994, Regulation of phospholipase A2 in human leukemia cell lines: its implication for intracellular signaling, *Proc Natl Acad Sci USA.* **91**:6274–6278.

Uckun F.M., Evans W.E., Forsyth C.J., Waddick K.G., Ahlgren L.T., Chelstrom L.M., Burkhardt A., Bolen J., Myers D.E., 1995, Biotherapy of B-cell precursor leukemia by targeting genistein to CD19-associated tyrosine kinases, *Science.* **267**:886–891.

Uckun F.M., Messinger Y., Chen C.L., O'Neill K., Myers D.E., Goldman F., Hurvitz C., Casper J.T., Levine A., 1999, Treatment of therapy-refractory B-lineage acute lymphoblastic leukemia with an apoptosis-inducing CD19-directed tyrosine kinase inhibitor. *Clin Cancer Res.* **5**:3906–3913.

Uckun F.M., Narla R.K., Jun X., Zeren T., Venkatatchalam T., Waddick K.G., Rostostev A., and Myers D.E., 1998, Cytotoxic activity of epidermal growth factor-genistein against breast cancer cells, *Clin Cancer Res.* **4**:901–912.

Upadhyay S., Neburi M., Chinni S.R., Alhasan S., Miller F., and Sarkar F.H., 2001, Differential sensitivity of normal and malignant breast epithelial cells to genistein is partly mediated by p21 (WAF1), *Clin Cancer Res.* **7**:1782–1789.

van Rijn J., van den Berg J.,1997, Flavonoids as enhancers of x-ray-induced cell damage in hepatoma cells., *Clin Cancer Res.* **3**:1775–1779.

Veldman C.M., Schmid C., 1998, Differential effects of fluroide and insulin-like growth factor I on sodium-dependent alanine and phosphate transport in a human osteoblast-like cell line, *Growth Horm IGF Res.* **8**:55–63.

Venkatakrishnan G., S.R., and Groopman J.E., 2000, Chemokine receptors CXCR-1/2 activate mitogen-activated protein kinase via the epidermal growth factor receptor in ovarian cancer cells, *J Biol Chem.* **275**:6868–6875.

Verma S.P., Salamone E., and Goldin B., 1997, Curcumin and genistein, plant natural products, show synergistic inhibitory effects on the growth of human breast cancer MCF-7 cells induced by estrogenic pesticides, *Biochem Biophys Res Commun.* **233**:692–696.

Versantvoort C.H., Rodes T., and Twentyman P.R., 1996, Acceleration of MRP-associated efflux of rhodamine 123 by genistein and related compounds, *Br J Cancer.* **74**:1949–1954.

Wang C., and Kurzer M.S., 1998, Effects of phytoestrogens on DNA synthesis in MCF-7 cells in the presence of estradiol or growth factors, *Nutr Cancer.* **31**:90–100.

Wang T.T., Sathyamoorthy N., Phang J.M., 1996. Molecular effects of genistein on estrogen receptor mediated pathways. *Carcinogenesis.* **17**:272–275.

Watanabe S., Yamaguchi M., Sobue T., Takahashi T., Miura T., Arai Y., Mazur W., Wahala K., and Adlercreutz H., 1998. Pharmacokinetics of soybean isoflavones in plasma, urine and feces of men after ingestion of 60 g baked soybean powder (kinako), *J Nutr.* **128**:1710–1715.

Weber G., Shen D.F., Li W., Yang H., Look K.Y., Abonyi M., and Prajda N., 2000, Signal transduction and biochemical targeting of ovarian carcinoma, *Eur J Gynaecol Oncol.* **21**:231–236.

Wegenka U.M., Lutticken C., Buschmann J., Yuan J., Lottspeich F., Muller-Esterl W., Schindler C., Roeb E., Heinrich P.C., and Horn F., 1994, The interleukin-6-activated acute-phase response factor is antigenically and functionally related to members of the signal transducer and activator of transcription (STAT) family, *Mol Cell Biol.* **14**:3186–3196.

Willard S.T., and Frawley L.S., 1998, Phytoestrogens have agonistic and combinatorial effects on estrogen-responsive gene expression in MCF-7 human breast cancer cells, *Endocrine.* **8**:117–121.

Xiang H., Schevzov G., Gunning P., Williams HM., and Silink M., 2002, A comparative study of growth-inhibitory effects of isoflavones and their metabolites on human breast and prostate cancer cell lines, *Nutr Cancer.* **42**:224–232.

Xu J., and Loo G., 2001, Different effects of genistein on molecular markers related to apoptosis in two pheno-typically dissimilar breast cancer cell lines, *J Cell Biochem.* **82**:78–88.

Yabunaka N., 1995, Studies on the regulation of glycolipid sulfotransferase activity in human renal cancer cells, *Hokkaido Igaku Zasshi.* **70**:289–300.

Yabunaka N., Honke K., Ishii A., Ogiso Y., Kuzumaki N., Agishi Y., and Makita A., 1997, Involvement of Ras in the expression of glycolipid sulfotransferase in human renal cancer cells, *Int J Cancer.* **71**:620–630.

Yamada A., Hara A., Inoue M., Kamizono S., Higuchi T., and Itoh K., 1997, Beta 2-integrin-mediated signal up-regulates counterecptor ICAM-1 expression on human monocytic cell line THP-1 through tyrosine phosphorylation, *Cell Immunol.* **178**:9–16.

Yamamoto S., Sobue T., Sasaki S., Kobayashi M., Arai Y., Uehara M., Adlercreutz H., Watanabe S., Takahashi T., Litoi Y., Iwase Y., Akabane M., and Tsugane S., 2001, Validity and reproducibility of a self-administered food-frequency questionnaire to assess isoflavone intake in a Japanese population in comparison with dietary records and blood and urine isoflavones, *J Nutr.* **131**:2741–2747.

Yamane M., and Abe A., 2000, Omega-hydroxylation activity toward leukotriene B (4) and polyunsaturated fatty acids in the human hepatoblastoma cell line, HepG2, and human lung adenocarcinoma cell line, A549, *J Biochem.* **128**:827–835.

Yamashita K., Suzuki M., Iwata H., Koike T., Hamaguchi M., Shinagawa A., Noguchi T., and Hayakawa K., 1996, Tyrosine phosphorylation is crucial for growth signaling by tissue inhibitors of metalloproteinases (TIMP-1 and TIMP-2), *FEBS Lett.* **396**:103–107.

Yan C., and Han R., 1999, Protein tyrosine kinase inhibitor genistein suppresses in vitro invasion of HT1080 human fibrosarcoma cells, *Zhonghua Zhong Liu Za Zhi.* **21**:171–174.

Yanagihara K., Ito A., Toge T., and Numoto M., 1993, Antiproliferative effects of isoflavones on human cancer cell lines established from the gstrointestinal tract, *Cancer Res.* **53**:5815–5821.

Yang C.C., Hsu C., Sheu J.C, Mai X.Y., and Yang S.D., 1998, Differential tyrosine phosphorylation/activation of oncogenic proline-directed protein kinase F (A)/GSK-3alpha in well and poorly differentiated human prostate carcinoma cells, *J Protein Chem.* **17**:329–335.

Yang E.B., Wang. D.F., Mack P., and Cheng L.Y., 1996, Genistein, a tyrosine kinase inhibitorm reduces EGF-induced EGF receptor internalization and degradation in human hepatoma HepG2 cells, *Biochem Biophs Res Commun.* **224**:309–317.

Yu J., Cheng Y., Xie L., and Zhang R., 1999, Effects of genistein and daidzein on membrane characteristics of HCT cells, *Nutr Cancer.* **33**:100–104.

Yu L., Blackburn G.L., and Zhou JR., 2003, Genistein and daidzein downregulate prostate androgen-regulated transcript-1(PART-1) gene expression induced by dihydrotestosterone in human prostate LNCaP cancer cells, *J Genistein Prostate Cancer.* **133**:389–392.

Yun J., Lee C.K., Chang I.M., Takatsu K., Hirano T., Min K.R., Lee M.K., and Kim Y., 2000, Differential inhibitory effects of sophoricoside analogs on bioactivity of several cytokines, *Life Sci.* **67**:2855–2863.

Zheng W., Dai Q., Custer L.J., Shu X.O., Wen W.Q., Jin F., Franke A.A., 1999, Urinary excretion of isoflavonoids and the risk of breast cancer, *Cancer Epidemiol Biomarkers Prev.* **8**:35–40.

Zhu Q., Meisinger J., Van Thiel D.H., Zhang Y., and Mobarhan S., 2002, Effects of soybean extract on morphology and survival of Caco-2, SW620, and HT-29 cells, *Nutr Cancer.* **42**:131–140.

# 10

# Chinese Medicine and Immunity

Haruhisa Wago and Hong Deng

## 1. Introduction

Traditional Chinese Medicine (TCM) has so far been known to be useful for cancer patients as both an anti-tumor and an immuno-enhancing drug[1,2]. If some kinds of crude drugs of TCM possess such biological activities, they would be of great help in cancer treatment. Recently, it was found that *Astragali radix* has immuno-enhancing and anti-tumor effects[3-5] and *Angelicae radix* also has anti-inflammatory, analgestic, interferon-inducing, and immunopotentiating effects[6-9]. In addition, *Zizyphi fructus* is reported to augment the natural killer cell activity[10], while *Cervi parvum cornu* is shown to exert anti-tumor effects by the inhibition of monoamine oxidase[11-13]. *Amomi semen* and *Rehmanniaee radix* are also known to have anti-tumor activity and to enhance the function of cytotoxic T lymphocyte with the increase of IL-2[14,15].

In this investigation, a mixture of these six kinds of crude drugs used for cancer patients in China was named Ekki-Youketsu-Fusei-Zai (EYFZ), and its biological activities including both immuno-enhancing and anti-tumor effects were examined with much attention to the *in vitro* functional activation of macrophages and T lymphocytes and also to the *in vivo* effects on natural killer (NK) cell activity and life-prolongation using tumor-bearing mice in order to elucidate the relationship of immuno-enhancement and anti-tumor influences from the viewpoints of alternative and complementary medicine[16-18].

## 2. Preparation of EYFZ

As mentioned before, EYFZ is a mixture of six kinds of crude drugs. All of these crude drugs were purchased from Tochimoto (Osaka, Japan). First, following the popular method for producing TCM in China, a mixture of *Astragali radix* (10.0 g), *Z. fructus* (5.0 g), *A. semen* (5.0 g), *Angelica radix* (8.0 g), *C. parvum cornu* (5.0 g) and *R. radix* (10.0 g) was added to 200 ml distilled water (DW), soaked for 20 min at room temperature, and then boiled for 30 min to extract the effective substances. Finally, the extracted solution

---

**Haruhisa Wago** • Department of Medical Technology, Saitama Medical School Junior College, Saitama 350-0495, Japan. Hong Deng, Electro-Chemical & Cancer Institute, Chofu, Tokyo 182-0022, Japan

*Complementary and Alternative Approaches to Biomedicine*, edited by Edwin L. Cooper and Nobuo Yamaguchi. Kluwer Academic/Plenum Publishers, 2004.

was centrifuged at 3000 rpm for 20 min, and the supernatant was further condensed to 43 ml (1 g crude drug/ml) by evaporation and diluted with DW to a desired concentration.

## 3. Activation of Murine Macrophages by EYFZ

We first examined the effect of EYFZ on the morphology and cytokine production of murine macrophage to elucidate the efficacy of EYFZ in the functional activation of immunocytes.

### 3.1. Macrophage Cell Line J774.1

Macrophages are known to be activated in a nonspecific manner and to be able to secrete various kinds of cytokines such as IL-1 $\alpha/\beta$, IL-6, IL-12, TNF-$\alpha$, IFN-$\alpha/\beta$ etc. for interaction with other immunocytes. They also may produce chemicals such as NO, $O_2^-$, prostaglandins etc., to kill the pathogens[19]. Additionally, macrophages are reported to be activated *in vitro* by a variety of agents, leading to an increase of metabolism, release of lysosomal enzyme, phagocytosis, and anti-tumor activity. These agents are generally immunopotentiators, some of which are called biological response modifiers (BRMs). Among BRMs, bacterial lipopolysaccharide (LPS) is one of the strongest activators of macrophages[20,21]. The, murine macrophage cell line J774.1 has been used in many investigations to verify the mechanisms underlying the LPS action on macrophages[22-26].

J774.1 cell line originates from BALB/c mice and can be used for the study of macrophage functions *in vitro* and has been reported to show the same cytokine production as the macrophages[27-29]. Thus, here in our experiment, we examined the effects of EYFZ on the morphological changes and cytokine production of J774.1 cells *in vitro* to elucidate the EYFZ action on the immune responses.

### 3.2. Morphological Changes of J774.1 Cells by EYFZ

J774.1 cells were suspended in culture medium at a concentration of $5 \times 10^4$ cells/ml, and 5 ml of cell suspension was plated in a 25 cm$^2$ plastic culture flask. The cells were incubated with 1.2 mg crude drug/ml of EYFZ, and the cell morphology was observed under a phase-contrast microscope and photographed to learn the morphological changes *in vitro*, implying the macrophage activation.

After incubation with EYFZ for 3, 6 and 12 hrs, the effect of EYFZ on the morphology was observed. Results showed that EYFZ at 1.2 mg crude drug/ml caused great changes in both size and shape. Particularly, J774.1 cells in the presence of EYFZ spread and elongated with numerous vacuoles and granules inside, although those without EYFZ mostly remained round. For example, the number of spread type cells with EYFZ after 12 hr-incubation significantly increased as compared with that of the control without EYFZ. This change corresponded to that with positive controls using LPS. Similarly, over 70% of J774.1 cells with EYFZ or LPS had a diameter over 21 µm, whereas about 60% of cells in the controls still had a diameter less than 21 µm (Fig. 10.1). These results showed that J774.1 cells changed greatly in shape and size in response to EYFZ, strongly suggesting the induction of functional activation of J774.1 macrophage cell line by EYFZ.

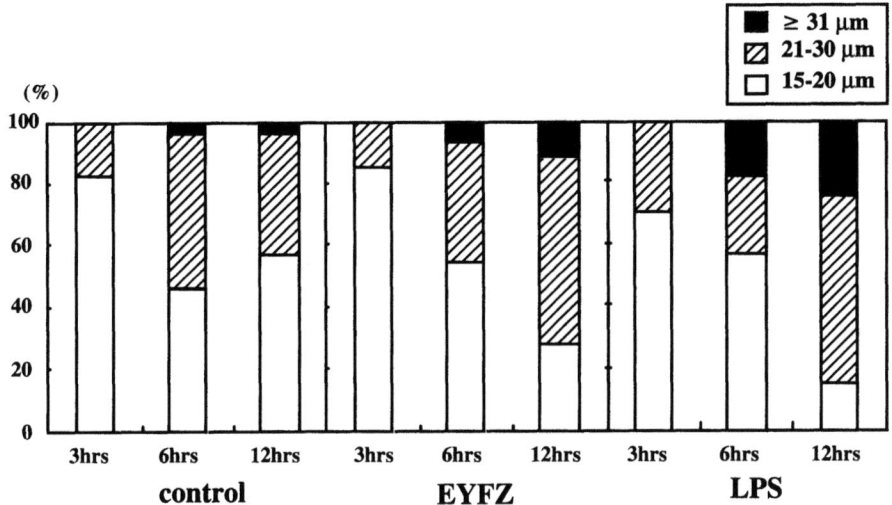

**Figure 10.1.** The time course of size change after EYFZ stimulation.
Effects of EYFZ and LPS (culture medium alone) on the size of J774.1 incubation for 3, 6, and 12 hrs. Open bars show the % of small cells (diameter was below 20 μm), hatched bars and solid bars show the % of large cells (diameters were over 21 μm to 30 μm and over 31 μm, respectively). (From Nakajima et al., J. Saitama Med. School, 28 : 117, 2001, with permission.)

### 3.3. Induction of Expression of Cytokine mRNAs by EYFZ

Macrophages are involved in almost all the steps of the immune responses in vertebrates and play a key role in the initial response of immunity before T and B cells are mobilized. Mechanisms by which macrophages act as effector cells in a defense system include the secretion of extracellular cytokines such as IL-1, IL-12, and TNF-α. Thus, we investigated whether or not EYFZ could stimulate the expression of these cytokine mRNAs from J774.1 using RT-PCR method.

J774.1 cells were cultured for 24 and 48 hrs in the presence of 1.2 mg crude drug/ml of EYFZ. After incubation, cells were harvested by scraping, and quickly frozen and stored at −80° C. Total cellular RNAs were extracted from $1 \times 10^6$ J774.1 cells for RT-PCR, and the cytokine mRNA expression of J774.1 cells treated with or without EYFZ was investigated. The gel electrophoretic patterns of RT-PCR products showed that the expression level of IL-12p35 and IL-12p40 was induced in J774.1 cells after 12 and 24 hrs treatment with EYFZ, though expressions of β-actin, a housekeeping gene, was almost constant (Fig. 10.2). There were no differences in other cytokine expression between treated and control cells.

### 3.4. Stimulation of Production of IL-12 by EYFZ

As EYFZ induced the expression of both IL-12p35 and IL-12p40 mRNAs, the IL-12 production by EYFZ was examined by ELISA method, because IL-12 is a heterodimetric cytokine composed of disulfide linked p35 and p40 subunits[30]. After J774.1 cells were incubated with various kinds of concentrations of EYFZ, IL-12 in the culture supernatants

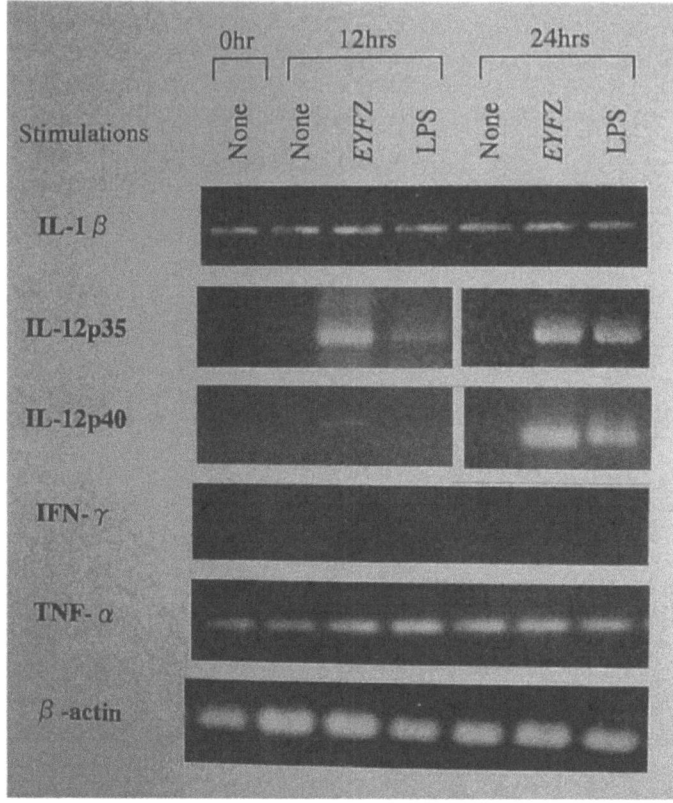

**Figure 10.2. The effect of EYFZ on the expression of cytokine mRNAs.**
J774.1 cells stimulated with EYFZ or LPS in culture were assayed for IL-1$\beta$, IL-12p35, IL-12p40, IFN-$\gamma$, TNF-$\alpha$, and $\beta$-actin mRNA by RT-PCR. The results are for a typical example of repeated experiments. (From Nakajima et al., J. Saitama Med. School, 28 : 117, 2001, with permission.)

were assayed by ELISA. Results clearly showed that IL-12 concentrations secreted from J774.1 cells in the presence of EYFZ were significantly higher than those of the control without EYFZ (Fig. 10.3).

## 3.5. Efficacy of EYFZ in Activation of Macrophages

The results showing that EYFZ induced morphological changes and IL-12 production of J774.1 cells strongly suggest that this Chinese medicine will be effective in the functional activation of murine macrophages. It is of great interest that EYFZ is capable of only IL-12 production. IL-12 has recently been brought into focus as an effector molecule to enhance the immunity by macrophages[31-33]. IL-12 exerts multiple effects on T and NK cells including the augmentation of IFN-$\gamma$ production, proliferation, and cytotoxic activity, and also plays an essential role in determining a Th1/Th2 balance[34-37]. In addition, IL-12 has been demonstrated to have anti-tumor effects[35], and is much more effective than other

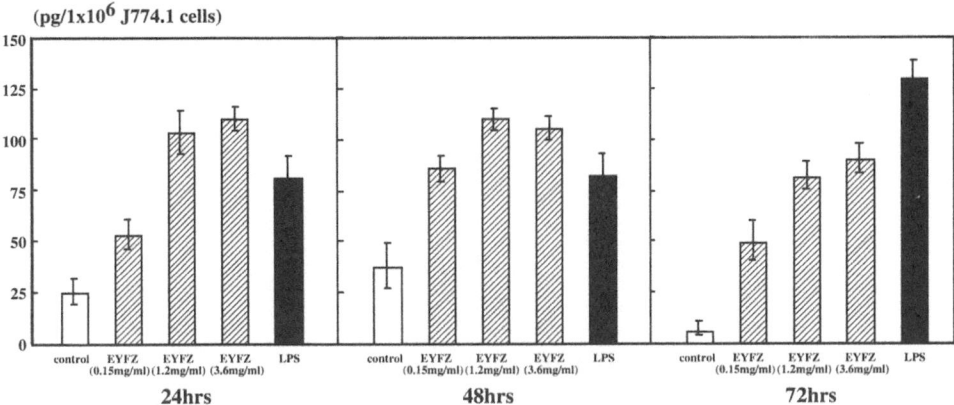

**Figure 10.3.** The secretion of IL-12 from J774.1 stimulated with EYFZ. The secretion of IL-12 was examined after incubation for 24, 48, and 72 hrs. J774.1 cells ($1 \times 10^4$ cells/ml) were cultured in the presence of EYFZ (0.15 mg/ml, 1.2 mg/ml and 3.6 mg crude drug/ml) (hatched bars), LPS (10 ng/ml) (solid bars), or control (culture medium only). Values represent the mean ±S.D. of four independent experiments. From Nakajima et al., J. Saitama Med. School, 28 : 117, 2001, with permission

cytokines. In this regard, EYFZ is considered to be an interesting Chinese medicine for its functional activation of macrophages.

## 4. Stimulation and Activation of Lymphocytes by EYFZ

We second examined the effects of EYFZ on human peripheral blood lymphocytes (PBL) to elucidate whether or not EYFZ could stimulate them to proliferate without mitogen, to express cytokine mRNA such as IFN-$\gamma$, IL-2, and IL-4 and to produce them *in vitro*.

### 4.1. Proliferation of PBL by EYFZ

PBL was obtained as follows: Human peripheral blood was drawn from healthy adult donors into a small tube with heparin, and diluted 1:2 with PBS. Cell suspensions were layered on a Ficoll Isopaque and centrifuged at 2000 rpm for 30 min, and then mononuclear cells were collected from the interface and washed three times with PBS. Cell concentration was adjusted to $1 \times 10^6$ cells/ml, and cells were suspended in RPMI 1640 medium. They were incubated for 2 hr at 37° C in a 5% $CO_2$ incubator to allow adherence. Then, nonadherent lymphocytes were gently collected, and finally PBL at a cell concentration of $5 \times 10^5$ was incubated with EYFZ for 72 hrs at 37° C in a 5% $CO_2$ incubator. The proliferation of PBL was examined by MTT assay using spectrophotometric analysis by an absorbance at 570 nm.

The result clearly showed that EYFZ directly stimulated PBL to proliferate without mitogen, and this proliferative effect was almost the same as that of the positive controls using ConA (Fig. 10.4). Since ConA has been used as the specific mitogen for T lymphocytes[38,39], EYFZ is suggested to stimulate peripheral T lymphocyte and exert a proliferative effect on them.

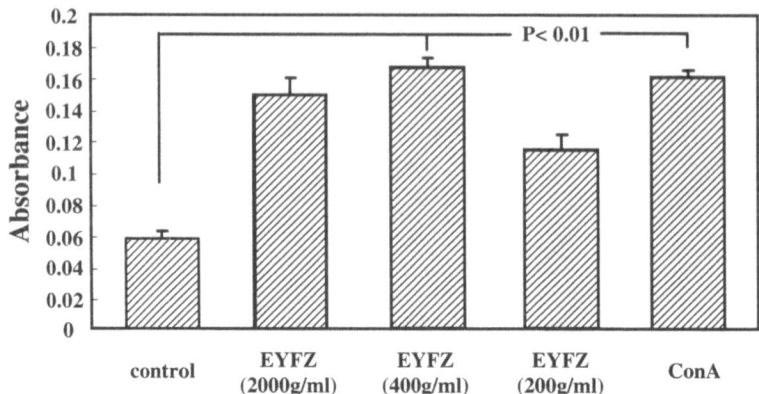

**Figure 10.4.** Effect of Ekki-Youketsu-Fusei-Zai (EYFZ) on proliferation of human peripheral blood lymphocytes (PBL) *in vitro*.
Human PBL ($5 \times 10^5$ cells/ml) were cultured with EYFZ (200 μg/ml, 400 μg/ml, 200 μg/ml), ConA (5 μg/ml) or culture medium alone in 96 well flat-bottomed plates. After 72 hr incubation, the proliferation of lymphocytes was measured by MTT assay. Each column and vertical bar represents the mean ±standard error of triplicates. (P 0.01 by Student's two tailed test) (control: culture medium without EYFZ or ConA) (From Deng et al., J. Saitama Med. School, 28 : 125, 2001, with permission)

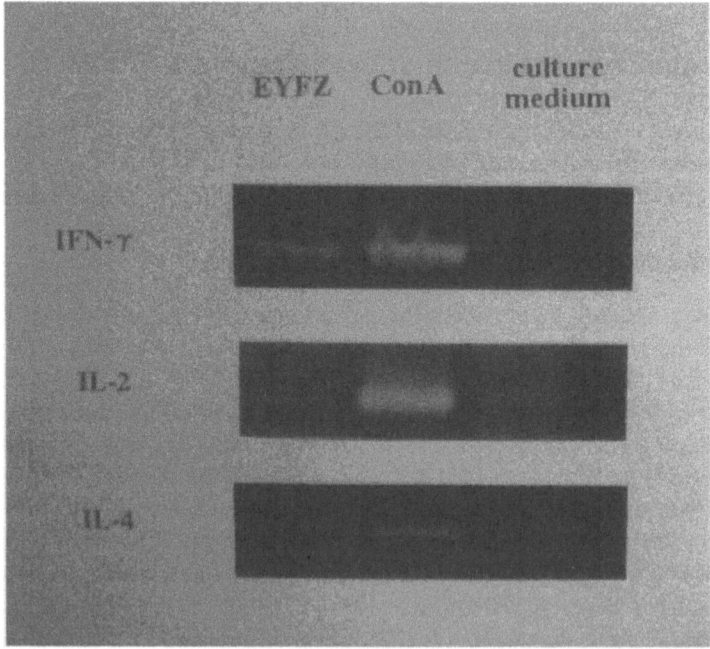

**Figure 10.5.** Effect of Ekki-Youketsu-Fusei-Zai (EYFZ) on expression of IFN-$\gamma$, IL-2 and IL-4 mRNA.
Human PBL ($5 \times 10^5$ cells/ml) were cultured with EYFZ (400 μg/ml), ConA (5 μg/ml) or culture medium alone in 25 cm$^2$ culture flask. After 72 hr incubation, RT-PCR analysis was performed to determine the expression of INF-$\gamma$, IL-2 and IL-4 mRNA. (From Deng et al., J. Saitama Med. School, 28 : 125, 2001, with permission)

## 4.2. Induction of Expression of Cytokine mRNA of T lymphocytes by EYFZ

Human PBLs were cultured with EYFZ for 72 hrs at 37° C in a 5% $CO_2$ incubator. After incubation, cells were used for the preparation of RNA. RT-PCR was performed to analyse the expression of cytokine mRNA by T lymphocytes. The result showed that EYFZ induced expression of IFN-$\gamma$ mRNA as compared with culture medium alone, but not expression of IL-2 and IL-4 (Fig. 10.5). ConA used as a positive control induced significant expression of IFN-$\gamma$, IL-2 and IL-4 mRNA more than in EYFZ-addition or in culture medium alone.

## 4.3. Stimulation of Production of IFN-$\gamma$ by EYFZ

Human PBLs were cultured for the analysis of production of IFN-$\gamma$, IL-2 and IL-4. After 72 hr incubation, the culture supernatant was collected and stored at $-80°$ C until use. Here, their cytokines were measured by ELISA to verify the effect of EYFZ on the cytokine production of PBLs. The result showed that EYFZ promoted significantly only the production of IFN-$\gamma$ in culture supernatant of PBLs as compared with that in the culture supernatant without EYFZ or ConA (Fig. 10.6). ConA used as a positive control significantly promoted the production of IFN-$\gamma$, IL-2 and IL-4 higher than in EYFZ-addition or in culture medium alone.

## 4.4. Efficacy of EYFZ in Stimulation of Human T Lymphocytes

Our results clearly demonstrated that EYFZ could induce only the expression of IFN-$\gamma$ mRNA and the production of IFN-$\gamma$ *in vitro*. The expression and production of IL-2 and IL-4 were not observed in our experiment. As has been reported, helper T cells are classified into two subsets (Th1 and Th2) on the basis of cytokines synthesized by T cells. Th1 cells secreting IL-2 and/or IFN-$\gamma$ causes cellular immunity, whereas Th2 cells secreting IL-4, IL-5, IL-6 and/or IL-10 are involved in humoral immunity[40]. In this regard, EYFZ is considered to participate in the enhancement of cellular immune response. A preliminary experiment using anti-CD4± and anti-CD8± monoclonal antibodies showed that CD4+ and CD8+ T cell subsets were at least associated with the IFN-$\gamma$ production in the presence of EYFZ (data not shown). Thus, EYFZ has a possibilty of possessing some effective components to induce only IFN-$\gamma$ from those T cells.

## 5. Anti-Tumor Activity of EYFZ and Its Effect on NK Cells *in Vivo*

Results of our *in vitro* experiments strongly showed that EYFZ induced the functional activation of macrophages and stimulated T lymphocytes to proliferate and activate. Then, we further examined anti-tumor effects of EYFZ on survival, tumor size, body weight, and natural killer (NK) cell activity in tumor-bearing mice by using a tumor cell line, colon-26.

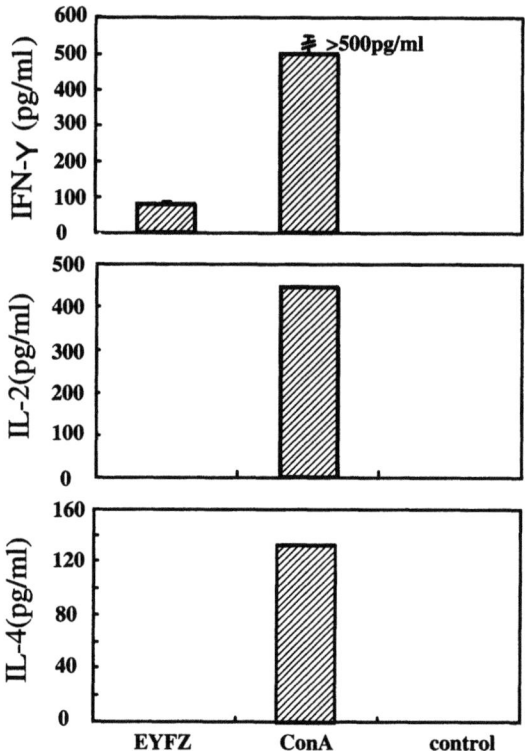

**Figure 10.6.** Effect of Ekki-Youketsu-Fusei-Zai (EYFZ) on production of IFN-$\gamma$, IL-2 and IL-4 in culture supernatant.
Human PBL were cultured under the same conditions as RT-PCR for IFN-$\gamma$, IL-2 and IL-4 mRNA expression. After 72 hr incubation, ELISA analysis was used to determine production of IFN-$\gamma$, IL-2 and IL-4. Values are mean ± standard error of triplicates and represent one of three independent experiments. ($P < 0.01$ by Student's two-tailed test) (From Deng et al., J. Saitama Med. School, 28 : 125, 2001, with permission)

## 5.1. The Life-Prolonging Effect of EYFZ

### 5.1.1. Tumor Cell Line Used in This Study and Its Implantation

Colon 26 cell line, colon-adenocarcinoma from BALB/c mice, was cultured *in vitro* with the basal medium RPMI 1640. This cell line has been successfully used as a model of tumor-bearing mice and cachexia[41,42]. A single cell suspension at a cell concentration of $5 \times 10^5$ was subcutaneously implanted into the back of a 6-week-old mouse, simultaneously starting the oral administration of EYFZ.

### 5.1.2. Evaluation of Survival, Tumor Size, and Body Weight of Tumor-Bearing Mice

The effect of oral administration of EYFZ was examined on the suvival of colon-26 implanted mice. When EYFZ was continuously administrated to the mice which had been implanted subcutaneously with colon 26 for 28 days, the life-prolonging effect was found.

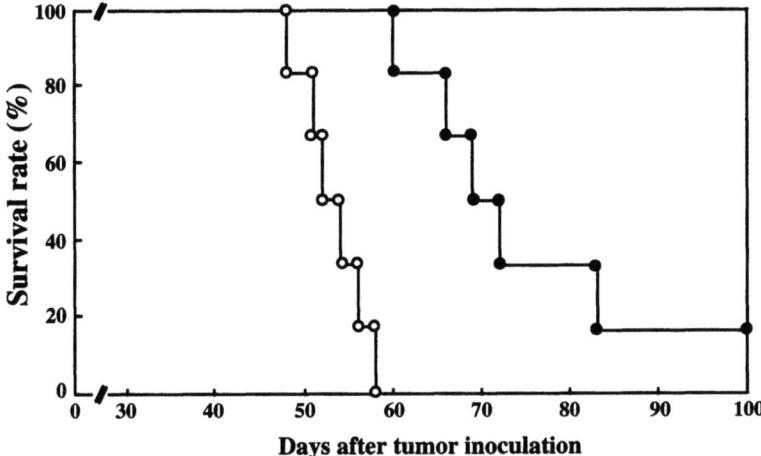

**Figure 10.7.** Effect of Ekki-Youketsu-Fusei-Zai (EYFZ) on survival of mice inoculated with colon-26 cells. Six BALB/c female mice per group were inoculated with colon-26 ($5 \times 10^5$ cells/mouse). Treatment group mice were orally administered with EYFZ (716.7 mg crude drug/kg) for 28 days just after the subcutaneous implantation of colon-26 cell line. The control group mice received only saline. Their survival rates are shown. (○): treatment group mice, (●): control group mice. $P < 0.01$; by log rank test. (From Deng et al., J.Saitama Med. School, 28 : 109, 2001, with permission)

All of the tumor-bearing mice in the control group died within 59 days, although the tumor-bearing mice treated with EYFZ died in 70 days on average, meaning a significant life-prolonging effect (Fig. 10.7).

On the other hand, results from observation for 34 days revealed that the tumor size in EYFZ—administrated mice was smaller than that in the control mice (Fig. 10.8). We further examined the effect of oral administration of EYFZ on the body weight in tumor-bearing

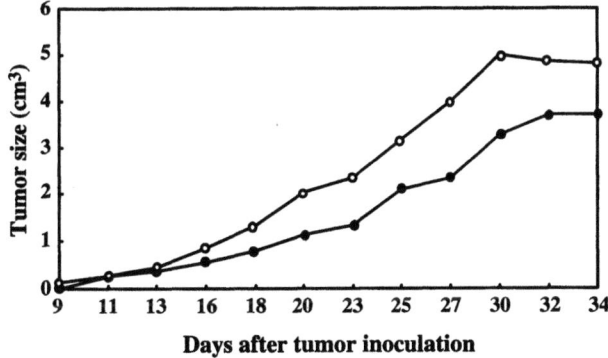

**Figure 10.8.** Effect of Ekki-Youketsu-Fusei-Zai (EYFZ) on the in vivo growth of colon-26 tumor cells. Mice treated with EYFZ (716.7 mg crude drug/kg) and control mice without EYFZ were inoculated with colon-26 cells according to the same protocol as in Fig. 10.7. Tumor group was measured after implantation and calculated as shown in Material and Methods. The tumor size was measured from day-9 to day-34 after transplantation. (μ): treatment group mice, (○): control group mice. (From Deng et al., J.Saitama Med. School, 28 : 109, 2001, with permission)

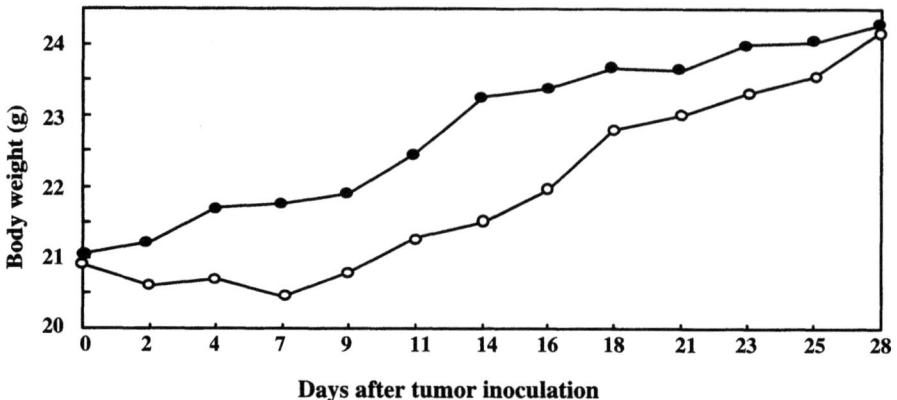

**Figure 10.9.** Effect of Ekki-Youketsu-Fusei-Zai (EYFZ) on body weight loss of mice inoculated with colon-26 cells.
Mice treated with EYFZ (716.7 mg crude drug/kg) and control mice without EYFZ were inoculated with colon-26 cells according to the same protocol as in Fig. 10.7. After implantation, the body weight was measured 2 or 3 times a week and finally on day-28. (●): treatment group mice, (○): control group mice. (From Deng et al., J.Saitama Med. School, 28 : 109, 2001, with permission)

mice. As a result, it was shown that the body weight of the EYFZ-treated mice was much greater than that of the control (Fig. 10.9).

## 5.2. Activation of NK Cells

### 5.2.1. Evaluation of NK Cell Activity

The spleen cell suspension was prepared by squeezing the spleen between two glass slides. Red blood cells were lysed and removed by osmotic treatment. After washing three times with serum-free basal medium, the cells were incubated at 37° C in a 5% $CO_2$ incubator for 2 hrs to remove the adherent cells. The non-adherent cells were collected as effector cells. Then, these effector cells ($5 \times 10^6$ cells/ml) were incubated with NK-sensitive target cells YAK-1 ($5 \times 10^4$ cells/ml), and incubated for 4 hrs at 37° C in a 5% $CO_2$ incubator. An effector—target ratio of 100:1 was considered optimum. After incubation, the supernatant was used for lactate dehydrogenase (LDH) assay to determine the cytotoxic activity of NK cells.

### 5.2.2. Augmentation of NK Cell Cytotoxicity by EYFZ

The effect of oral administration of EYFZ was examined on splenic NK cell cytotoxicity in tumor-bearing mice. Results showed that NK cell cytotoxicity of splenic cells in orally EYFZ-administrated mice was significantly higher than that in the control 17 and 24 days after the onset of transplantation and oral administration (Fig. 10.10).

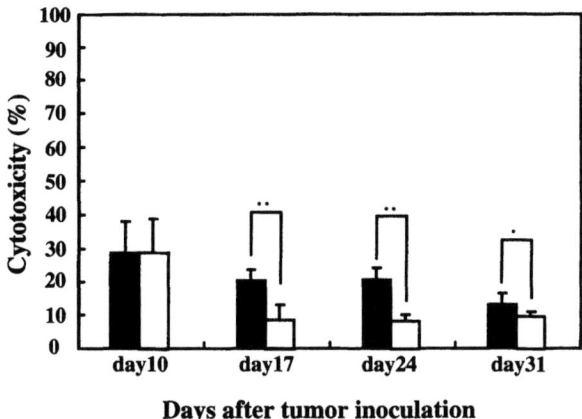

**Figure 10.10.** Effect of Ekki-Youketsu-Fusei-Zai (EYFZ) on NK cell activity of mice inoculated with colon-26 cells.
Five BALB/c female mice per group were inoculated s.c with colon 26 cells ($5 \times 10^5$ cells/mouse). Treatment mice were orally administered with EYFZ (716.7 mg crude drug/kg) for 28 days just after the subcutaneous implantation of colon-26 cell line. Control group mice received only saline. Each column and vertical bar represents the mean± standard deviation of 5 mice on day-10, day-17, day-24 and day-31 after inoculation. (■): treatment group mice, (□): control group mice. *P™Student's two-tailed test. (From Deng et al., J. Saitama Med. School, 28 : 109, 2001, with permission)

## 6. Discussion

Traditional Chinese Medicine (TCM), one of BRMs, is an defined crude extract from a mixture of many plants, animals and minerals, and the prescription is defined for each TCM. As mentioned before, there have been many investigations reporting the biological actions of TCMs on immune responses in mammals[43-45]. Our EYFZ, a TCM, was a mixure composed of extracts of six kinds of crude drugs, although its immunological activity has not yet been demonstrated.

From our studies, it was shown that EYFZ 1) induced the functional activation of macrophage cell line J774.1 and promoted the secretion of IL-12 from this cell line, 2) had some significant stimulating effects on proliferation of human lymphocytes and induction of IFN-$\gamma$ by T cell subsets, 3) had a life-prolonging effect on tumor-bearing mice and an inhibitory influence on tumor size, and 4) augmented the NK cytotoxic activity of splenic cells in tumor-bearing mice. Therefore, these results strongly suggest that this EYFZ is of great help for cancer patients as an anti-tumor drug from the viewpoint of alternative and complementary medicine.

It was noted that EYFZ was able to induce IL-12 production by J774.1, because IL-12 has recently been brought into focus as an effector molecule to enhance immunity by murine macrophages[31-33]. IL-12 also exerts multiple effects on T and NK cells including the augmentation of IFN- production, proliferation and cytotoxic activity. This phenomenon was actually demonstrated in the present study: EYFZ induced only the expression of IFN-mRNA and production of IFN-in culture supernatant. Thus, EYFZ is suggested to induce a strong Th1 response, leading to an increase of cellular immunity. NK cell activity in tumor-bearing mice was augmented by oral administration of EYFZ, implying the absorption of some effective components enhancing the splenic NK cell function from the digestive organs.

NK cells exhibit spontaneous cytotoxic activity in a non-major histocompatibility complex (MHC) restricted manner against virus-infected or cancer cells *in vivo* and their activity can be augmented by IFN-$\gamma$[46-48]. Therefore, EYFZ seems to have an anti-tumor effect on colon-26 implanted mice via augmentation of NK cell activity, which is also induced by IFN-$\gamma$ produced by T lymphocytes. EYFZ does not show direct cytotoxicity on some tumor cell lines including colon-26 and A549 (data not shown). Thus, it is suggested that the efficacy of EYFZ on the anti-tumor state is attributable to the enhancement of the host defense or immune system where macrophages, T lymphocytes and NK cells play an essential role.

## 7. Acknowledgments

We thank Dr. Kenichiro Hasumi of Electro-Chemical and Cancer Research Institute for his kind help and valuable suggestions in this investigation.

## 8. References

1. I. Adachi and T. Watanabe, *Gan To Kagaku.* **16**, 1538 (1989).
2. Y.S. Lee, I.S. Chung, I.R. Lee, K.H. Kin, W.S. Hong an Y.S. Yun, *Anticancer Res.* **17**, 323 (1997).
3. K.S. Zhao, C. Mancini and G. Doria, *Immunophamacol.* **20**, 225 (1990).
4. Y. Yoshida, M.Q. Wang, J.N. Liu, B.E. Shan and U. Yamashita, *Int. J. Immunophamacol.* **19**, 359 (1997).
5. B.H. Lau, H.C. Botolazzo and P.D. Lui, *Cancer Biotherpy.* **9**, 153 (1994).
6. S. Kojima, K. Inaba, S. Kobayashi and M. Kimura, *Biol. Pharm. Bull.* **19**, 47 (1996).
7. S. Tanaka, Y. Ikeshiro, M. Tabata and M. Konoshima, *Arzneimittelforschung.* **27**, 2039 (1977).
8. Y. Kumazawa, Y. Nakatsuru, H. Fujisawa, C. Nishimura, K. Mizunoe and Y. Otsuka, *J. Pharmacobiodyn.* **8**, 417 (1985).
9. N.L. Wang, H. Kiyohara, T. Matsumoto, H. Otsuka, M. Hirano and H. Yamada, *Planta Med.* **60**, 425 (1994).
10. Y. Yamaoka, T. Kawakita, M. Kaneko and K. Nomoto, *Biol. Pharm. Bull.* **19**, 936 (1996).
11. B.X. Wang, X.H. Zhao, S.B. Qi, S. Kaneko, M. Hattori and T. Namba, *Chem. Pharm. Bull*, **36**, 2587 (1988).
12. B.X. Wang, A.J. Liu, Q.G. Wang, G.R. Wei, J.C. Chui and N. Yang, *Chinese Dispatches Pharmacol.* **3**, 9 (1985).
13. X.B. Sun and C.C. Zhou, *Chinese Patent Pedicine Res.* **2**, 24 (1986).
14. M. Hamada, Y. Fujii, H. Yamamoto, Y. Miyazawa, S.M. Shui and Y.C. Tung, *J. Ethnopharmacol.* **24**, 311 (1988).
15. L.Z. Chen and X.W. Zhou, *Chung Kuo Yao Li Hsuch Pao.* **16**, 337 (1995).
16. H. Deng, K. Nakajima, X. Ma, K. Hasumi, T. Akatsuka and H. Wago, *J. Saitama Med. School.* **28**, 109 (2001).
17. K. Nakajima, H. Deng, X. Ma, K. Hasumi, T. Akatsuka and H. Wago, *J. Saitama Med. School.* **28**, 117 (2001).
18. H. Deng, K. Nakajima, X. Ma, K. Hasumi, T. Akatsuka and H. Wago, *J. Saitama Med. School.* **28**, 125 (2001).
19. C.F. Nathan, *Clin. Invest.* **79**, 319 (1987).
20. P. Alexander and R. Evans, *Nat. New Biol.* **232**, 76 (1971).
21. R.B. Johnston, C.A. Godzik and Z.A. Cohn, *J. Exp. Med.* **148**, 115 (1978).
22. F. Amano and Y. Akamatsu, *Infect Immun.* **59**, 2126 (1991).
23. T. Noda and F. Amano, *Biol. Pharm. Bull.* **21**, 673 (1998).
24. D. Salvemini, A. Pistelli, V. Mollace, E. Anggard and J. Vane, *Biochem. Pharmacol.* **44**, 17 (1992).
25. F. Kura, K. Suzuki, H. Watanabe, Y. Akamatsu and F. Amano, *Infect Immun.* **62**, 5419 (1994).

26. M. Fujikura, M. Muroi, Y. Muroi, N. Ito and T. Suzuki, *J. Biol. Chem.* **268**, 14898 (1993).
27. J. Klostergaard, W.A. Foster and M.E. Leroux, *J. Biol. Response Mod.* **6**, 313 (1987).
28. K. Kawakami, M.H. Qureshi, Y. Koguchi, K. Nakajima and A. Saito, *FEMS Microbiol. Lett.* **175**, 87 (1999).
29. M.G. Lei and D.C. Morrison, *Infect Immun.* **68**, 5084 (2000).
30. R. Manetti, P. Parronchi, M.G. Giudizi, M.P. Piccinni, E. Maggi and G. Trinchieri, *J. Exp. Med.* **177**, 1199 (1993).
31. T. Kato, R. Hakamada, H. Yamane and H. Nariuchi, *J. Immunol.* **156**, 3932 (1996).
32. U. Shu, M. Kiniwa, C.Y. Wu, C. Maliszewski, N. Vezzio and J. Hakimi, *Eurp. J. Immunol.* **25**, 1125 (1995).
33. T. Kato, H. Yamane and H. Nariuchi, *Cell. Immunol.* **181**, 59 (1997).
34. C.S. Hsieh, S.E. Macatonia, C.S. Tripp, S.F. Wolf, A.O'Garra and K.M. Murphy, *Science.* **260**, 547 (1993).
35. E.J. Roy, U. Gawlick, B.A. Orr, L.A. Rund and A.G. Kranz, *J. Immunol.* **165**, 7293 (2000).
36. M.J. Smith, M. Taniguchi and S.E. Street, *J. Immunol.* **165**, 2665 (2000).
37. K. Takeda, Y. Hayakawa, M. Atsuta, S. Hong, L. van Kaer and K. Kobayashi, *Int. Immunol.* **12**, 909 (2000).
38. R.J. Armitage, A.E. Namen, H.M. Sassenfeld and K.H. Grabstein, *J. Immunol.* **144**, 938 (1990).
39. A.L. Causey, R.M. Wooten, L.W. Clem and J.E. Bly, *J. Immunol. Methods.* **175**, 115 (1994).
40. T.R. Mosmann and R.L. Coffman, *Ann. Rev. Immunol.* **7**, 145 (1989).
41. Y. Tanaka, H. Eda, T. Tanaka, T. Udagawa, T. Ichikawa and I. Horii, *Cancer Res.* **50**, 2290 (1990).
42. T.H. Corbett, D.P. Griswold, B.J. Roberts, J.C. Peckham and F.M. Schabe, *Cancer Res.* **35**, 2434 (1975).
43. S. Sakaguchi, E. Tsutsumi and K. Yokota, *Biol. Pharm. Bull.* **16**, 782 (1993).
44. S. Sakaguchi, S. Furusawa, K. Yokota, K. Sasaki and Y. Takayanagi, *Biol. Pharm. Bull.* **18**, 621 (1995).
45. K. Terasaki, M. Nose and Y. Ogiwara, *Biol. Pharm. Bull.* **20**, 809 (1997).
46. W.D. DeWys, D. Gegg, P.T. Lavin, P.R. Band J.M. Bennett, *Am. J. Med.* **69**, 491 (1980).
47. K. Soda, M. Kawakami, A. Kashii and M. Miyata, *Jpn. J. Cancer Res.* **85**, 1124 (1994).
48. G. Trinchieri, *Adv. Immunol.* **47**, 187 (1989).

# 11

# Testing Efficacy of Natural Anxiolytic Compounds

## A. A. Roberts

## 1. Introduction

*Anxiety disorders* are a group of mental disorders that range in their severity from occasional, brief episodes of relatively benign nervous tension to severe, recurrent and disabling panic attacks that interfere with activities of daily living. In addition to the suffering of the affected individual, anxiety disorders have larger social and economic ramifications, such as loss of workplace productivity.[1] Treatment approaches to anxiety disorders include psychoanalytic, cognitive, and pharmacologic therapies. At present, state of the art Western medical therapies for anxiety rely heavily on *anxiolytic* (anti-anxiety) pharmaceuticals, some of which have sedative and cognitive side effects. Since the 1960's, when the prototypic anxiolytic drug diazepam (valium) was discovered serendipitously by L. Sternbach and E. Reeder[2], anti-anxiety drugs have undergone refinement with the goal of reducing undesirable sedative and amnestic side effects. Advances in the fields of neurobiology and psychology have yielded insights into the neurotransmitter systems involved in fear and anxiety responses that have facilitated the development of more selective anxiolytic drugs.[3-5]

Anxiety disorders are a common complaint by patients presenting to psychiatrists, psychologists and other health care professionals. Many individuals who suffer from anxiety disorders seek treatment from health care practitioners outside the realm of psychiatry.[1] Alternative providers include family physicians, clinical psychologists, counselors, and alternative medicine providers of various sorts. In addition, there is a growing market catering to self-adminstration of herbal and other dietary supplements for their anxiolytic effects. Anxiolytic herbal supplements include kava kava, valerian root, passionflower, and others. The popular use of plant extracts in the self-medication of anxiety warrants thorough scientific evaluation of these compounds to determine their individual efficacies, potencies, physiologic mechanisms, and safety. Complete characterization of herbal constituents is critical for predicting and treating potential toxicities, such as the acute hepatitis that has been attributed to ingestion of the anxiolytic herb, kava kava.[6] Individual preparations should

---

**A. A. Roberts** • M.D. Ph.D., Department of Pathology and Laboratory Medicine, University of California, A7-149 Center for Health Sciences, Los Angeles, CA 90095.
*Complementary and Alternative Approaches to Biomedicine*, edited by Edwin L. Cooper and Nobuo Yamaguchi. Kluwer Academic/Plenum Publishers, 2004.

be evaluated independently as their constituents may vary. Clearly the safe use of plant-derived medicines requires a detailed understanding of their pharmacologic mechanisms based on *in vitro* and *in vivo* test results.

## 2. Classification of Anxiety Disorders

The Diagnostic and Statistical Manual of Mental Disorders (DSM-IVTR) includes the following conditions under the heading of anxiety disorders: acute stress disorder; agoraphobia; generalized anxiety disorder; obsessive-compulsive disorder; panic disorder; post-traumatic stress disorder; social phobia; and specific phobia.[7] These disorders vary in their severity and symptomatology, as well as treatment responses. Acute stress disorder is manifested as panic attacks, activations of the sympathetic nervous system in association with a situational trigger that produce a physiologic and behavioral "fight or flight" response. Generalized anxiety disorder typically involves chronic low-level nervousness and worry that has mild to moderate effects on activities of daily living. Anxiety is also a component of some depressive disorders.

The DSM classification of anxiety disorders has been largely based on patient populations in the United States[8]; however, anxiety disorders also pose a significant global problem.[9] The World Health Organization Global Burden of Disease Survey has predicted that the disability costs of major depression will be second only to ischemic heart disease by the year 2020.[8] Anxiety disorders may present in a culturally specific manner, and individuals from different ethnic backgrounds may present with different complaints. For example, Chinese-American patients with panic disorder may complain of "hot and cold" symptoms (*pa-leng*), whereas Japanese patients with social anxiety have described "fear of losing face and facing situations".[10] While understanding cultural specifics is helpful in diagnosis and treatment, much has been gained by focusing on the common neurobiological basis of anxiety disorders. It is now understood that anxiety disorders are mediated by autonomic and neuroendocrine systems that are under the control of the central nervous system (CNS). The neural pathways mediating anxiety have been characterized, and this information has been the impetus for research and development of anxiolytic (anti-anxiety) drugs.

## 3. Gaba$_A$ Receptors and Anxiety

The CNS mechanisms underlying anxiety and stress-related behaviors have been fairly well characterized. Ample evidence supports a neurally-mediated, physiologic and behavioral "stress response" that is under the control of the hypothalamus and the CNS inhibitory neurotransmitter, g-aminobutyric acid (GABA).[4,5,11,12] GABA acts throughout the central nervous system as a dampening influence on neuronal activity helping to protect neurons from toxic overstimulation.[13] The CNS actions of GABA are largely mediated by type A, GABA (GABA$_A$) receptors, which are also the major CNS targets for several classes of anxiolytic and sedative drugs. Drugs that potentiate the activity of GABA$_A$ receptors, such as diazepam (Valium), barbiturates, and alcohol, inhibit anxiety- or fear-driven behavioral and physiologic responses to stressful stimuli. These GABA enhancing drugs have anxiolytic properties, characterized by a reduction in stressor-induced sympathetic nervous system activity, subjective feelings of anxiety, and anxiety-driven behavioral responses.

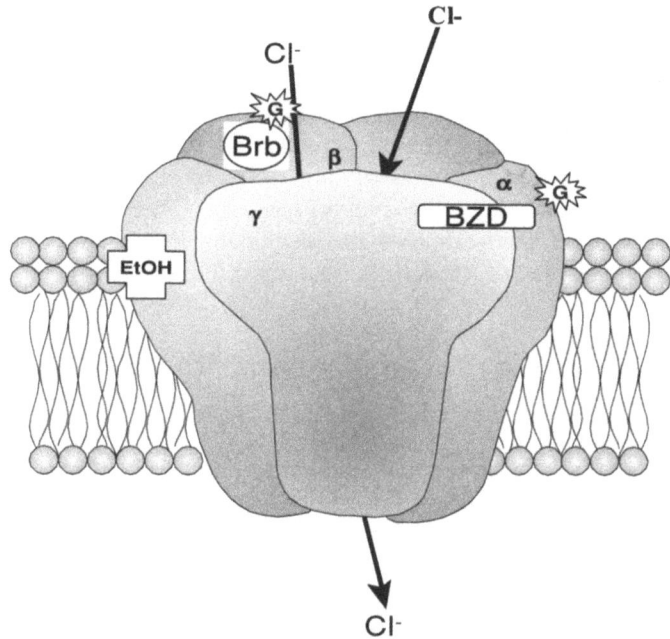

**Figure 11.1.** Basic molecular constituents of the type A, g-aminobutyric acid GABA$_A$ receptor. This schematic represents the transmembrane pentameric arrangement of GABA$_A$ receptor subunits. Five protein subunits surround and gate a central chloride channel. The benzodiazepine binding site (**BZD**) is shown on the extracellular surface spanning the $\alpha$- and $\gamma$-subunit interface. The barbiturate binding site (**Brb**) is shown on the $\beta$-subunit. Binding sites for $\gamma$-aminobutyric acid (**G**) are shown on the b- and a-subunits. The binding site for ethanol (**EtOH**) is depicted at the membrane interface.

There are a wide range of potential therapeutic applications for anxiolytic medications, from anticipatory anxiety in victims of acute panic attacks to the chronic anxious thoughts or feelings of generalized anxiety disorder.

GABA$_A$ receptors are trans-membrane receptors that occur widely throughout the CNS. They are composed of five transmembrane protein subunits that surround and gate a central, chloride-conducting ion channel (Figure 11.1). GABA$_A$ receptor subunit polypeptides are a large multigene family containing at least 16†different members including 6$\alpha$, 3$\beta$, 3$\gamma$, $\delta$, $\varepsilon$, $\theta$ and $\pi$ subunits.[14] GABA activity depends on the identity of the $\alpha$ subunit(s), but the BZD binding site occurs at the interface of an $\alpha$ and a $\gamma$ subunit, and the identity of both subunits influences the activity of BZD site ligands.[14-16] The predominant GABA$_A$ receptor subtype in the rat brain has a subunit composition of $\alpha_1\beta_{2,3}\gamma_2$, whereas together $\alpha_2\beta_{2,3}\gamma_2$ and $\alpha_3\beta_{2,3}\gamma_2$ constitute the next most prevalent subtypes[17]. The anxiolytic, sedative, and amnestic properties of diazepam are mediated by distinct subtypes of GABA$_A$ receptors, and subtype-specific drugs have been developed with the goal of reducing undesirable sedative and amnestic effects, whilst maximizing anxiolytic efficacy[18]. BZD agonists are compounds that bind at the BZD modulatory site and potentiate GABA activity. BZD inverse agonists decrease the actions of GABA and BZD antagonists have no effect on GABA activity although they bind the BZD site and competitively inhibit BZD agonist or inverse agonists. Select BZD site ligands have anxiolytic effects that are equivalent to the classical, full agonist BZDs, like diazepam or clordiazepoxine, but reduced sedative and amnestic effects.[19] Some authors have attributed these selective anxiolytic effects to partial agonist activity of BZD ligands at $GABA_A$ receptors, i.e. ligands that bind with equal affinity but result in lesser activation of GABA activity than full agonists.[20] Molecular studies indicate that anxioselective benzodiazepines have high affinity for $\alpha$2- and/or $\alpha$3-containing $GABA_A$ receptor subtypes and lower affinity for $\alpha$1-containing receptor subtype(s).[20] The $\alpha$1 subunit containing $GABA_A$ receptor subtype has a high affinity for drugs with sedative effects, such as zolpidem, whereas the $\alpha$2 and $\alpha$3 containing subtypes have a high affinity for anxioselective drugs, such as bretazenil, imidazenil, abecarnil, and others.[20,21]

## 4. Behavioral Models for Testing Anxiolytic Compounds

Anxiolytic agents are evaluated by behavioral tests using animals that are exposed to environmental stressors, such as an unfamiliar environment, in controlled laboratory conditions. The best experimental stressor paradigms allow selective assessment of anxiolytic drug effects and have internal controls for arousal level and motor activity built into the experimental model. For example, the *light/dark box test* uses an apparatus consisting of two compartments, one dark and one light, separated by a divider with a small opening at the base.[22] Mice are placed in the center of the light area, facing away from the dark one, and permitted to explore for a short time (usually five minutes). The investigator measures the number of transfers between compartments and the percentage of time spent in the light side. The percentage of time spent in the light compartment, but not the number of transfers, is the best indicator of anxiolytic activity, whereas the number of transfers reflects both anxiety and exploration, which is influenced by arousal state and motor function.[23,24] A selective anxiolytic drug will increase time in the light without significantly affecting transfers.

The elevated plus-maze test is a rodent test for assessing anxiolytic agents.[25] The maze is elevated to table height and consists of two open arms, and two closed arms (with an open roof), arranged so that the two arms of each type are opposite each other). Similar to the light/dark box test, the elevated plus maze manifests anxiolytic effects as the increase in the time spent in the open compartment(s). A seated observer records number of entries into and time spent in open and closed arms, as well as the total number of arm entries, over a period of five minutes. Increased time number of entries into, and time spent in, open arms indicates an anxiolytic effect of the drug, whereas, similar to the light/dark box test, the total entries more closely reflects the arousal level and motor activity of the animal.

The *staircase test* is another behavioral paradigm used to test anxiolytic properties of exogenous compounds administered prior to testing.[26] In this behavioral test, a mouse is placed in an enclosed staircase consisting of five opaque plastic steps, and the number of steps climbed and rearings are recorded over a five minute time period. The numbers of rears and steps climbed have been shown to vary independently, and are thought to reflect anxiolytic and sedative effects, respectively.[27] The ratio of steps: rears is usually taken as an index of anxiolytic activity.[26]

Naturally, the power of behavioral studies increases with the size of the experimental group; however, four to six animals per experimental group are usually adequate to ensure statistically significant data. Appropriate external controls include a group of animals receiving sham drug administration at the same time point as the experimental group, typically composed of the vehicle that was used as a delivery substance for the drug. In the event that the animal is to be sacrificed for physiologic assays subsequent to behavioral testing, age and gender matched controls that do not receive any intervention are also indicated. Ideally, the experimental groups should all contain animals from at least three different litters and each of these litters should be represented in every experimental group. This design minimizes baseline differences between groups that are independent of the drug or testing effects.

## 5. Natural Anxiolytic Compounds Acting at Gaba$_A$ Receptors

Studies utilizing a variety of testing paradigms have demonstrated that stressor-responsive behaviors and GABA$_A$ receptor activity are influenced by prior administration of modulatory GABA$_A$ receptor ligands that act at the BZD binding site.[22,23,28] Classical BZDs, such as diazepam and chlordiazepoxide, are the prototypical class of anxiolytic drugs, and set the standard for performance of anxiolytic drugs. However, some patients do not respond to conventional treatment and the risk of tolerance and dependence is significant. The sedative and amnestic effects cause some patients experience unwanted drowsiness or cognitive impairment secondary to BZDs. These unwanted side effects drive some consumers to avoid pharmaceutical prescriptions in favor of "natural" compounds that have the advantage of not requiring a prescription. An added appeal is the observation that herbal preparations are often, although by no means always, cheaper than their drugstore counterparts. A number of natural compounds have now been identified that have some affinity for GABA$_A$ receptors, some of which have demonstrated anxiolytic effects.

Although it has undergone a reemergence of late, the medicinal use of plants for the treatment of anxiety dates back at least to the turn of the century in the United States and probably farther.[29] Passionflower (*Passiflora incarnata* and related species) is a tropical climbing vine that acts as a sedative/tranquillizer with anxiolytic properties.[30] Traditionally, the aerial parts have been administered predominantly as a liquid (alcoholic) tincture.[29] The major bioactive compounds in passionflower are flavonoids, indole alkaloids, and cyanogenic glycosides.[29] Flavonoids are low molecular weight, polyphenolic secondary metabolites of vascular plants that give rise to the red, blue, and purple anthrocyanin pigments apparent in the fall foliage of temperate climates (figure 11.2.[31] Beyond their aesthetic appeal,

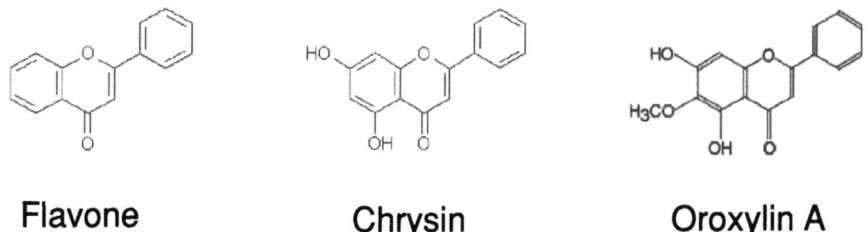

**Figure 11.2.** Chemical structures of flavonoids.
Basic flavone structure illustrated on left, anxiolytic flavone *chrysin* in center, and benzodiazepine site antagonist *oroxylin A* on right.

the flavonoids are also a marvelous example of biological diversity within and beyond the plant kingdom, and they play a critical role in innate immune function and stress protection across a wide range of plant and animal species. For example, flavonoids have important ultraviolet-absorbing and antioxidant properties that protect plants and make these compounds attractive for development as anti-cancer therapies in humans.[32] In addition, some flavonoids exhibit neuroprotective properties, partly due to their antioxidant properties.[33]

Flavonoids have complex actions in the CNS including analgesia, including effects on motor activity and arousal[34,35]. Anticonvulsant flavonoids have also been identified,[36,37] some of which have sedative and anxiolytic effects as well[37,38]. Anxiolytic flavonoids, such as chrysin, apigenin, and 6,3'-dinitroflavone, potentiate the action of GABA by binding to the BZD site. These compounds have demonstrated anxiolytic effects in behavioral testing and are under investigation for the treatment of anxiety[36,38,39]. C*hrysin* (5,7-dihydroxyflavone) is a naturally occurring flavonoid that has been isolated from passionflower and demonstrated to have potent anxiolytic effects in behavioral studies with minimal sedative effects[40]. In addition, synthetic derivatives of flavonoids have also been identified that have potent anxiolytic effects without myorelaxant, amnestic or sedative actions[41].

## 6. Behavioral Tests of Natural Anxiolytic Compounds

Evaluating the medical utility of naturally occurring flavonoids can be rather daunting, due to the diversity of these compounds and the presence of multiple flavonoids and other bioactive compounds within individual plants. Often, old folk remedies or natural medicine practices provide logical starting points for the identification of novel biotherapeutics; however, the task remains of sorting through the various candidate compounds to elicit the active ingredient that is producing the desired treatment effect. Prior to extensive purification and isolation steps, some behavioral testing can be done on plant extracts, which are most analogous to the traditional herbal remedies.

Soulimani, *et al.*[29] administered lyophilized hydroalcoholic and aqueous extracts of *Passiflora incarnata* prior to administering behavioral tests of anxiety to mice. They found an anxiolytic effect of the hydroalcoholic extract at one dose in the light/dark box test and staircase test that did not correspond to a sedative effect, whereas the aqueous extract had apparent sedative effects. Although it would be more convincing if a dose-response curve had been demonstrated, this study did not use purified constituents, and thus it is possible to envision an inhibitory compound that interferes at higher doses. Furthermore, this level of study can help to focus investigators on a subset of bioactive compounds that are soluble in aqueous/alcoholic extracts.

Species differences in anxiolytic potency of plants reflect differences in flavonoid content. For example, Petry *et al.* compared the effects of pre-treatment with diazepam (1 mg/kg) and hydroethanol extracts of *Passiflora alata* and *Passiflora edulis* (25, 50, 100 and 150 mg/kg) on elevated plus-maze behavior in rats.[42] They found that animals treated with *P. alata* (100 and 150 mg/kg) and *P. edulis* (50, 100 and 150 mg/kg) spent more time and entered more frequently into open arms and less into closed arms without altering total entries, indicating an anxiolytic effect at physiologic doses. Additionally, *P. edulis* extract caused an increase in the total number of entries at doses of 50 and 100 mg/kg, indicating elevated motor activity and suggesting a lesser sedative effect than diazepam. Anxiolytic effects were detected in *P. edulis* at half the dose (50 mg/kg) of *P. alata* (100 mg/kg), consistent with the finding of almost twice the flavonoids in hydroethanol extracts of *P. edulis* leaves than *P. alata*.

Investigational studies using plant extracts are also helpful in identifying the portion of the plant that has the highest concentration of the active compound for efficient purification. Dahwan et al. have demonstrated that alternative methods of extraction of a *Passiflora incarnata*, differ in their anxiolytic properties and that the potency is dependent which part of the plant is used.[43] Using the elevated plus-maze model in mice, they demonstrated that methanol extracts of leaves, stems, flowers, and whole plants conferred anxiolytic effects at different potencies, whereas the roots were practically devoid of anxiolytic effects. Thus, inclusion of roots or flowers can yield a significant decrease in the anxiolytic potency, and the authors recommend separation of these parts prior to any pharmacological studies on *P. incarnate*. More recently, these same authors demonstrated considerable differences in the potency of tinctures prepared from *Passiflora incarnata* by five reputable manufacturers of homeopathic medicines.[44]

The use of whole plant preparations instead of isolated chemical components is cheaper and more in line with traditional preparations than purified individual chemical constituents. The advantages lie in ease of preparation, lower secondary contaminants due to fewer chemical extractions, and the possibly synergism between multiple active chemical constituents. Disadvantages to crude herbal extracts include compositional variability that makes standardization difficult and possible antagonism between multiple active chemical components. In order to standardize the potency of flavonoids and other herbal medicinal products, a measure of therapeutic potency is needed, particularly if the primary active constituent has not been identified. Bioequivalence studies are *in vitro*, animal, or clinical studies that are done to establish equivalent therapeutic doses of different herbal preparations.[45] Passionflower preparations can be standardized for anxiolytic effect based on their potency in animal testing paradigms, such as the open arm plus maze, staircase test, and light-dark box test, or by clinical trials. Once individual flavonoids have been isolated from passionflower and other anxiolytic herbs, standardization can be based on pharmaceutical equivalence.

The anxioselective properties of the flavonoid chrysin (isolated from passionflower) have been demonstrated in several different studies, and its anxiolytic properties have been linked to activity at the BZD binding site of the $GABA_A$ receptor. Medina et al. of Argentina demonstrated antiseizure activity of chrysin in a mouse model in which tonic-clonic seizures were induced by pentylenetetrazol.[36] Concomitant administration of Ro 15-1788, an antagonist at the benzodiazepine site of the $GABA_A$ receptor, blocked the antiseizure effect, indicating that chrysin acts via BZD binding sites on $GABA_A$ receptors. Several years later, the same group published a study comparing the behavioral effects of chrysin to those of diazepam in the elevated plus-maze test of anxiety.[40] They found that moderate doses of either diazepam or chrysin increased the number of entries into and time spent on open arms, consistent with an anxiolytic action of both compounds. Pretreatment with Ro 15-1788 abolished the effects of diazepam and chrysin on the elevated plus-maze, indicating that both compounds act via benzodiazepine binding sites on $GABA_A$ receptors. Interestingly, in tests of arousal and motor activity, such as the holeboard and the horizontal wire test, high doses of diazepam but not of chrysin had sedative and myorelaxant effects. These results suggest that chrysin acts as an anxiolytic without inducing sedation as a side effect, enhancing its desirability as an anxiolytic medication. As is the case for a number of BZD ligands, this physiologic difference may reflect differing specificities of chrysin and diazepam for $GABA_A$ receptor subtypes. Subunit specificity can be determined by a number of different methods including affinity chromatography followed by immunoprecipitation and *in vitro* expression studies.

## 7. In Vitro Studies of Natural Anxiolytic Compounds

Once an anxiolytic compound has established efficacy in animal behavioral models, it is necessary to establish the mechanism of action. A limited number of studies have examined *in vitro* activity of select flavonoids at $GABA_A$ receptors by neurochemical and neurophysiologic testing. A variety of flavonoids from different plant sources have been shown to have affinity for the BZD site on $GABA_A$ receptors.[39,41] For example, Hui et al. isolated four flavonoids from Scutellaria baicalensis. (wogonin, baicalin, baicalein, and scutellarein) that have affinity for the BZD site of the $GABA_A$ receptor; All four compounds contained the flavonoid phenylbenzopyrone nucleus.[46]

*In vitro* electrophysiologic studies are few in number and have yielded some unexpected results. For example, Goutman et al. demonstrated dose-dependent inhibition of $\alpha1\beta1\gamma2s$ $GABA_A$ receptor currents by the anxiolytic flavonoid, chrysin and apigenin among others[47]. Another anxiolytic flavone, wogonin, enhanced the GABA-activated currents in rat dorsal root ganglion neurons, and in Xenopus laevis oocytes expressing $\alpha1\beta1\gamma2s$ $GABA_A$ receptors. The effect was partially reversed by co-application of BZD site antagonist Ro15-1788.[48] These results are somewhat preliminary, and further in vitro testing will be necessary to determine the activity of anxiolytic flavonoids at $\alpha2$ and $\alpha3$-containing $GABA_A$ receptors, the subtypes that have been linked to the anxiolytic properties of diazepam.

Some synthetic flavonoids have exhibited antagonism of diazepam. For example, Viola et al. demonstrated that 6-chloro-3'-nitroflavone is a functional antagonist at the BZD site.[49] This flavone derivative was devoid of anxiolytic, anticonvulsant, sedative, or myorelaxant actions in mice or amnestic effects in rats but was able to block these actions of diazepam.[49] The results of in vitro studies were consistent with antagonistic activity, demonstrating that 6-chloro-3'-nitroflavone blocks potentiation of $\alpha x\beta1\gamma2s$ $GABA_A$ receptor activity by diazepam.[49] Huen et al. also demonstrated antagonism at the BZD site by the flavonoid oroxylin A (5,7-dihydroxy-6-methoxyflavone), which selectively abolished the anxiolytic, myorelaxant and motor incoordination, but not the sedative and anticonvulsant effects elicited by diazepam.[50]

## 8. Conclusion

Over the past decade or two, alternative medical therapies have moved into mainstream consciousness and gained popular use. A substantial literature has accumulated concerning therapeutic applications, safety, and mechanisms of natural medicinal preparations from plants, from crude extracts to their refined phytochemical constituents. Select herbal preparations have been used in traditional, non-allopathic treatments for insomnia and nervousness, indicating a potential for anxiolytic therapy. Classic benzodiazepines, such as diazepam, act by potentiating the inhibitory neurotransmitter GABA and are the most widely-prescribed anxiolytic drugs, although they have undesirable sedative and behavioral side effects. Certain flavonoids with affinity for the BZD site on $GABA_A$ receptors have anxiolytic effects in humans and in animal models that warrant further characterization. A select group of flavonoids, including chrysin from *Passiflora*, and wogonin from *Scutellaria*, are BZD site ligands that have anxiolytic activity in behavioral testing without significant sedative or amnestic effects. Anxiolytic herbal preparations containing flavonoids are already in popular use, therefore, further characterization of their mechanisms, applications, and safety is warranted. Individual preparations should be standardized based on their pharmacokinetic characteristics.

## 9. References

1. J.P. Lepine, The epidemiology of anxiety disorders: prevalence and societal costs. *J Clin Psychiatry*, 2002. **63**(Suppl 14): p. 4–8.
2. L. Sternbach, et al., Quinazolines and 1,4-benzodiazepines. XXV. Structure-activity relationships of aminoalkyl-subsituted 1,4-benzodiazepin-2-ones. *J Med Chem*, 1965. **8**(6): p. 815–821.
3. P. Sah, et al., The amygdaloid complex: anatomy and physiology. *Physiol Rev*, 2003. **83**(3): p. 803–834.
4. A. Shekhar, L. Sims, and R. Bowsher, GABA receptors in the region of the dorsomedial hypothalamus of rats regulate anxiety in the elevated plus-maze test. II. Physiological measures. *Brain Res*, 1993. **627**(1): p. 17–24.
5. A. Shekhar and J. Katner, Dorsomedial hypothalamic GABA regulates anxiety in the social interaction test. *Pharmacol Biochem Behav*, 1995. **50**(2): p. 253–258.
6. E. Ernst, Safety concerns about kava. *Lancet*, 2002. **359**(9320): p. 1865.
7. M.B. First, *Diagnostic and Statistical Manual—Text Revision (DSM-IV-TR$^{TM}$, 2000)*. 2000, Washington, D.C.: American Psychiatric Association.
8. S. Lee, Socio-cultural and global health perspectives for the development of future psychiatric diagnostic systems. *Psychopathology. 2002 Mar–Jun;35(2–3):152–7.*, 2002. **35**((2–3)): p. 152–157.
9. L. Y, The burden of depression and anxiety in general medicine. *J Clin Psychiatry*, 2001. **62**(8): p. 4–9.
10. J. Chen, L. Reich, and H. Chung, Anxiety disorders. *West J Med*, 2002. **176**(4): p. 249–253.
11. E. De Souza, Neuroendocrine effects of benzodiazepines. *J Psychiatr Res*, 1990. **24**(Suppl 2): p. 111–119.
12. R. Lydiard, The role of GABA in anxiety disorders. *J Clin Psychiatry*, 2003. **64**(Suppl 3): p. 21–27.
13. P. Dodd, Excited to death: different ways to lose your neurones. *Biogerontology*, 2002. **3**(1–2): p. 51–56.
14. P. Whiting, et al., Molecular and functional diversity of the expanding GABA-A receptor gene family. *Ann N Y Acad Sci*, 1999. **868**: p. 645–653.
15. D. Pritchett, H. Luddens, and P. Seeburg, Type I and type II $GABA_A$-benzodiazepine receptors produced in transfected cells. *Science. 1989 Sep 22;245(4924):1389-92*, 1989. **245**(4924): p. 1389–1392.
16. G. Smith and R. Olsen, Functional domains of $GABA_A$ receptors. *Trends Pharmacol Sci*, 1995. **16**(5): p. 162–168.
17. R. McKernan and P. Whiting, Which $GABA_A$-receptor subtypes really occur in the brain? *Trends Neurosci*, 1996. **19**(4): p. 139–143.
18. E. Korpi, et al., GABA(A)-receptor subtypes: clinical efficacy and selectivity of benzodiazepine site ligands. *Ann Med.*, 1997. **29**(4): p. 275–282.
19. S. Stahl, Selective actions on sleep or anxiety by exploiting GABA-A/benzodiazepine receptor subtypes. *J Clin Psychiatry*, 2002. **63**(3): p. 179–180.
20. J. Atack, Anxioselective Compounds Acting at the $GABA_A$ Receptor Benzodiazepine Binding Site. *Curr Drug Target CNS Neurol Disord*, 2003. **2**(4): p. 213–232.

21. G. Griebel, et al., SL651498: an anxioselective compound with functional selectivity for alpha2- and alpha3-containing gamma-aminobutyric acid(A) (GABA(A)) receptors. *J Pharmacol Exp Ther*, 2001. **298**(2): p. 753–768.
22. J. Crawley and F. Goodwin, Preliminary report of a simple animal behavior model for the anxiolytic effects of benzodiazepines. *Pharmacol Biochem Behav*, 1980. **13**(2): p. 167–170.
23. R. Young and D. Johnson, A fully automated light/dark apparatus useful for comparing anxiolytic agents. *Pharmacol Biochem Behav*, 1991. **40**(4): p. 739–743.
24. E. Lepicard, et al., Differences in anxiety-related behavior and response to diazepam in BALB/cByJ and C57BL/6J strains of mice. *Pharmacol Biochem Behav*, 2000. **67**(4): p. 739–748.
25. S. Pellow, et al., Validation of open:closed arm entries in an elevated plus-maze as a measure of anxiety in the rat. *J Neurosci Methods*, 1985. **14**(3): p. 149–167.
26. J. Simiand, P. Keane, and M. Morre, The staircase test in mice: a simple and efficient procedure for primary screening of anxiolytic agents. *Psychopharmacology*, 1984. **84**(1): p. 48–53.
27. L. Steru, et al., Comparing benzodiazepines using the staircase test in mice. *Neurosci Biobehav Rev*, 1985. **9**: p. 45–54.
28. S. File, What can be learned from the effects of benzodiazepines on exploratory behavior? *Neurosci Biobehav Rev*, 1985. **9**: p. 45–54.
29. R. Soulimani, et al., Behavioral effects of passiflora incarnata and its indole alkaloid and flavenoid derivatives and maltol in the mouse. *J. Ethnopharmacol*, 1997. **57**: p. 11–20.
30. S. Akhondzadeh, et al., Passionflower in the treatment of opiates withdrawal: a double-blind randomized controlled trial. *J Clinical Pharmacy and Therapeutics*, 2001. **26**: p. 369–373.
31. T. Field, D. Lee, and N. Holbrook, Why leaves turn red in autumn. The role of anthocyanins in senescing leaves of red-osier dogwood. *Plant Physiol*, 2001. **127**: p. 566–574.
32. B. Winkel-Shirley, Biosynthesis of flavonoids and effects of stress. *Curr Opin Plant Biol*, 2002. **5**: p. 218–223.
33. H. Ha, et al., Quercetin attenuates oxygen-glucose deprivation- and excitotoxin-induced neurotoxicity in primary cortical cell cultures. *Biol Pharm Bull*, 2003. **26**(4): p. 544–546.
34. a.M.A. Speroni E, Neuropharmacological activity of extracts from Passiflora incarnata. *Planta Med*, 1988. **54**: p. 488–491.
35. C.S. Picq M, Prigent AF, Effect of two flavonoid compounds on central nervous system. Analgesic activity. *Life Sci*, 1991. **49**(26): p. 1979–1988.
36. J. Medina, et al., Chrysin (5,7-di-OH-flavone), a naturally-occurring ligand for benzodiazepine receptors, with anticonvulsant properties. *Biochem Pharmacol*, 1990. **40**(10): p. 2227–2231.
37. E. Nogueira and V. Vassilieff, Hypnotic, anticonvulsant and muscle relaxant effects of Rubus brasiliensis. Involvement of GABA(A)-system. *J Ethnopharmacol*, 2000. **70**(3): p. 275–280.
38. A. Paladini, et al., Flavonoids and the central nervous system: from forgotten factors to potent anxiolytic compounds. *J Pharm Pharmacol*, 1999. **51**(5): p. 519–526.
39. M. Marder and A. Paladini, GABA(A)-receptor ligands of flavonoid structure. *Curr Top Med Chem*, 2002. **2**(8): p. 853–867.
40. C. Wolfman, et al., Possible anxiolytic effects of chrysin, a central benzodiazepine receptor ligand isolated from Passiflora coerulea. *Pharmacol Biochem Behav*, 1994. **47**(1): p. 1–4.
41. J. Medina, et al., Overview–flavonoids: a new family of benzodiazepine receptor ligands. *Neurochem Res*, 1997. **22**(4): p. 419–425.
42. R.F. Petry RD, de-Paris F, Gosmann G, Salgueiro JB, Quevedo J, Kapczinski F, Ortega GG, Schenkel EP, Comparative pharmacological study of hydroethanol extracts of Passiflora alata and Passiflora edulis leaves. *Phytother Res*, 2001. **15**(2): p. 162–164.
43. K. Dhawan, S. Kumar, and A. Sharma, Comparative anxiolytic activity profile of various preparations of Passiflora incarnata linneaus: a comment on medicinal plants' standardization. *J Altern Complement Med*, 2002. **8**(3): p. 283–291.
44. K. Dhawan, S. Kumar, and A. Sharma, Anxiolytic activity of aerial and underground parts of Passiflora incarnata. *Fitoterapia*, 2001. **72**(8): p. 922–926.
45. D. Loew and M. Kaszkin, Approaching the problem of bioequivalence of herbal medicinal products. *Phytotherapy research*, 2002. **16**: p. 705–711.
46. K. Hui, X. Wang, and H. Xue, Interaction of flavones from the roots of Scutellaria baicalensis with the benzodiazepine site. *Planta Med. 2000 Feb;66(1):91-3*, 2000.

47. J. Goutman, et al., Flavonoid modulation of ionic currents mediated by GABA(A) and GABA(C) receptors. *Eur J Pharmacol*, 2003. **461**((2–3)): p. 79–87.
48. K. Hui, et al., Anxiolytic effect of wogonin, a benzodiazepine receptor ligand isolated from Scutellaria baicalensis Georgi. *Biochem Pharmacol*, 2002. **64**(9): p. 1415–1424.
49. H. Viola, et al., 6-Chloro-3′-nitroflavone is a potent ligand for the benzodiazepine binding site of the GABA(A) receptor devoid of intrinsic activity. *Pharmacol Biochem Behav*, 2000. **65**(2): p. 313–320.
50. M. Huen, et al., 5,7-Dihydroxy-6-methoxyflavone, a benzodiazepine site ligand isolated from Scutellaria baicalensis Georgi, with selective antagonistic properties. *Biochem Pharmacol*, 2003. **66**(1): p. 125–132.

# 12

# Alternative Approaches to Pain Relief during Labor and Delivery

## Michel Tournaire

## 1. Introduction

Even though delivery is a natural phenomenon, it has been demonstrated that the accompanying pain is considered severe or extreme in more than half of the cases. Besides conventional approaches, such as epidural analgesia, many complementary or alternative methods have been reported to reduce pain during labor and delivery. Complementary or Alternative Medicine (CAM) can be defined as theories or practices that are not part of the dominant or conventional medical system. Some of them have been reclassified as part of conventional medicine when supported by clinical experience or scientific data[1].

These methods are popular because they emphasize the individual personality, and the interaction between mind, body and environment[2]. They are attractive to people who want to be more involved in their own care and feel that such therapies are more in harmony with their personal philosophies. The conventional medical community usually offers traditional choices of analgesia, such as epidural and intravenous drugs. Patients may have access to alternative methods but will generally be obliged to do the relevant research themselves beforehand. Those seeking alternatives are not necessarily dissatisfied with conventional medicine, but attempt to supplement rather than replace traditional care. Quite often users of complementary medicine do not inform the practitioners in charge of their pregnancy and delivery. There are also different expectations for the management of pain during labor according to the category of professionals. Physicians are expected to provide pharmacological therapy, whereas midwives, nurses and other auxiliaries are required to assist patients with psychological methods and in fact use alternative approaches more often. The theoretical bases for many alternative methods derive from Eastern tradition or philosophy.

After a description of labor pain we shall mention the conventional treatments and describe the different complementary methods applicable to labor pain.

---

**Michel Tournaire** • Obstetrics and Gynecology Department, Saint Vincent de Paul Hospital, University of Paris, 75014 Paris, France.
*Complementary and Alternative Approaches to Biomedicine*, edited by Edwin L. Cooper and Nobuo Yamaguchi. Kluwer Academic/Plenum Publishers, 2004.

## 2. Nature of Pain during Labor and Delivery

A scientific definition of pain is "an unpleasant sensory and emotional experience associated with actual or potential tissue damage"[3]. Acute pain such as labor pain has two dimensions: a sensory or physical dimension, with the transmission of information, the pain stimuli, to the brain, and an affective dimension due to interpretation of these stimuli through the interaction of a wide variety of emotional, social, cultural and cognitive variables unique to the individual (Figure 12.1).

For the management of pain, conventional medicine focuses more on the physical side, while alternative methods deal mainly with emotional considerations. In the laboring patient, the two stages of labor correspond to different types of pain and routes of transmission. During dilation (first stage), visceral pain predominates, due to mechanical distention of the cervix and of the lower part of the uterus. These stimuli are transmitted to the spinal cord at the level of the tenth thoracic to the first lumbar root. Uterine contractions may be felt as back pain because the nerves that supply the uterus also supply the skin on the lower back or lumbosacral area. During the descent phase (second stage), pain is also caused by distension and stretching of the pelvic floor and perineum. These stimuli are transmitted via the pudendal nerve to the second to fourth sacral nerves.

Even though pain is a personal experience, it can be analyzed by means of quantitative pain measures. Verbal reports using standardized instruments, such as the McGill Pain Questionnaire and the Visual Analogue Scale, have been the most common methods of pain assessment both in clinical practice and research[4].

A Canadian study comparing different pain syndromes found that average labor pain scores were higher in both nulliparous (first delivery) and multiparous women than the average scores previously recorded for outpatients with sciatic pain, toothache and fracture pain[5–6] (Figure 12.2). However, whereas the average score is higher, its exact value differs greatly from one woman to another. Bonica found that labor pain was mild in 15% of cases, moderate in 35%, severe in 30%, and extreme in 20%[7].

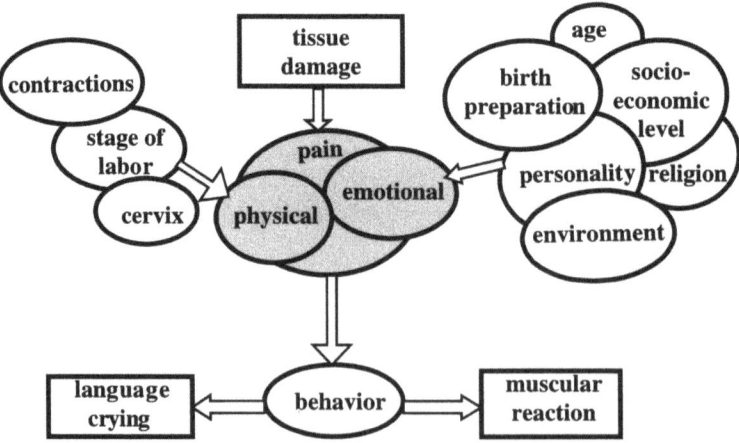

**Figure 12.1.** Components of pain (6)

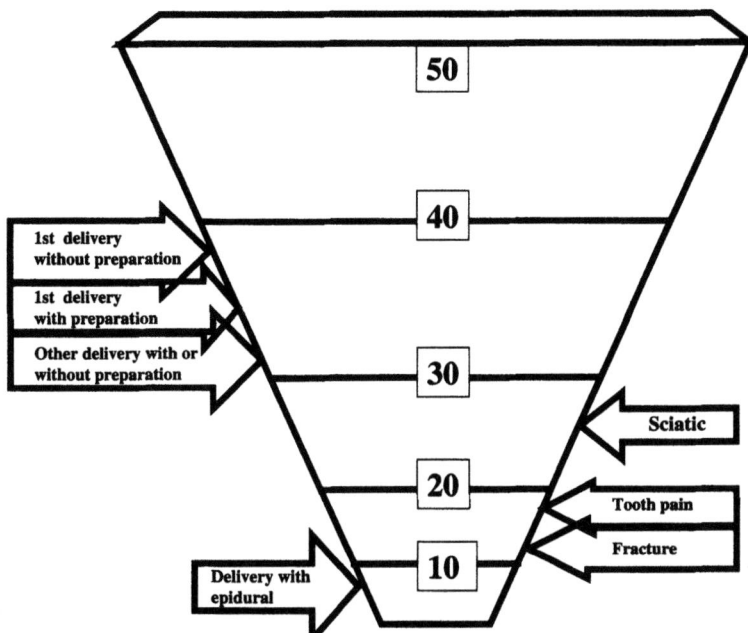

**Figure 12.2.** Pain scores[4–5–6]

Some factors are associated with increased pain: first delivery, history of dysmenorrhea (painful periods), fear of pain, and religious practice. Some factors diminish pain: childbirth preparation classes, complications during pregnancy, wish to breast feed, high socio-economic status, older age[5].

To evaluate the efficiency of the different therapies, we have applied conventional scientific methods to published studies. In other words, do these studies report a statistically significant reduction in labor pain? As we shall see, few publications in the field of CAM medicine meet these standards. However, we should consider that pain, which can now be quantified, is only one component of a woman's overall experience of labor and birthing. Personal satisfaction is not always correlated with the level of pain and although difficult to quantify should be included in the evaluations. Note that the term analgesia means pain relief without total loss of sensation, while anesthesia is defined as pain relief with total loss of sensation.

## 3. Conventional Treatments

### 3.1. Regional Analgesia: Epidural

An epidural involves the introduction of a local anesthetic agent to the sensitive nerves conducting the pain messages on their way to the spine. A catheter (fine flexible tube) is usually placed in the epidural space, allowing intermittent or continuous infusion throughout the delivery.

The epidural is the most efficient way of reducing labor pain (Figure 12.2). Eighty-five to 95 per cent of women report complete relief of pain during the two phases of delivery: cervical dilation and descent of the baby[8]. Complete failure is rare and usually due to technical problems, as when the epidural space cannot be reached with the catheter. Delivery pain relief can be partial. The painful feeling of contractions persists, but at a lower intensity. Sometimes the area of analgesia is incomplete. For example, the pain can be felt laterally in half of the abdomen. When the lower nerves are not, or are insufficiently, dulled, pain may develop during the second phase of labor. One of the main advantages of the epidural is that it is efficient regardless of the cultural context, with few side effects. But it is not always available.

## 3.2. Injected Drugs

Morphine-like drugs (opioids) can be given continuously or in intermittent doses at the patient's request or via patient-controlled administration. Recent reports suggest that the analgesic effect of these agents in labor is limited and that the primary mechanism of action is heavy sedation, which means that consciousness is reduced during delivery. Such drugs may also have some effect on the newborn, with rare but possible breathing difficulties that may require assistance. Few studies have dealt with efficiency. Most were done in the 1960s and provide information on patient satisfaction (generally good in about half of the cases), but without quantitative evaluation of pain reduction.

## 3.3. Nitrous Oxide

Nitrous oxide gas is given for inhalation at subanesthetic concentrations. Despite being used for more than 100 years, there is no clear quantitative evidence of the efficacy of nitrous oxide in relieving labor pain. The subjective feelings of mothers giving birth suggest, however, that nitrous oxide is beneficial in many cases. Many women report significant analgesia with it, and many would choose it again for another delivery.

## 4. Alternative Approaches

Complementary and alternative methods applicable to labor pain can be divided into mind-body interventions, alternative systems of medical practice, manual healing, bioelectromagnetic and physical methods, and alternative medication[1].

## 4.1. Mind-Body Interventions

Mind-body interventions are based on the interconnectedness of mind and body and on the power of each to affect the other. Many mind-body interventions are applied to chronic illness, but this technique also appears to be applicable to the acute situation of delivery.

### 4.1.1. Psychoprophylactic Methods

Grantly Dick-Read introduced "natural childbirth" in 1933. He believed that childbirth pain was a pathologic response produced by fear, apprehension and tension. He felt it essential to teach women the anatomic and physiologic facts of childbirth, and to instruct them in physical and mental relaxation. Both approaches are alleged to diminish pain by familiarizing the pregnant woman with the process of childbirth and by creating an atmosphere of confidence.

Fernand Lamaze introduced his method in France in 1951 after a visit to Russia. This method was first optimistically called "painless childbirth", but later the more appropriate term "fearless childbirth" was applied. It is based on the Pavlovian concept of conditioned reflex training. By focusing on certain breathing patterns or concentration points such as a mark on a nearby wall, it should be possible to block pain messages to the brain.

Bradley's method emphasizes natural childbirth, with the parents working as a team. Students of this method are taught about deep abdominal breathing and an understanding of the labor and delivery process. Rather than trying to block out pain, Bradley's method encourages concentrated awareness that works through the pain.

These methods are expected to provide better information about the process of delivery, reduce fear, give greater satisfaction with a sense of achievement and happiness, and create a better child-mother relationship. The importance of a good relationship between the patient and the caregiving team is also emphasized. Of these approaches, we have only found an evaluation of the Lamaze method: a study by Melzack in 1984[5] using the McGill pain scale found a slight decrease in average pain score in patients using the Lamaze method, but this was not statistically significantly different from the control group (Figure 12.2).

This study also showed that the average pain score is slightly higher at the first delivery, compared with subsequent deliveries, but the difference is not statistically significant.

### 4.1.2. Leboyer's Method

Frédéric Leboyer described his method in France in 1974 in his book "Birth without violence". Inspired by Indian yoga, this method focuses on providing a better welcome for the newborn. In contrast to the usual environment, with too much light and noise stressing the baby, Leboyer proposes calm for mother, father and professionals, and darkness, little noise and a warm bath for the newborn. For the comfort of the mother during delivery, Leboyer considers that serenity obtained through attention to the baby raises the pain threshold. There has been no specific evaluation of the effects on labor pain. However, couples express a high degree of satisfaction. Although Leboyer's method is rarely used now as described in the 1970s, many birth practitioners still consider it to have a positive effect, with gentle and heightened attention for the newborn.

### 4.1.3. Hypnosis

The word hypnosis originates from the Greek "hypnos" meaning "sleep". In fact it is not sleep but a state of focused concentration in which the patient can be relatively unaware, but not completely blind to her surroundings[2]. During hypnosis, suggestions may be made, focusing on diminishing awareness of pain, fear and anxiety. The woman is prepared with initial hypnotic experiences which include three steps: absorption of the words or images presented by the therapist, dissociation, a suspension of critical judgment, and responsiveness. A few comparative studies have evaluated the efficiency of hypnosis.

In 1962, Davidson et al.[9] compared 70 patients who had six lessons of autohypnosis with 70 who had received Dick-Read's training and with 71 who had no special antenatal training. The study was not randomized as patients were allowed to choose their group. A statistically significant reduction in the duration of the first stage of labor was found in the hypnosis group as compared to the two other groups. Autohypnosis was effective on labor pain: 59% of this group required no analgesia, compared with 1.4% in the control group. All patients required analgesia in the Dick-Read training group. The subjective impression of labor was much more pleasant in the autohypnosis group. In 1990, Harmon et al.[10] completed a randomized study showing shorter stage 1 labor, less medication and higher pain thresholds in the hypnosis group than in the control group. In 1995, Mairs[11] compared 29 primigravida women who chose to join "hypnosis for childbirth" classes and 29 in a control group. The trained group reported statistically significantly lower ratings of both pain and anxiety. However, there was no statistically significant difference between the two groups in their drug usage during labor.

A few negative effects of hypnosis have been reported, including mild dizziness, nausea, and headache. These seem to be associated with failure to dehypnotize the patient properly. Caution should be used in patients vulnerable to psychotic decompensation.

To summarize, hypnosis seems to reduce fear, tension and pain during labor and to raise the pain threshold. It reduces the need for chemical analgesia. Patients have a greater sense of control over painful contractions. Hypnosis, therefore, can be considered as a helpful adjunct during the course of labor and delivery.

### *4.1.4. Biofeedback*

Biofeedback uses monitoring instruments to provide feedback to patients, i.e. physiological information of which they are normally unaware. Electrodes feed information to a monitoring box which registers the results by a sound or a visual meter which varies as the monitored function increases or decreases. For women in labor, several biofeedback-assisted relaxation techniques have been introduced.

Duchene[12] completed in 1989 a prospective randomized trial in which tension of the abdominal muscles was monitored. As uterine contractions occurred the women focused on relaxing the abdominal muscles. The reports of pain using visual analogue scales and verbal description scales showed significantly lower pain values in the biofeedback group and less medication. In 1992, Bernat et al.[13] used a fingertip thermometer. When the patient relaxes, vasodilation occurs and the finger temperature increases. However, none of the experimental subjects attempted to use fingertip temperature control as a coping technique during labor. The authors concluded that a lack of hospital staff support may have contributed to this study's outcome.

In conclusion, biofeedback-assisted relaxation techniques applied to pain control yield contradictory results. Their efficiency is certainly contingent on strong support from caregivers to facilitate the use of the technique.

### 4.1.5. Yoga

Yoga, a method of Indian origin, proposes control of mind and body. Between the different types of yoga, "energy yoga" can be applied to pregnancy and delivery. Through special training of breathing, it achieves changes in levels of consciousness, relaxation, receptivity to the world and inner peace. According to professionals who use this technique for delivery, yoga shortens the duration of labor, decreases pain and reduces the need for analgesic medication. However, we have not found any scientific confirmation of these assertions.

### 4.1.6. Sophrology

The word sophrology derives from two Greek words, "sos" harmony or serenity and "phren" conscience or spirit. This technique derived from Indian yoga was introduced in Europe during the 1960s. Its purpose is to improve the control of body and spirit through three degrees of dynamic relaxation: concentration, contemplation and meditation. Applied to obstetrics, better control of the delivery process is expected. Patients individually report a high degree of satisfaction with this experience of relaxation during prenatal classes and delivery, but there is no controlled evaluation in the literature.

### 4.1.7. Haptonomy

Derived from the Greek words "hapsis" affectivity and "nomos" knowledge, haptonomy can be defined as the science of affectivity. This approach was proposed by Frans Vedman[14] in the Netherlands during the 1940s. Specific zones of affectivity are reported to improve the contact between father, mother and baby and to help to share emotions. In practice, haptonomy is appreciated by couples during pregnancy, but it seems to be used irregularly during labor, particularly because teams in charge of delivery are not always aware of its existence. Practitioners expect a quicker and easier delivery as well as a better relationship between parents and newborn. There is no published evaluation of haptonomy and in fact such an assessment should not be expected because, as the specialists of this method say, "affectivity cannot be put into numbers".

### 4.1.8. Music Therapy

Music addresses many of the physical and psychological needs of patients. In obstetrics, a slow and restful type of music may be used as a sedative to promote relaxation during the early stage of labor. Music with a steady beat may be used as a stimulant to promote movement during the latter stages. The literature findings are discordant.

In Austin, Texas, a music program is used during the third trimester of pregnancy[15]. The prospective mother and her partner are allowed to select the kind of music they like for the different stages of labor. After a study of 30 deliveries, only one half of the women who had listened to music required analgesia. However, in a randomized study Durham et al.[16] could not demonstrate the value of music in reducing the need for analgesic medication. But the subjective sense of satisfaction appeared to be higher in the group that listened to music. The beneficial effects of music therapy have not made much of an impression on clinical medicine. Further scientific studies are needed.

## 5. Alternative Systems of Medical Practice

### 5.1. Acupuncture

Acupuncture has been used in China for more than 2000 years. Specific anatomic parts of the body are stimulated for therapeutic purposes. This can be done in the usual way with needles, but practitioners may also use heat, pressure, impulse of magnetic energy, burning by a preparation of the herb *Artemia vulgaris*, electrical stimulation or surface electrodes at acupuncture loci[2].

Acupuncture is based on the balance between Yin and Yang. Treatment is aimed at reconstituting the normal movement between these two opposites. The meridians are considered as energy channels. Most treatments of obstetrical and gynecological problems involve the use of points on differents meridians: spleen-pancreas located on the inside of the ankle bone, conception, governing or penetrating vessel.

Acupuncture may produce effects through several different mechanisms. One hypothesis is that acupuncture points have electrical properties that, when stimulated, may alter the level of chemical neurotransmitters in the body. Another hypothesis is that endorphins are released due to activation of the hypothalamus. The effects of acupuncture have also been attributed to alterations in the natural electrical currents or electromagnetic fields in the body.

The use of acupuncture for pain relief has given equivocal results. Wallis et al.[17] in San Francisco in 1974 found that 19 of 21 patients had inadequate pain relief based on pain scores and none of the 21 subjects was judged by investigators to have adequate analgesia. Abouleish et al.[18] in the USA in 1970 used electroacupuncture and described relief of pre-existing pain in 7 of 12 participants. The investigators, who found the technique time-consuming, cited some disadvantages: inconsistency, unpredictability and lack of completeness. Hyodo et al.[19] in Japan in 1977 in a study of subjective and objective relief of labor pain in 32 women described an improvement in approximately 60% of 16 primiparous women and 90% of 16 multiparous women. However, all patients received systemic sedation. The authors concluded that acupuncture is useful for delivery because of its safety, despite erratic and less potent results than conventional analgesic techniques. In Nigeria in 1986, 19 of 30 women (63.3%) given sacral acupuncture by Umeh[20] indicated that they had adequate pain relief by responses on a visual scale and did not request another form of analgesia. Yanai et al.[21] in Israel in 1987 evaluated electroacupuncture during the labor of 16 parturients. Fifty-six percent of the women reported mild to good pain relief and 81% described increased relaxation. The perceived positive effects led the authors to believe that acupuncture should be pursued as an additional method of pain control. In 1999, Lyrenas et al.[22] in Sweden studied 31 primiparous women who had received repeated acupuncture compared with untreated women. Pain assessed on a visual analogue scale was not reduced in women treated with acupuncture, and the acupuncture did not reduce the need for analgesics during labor. Finally, in 2002 Ramnero et al.[23] in Sweden reported a randomized, controlled study in 90 parturients, 46 of whom received acupuncture during labor as a complement or alternative to conventional analgesia. Acupuncture significantly reduced the need for epidural analgesia (12% vs. 22%). Patients in the acupuncture group reported a significantly greater degree of relaxation compared with the control group. The authors considered these results to suggest that acupuncture could be a good alternative or complement for women who seek an alternative to pharmacological analgesia in childbirth, but further trials are required to clarify whether the main effect of acupuncture during labor is analgesic or relaxing.

To summarize, acupuncture studies are difficult to conduct and analyze for several reasons, including lack of standardization with use of multiple acupuncture points, and the difficulty of choosing a control group. Within the control group the needles might be correctly placed but not stimulated, or needles could be placed in inappropriate sites. There were no reported complications in any of the studies but there is a potential risk of infection. To achieve a good analgesic effect during labor, a relatively long induction period may be required. It is difficult for a woman in labor to remain still for 15 to 30 minutes, and some patients felt discomfort because of the restrictions in movement.

Overall, acupuncture may have a beneficial therapeutic effect on labor pain but better designed studies need to be completed with, if possible, standardization of the points used, and better control groups. It should also be recognized that the procedure is time-consuming, and that the required training of patients and personnel alike may be considerable.

## 5.2. Acupressure Systems

Acupressure is a descendant of Chinese manipulative therapy in which points are stimulated by pressure, using hands, fingers and thumbs[24]. Some midwives use acupressure to release the pain of labor. Pressure is applied simultaneously to both sides of the spine in the lower back. Pressure against spots that are sensitive can be particularly efficacious. Force is initially applied during contractions and then continuously.

## 5.3. Homeopathy

Homeopathy involves the use of diluted substances that cause symptoms in their undiluted form. According to homeopathic theory, remedies stimulate the self-healing mechanism. The amount of medicine prescribed is so small that it often cannot be measured in molecular amounts[25]. We found no studies evaluating the effect of homeopathic treatment on labor pain. Smith[26] has reviewed cervical ripening and labor induction by "caulophyllum". There were no differences between the homeopathy and control groups in a randomized, controlled trial involving 40 women.

## 5.4. Manual Healing

Manual healing methods used today during delivery include therapeutic touch and massage therapy.

### 5.4.1. Therapeutic Touch

The purpose of therapeutic touch in labor is to communicate caring and reassurance. Painful contractions of the uterus can be treated by the application of pressure with the hands to the woman's back, abdomen, hips, thighs, sacrum or perineum. Whether touch is perceived as positive or not is dependent on who is touching the patient: in one study, touching was perceived positively by 94% of patients when they were touched by a relative or friend, 86% by their husbands, 73% by a nurse, and 21% by a physician[27]. Anxiety is reported to be reduced in patients who receive reassuring touch. In a retrospective study of 30 patients, 77% experienced "less pain" when they were touched during labor and 40% reported less need for pain medication.

### 5.4.2. Massage Therapy

The practice of massage varies from the tickling massage of "kung fu" to a firm massage. The effect of a gentle massage of the periumbilical area by their partner was studied in nine women compared with six who received no massage[28]. There was no significant difference between the two groups in pain evaluated by means of visual scale, and no difference in the time of use of epidural analgesia for labor.

## 5.5. Bioelectromagnetic Applications and Physical Methods

### 5.5.1. Transcutaneous Electrical Nerve Stimulation

TENS involves administration of low voltage electrical stimuli through flat electrodes applied to the skin (2). TENS units consist of a stimulator and two pairs of electrodes. The upper electrodes are taped at the level of the tenth thoracic to the first lumbar root and the lower pair at the level of the second to fourth sacral nerves. The stimulator has two channels for the two pairs of electrodes. Initial reports were encouraging. Augustinson et al.[29] (1977) found that among 147 women, 44% rated pain relief as good or very good and 44% as moderate. Bundsen et al.[30] found TENS to be especially beneficial for labor pain localized in the back. However, a meta-analysis by Carroll et al.[31] of 10 randomized, controlled trials in 877 women, 436 receiving TENS and 441 as controls, revealed no significant difference in pain and the use of additional analgesic interventions was not different between the two groups.

### 5.5.2. Sterile Water Blocks

Counter-irritation is the process by which localized pain may be relieved by irritating the skin in the same dermatomal distribution. For example, the uterus is supplied by the lower thoracic spinal cord segments. Some of these receive stimuli from the skin of the lower back and the sacrum. Labor analgesia may be produced by counter-irritation of this area. Irritation can be achieved by intracutaneous injection of sterile water papules over the sacrum with a fine needle. Lytzen et al.[32] in Sweden noted instant and complete relief of lower back pain in the first stage of labor in 83 women. Pain relief lasted as long as three hours. In some women the block was repeated. Half of the women required no other form of analgesia. Labrecque et al.[33] compared water blocks with TENS for the treatment of lower back pain during labor. Women who received the sterile water injections rated the intensity of pain lower than did women in the TENS group. Intracutaneous water injections are associated with a sharp injection pain that lasts between 20 and 30 seconds which some women find less acceptable than lower back pain. This method may be an alternative for women who have lower back pain during labor but wish to avoid epidural analgesia.

### 5.5.3. Hydrotherapy

The popularity of undergoing part of labor in water has increased dramatically around the world. The expected benefits include pain relief and decreased use of analgesia and anesthesia. Several studies have reported use of analgesia for women undergoing labor in water, but others have found no difference from control groups. In a 1987 non-randomized, prospective, controlled study, Lenstrup et al.[34] evaluated the effect of a warm bathtub on 88 parturients, and found that cervical dilation rate and pain relief could be improved in patients who had a bath during the first stage of labor. In 1988, independent midwives used hydrotherapy in which a clean bathtub was filled with warm water[24]. The clinical impression of practitioners who use hydrotherapy is that their patients experience shorter and less painful labor. Burn et al.[35] studied 302 women who used a labor pool. Fifty percent of the primigravidas in the pool group used pain medication, compared with 76% in the control group. Rush et al.[36] found in a randomized study in 785 women that the tub group required fewer pharmacological agents than the control group (66% vs. 59%, P = 0.06). Cammu et al.[37] in a prospective randomized trial using a visual analogue scale showed that absolute values of labor pain were not significantly different between hydrotherapy and control groups.

Maternal satisfaction with this birth experience has been measured and women report increased levels of satisfaction, self-esteem, pain relief and relaxation with immersion. Ruptured membranes have been discussed as a potential problem in the use of hydrotherapy, although in the study of Lenstrup et al. they were not considered a contraindication. Odent et al.[38] reported no infectious complications in patients who gave birth in water, even if the membranes were already ruptured.

## 6. Alternative Medications

### 6.1. Herbal Medicine

Herbal medicine is described as the use of plant materials in medicine and food for therapeutic purposes. Various herbal remedies are used during the prenatal period to "prepare" the uterus and cervix for childbirth and ease pain during labor and delivery.

In a study of the practice of a group of independent midwives in Utah[24], specific herbs were used because of their perceived actions and properties, in particular a "5-week formula" which is a combination of ten herbs used during the last five or six weeks of pregnancy. This is said to facilitate birth. Some herbal remedies are used as the principal method of managing pain and enhancing endurance during delivery. Practitioners observed that these herbal formulas had a calming and relaxing effect. Labor pain can also be treated specifically with motherwort. The effect of raspberry leaf in facilitating labor in 192 multiparous women was studied by Simpson et al.[39] in a double-blind, randomized, placebo-controlled trial in Australia. Raspberry leaf was consumed in tablet form from 32 weeks of gestation until labor. Contrary to popular belief, it did not shorten the first stage of labor but rather the second (mean difference 9.59 minutes), and also lowered the rate of forceps deliveries (19.3% vs. 30.4%). The difficulty with herbal remedies is that few have undergone scientific scrutiny, chemical isolation, or extraction to identify the pharmacologically active agent or to enable toxicity testing.

## 6.2. Aromatherapy

Aromatherapy uses essential oils extracted from aromatic botanical sources to treat and balance the mind, body and spirit[25]. It combines the physiological effects of massage with the use of essential oils. One of the purposes of this method is to relieve anxiety and stress and to help relaxation. Massage around the lower back with jasmine, juniper, geranium, clary sage, rose and lavender have been reported to provide subjective benefit in labor.

## 7. Conclusion

Complementary and alternative medicine can be defined as methods that are not currently part of the dominant or conventional medical system. CAM exists because conventional medicine can be limited in its ability to provide relief and to meet patients' needs. CAM and conventional medicine share the responsibility for applying evidence-based practice and for seeking scientific proof to justify a planned intervention, as well as the obligation to avoid harmful or useless practices.

For labor pain, most studies demonstrate the greatest benefit during the beginning of the dilation phase. When women enter the active phase of dilation or during delivery itself, there is more need for additional conventional analgesics. This suggests that complementary medicine may be useful for the early onset of pain or as a distracter, diverting women's attention from the source of pain. In some cases the number of parturients who successfully use alternative methods is greater than what would be expected from a placebo effect. In a few cases the amount of pain medication was reduced but this was not consistently true. The degree of success of a method is correlated with the availability of support staff in both educational and trial phases of the studies, and necessarily in clinical practice. Whereas physicians do not need to be experts in the management of alternative therapies, they should at least possess some basic knowledge of complementary medicine. In the future, the demand for complementary medicine will probably continue to rise. Care providers have to facilitate informed choices through discussion of their own experience and knowledge. One of the difficulties for the physician is to identify studies sufficiently well-designed to help them guide their patients.

## 8. Key Words

Labor, delivery, pain, hypnosis, biofeedback, yoga, sophrology, stimulation, acupuncture, transcutaneous electrical nerve stimulation, sterile water blocks, manual healing, hydrotherapy, homeopathy.

## 9. References

1. R. A. Chez, and W. B. Jonas, The challenge of complementary and alternative medicine, *Am. J. Obstet. Gynecol.* **177**, 1156–61 (1997).
2. B. A. Gentz, Alternative therapies for the management of pain in labor and delivery, *Clinical Obstet. Gynecol.* **44**, 704–735 (2001).

3. N. K. Lowe, The nature of labor pain, *Am. J. Obstet.* **186**, 16–24 (2002).
4. R. Melzack, The McGill Pain Questionnaire: major properties and scoring methods, Pain, **1**, 277–299 (1975).
5. R. Melzack, The myth of painless childbirth, Pain, **19**, 321–337 (1984).
6. H. Cardin, M. T. Moisson Tardieu, and M. Tournaire, *La péridurale*, Balland Paris (1986).
7. J. J. Bonica, *Textbook of pain*, Churchill-Livingstone, Edinburgh (1984).
8. P. Brownridge, and S. Cohen, Neural blockade for obstetrics and gynecologic surgery. In *Neural Blockade in Clinical Anesthesia and Management of Pain*. J. P. Lippincott Company, Philadelphia (1988).
9. J. A. Davidson, An assessment of the value of hypnosis in pregnancy and labor, *Br. Med. J.* **2**, 951–953 (1962).
10. T. M. Harmon, M. J. Hynan, and T. E. Tyre, Improved obstetric outcomes using hypnotic analgesia and skill mastery combined with childbirth education, *J. Consult. Clin. Psychol.* **58**, 525–530 (1990).
11. D. Mairs, Hypnosis and Pain in Childbirth, Contemp Hypnosis, **12**, 111–118 (1995).
12. P. Duchene, Effects of biofeedback on childbirth pain, *J. Pain Symptom Manage*, **4**, 117–123 (1989).
13. S. H. Bernat, P. J. Wooldridge, M. Marcki et al., Biofeedback-assisted relaxation to reduce stress in labor, *J. Obstet. Gynecol. Neonatl. Nurs* **21**, 295–303 (1992).
14. F. Veldman, Haptonomy science de l'affectivité, *Presse Universitaire de France*, Paris (1989).
15. C. Marwick, Leaving concert hall for clinic, therapists now test music's "charms", *Jama,* **275**, 267–268 (1996).
16. L. Durham, and M. Collins, The effect of music as a conditioning aid in prepared childbirth education, *J. Obstet. Gynecol. Neonatal Nurs.* **15**, 268–270 (1986).
17. L. Wallis, S. M. Shnider, R. J. Palahniuk et al., An evaluation of acupuncture analgesia in obstetrics, *Anesthesiology*, **41**, 596–601 (1974).
18. E. Abouleish, and R. Depp, Acupuncture in obstetrics, *Anesth. Analg.* **54**, 82–88 (1975).
19. M. Hyodo, and O. Gega, Use of Acupuncture Anesthesia for Normal Delivery, *Am. J. Chin. Med.* **5**, 63–69 (1977).
20. B. Umeh, Sacral acupuncture for pain relief in labor: initial clinical experience in Nigerian women, *Acupunct. Electrother Res.* **11**, 147–151 (1986).
21. N. Yanai, E. Shalev, E. Yagudin, and H. Zuckerman, The use of acupuncture during labor, *Am. J. Acupunct.* **15**, 311–312 (1987).
22. S. Lyrenas, H. Lutsch, J. Hetta et al., Acupuncture before delivery: effect on Pain Perception and the Need for Analgesics, *Gynecol. Obstetric. Invest.* **29**, 118–124 (1999).
23. A. Ramnero, V. Hanson, and M. Kihlgren, Acupuncture treatment during labour. A randomised controlled trial, *B.J.O.G.* **109**, 637–644 (2002).
24. C. Sakala, Content of care by independent midwives: assistance with pain in labor and birth, *Soc. Sci. Med.* **26**, 1141–58 (1988).
25. A. D. Allaire, Complementary and alternative medicine in the labor and delivery suite, *Clinical Obstet. Gynecol.* **44**, 681–691 (2001).
26. C. A. Smith, Homeopathy for induction of labour, *Cochrane Database Syst. Rev.* CD 003399 (2001).
27. K. S. Penny, Postpartum perceptions of touch received during labor, *Res. Nurs. Health* **2**, 9–16 (1979).
28. L. W. Hedstrom, and N. Newton, Touch in labor: a comparison of cultures and eras. Birth **13**, 181–186 (1986).
29. L. E. Augustinsson, P. Bohlin, P. Bundsen et al., Pain relief during delivery by transcutaneous electrical nerve stimulation. Pain **4**, 59–65 (1977).
30. P. Bundsen, L. E. Peterson, and U. Selstam, Pain relief labor by transcutaneous electrical nerve stimulation, a prospective matched study, *Acta Obstet. Gynecol. Scand.* **60**, 459–168 (1981).
31. D. Carroll, R. A. Moore, M. R. Tramer et al., Transcutaneous electrical nerve stimulation does not relieve labor pain: updated systematic review, *Contemp, Rev. Obstet. Gynecol.* **9**, 195–205 (1997).
32. T. Lytzen, L. Cederberg, and J. Moller Nielsen, Relief of low back pain in labor by using intracutaneous nerve stimulation (INS) with sterile water papules, *Acta. Obstet. Gynecol. Scand.* **6**, 341–43 (1989)
33. M. Labrecque, A. Nouwen, M. Bergeron. et al., A randomized controlled trial of nonpharmacologic approaches for relief of low back pain during labor, *J. Fam. Pract.* **48**, 259–263 (1999)
34. Lenstrup, A. Schantz, A. Berget, E. Feder, H. Roseno, and J. Hertel, Warm tub bath during delivery, *Acta. Obstet. Gynecol. Scand.* **66**, 709–712 (1987).
35. E. Burn, and K. Greenish, Water birth. Pooling information, *Nurs. Times* **89**, 47–49 (1993).

36. J. Rush, S. Burlock, R. Lambert et al., The effects of whirlpool baths in labor: A randomized controlled trial, *Birth* **3**, 136–143 (1996).
37. H. Cammu, K. Classen, I. Van Wettere et al., To bathe or not to bathe during the first stage of labor, *Acta Obstet. Gynecol. Scand.* **73**, 468–472 (1994).
38. M. Odent, Birth under water. *Lancet* **2**(8365–66), 1476–1477 (1983).
39. M. Simpson, M. Parsons, J. Grenwood, and K. Wade, Raspberry leaf in pregnancy: its safety and efficacy in labor, *J. Midwifery Womens Health* **46**, 51–59 (2001).

# 13

# Alternative and Comparative Medicine in Dentistry

Kazuo Komiyama, Kazuyosi Koike, Masahiro Okaue, Takahiro Kaneko, Mitsuhiko Matsumoto

## 1. Introduction

In Japan, Oriental herbal medicines (Kampo) are clinically designated as drugs that can be prescribed by any doctor. The quality of the materials used in Kampo are well controlled by pharmaceutical companies. One of the leading Japanese companies, Tsumura Pharmaceuticals Co. Japan, classifies the herbal materials according to the sub-species of the plant and the place it was collected or grown. Kampo is no longer an old or mysterious prescription and its use must be re-evaluated and considered as a choice for therapy today.

In Kampo therapy, different drugs are used for the same disease, depending on the individual's constitution and the state of the disease. The therapeutic effects of certain Kampos differ markedly among patients according to their different physical constitutions and responsiveness to the drugs. Since genetic difference should also be a determinant of a patient's responsiveness to Kampos, it is plausible that a certain Kampo may exert extremely different actions in a host-dependent manner. To our knowledge, however, there have been few studies on Kampo from such a perspective.

In order to conduct such a study and acquire a mass of clinical data under strict scientific methods, prescriptions for Kampo medicine would need to be standardized. In this paper, we describe the background of alternative and comparative medicine in dentistry and introduce our clinical trails using Kampo for various disturbances in saliva secretion as well as a review of literature on the subject.

---

**Kazuo Komiyama** • Departments of Pathology; Division of Immunology and Pathobiology, Dental Research Center, Nihon University School of Dentistry  **Kazuyosi Koike** • Oral diagnosis  **Masahiro Okaue** • Oral and Maxillo-facial Surgery  **Takahiro Kaneko** • Oral and Maxillo-facial Surgery, Saitama Medical Center, Saitama Medical School  **Mitsuhiko Matsumoto** • Oral and Maxillo-facial Surgery.
Correspondence: Kazuo Komiyama, Department of Pathology, Nihon University School of Dentistry, 1-8-13, Kanda-Surugadai, Chiyoda-ku, Tokyo 101-8310, Japan. E-mail:komiyama@dent.nihon-u.ac.jp

*Complementary and Alternative Approaches to Biomedicine*, edited by Edwin L. Cooper and Nobuo Yamaguchi. Kluwer Academic/Plenum Publishers, 2004.

## 2. Alternative and Comparative Medicine in Japanese Dentistry

The history of dentistry is quite old, osteomyelitis of the jaw bone, exostosis and caries have been identified in the teeth and skull of Stone Age man [1]. Many of the skulls had teeth worn down to such an extent, due to masticating hard food, that the pulp had become exposed, causing extensive suppurative destruction. Dentists taking a special position in the special medical profession, are described on "papyrus" [1]. In Japan, dental treatment during the Edo period was aimed primarily at curing toothache, by means of medication, moxibustion, acupuncture, hot irons, charms and incantations. Tooth extractions were also performed.

An example of a Kampo approach to gingivitis and/or oral pain is as follows: sea kelp, the leaf of Nandina domestica, and the meat of pickled Japanese plums are rolled together into a ball. This conglomeration was then placed in a small silk pouch and kept in the mouth at the site of distress.

Worthy of mention when discussing dental treatments during the Edo period are wooden denture plates. These dentures are similar to present day suction plate dentures. Japan preceded the West by more than 300 years in the development of plate dentures. These dentures were designed to have both aesthetic and functional aspects. Japanese dentures

**Figure 13.1.** A dentist extracting the teeth. (By permission from the Dental Historic Museum at Matsudo, Nihon University)

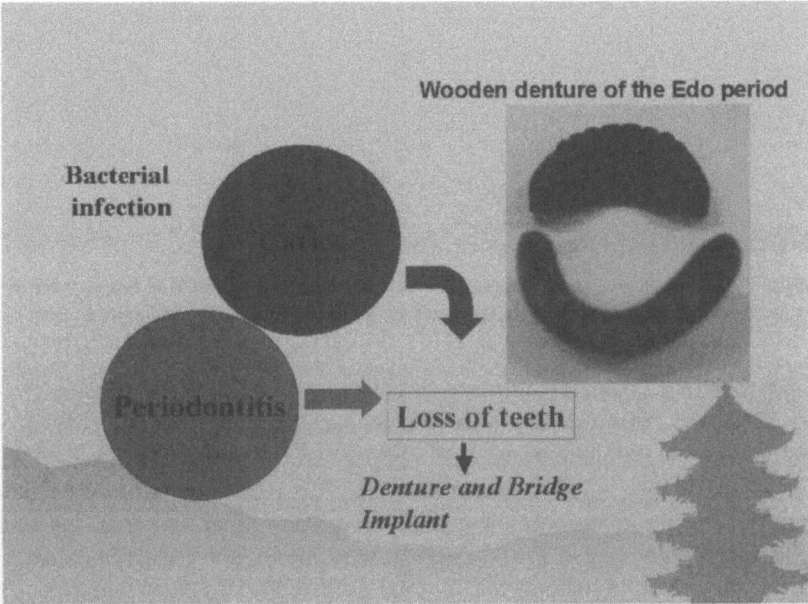

**Figure 13.2.** Dental medicine is almost all of an alternative nature. The causes for missing teeth can be controlled by brushing. Missing teeth are replaced by bridges, implants and dentures. A wooden denture of the Edo period is reproduced here with permission from Dental Museum at Matsudo, Nihon University

were a natural looking product to replace teeth that had been lost. Surprisingly enough, these dentures enabled people to chew quite adequately.

These dentures were made of a variety of materials. Pagodite (talc), ivory, natural teeth, animal bones, and other substances were used for anterior teeth. Copper rivets and nails were pounded into the molars to create a chewing surface. Boxwood, cherry wood, and apricot wood were used to form the denture base. The teeth were tied to the base with "shamisen" (Japanese balalaika) string [2]. Since nathology and occlusion theories were not developed at the time, the technology was aimed at speaking and eating. Dental technology of today is based on modern medicine: anatomy, physiology and occlusions.

### Xerostomia: a coming issue in dentistry

Saliva plays an extremely important role in maintaining oral physiological conditions. Saliva keeps the mouth moist and helps the tongue to move. Furthermore, saliva is required when swallowing food. Saliva not only has these functions, but it also acts as an agent for bio-defense molecules against various pathogens. Saliva contains numerous anti-bacterial molecules: lysozyme, secretory IgA, hormones, and enzymes. Saliva dilutes pathogenic substances and chemicals and protects oral mucosa and digestive tracts.

Saliva is synthesized from the major salivary glands: the parotid gland, sub-mandibular gland, and sublingual gland, and is additionally secreted by minor salivary glands of mouth

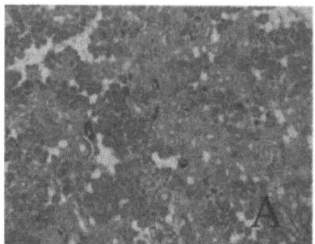 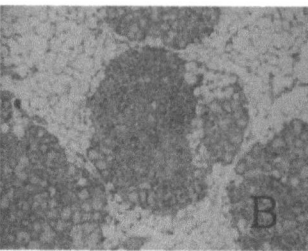 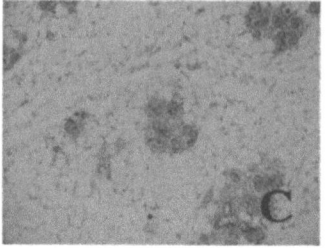

**Figure 13.3.** Loss of salivary glands in patients with Xerostomia. The number of lobules were decreased in elderly patients and those with Sjogren's syndrome. A: normal adult. B: Sjogren syndrome. C: aged person

submucosa. The total secreted volume of an adult is 1.5 L per day on average. The volume of saliva differs greatly from person to person, and a reduction in the volume of saliva is caused by various problems in the oral mucosa and functions. The number of salivary glands is reduced by aging and disease conditions such as autoimmune disorders like Sjogren syndrome. In cancer patients receiving radiation therapy of an oral area, the acinic cells of the glands undergo apoptosis and necrosis that causes a decrease in saliva flow and Xerostomia. Moreover, in patients with a total neck resection, the submandibular glands are totally removed. In these patients, artificial saliva is administrated to help them swallow food. Another factor of Xerostomia is age. The number of salivary gland-lobules decrease with aging.

## 3. Clinical Trials of Kampo on Patients With Xerostomia

### Patients and Methods

#### Patients

A total of 50 patients (Female: 35, Male: 15) participated in the present study, including 10 with primary Sjogren's, 15 with dry mouth due to receiving radiation, 5 with total neck resection (including sub-mandible glands), 20 with dry-mouth and no disease conditions (probably aging).

#### Treatment protocol

All the patients received a general physical examination and a rheumatological evaluation before entering the study. Bakumondo-to (Tsumura Pharmaceuticals Co) was administered in a 6 g pack of granules of ophiogonins extract formula, three times a day after meals.

Bakumondo-to contains the following six herbs: Ophiopogonis tuber (*Opiopogon japonicus ker-gawler Lilliacese*), Ginsen radix (*Panax ginseng C.A. Meyer Araliaces*), Glycyrrhizae Radix (*Glycyrrhiza glabra L. leguminosae*), Oryzae Fructus (*Oryzae sativa L. Graminease*), pinelliae Tuber (*Pinellia Teranata Breitenbach Arecase*) and Zizyphi Fructus (*Zizyphus jujuba Miller Var. Rhamnacease*). The patient's subjective symptoms were examined before, and 4 weeks after beginning the therapy. To assess the effects of Bakumondo-to on saliva secretion, gum tests were performed in all patients before and after the therapy. The total saliva was pooled in a 50 ml polystyrene tube for 10 minutes.

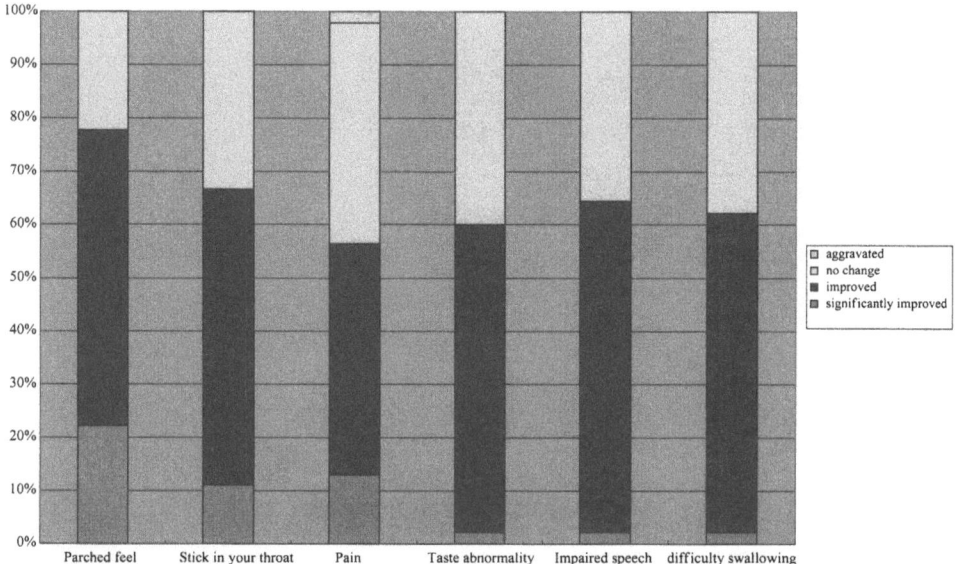

**Figure 13.4.** The effects of subjective symptoms of patients administered with Bakumondo-to for 4 weeks.

### Results and discussion

After 2 weeks of Bakumondo-to therapy, 87.0% (39 out of 45) of the patients participating in this study showed an improvement in their clinical condition The subjective symptoms of patients improved: 77.7% stated that the parched feeling in their mouths had improved; 66.6%, stickiness in mouth, 57%, pain of the oral mucosa, 59.9%, taste abnormality, 64.4%, impaired speech, and 62.2%, difficulty in swallowing (Fig. 13.4).

The mean volume of saliva secreted (Gum test) increased significantly from $7.5 \pm 5.3$ to $11.6 \pm 5.8$ ($P < 0.005$) (Fig. 13.5). No apparent clinical effects were observed other than dry-mouth and the symptoms in this study. However, a week after discontinuing the administration of Bakumondo-to, most of the patients reverted to pre-treatment conditions. No serious adverse reactions were observed in the patients, except for one who complained of a mildly abnormal sensation in the oral cavity. This side effect disappeared within a few days without any additional therapy when Bakumondo-to administration was discontinued.

The mechanisms of Bakumondo-to that caused this positive clinical effect in many patients are not clear. Pathological examination showed that the salivary glands had not regenerated and the number of glands had not increased in the participants. However, similar results have been obtained in other studies. Ohno, S. et al. [7] examined the effects of a Kampo prescription in 40 patients with Sjogren's syndrome (referred to as SJS herein after). The Kampo prescription "Hochu-ekki-to" was administered to 32 SJS patients as a control group. These Kampo prescriptions were granule extracts prepared by Tsumura Pharmaceuticals Co. Each participant took the extract continuously for four weeks. A gum test was carried out prior to the start of the experiments to avoid learned responses. The saliva levels were based on the measured values at the second gum test and subsequent tests.

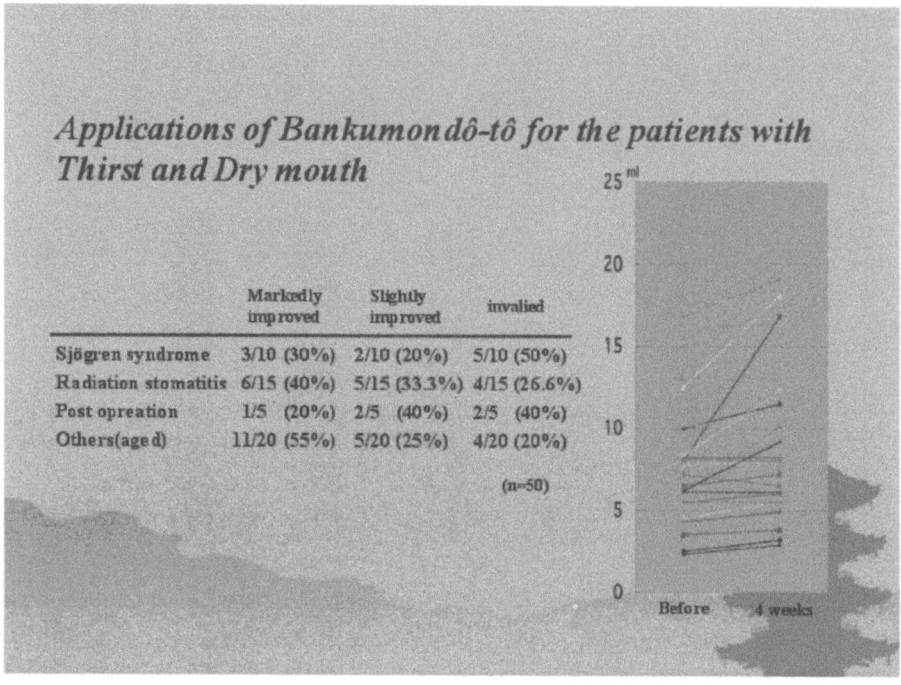

**Figure 13.5.** Improved saliva flow after Bakumondo-to treatment. The total saliva was collected 10 min before and after administering Bakumondo-to

The results indicated that Bakumondo-to increased saliva secretion with a post-administration level of $11.4 \pm 1.4$ ml, compared with the pre-administration value of $8.2 \pm 1.1$ ml ($p < 0.01$). The increase in saliva secretion was $3.16 \pm 0.78$ ml in the group given Bakumondo-to, but no increase was observed in the group given Hochu-ekki-to; showing a significant difference ($p < 0.005$). Gradual increases in saliva secretion were observed in patients who took Bakumondo-to over a long term ($r = 0.7290$). The patients were classified in stages according to sialograms. A marked increase in saliva secretion was found in those at light stages. A significantly large number of patients who showed a reaction to Bakumondo-to were diagnosed, according to Oriental medicine, as "Kyo-sho" of Yin-Yang relativity of the lower section of body.

To investigate its effective components, Yamamoto, S. et al. [5] administered Bakumondo-to to healthy subjects and patients with bronchial asthma; and attempts were made to identify the components excreted in urine. Of the 5 substances detected, only 3 were identified: medicarpin and liquititigenin, a component of Kanzo (Glycyrrihizae radix) and davidigenin, which is a metabolic product of Kanzo (glycrrhizase radix) flavonoid. Investigation of the speed of secretion into the urine showed different characteristics for each susbstance. The secretion of medicarpin was almost complete as early as 3 hr after administration. In contrast, no davidigenin was excreted for 3 hr, excretion starting from between 3 and 6 hrs after administration and continuing for 24 hrs. Furthermore, the dynamics and volume of excretion varied from subject to subject, leading to the assumption that Kampo efficacy is closely related to individual differences. To elucidate the mechanism of saliva secretion increase by Bakumondo-to, Wakui et al. [6] applied Bakumondo-to directly to isolated rat submandibule acinus cells.

The data of microfluorometry showed an increased intercellular $Ca^{2+}$ level following extracellular administration of Bakumondo-to. Measurement of membrane currents using the whole-cell patch-clamp method revealed that Bakumondo-to evoked fluctuation in the $K^+$ and $Cl^-$ currents. These changes in the ionic membrane currents were intracellular $Ca^{2+}$ dependent and were also produced by extracellular administration of acethlcholine (Ach), a physiological agonist, intracellular administration of GTP-$\gamma$S, an activator of GTP-binding proteins, and inositol triphosphate ($IP^3$). The fluctuation of the membrane currents was ascribed to the repeated elevation of intracellular $Ca^{2+}$ by stimulation. Administration of atropine markedly inhibited the Bakumondo-to evoked responses of the membrane currents. These results suggest that Bakumonndo-to activates muscarinic receptors on the membrane of submandibular acinar cells and the subsequent mobilization of intracellular $Ca^{2+}$ by $IP^3$ causes secretion of saliva.

The above data clearly demonstrate that Bakumondo-to is helpful for patients with disturbed saliva flow such as xerostomia. The effect of Bakumonodo-to on increasing saliva secretion lasted a relatively short time, with the patient's condition soon reverting after discontinuing its administration. Further study is required with large scale and long-term administration as well as experiments investigating the role of the 6 different herbs in Bakumondo-to.

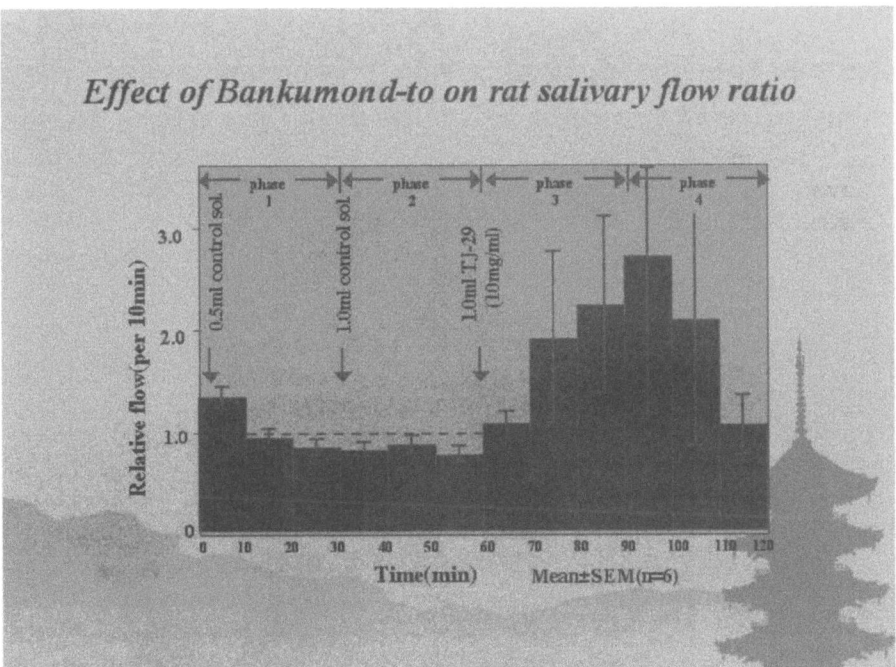

**Figure 13.6.** Effects of Bakumondo-to on salivary flow. Saliva was collected every 10 min by cannulation of submandibular glands. Bakumondo-to was administrated through the femoral vein (10 mg/ml), from ref. 6

## 4. References

1. Weinberger, B. W. (1982) The teeth of prehistoric man and his descendants: In An Introduction to the History of Dentistry, The C. V. Mosby Co. pp. 13–77.
2. Sen Nakahara et al. Dentistry in Japan: In Manners and customs of dentistry in ukiyoe, Ishi Yaku Printe Ltd. pp. 133–134.
3. Ohono, S. et al. (1995) Effects of Bakumonodo-to on saliva secretion in patients with Sjogren syndrome. 33–40.
4. Wakui et al. (1992) Mechanism of saliva secretion by Bakumonodo-to. Kampo and Imumo-Allergy 6, 64–73. (Japanese)
5. Yamamoto, S. et al. (1994) Effective components in Bakumonodo-to (Mai men dong-tang).
6. Kampo and Imumo-Allergy 8, 89–97. (Japanese)

# III

# Physical Intervention: Touch, Hydrotherapy, Sound

# 14

# Methodological Concerns when Designing Trials for the Efficacy of Acupuncture for the Treatment of Pain

## P.J. White

## 1. Introduction

It has taken some considerable time for acupuncture to gain any form of respectability within western medicine. Indeed there is still some resistance from many quarters where acupuncture is still viewed with a large dose of scepticism. Not least is the problem that many of the concepts inherent within traditional Chinese medicine (TCM) do not translate well into western scientific thinking[1]. It seems churlish however to discount this modality because of an inability to either grasp its basic tenets or due to perhaps an inadequacy of western science to adequately measure the forces that TCM maintains exists. It must be remembered that TCM has evolved within a framework of pure observation of cause and effect and has anecdotally been shown to be effective over millennia. Adherents may maintain that the inability to provide 'proof' does not diminish its usefulness as an effective treatment. It is perhaps this very attitude, which relies on anecdotal evidence rather than scientific evidence, which has alienated many in the western medical profession[2]. The idea that acupuncture is simply 'mumbo jumbo' is however gradually subsiding within the west. This is occurring in the wake of ongoing scientific investigation and credible hypotheses as to its mechanism as well as the emergence of guidelines designed to assist with the design and reporting of trials, both in general and specifically for acupuncture[3–5].

The subject of what constitutes methodologically rigorous research into the efficacy of acupuncture has long been a topic for heated debate. Much has been written on the subject of acupuncture and a Medline search revealed a mean publication rate of 240 articles per year between 1976 and 1996[6]. Despite this, its efficacy in the context of pain relief is still a contentious issue. Trial methodologies have frequently been severely flawed with problems such as insufficient subject recruitment, non-blinding of patients, non-randomisation, inappropriate control or inappropriate treatment being fairly common[7]. It is therefore not surprising that results obtained from studies have often been inconclusive and contradictory.

---

*Complementary and Alternative Approaches to Biomedicine*, edited by Edwin L. Cooper and Nobuo Yamaguchi. Kluwer Academic/Plenum Publishers, 2004.

The model of the randomised, double blind placebo controlled trial has historically evolved within the pharmaceutical industry where this method of testing is ideally suited and has been accepted as the 'gold standard'[8], although its use and in particular the practice of blinding has been contested[9]. The randomised, controlled, parallel arm trial methodology has extended its influence away from the pharmaceutical industry and has largely been adopted as standard practice for clinical research within complementary medicine. Whilst the search for definitive answers through the use of this method is to be wholly applauded, it does however pose particular problems for manual therapies, such as acupuncture, where it becomes extremely difficult to conduct a double blind trial. The practitioner may unintentionally communicate their expectations or preferences to the patient and this may be a major source of bias[10;11]. Another problem, which is particularly germane within complementary medicine, is that most disciplines within this field would purport to be holistic in nature (acupuncture is no exception to this). The utilisation of the randomised controlled trial is very reductionist in nature[12] and seeks to quantify or justify one very small facet of treatment and again this may prove to be an ongoing problem for acupuncture research.

Another alternative to the parallel arm controlled trial might be a crossover design. In this case, patients each receive two different types of treatment i.e. one course after another. This would certainly be acceptable from an 'intention to treat' perspective as everybody would ultimately receive both treatments. There are however problems associated with this and particularly so when conducting an acupuncture trial. Patients may favour a particular intervention and if the interventions are overtly different to each other (e.g. mock TENS as a control), patients may then be reluctant to change to the alternative treatment. Similarly, a certain amount of objectivity may be lost, as the patient is obviously not blinded to the nature of the treatment even if they are blinded to the fact of it being an inactive control. Secondly it is not known how long any beneficial effects of acupuncture may last as it has a variable and unpredictable time based response. Therefore, it has not been possible to clearly define a suitable 'wash out' period for this treatment. A recent trial involving acupuncture for chronic neck pain has suggested a treatment effect lasting a year or more[13]. It would therefore be impossible to clarify the relative merits of either treatment in a crossover design as one may 'contaminate' the other unless there were a very long period between treatments[14]. Clearly this would not be practical within time constraints often imposed by funding bodies.

There has been much debate as to what constitutes appropriate methodology and the main issues are now discussed.

## 2. Blinding

The nature of acupuncture is such that it is a 'fluid' and dynamic treatment modality. This requires that the practitioner is able to react to changes in the patient's condition and therefore make 'on the spot' decisions as to how best to proceed. It may, for example, be required to change the points selected either prior to or even during an individual treatment session in the light of the patient's response[15]. This poses problems for acupuncture research. Clearly this necessitates the practitioner having knowledge of the patient's condition in order to facilitate adequate treatment. This in turn implies, as with any invasive technique, that double blinding is impractical for effective acupuncture studies[16;17]. Pure double blinding has been attempted[18] whereby a third party told the acupuncturist which area needed treating. The practitioner was not allowed to speak to the patient and in 50% of cases they were given an inappropriate area to treat. Whilst this was indeed double blinding, this approach will inevitably lead to problems with the treatment itself. Clearly the practitioner was not able to react to changes in the patient's condition, neither was he/she able to adequately palpate for tender (or ahshi) points both of which may have led to inferior treatment being provided. Lastly it might not, under these circumstances, have been possible to ensure that even points used at the inappropriate area, would have no effect on the real condition either through the mechanism of diffuse noxious inhibitory control DNIC[19] or simply because acupuncture points often have multiple effects. Unfortunately if there is no blinding at all the expectations of both the patient and the therapist may have a strong influence on the results of a trial. The best option to ensure maximum objectivity therefore is to blind the patient, use patient centred outcome measures and to use independent blinded assessors where possible[11;20]. This would maintain objectivity in reporting and gathering of data but of course does not solve the problem of therapist bias. In this case, it is imperative to ensure standardisation of treatment and contact with the patient as much as possible and this is attainable by rigorous protocol design and application. Furthermore, anonomisation of patient records and blind analysis of results will also aid in reducing bias as a confounding factor.

## 3. Homogeneity and Treatment Protocol

Conditions examined in a trial must be limited to one specific pathology or painful site as efficacy may vary from condition to condition[21;22]. In the same way, acute pain may react very differently to chronic pain and it is essential that trial methodology reflect this in the inclusion and exclusion criteria. If many painful sites are included in a single trial, the results would be very difficult to interpret and it would be all too easy to draw erroneous conclusions.

It must be born in mind when examining the results of trials, that the treatment needs to be appropriate and adequate given the current state of knowledge[23] and must be such that it can be considered as being adequate by general consensus of those who use acupuncture. If a trial is undertaken therefore, it is imperative that the investigator has a good working knowledge of the complexities of the treatment regime used so that informed and logical decisions can be made as to its adequacy and therefore external validity. The question of what is 'adequate' is of course a matter for debate and may vary widely from practitioner to practitioner but this important problem needs to be addressed. A method for developing treatment protocol 'The BRITS method' has been suggested as a way forward to improving the quality of treatment in acupuncture trials[24]. It recommends among other things, a comprehensive review of literature in order to ascertain an optimum consensus of treatment protocol. As Birch[23] stated 'a clinical trial will be a fair test of acupuncture only if an adequate treatment is administered.' If the number of needles used is too few, this could seriously impair efficacy, as often, the combination of acupuncture points used may give a much more powerful effect than each point used individually[25]. Treatment regimes in a trial must therefore reflect clinical practice if we are to arrive at a realistic evaluation of its clinical effects.

Indeed it has been pointed out that whilst much critical research has focussed on the adequacy of the scientific protocol, the adequacy of the treatment itself must also be evaluated[26]. Ezzo et al[27] examined the aspects of acupuncture treatment i.e. how many treatment sessions and how many needles needed to be used in order to provide an adequate treatment. They concluded that treatments comprising of at least six sessions and using at least six needles each time were consistently associated with a positive result. One suggestion was that this might be due to a cumulative effect of treatment[28]. This would of course have serious implications when designing trials as those with too few sessions could be considered as giving a sub-optimal dose and therefore poor treatment. Clearly this could therefore result in a type II error.

The use of formulaic acupuncture at all has been disputed[29]. It is argued that traditional Chinese acupuncture works because it is individualised, therefore any attempt to standardise the treatment and adhere to a narrow pre-set selection of points for treatment does not constitute adequate therapy and could seriously restrict the efficacy of acupuncture[30]. It should be pointed out however, that as yet, there is no evidence to suggest that one form of acupuncture (western formulaic style or traditional Chinese) is superior to the other. Indeed there is evidence that points to poor inter-practitioner reliability when diagnosing using traditional Chinese methods[31]. The treatment protocol of any acupuncture research program must not be too narrow and should therefore contain sufficient flexibility to allow the practitioner to be able to select points that will most benefit the patient. Not to do so would constitute negligent research through poor treatment and this may negate any possible therapeutic advantages for acupuncture (again, a type II error). This could certainly be a large problem where 'eastern' philosophy has been adapted to try and fit into a 'western' experimental paradigm. Perhaps this is the wrong way around and experimental design should be adapted to cater for the flexibility inherent within acupuncture[32].

## 4. Randomisation

Another potential source of bias is that of the make up of the groups being treated within a controlled trial and particularly so in a group comparative trial. It is vital that the patient makeup of each group is as far as possible identical in every aspect. The only way that this can be achieved is by randomly allocating patients to each group[17]. If this is not done then clearly the therapist could allocate patients to each group on the basis of patient preference or on criteria such as whether it was felt a subject would obtain better results from one form of treatment compared to another, this could be a major source of bias. It is thought that such bias can lead to gross overestimation of the effectiveness of treatment[33]. Randomisation therefore ensures that both known and unknown prognostic factors are equally distributed[17]. The nature of randomisation however is such that groups could still be unbalanced in terms of certain factors. For example, it may become evident that one group is comprised of predominantly female subjects compared to male in the other and this could happen purely by chance. This problem is resolved by stratification e.g. factors such as age, gender or indeed any prognostic factors are also balanced across the groups.

## 5. Sample Size

In any trial it is of vital importance that adequate numbers of subjects are recruited. If this is not done then it may not be possible to generalise results to the population as a whole as there would be a much greater possibility that results obtained were through chance alone and therefore erroneous. If the expected treatment benefit is small compared to a placebo then subject numbers will need to be greater in order to detect a truly significant difference. Lewith and Machin[20] give a good example of a study involving 40 patients i.e. 20 having real acupuncture and 20 having sham acupuncture (use of 'non' acupuncture points). It was assumed that real acupuncture would have a response rate of 70% compared to 50% for sham acupuncture. Lewith and Machin point out that with this number of subjects there was only a 36% chance of detecting a significant difference between the two treatments (assuming $p = 5\%$). They further calculated that 130 patients would need to have been recruited to each group in order to adequately and confidently detect a significant difference. This example illustrates the importance of recruiting sufficient subjects in order to be confident about the results. It also shows that these numbers must be ascertained as the result of a power calculation. Unfortunately, many studies all too often failed to do this and as a consequence too few subjects were recruited, thus degrading results.

## 6. Control Options

Several options for controls are available, however these enable researchers to measure different things. The choice of control therefore is of prime importance[34]. For example, a waiting list control may be utilised and this may be useful when measuring chronic conditions that are not expected to change in the short term. In this scenario, patients are actively treated and compared to other similar patients who are still waiting to be treated. This circumvents the sometimes difficult issue of 'intention to treat' as eventually all patients are treated. Also it does not subject the untreated patient to any potential harm and gives valuable information on the influence of regression to the mean. It does not however control for the purely placebo effects of a particular treatment.

Another option for control might be to use another already proven and accepted treatment. Here both groups of patients would be receiving treatment and this may be viewed as more ethically acceptable[32]. Once again however, there is no control for the placebo effect of treatment and so this method may be more appropriate to compare relative efficacy of two already validated forms of treatment rather than when trying to ascertain the efficacy of treatment in its own right.

The question of whether a treatment is more effective than a placebo is an important one to address however. It is thought that all treatments are comprised of two elements i.e. the real physiological effect and also the placebo or non-specific effects associated with it. This placebo effect may be propagated by the process and rituals involved in treatment administration, patient expectation, suggestibility and conditioning, all of which may have considerable effect on treatment efficacy[34]. It can therefore be difficult to attribute any degree of improvement in the patient's condition to either the treatment itself or to the non-specific effect of the treatment mechanism. Teasing apart these two elements is crucial in order to "increase our understanding about the psychological and psycho physiological parameters involved in medical treatment."[34]. It is also quite simply not good medical practice to continue to treat patients on the basis of ignorance[35]. This is thought to be particularly relevant with modalities such as acupuncture where a) there is a great deal of physical stimulation and contact with the therapist and b) the patient is pierced with a needle, which is thought to have a profound enhanced placebo effect[36;37].

## 7. Placebo Controls

When conducting a trial into an unproven treatment modality therefore, it is necessary to use a placebo control in order to allow realistic evaluation as to its efficacy and to be able to adequately assess the contribution of the placebo effect of the treatment[32;38]. Several 'placebo' control options have been put forward and used in the context of acupuncture research and these will now be examined.

### 7.1. Sham Acupuncture

Many of the early studies used the method of 'sham acupuncture' as the placebo control. This involves the insertion of needles, to normal depth, into non-acupuncture points. That is at locations in the body that are thought to be devoid of any known therapeutic value in terms of acupuncture. At first glance this would appear to be the most obvious and logical approach to control for the placebo effect. Patients in both groups (active and placebo) would receive similar stimulation in so much as they would both be able to see the needles in place and may experience, to a certain level, similar tactile sensations. However, this technique may not be as 'inert' as initially thought for several reasons. Apart from the points known to exist on the mapped meridians, there are also a great number of 'extra' points, which often vary from text to text. It may therefore be difficult to guarantee that any point chosen is in fact truly 'inert' from an acupuncture point of view[38]. Secondly, work conducted by Le Bars on diffuse noxious inhibitory control[19] suggests that any noxious stimulus e.g. the insertion of a needle at a non-acupuncture point, may have a physiological affect and can attenuate pain. This is thought to have a significant effect on between 40–50% of patients (although some sources would put this as high as 60%[39]) which although probably less effective that real treatment, is thought to be more effective than a true placebo[20]. Next, if it is accepted that there may be some electro-chemical action on the body by inserting a needle into an acupuncture point[40;41], it would be difficult to adequately assess how far away from a true acupuncture point the needle needs to be before it no longer has any 'electrical' influence over the point[8]. Clearly a study involving sham acupuncture will not control for a placebo effect[7;11;29]. Indeed in reality such a study would be comparing the relative efficacy of point location[22] or indeed perhaps comparing acupuncture against a slightly less effective form of acupuncture. Lastly there is also an ethical issue associated with this form of control. It may be viewed as ethically unacceptable to subject patients to an invasive technique such as acupuncture if it is thought that there may be little or no benefit to the patient and this may effectively expose the patients to unnecessary harm[38]. The use of sham acupuncture as a placebo cannot be justified and has indeed been the source of much confusion in the field of acupuncture research. This has been particularly relevant where results of trials using sham acupuncture have been reviewed by non-acupuncturists or those not fully understanding the complexities around this issue thus leading to false conclusions and a high probability of a type II error.

## 7.2. Pseudo Sham Acupuncture.

In this control, real acupuncture points are used but are chosen in such a way as to have no effect on the condition being treated. This would have the advantage of causing the patient to experience the normal feelings of needle sensation (deqi) and would therefore feel the same as active acupuncture. This type of control however has several important failings. Once again the patient is subjected to a potential source of harm through the use of other non-tested acupoints and with no intention to heal. Within acupuncture therapy there are so many options of point combinations that it may not be possible to guarantee that the combination of control points have no effect at all on the condition to be treated. Even if this could be guaranteed, the points chosen may still have some effect on the condition however, through the mechanism of DNIC. Once again this control would tend to be useful for comparing the relative efficacy of point selection and would not give any indication as to the true placebo effect.

## 7.3. Superficial Needling (or Minimal Acupuncture)

Another control option is that of superficial or minimal needling. In this case, needles (usually shorter than normal) are inserted only 1–2 mm into the skin and usually at non-acupuncture points. This has the advantage that, to the patient, it appears that normal acupuncture is occurring. Superficial needling is thought to be less effective than sham acupuncture[42]. It has been argued however that even with minimal needling, there is still some small noxious stimulus and this may therefore have a physiological effect albeit fairly small[43]. This particular control option is also very similar to Japanese acupuncture where superficial needling is often employed as a therapeutic technique. Clearly, if efficacy for this style of acupuncture were proven then superficial acupuncture would not be appropriate for use as a placebo control.

## 7.4. Dummy Needling

There have been several methods attempted in order to simulate acupuncture treatment without actually piercing the skin. An experiment was carried out using a plastic guide tube and a cocktail stick[44] where the end of the tube was placed over a bony prominence and the stick simply slid inside and tapped so that a sensation could be felt. The researchers went to elaborate lengths to ensure that the patient was unable to see what was happening and this often proved difficult. It is felt however, that preventing patients from being able to see the procedure may in itself cause a certain amount of suspicion. This technique would therefore be very difficult to routinely use on any points other than on the back and this would seriously limit the number of point options available to the practitioner.

In order to give maximum realism, points used as placebo treatment should be the same as those used in the active treatment group but no penetration should occur. Work has been carried out on a placebo needle, which works in the same way as a stage dagger[45]. As the needle is pushed against the skin, it causes a pricking sensation but as more pressure is applied, the shaft of the needle disappears into the handle. This gives the impression that the needle is actually entering the skin. The needle is held in position by a small adhesive plastic ring, which can also be used with the real needles so as to aid consistency. This system was tested on 60 volunteers who were subjected to both real and placebo needling. Streitberger's validation results were very promising, however independent validation has not yet been carried out and this would need to be performed prior to wide acceptance of these needles. It has also been pointed out that the needles might not be appropriate for use in certain areas such as the scalp, toes and fingers, and that there may be some limitation in the angle of needle insertion[46]. This particular control technique, although as yet unproven, may represent the best option thus far for a truly realistic acupuncture placebo control.

## 7.5. Decommissioned Electrical Unit

The use of a decommissioned electrical stimulation unit such as a transcutaneous electrical nerve stimulation (TENS) machine or similar will provide a completely inactive control treatment[38;43]. Usually the cables joining the patient to the machine are altered in such a way as to ensure that there is no real electrical connection. The patient therefore receives a visual and/or auditory stimulus but no actual current or electrical stimulus of any kind. This of course means that the intervention is completely safe although apart from any potential placebo effect, there is no intention to treat. The obvious disadvantage of such a control is that firstly it lacks any tactile stimulation unlike the 'deqi' felt with acupuncture. This, it is suggested, could have an effect on patient expectation where a strong sensation may elicit a much more powerful suggestion and could therefore be viewed as more beneficial by the patient thus increasing their expectation of effectiveness[47]. Secondly, it does not even remotely resemble needling and this might also have an effect on expectation, particularly if, as has been suggested, needling has an enhanced non-specific effect[37]. Despite this, the placebo effect of this type of control is thought to be in the order of 30%, which is comparable to placebo medication used in drug trials[38]. The expectation of effectiveness of acupuncture and a decommissioned electrical unit and therefore their relative suggestibility, have been shown to be equal[47;48]. The decommissioned electrical unit has therefore been demonstrated to be a valid placebo control and has been utilised in several studies. In Richardson's review of acupuncture for pain relief [49] six out of twenty eight studies (21%) used sham TENS and in Lewith and Vincent's paper on chronic pain studies [22], seven out of twelve studies (58%), with adequate controls, used mock TENS.

## 8. Ensuring the Relative Credibility of Active and Control Interventions

The outcome and integrity of any clinical trial may well be dependent upon the credibility of the control[16;50]. It is felt that the level of suggestion offered by the practitioner may have a considerable effect on the credibility of any sham treatment that may be used as a placebo control[11]. Where the control differs substantially from the treatment actually being examined, it is of prime importance that this credibility must be routinely assessed. If this is not assessed and a study shows that there is a difference between acupuncture and a placebo, in favour of the active treatment, it could be argued that perhaps acupuncture is simply a more convincing placebo and therefore has a greater effect. If the two interventions used in a trial can be demonstrated to have equal credibility and therefore 'placebo' power, then the true placebo may legitimately be used as a control. The credibility of controls can be ascertained by comparing the expectancy of effectiveness of the real versus placebo treatment. This issue has been addressed by Borkovec and Nau who have devised and tested a 'credibility rating'[50]. It has four simple questions and is therefore quick to complete:

1. How confident do you feel that this treatment can alleviate your complaint?
2. How confident would you be in recommending this treatment to a friend who suffered from similar complaints?
3. How logical does this treatment seem to you?
4. How successful do you think this treatment would be in alleviating other complaints?

The psychometric properties of this measure have been well established and it has been shown to have test-retest reliability, internal consistency, repeatability[43;51] and it has also been used in many trials

## 9. Summary

The aim of all clinical research must surely be to assess an intervention or process or relationship between variables, ultimately in order to be able to improve clinical practice and healthcare for patients/consumers. As such, evidence gathered must have external validity (generalisability)[52] and it must be relevant to clinical practice. It can be seen therefore that research into acupuncture for pain relief is an inherently difficult task, which requires much sensitivity and careful planning in order to minimise error when drawing conclusions. The researcher needs to adequately address problems such as sample size, blinding, randomisation and appropriate use of control and treatment. It is also most important to be able to clarify that the placebo (if used) is credible to the patient.

The field of acupuncture has been overwhelmed with hundreds of sub-standard trials and much doubt still surrounds the issue of efficacy. Few good quality rigorous trials exist to support the use of acupuncture for pain relief in general[21]. It has also been noted that the question of long-term efficacy has not been assessed by the many trials[29] and this needs to be addressed prior to its acceptance. Further well conducted rigorous trials are therefore needed, utilising the lessons learnt through the substantial number of previous trials, in order to adequately assess the efficacy of acupuncture for painful conditions. Such trials must therefore:

- Utilise a placebo control
- Be randomised and stratified
- Ensure patient blinding
- Use self completed outcome measures and blinded assessors
- Have sufficient and well validated outcome measures
- Check credibility of treatment
- Have a well defined entry criteria which includes a single condition only
- Have a long term follow up

- Enrol sufficient subjects, the number to be determined by a power calculation
- Provide adequate acupuncture treatment as would be found in clinical practice
- Be group comparative so as to negate any carry-over effects

In this way good quality trials can be undertaken which will yield reliable results and therefore add meaningful data to the body of knowledge appertaining to the effectiveness of acupuncture.

## 10. References

1. Salzberg C, Miller A, Johnson LK. Acupuncture: history, clinical uses, and proposed physiology. *Physical Medicine and Rehabilitation Clinics of North America* 1995;**6**:905–16.
2. Lynn J. Using complementary therapies: acupuncture. *Professional Nurse* 1996;**11**:722–4.
3. White A, Filshie J, Cummings M. Clinical trials of acupuncture: consensus recommendations for optimal treatment, sham controls and blinding. *Complement Ther.Med* 2001;**9**:237–45.
4. Moher D, Schulz K, Altman G. The CONSORT statement: revised recommendations for improving the quality of reports of parallel group randomised trials. *The Lancet* 2001;**357**:1191–4.
5. MacPherson H, White A, Cummings M, Jobst K, Rose K, Niemtzow R. Standards for reporting interventions in controlled trials of acupuncture: The STRICTA recommendations. *Acupuncture in Medicine* 2002;**20**: 22–5.
6. Barnes J, Abbot NC, Harkness EF, Ernst E. Articles on complementary medicine in the mainstream medical literature: an investigation of MEDLINE, 1966 through 1996. *Arch Intern.Med* 1999;**159**:1721–5.
7. Araujo M. Does the Choice of Placebo Determine the Results of Clinical Studies on Acupuncture. *Research in Complementary Medicine* 1998;**5**:8–11.
8. Newham DJ. Methodological perspectives. *Disabil.Rehabil* 1999;**21**:134–6.
9. Kiene H. A Critique of the Double-Blind Clinical Trial—Part 1. *Alternative Therapies in Health and Medicine* 1996;**2**:74–80.
10. Gaus W,.Hogel J. Studies on the efficacy of unconventional therapies. Problems and designs. *Arzneimittel Forschung/Drug Research* 1995;**45**:88–92.
11. Margolin A, Avants SK, Kleber HD. Investigating alternative medicine therapies in randomized controlled trials. *JAMA* 1998;**280**:1626–8.
12. Patel MS. Problems in the evaluation of alternative medicine. *Soc.Sci.Med.* 1987;**25**:669–78.
13. White, P. A study for the efficacy of a western acupuncture protocol for the treatment of chronic mechanical neck pain. 2002. University of Southampton. Ref Type: Thesis/Dissertation.
14. Lewith GT. Can we assess the effects of acupuncture? [editorial]. *Br.Med.J.Clin.Res.Ed* 1984;**288**:1475–6.
15. Harden RN. The pitfalls of clinical acupuncture research: can east satisfy west? *Arthritis Care and Research* 1994;**7**:115–7.
16. Zaslawski C, Rogers C, Garvey M, Ryan D, Yang CX, Zhang SP. Strategies to maintain the credibility of sham acupuncture used as a control treatment in clinical trials [see comments]. *J.Altern.Complement Med.* 1997;**3**:257–66.
17. Ernst E,.White AR. A review of problems in clinical acupuncture research. *Am.J.Chin Med.* 1997;**25**:3–11.
18. Godfrey C,.Morgan P. A controlled trial of the theory of acupuncture in musculoskeletal pain. *J.Rheumatology* 1978;**5**:121–4.
19. Le Bars D, Villanueva L, Willer J, Bouhassira D. Diffuse Noxious Inhibitory Controls (DNIC) in Animals and Man. *Acupuncture in Medicine* 1991;**9**:47–56.
20. Lewith GT,.Machin D. On the evaluation of the clinical effects of acupuncture. *Pain* 1983;**16**:111–27.
21. Ernst E. Is acupuncture effective for pain control? [letter]. *J.Pain Symptom.Manage.* 1994;**9**:72–4.
22. Lewith G,.Vincent C. Evaluation of the Clinical Effects of Acupuncture. A problem Reassessed and a Framework for Future Research. *Pain Forum* 1995;**4**:29–39.
23. Birch S. Issues to consider in determining an adequate treatment in a clinical trial of acupuncture. *Complementary Therapies in Medicine* 1997;**5**:8–12.
24. Birch S. Testing the clinical specificity of needle sites in controlled clinical trials of acupuncture. *Proceedings of the second annual meeting, society for acupuncture research* 1995;274–94.

25. Hopwood V, Lovesey M, Mokone S. Acupuncture and Related Techniques in Physical Therapy. Churchill Livingstone, 1997.
26. White AR,.Ernst E. A trial method for assessing the adequacy of acupuncture treatments. *Altern.Ther.Health Med.* 1998;4:66–71.
27. Ezzo J, Berman BM, Hadhazy V, Jadad A, Lao L, Singh BB. Is acupuncture effective for the treatment of chronic pain? A systematic review. *Pain* 2000;86:217–25.
28. Pomeranz B. Bruce Pomeranz, PHD. Acupuncture and the raison d'etre for alternative medicine [interview by Bonnie Horrigan]. *Altern.Ther.Health Med.* 1996;2:85–91.
29. Thomas M. Acupuncture studies on pain. *Acupuncture in Medicine* 1997;15:23–31.
30. Hammerschlag R. Methodological and ethical issues in clinical trials of acupuncture. *J.Altern.Complement Med.* 1998;4:159–71.
31. Sherman K, Cherkin D, Hogeboom C. The diagnosis and treatment of patients with chronic low-back pain by traditional Chinese medical acupuncturists. *J.Altern.Complement Med.* 2001;7:641–50.
32. Hammerschlag R,.Morris MM. Clinical trials comparing acupuncture with biomedical standard care: a criteria-based evaluation of research design and reporting [corrected] [published erratum appears in COMPLEMENTARY THER MED 1997 Dec; 5(4): 253]. *Complementary Therapies in Medicine* 1997;5:133–40.
33. Schulz K, Chalmers I, Hayes R, Altman D. Empirical Evidence of Bias. *JAMA* 1995;273:408–12.
34. Peck C,.Coleman G. Implications of Placebo Theory for Clinical Research and Practice in Pain Management. *Theoretical Medicine* 1991;12:247–70.
35. Mackay M. Acupuncture and placebos [letter]. *N.Z.Med.J.* 1985;98:606–7.
36. Walach H, Maidhof C. Is the Placebo Effect Dependent on Time? A Meta-Analysis. In Kirsch D, ed. *How Expectancies Shape Experience*, pp. 321–32. Washington DC: American Psychological Association, 1999.
37. Kaptchuk TJ, Goldman P, Stone D, Stason W. Do medical devices have enhanced placebo effects? *J.Clin.Epidemiol.* 2000;53:786–92.
38. de la Torre CS. The choice of control groups in invasive clinical trials such as acupuncture. *Frontier Perspectives* 1993;3:33–7.
39. Engelbart JH,.Kloppenburg GV. The treatment of pain by acupuncture. The riddle of acupuncture and the place of the acupuncturist in the pain-team. *Acta Anaesthesiol.Belg.* 1981;32:33–43.
40. Stux G,.Pomeranz B. Basics of acupuncture. Springer Verlag New York 1995.
41. Sims J. The mechanism of acupuncture analgesia: a review. *Complementary Therapies in Medicine* 1997;5:102–11.
42. Vincent C,.Lewith G. Beijing to Belgrade—Making Sense of Acupuncture Research. *APS Journal* 1994;3:89–91.
43. Vincent C,.Lewith G. Placebo controls for acupuncture studies. *Journal of the Royal Society of Medicine* 1995;88:199–202.
44. White AR, Eddleston C, Hardie R, Resch KL, Ernst E. A pilot study of acupuncture for tension headache, using a novel placebo. *Acupuncture in Medicine* 1996;14:11–5.
45. Streitberger K,.Kleinhenz J. Introducing a placebo needle into acupuncture research. *Lancet* 1998;352:364–5.
46. Kaptchuk TJ. Placebo needle for acupuncture [letter]. *Lancet* 1998;352:992.
47. Petrie J,.Hazleman B. Credibility of placebo transcutaneous nerve stimulation and acupuncture. *Clinical and Experimental Rheumatology.* 1985;3:151–3.
48. Wood R,.Lewith G. The Credibility of Placebo Controls in Acupuncture Studies. *Complementary Therapies in Medicine* 1998;6:79–82.
49. Richardson PH,.Vincent CA. Acupuncture for the treatment of pain: a review of evaluative research. *Pain* 1986;24:15–40.
50. Borkovec T,.Nau S. Credibility of Analogue Therapy Rationales. *J.Behav.Ther.and Exp.Psychiat* 1972;3:257–60.
51. Vincent C. Credibility Assessment in Trials of Acupuncture. *Comp.Med.Res.* 1990;4:8–11.
52. Downs S,.Black N. The feasability of creating a checklist for the assessment of the methodological quality both of randomised and non randomised studies of health care interventions. *J.Epidemiol.Community Health* 1998;52:377–84.

# 15

# Treatment for Atopic Dermatitis by Acupuncture

**M. Fukuda, N. Kawada, H. Kawamura and T. Abo**

## 1. Introduction

The onset of atopic dermatitis usually occurs at infancy or during childhood and in some cases spontaneously cures before adolescence[1,2]. Generally speaking, children have a strong immune system (i.e., a high level of lymphocytes)[3] while the immune system of adults gradually weakens[4]. In Japan, the level of granulocytes surpasses that of lymphocytes from 15 to 20 years of age (our unpublished data). We speculate that this age-associated change of the immune system might be responsible for the spontaneous cure of atopic dermatitis. In other words, many cases of atopic dermatitis naturally subside due to an age-dependant decrease in the number of T cells.

In spite of the usual course described above, some patients with atopic dermatitis do not cure spontaneously, but continue to suffer from severe symptoms such as inflamed reddish facial skin[1,2]. Such patients have a long history of using topical ointment containing steroid hormones. Are these severe cases of atopic dermatitis inevitable because of their atopic nature or could they be curable by some appropriate therapy? We speculate that when steroid hormones contained in ointment remain long enough in the dermal tissue of the skin, they become oxidized cholesterols. Steroid hormones have a cholesterol structure, and the skin of patients with atopic dermatitis who use ointment containing such hormones begins to give off a bad odor. This would seem to indicate that the oxidized substance stimulates tissue, causing granulocyte infiltration. We confirmed this phenomenon in an animal model[5] in which oxidized cholesterols directly induced the accumulation of granulocytes at corresponding sites (our unpublished observation). Such granulocytes are further activated by many inflammatory cytokines, (e.g., TNF$\alpha$, IFN$\gamma$, G-CSF, etc.), as well as by sympathetic nerve stimulation after interaction with resident bacteria.

---

**M. Fukuda** • Fukuda-iin, Niigata   **N. Kawada** • Department of Obstetrics and Gynecology, Uhrin Hospital, Fukushima   **H. Kawamura** • Department of Immunology, Niigata University School of Medicine, Niigata 951-8510, Japan   **T. Abo** • Department of Immunology, Niigata University School of Medicine, Niigata 951-8510, Japan

T. Abo address. Department of Immunology, Niigata University School of Medicine, Niigata 951-8510, Japan Fax: +81-25-227-0766. E mail: immunol2@med.niigata-u.ac.jp

*Complementary and Alternative Approaches to Biomedicine*, edited by Edwin L. Cooper and Nobuo Yamaguchi. Kluwer Academic/Plenum Publishers, 2004.

During the past four years, we have attempted to eliminate such oxidized substances from the skin of patients with severe atopic dermatitis and have been successful in almost all such cases (> 95%). Since our patients suffered from withdrawal from steroid hormones during therapy, we applied acupuncture. If we can understand the actual adverse reaction to steroid hormones, we, as well as other clinicians, will be able to successfully treat many patients with severe atopic dermatitis, even those addicted to steroid hormones. Although the precise reasons are not yet determined, the granulocytosis seen in patients with severe atopic dermatitis and with the withdrawal syndrome during therapy is interesting. At least in severe cases of atopic dermatitis, we have to consider the function of granulocytes as well as the functions of IgE, T cells, and eosinophils[6,7].

## 2. Experimental Procedure

Patients with severe atopic dermatitis (total n = 89, male n = 45, and female n = 44; ages 12 to 28, average 19.9 ± 6.0) were treated. The severity of the atopic dermatitis was estimated according to a previously reported system of grading[8]. As indicated by the authors in that paper[8] and others[9], it is not easy to measure the disease activity in atopic dermatitis. The simplest method categorizes such patients into "mild", "moderate" and "severe" atopic dermatitis groups[8]. Since almost all patients (approximately 100%) in this study had used ointment containing steroid hormones for several years, at least 80% of the patients with atopic dermatitis were classified as "severe" cases. Age-matched healthy controls (n = 100) were also selected to compare their leukocyte pattern with that of the patients.

The number of white blood cells (total leukocytes) was counted in the blood of healthy controls and patients with severe atopic dermatitis. The proportions of granulocytes (including neutrophils and basophils), eosinophils, and lymphocytes were determined by May-Grünwald-Giemsa staining. These data were obtained from the blood of the controls and patients during therapy.

Urine was collected from 13 patients with atopic dermatitis (before therapy) and 20 age-matched healthy controls for 24 hrs, and the concentrations (/liter) of these substances were measured by the HPLC method[10].

All parameters were measured in the laboratory of SRL (Tokyo, Japan, http://www.SRL-inc.co.jp/).

In addition to psychological support for the patients, we applied acupuncture. Briefly, fingers and toes were punctured with 26-gauge needles; more precisely, sites lateral to the nails were punctured, resulting in some bleeding. This stimulation induced parasympathetic nerve activation, e.g., recovery of peripheral circulation, a decrease in blood pressure, and relief from itching, etc. In some cases, severe withdrawal occurred during therapy and some elevation of serum transaminases, indicating hepatic failure, was seen. In such severe cases (less than 10% of the patients treated), transfusion of physiological saline (500 ml containing a diuretic) was intravenously administered once a day (up to several times).

Differences between the results obtained from healthy control subjects and patients with atopic dermatitis were analyzed using Student's *t*-test.

## 3. Results and Discussion

To determine the immunologic states of patients with severe atopic dermatitis (n = 89), the number of leukocytes and the percentage of the leukocyte population were

**Table 15.1.** A comparison of the distribution of leukocytes between healthy subjects and patients with atopic dermatitis.

| Parameter | Healthy subjects (n = 100) | Patients with severe atopic dermatitis (n = 89) | | |
|---|---|---|---|---|
| | | Before therapy | 2 wks after therapy | At discharge |
| Number of leukocytes | $6,500 \pm 1,180$ | $7,700 \pm 1,800^*$ | $8,500 \pm 2,400^*$ | $7,500 \pm 1,800^*$ |
| % Granulocytes[a] | $57.0 \pm 9.2$ | $59.5 \pm 10.7$ | $63.3 \pm 10.9^*$ | $52.0 \pm 10.3$ |
| % Eosinophils | $1.8 \pm 1.0$ | $11.8 \pm 8.4^*$ | $13.1 \pm 8.1^*$ | $13.7 \pm 8.5^*$ |
| % Lymphocytes | $38.9 \pm 5.2$ | $27.8 \pm 8.6^*$ | $22.6 \pm 8.8^*$ | $34.2 \pm 9.4$ |

[a] Neurophils and basophils (not including eosinophils)
*$p < 0.05$

enumerated in the peripheral blood (Table 15.1). These patients suffered from severe atopic dermatitis of the face and other sites of the body. All patients had used ointment containing steroid hormones for several months to several years. Before visiting our hospital, they had been treated at several other hospitals. For a comparison of the leukocyte pattern, age-matched healthy subjects (n = 100) were examined in parallel. Before therapy, these patients showed elevated levels of the number of leukocytes and the proportion of eosinophils, and a decreased level of the proportion of lymphocytes ($P < 0.05$). As for the absolute number of leukocytes, it was estimated that the numbers of granulocytes and eosinophils, but not of lymphocytes, increased ($P < 0.05$).

We previously reported that predominance of the sympathetic nervous system induces an elevation of granulocytes in the peripheral blood and other tissues of humans[3, 10, 11] and mice[12-14]. Since this is the case, it was speculated that the sympathetic nervous system of patients with severe atopic dermatitis who showed elevated levels of granulocytes might be predominant. This was obvious from the fact that the majority of the patients suffered from tachycardia and insomnia. We also examined the urine level of catecholamines as well as of metabolites of innate steroid hormones (Fig. 15.1). The patients showed elevated levels (per day) of free catecholamines and VMA (a metabolite of catecholamines) ($P < 0.05$). The level of 17-KS was low ($P < 0.05$) while that of 17-OHCS remained normal ($P > 0.05$).

To examine the possibility that sympathetic nerve activation was induced by oxidized cholesterols derived from stagnant steroids, we measured the serum levels of lipids and cholesterols (Fig. 15.2). Among tested parameters, the levels of total lipid, total cholesterol, $\beta$-lipoprotein, cholesterol ester, and free cholesterol were comparable between the controls and patients. However, the level of lipid peroxide (which is known to include oxidized cholesterols) had increased while that of total bile acid had decreased in the patients ($P < 0.01$).

We began to treat patients according to our hypothesis on the accumulation of oxidized cholesterols in the skin. This treatment included 1) the cessation of ointment usage, 2) acupuncture, and 3) psychological support. Two weeks after commencement of the therapy, the leukocyte pattern (the increase in the number of granulocytes and eosinophils) seen in patients with severe atopic dermatitis became much more prominent (see Table 15.1, middle column). Namely, one of the major signs of the withdrawal syndrome was the worsening of the leukocyte pattern, as well as the worsening of inflammation. These symptoms, however, gradually subsided (Table 15.1 and Fig. 15.3). At the time of discharge from the hospital

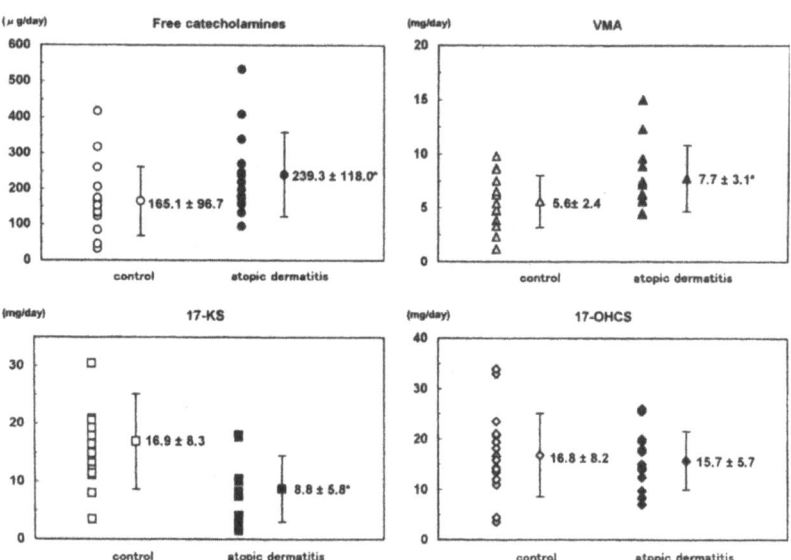

**Figure 15.1.** Excretion of free catecholamines, VMA, 17-KS, and 17-OHCS into the urine (per day) controls and the patients. Urine was collected from patients with severe atopic dermatitis (n = 13) and age-matched healthy controls (n = 20). *P < 0.05

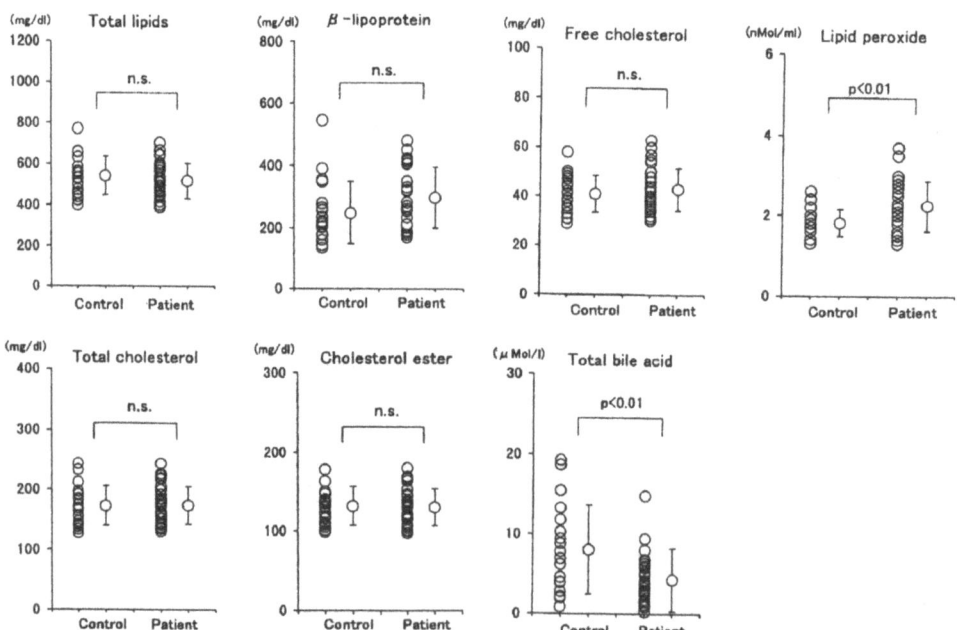

**Figure 15.2.** A comparison of the serum levels of lipids and cholesterols between controls and patients with severe atopic dermatitis. Blood samples were collected from severe atopic dermatitis (n = 29) and age-matched healthy controls (n = 19) and the serum levels of lipids and cholesterols were measured. The level of lipid peroxide increased but that of total bile acid decreased in the patients (P < 0.01)

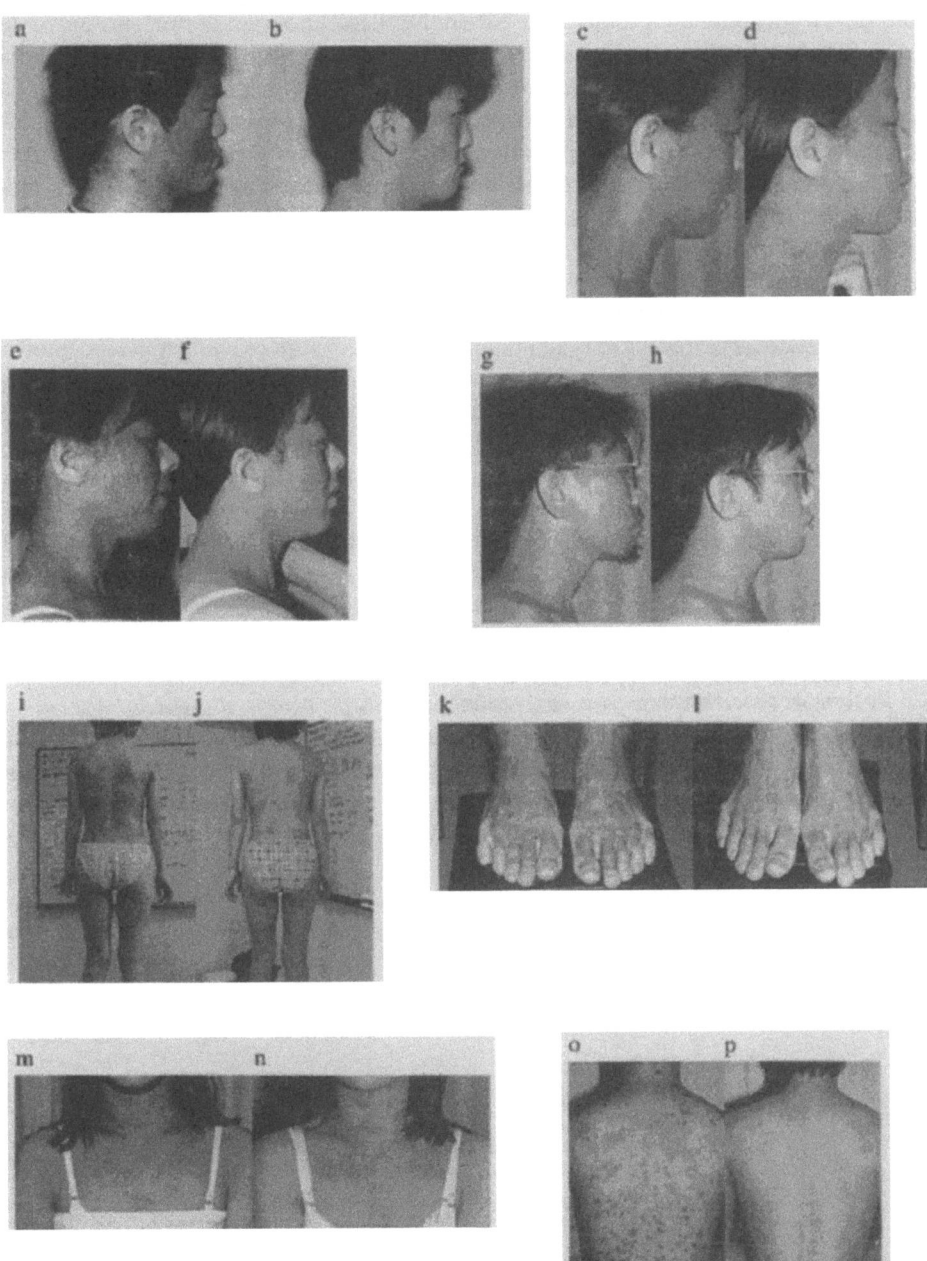

**Figure 15.3.** Successful treatment of severe atopic dermatitis by acupuncture. Right: before treatment (a, c, e, g, i, k, m, and o), Left: after treatment (a→b, c→d, e→f, g→h, i→j, k→l, m→n, and o→p). After cessation of the use of ointment containing steroid hormones, the patients suffered from the withdrawal syndrome. This was, however, countered by acupuncture, resulting in successful treatment within one month to several months

(Table 15.1, right column), the leukocyte pattern had become almost normal, except for the eosinophil value. As shown in Fig. 15.3, the severe atopic dermatitis improved remarkably in all patients. The time intervals before and after the present treatment ranged from one to five months. At the beginning of therapy, however, some patients (approximately 5%) dropped out due to the severe withdrawal syndrome. Some patients (approximately 20%) showed a sporadic exacerbation of atopic dermatitis even after discharge. However, these symptoms gradually disappeared with time. We recommended exercise for these patients since appropriate exercise seems to be very effective for eliminating the exacerbation.

In a series of recent studies, we reported that immunologic states are under the regulation of the autonomic nervous system[3, 10–15]. This is due to the fact that granulocytes carry surface adrenergic receptors[16] and lymphocytes carry surface cholinergic receptors[15]. If the sympathetic nervous system is stimulated, the number and function of granulocytes are activated. Inversely, if the parasympathetic nervous system is stimulated, the number and function of lymphocytes such as T and B cells are activated. A rapid change in the distribution of leukocytes arises from the fact that the life span of leukocytes is very short, e.g., granulocytes live only 2 days after maturation[13]. Moreover, sympathetic nerve stimulation induces the acceleration of granulocyte trafficking, i.e., the bone marrow (the pool organ of granulocytes)→the circulation→the mucosal tissues. As a consequence, sympathetic nervous system stimulation increases the number of granulocytes in the blood. This includes granulocytosis seen in the daytime (the circadian rhythm)[11], resulting from physical stress[13], and neonatal granulocytosis induced by commencement of pulmonary respiration at birth[3].

In light of these findings, we analyzed the leukocyte pattern in patients with severe atopic dermatitis, whose disease had been diagnosed as incurable at other hospitals. In contrast to what we anticipated, the level of granulocytes was found to be elevated while that of lymphocytes was low in these patients. Namely, they were in an immunosuppressive state. There have been similar reports by other investigators[17–20] about this immunosuppressive state in patients with severe atopic dermatitis.

We postulated that some oxidized substances (e.g., oxidized cholesterols resulting from steroid hormones used as an ointment) had induced sympathetic nervous system dominance (Fig. 15.4). Indeed, almost all of these young patients had symptoms such as tachycardia, hypertension, insomnia, anxiety, fatigue, etc. This postulation was confirmed

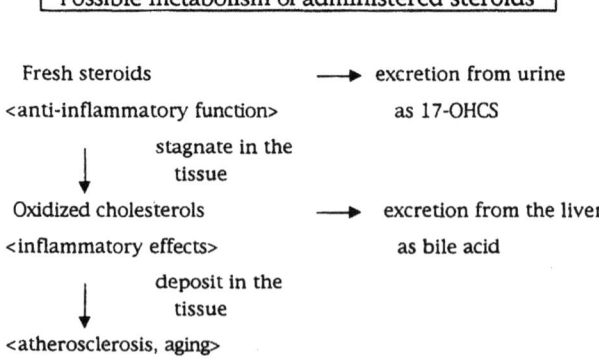

**Figure 15.4.** Fate of administered steroids in patients with atopic dermatitis

by the results of urine analysis, namely, the levels of free catecholamines and VMA were elevated in the urine of the patients. For an unknown reason, the level of 17-KS was low in these patients. The elevated serum level of lipid peroxide (which includes oxidized cholesterols) was also found to be elevated in the patients. At present, we do not know why the patients showed a decreased serum level of total bile acid. As mentioned below, one possibility is the acceleration of oxidized cholesterol secretion as bile acid.

According to our hypothesis, we discontinued the use of the ointment containing steroid hormones and waited for the oxidized cholesterols to be excreted from the skin. As well established, the usual routes for the excretion of metabolized steroid hormones are the urine (as 17-OHCS) and bile (as bile acid because of high oxidization). If they remain in the tissue for long, however, it is speculated that they become oxidized cholesterols which then induce granulocyte-associated inflammation. Finally, they may induce atherosclerosis and aging, as seen in some patients, if not all, with asthma and autoimmune diseases such as scleroderma, who have used steroid hormones for a long time (i.e., several years or longer).

Our successful treatment of severe atopic dermatitis supports our hypothesis. During the therapy, patients experienced the excretion of a purulent substance from the skin. To overcome the withdrawal syndrome from steroid hormones, we applied acupuncture as well as giving psychological support. However, when we applied acupuncture without the cessation of steroid hormones, the effects were limited or nil (our unpublished observation). Acupuncture induces the reflex of parasympathetic nerve stimulation[16, 21, 22] and results in the recovery from circulation failure or other ailments (our unpublished observations).

As shown by the present data, many of these patients were genetically predisposed to atopic dermatitis as reflected by the high level of eosinophils before and after therapy. In this regard, some patients suffered a recurrence of disease. However, possibly because of their ages (older than 16 years), the recurrence ratio was not so high (less than 10%). In the case of patients with atopic dermatitis who had not yet used ointment containing steroid hormones, two or three weeks were sufficient to achieve successful treatment by acupuncture. In other words, severe cases of atopic dermatitis might result from long use of steroid hormones.

With the present concept, we can easily understand the reason for the difficulty in reducing doses of steroid hormones in many diseased persons who have used such hormones for a long time. We call it steroid hormone addiction. Increased amounts of steroid hormones are always required for neutralization (but still temporary) of the stagnated, oxidized substance.

Our hypothesis still remains a speculation, and we also have to consider many other reasons. Not only T cells, IgE, and eosinophils but also granulocytes might be highly associated with severity of atopic dermatitis. Thus, granulocytosis always accompanied severe cases of atopic dermatitis and the withdrawal syndrome. Although we do not yet know the precise mechanisms underlying granulocytosis, one of the candidates may be stagnant steroid hormones which then induce sympathetic nerve activation.

## 4. Conclusions

Patients with severe atopic dermatitis are sometimes seen even after 15–20 years of age, despite the fact that this is the age when atopic dermatitis subsides in usual cases.

We consider the possibility that steroid hormones administered as ointment stagnate in the skin and become oxidized cholesterols. Such substances may induce circulation failure and granulocyte-associated inflammation. We herein report the withdrawal from the above ointment in such patients, resulting in successful treatment within several months. Since they suffered from the withdrawal syndrome for the first two or three weeks, acupuncture was performed. Before therapy, these patients showed elevated levels of granulocytes and eosinophils, and an inverse decreased level of lymphocytes in the blood. The number and proportion of granulocytes are known to increase by sympathetic nerve stimulation. This, therefore, indicated that these patients were in a dominant state of the sympathetic nervous system. During the therapy, this leukocyte pattern became much worse due to the withdrawal syndrome. However, in parallel with the amelioration of inflammation, normalization of the leukocyte pattern was observed. These results revealed that, by some yet-undetermined reasons, patients with atopic dermatitis who had been treated by steroid ointment for a long time, showed an unusual pattern of leukocytes (i.e., granulocytosis) and that this leukocyte pattern was normalized by the withdrawal of steroid hormones in parallel with the amelioration of the disease.

*Note*

Subsequent to collection of the data on the patients herein reported, we achieved successful treatment of 500 other patients with severe atopic dermatitis.

## 5. Acknowledgments

We thank a grant-in-aid for Scientific Research from the Ministry of Education, Science and Culture, Japan. We also thank Mrs. Yuko Kaneko for preparation of the manuscript.

## 6. References

1. R.J. Kaplan and E.W. Rosenberg, *Postgrad. Med.* **64**, 52 (1978).
2. G. Rajka, *Clin. Rev. Allergy* **4**, 3 (1986).
3. T. Kawamura, S. Toyabe, T. Moroda, T. Iiai, H. Takahashi-Iwanaga, M. Fukuda, H. Watanabe, H. Sekikawa, S. Seki and T. Abo, *Hepatology* **26**, 1567 (1997).
4. C. Miyaji, H. Watanabe, M. Minagawa, H. Toma, Y. Nohara, H. Nozaki, Y. Sato and T. Abo, *J. Clin. Immunol.* **17**, 420 (1997).
5. S. Maruyama, M. Minagawa, T. Shimizu, H. Oya, S. Yamamoto, N. Musha, W. Abo, A. Weerasinghe, K. Hatakeyama and T. Abo, *Cell Immunol.* **194**, 28 (1999).
6. J.D. Bos, E.J.M. Van Leent and J.H.S. Smitt, *Exp. Dermatol.* **7**, 132 (1998).
7. T. Werfel and A. Kapp, *Curr. Probl. Dermatol.* **28**, 29 (1999).
8. G. Rajka and T. Langeland, *Acta. Derm. Venereol.* **144**, 13 (1989).
9. A.Y. Finlay, *British J. Dermatol.* **135**, 509 (1996).
10. M. Minagawa, J. Narita, T. Tada, S. Maruyama, T. Shimizu, M. Bannai, H. Oya, K. Hatakeyama and T. Abo, *Cell. Immunol.* **196**, 1 (1999).
11. S. Suzuki, S. Toyabe, T. Moroda, T. Tada, A. Tsukahara, T. Iiai, M. Minagawa, S. Maruyama, K. Hatakeyama, K. Endo and T. Abo, *Clin. Exp. Immunol.* **110**, 500 (1997).
12. H. Kawamura, T. Kawamura, Y. Kokai, M. Mori, A. Matsuura, H. Oya, S. Honda, S. Suzuki, A. Weerasinghe, H. Watanabe and T. Abo, *J. Immunol.* **162**, 5957 (1999).
13. T. Moroda, T. Iiai, A. Tsukahara, M. Fukuda, S. Suzuki, T. Tada, K. Hatakeyama and T. Abo, *Biomed. Res.* **18**, 423 (1997).

14. S. Yamamura, K. Arai, S. Toyabe, E.H. Takahashi and T. Abo, *Cell. Immunol.* **173**, 303 (1996).
15. S. Toyabe, T. Iiai, M. Fukuda, T. Kawamura, S. Suzuki, M. Uchiyama and T. Abo, *Immunology* **92**, 201 (1997).
16. K. Nishijo, H. Mori, K. Yoshikawa and K. Yazawa, *Neuroscience Letters* **227**, 165 (1997).
17. M.J. Lesko, R.S. Lever, R.M. Mackie and D.M. Parrott, *Clin. Exp. Allergy* **19**, 633 (1989).
18. J.L. Rogge and J.M. Hanifin, *Arch. Dermatol.* **112**, 1391 (1976).
19. B. Vandercam, J.M. Lachapelle, P. Janssens, D. Tennstedt and M. Lambert, *Dermatology* **194**, 180 (1997).
20. C. Walker, M.K. Kagi, P. Ingold, P. Braun, K. Blaser, C.A. Bruijnzeel-Koomen and B. Wuthrich, *Clin. Exp. Allergy* **23**, 145 (1993).
21. K.C. Tam and H.H. Yiu, *Am. J. Chin. Med.* **3**, 369 (1975).
22. T. Williams, K. Mueller and M.W. Cornwall, *Phys. Ther.* **71**, 523 (1991).

# 16

# Hydrotherapy can Modulate Peripheral Leukocytes: An Approach to Alternative Medicine

N. Yamaguchi, S. Shimizu and H. Izumi

## 1. INTRODUCTION

Along with treating illness, one of the purposes of alternative medicine is to promote the quality of life (QOL) of healthy people. In Japan, centuries of tradition have shown that alternative therapies like hot spring hydrotherapy, acupuncture, and herbal medicine enhance the QOL of empirically healthy individuals. Evidence has been accumulating that this may be the result of immune system regulation. The scientific basis, however, has not yet been established.

In our investigation, we measured how the number of CD positive cells and cytokine producing cells changed before and after hot spring hydrotherapy. The results showed that these subsets could reflect the number and function of immuno-competent cells. For example, in an individual with a low CD16 cell number, the number increased after treatment while it decreased in another individual with a high CD16 cell number. Our results led us to believe that leukocyte subsets could be an interesting indicator for the evaluation of alternative therapies.

Many systems are in place to evaluate Western therapies that aim at healing the symptoms of an illness. However, when the purpose of a therapy is to enhance the QOL of healthy people, such as some alternative medical therapies, no widely-accepted evaluation system has been established. To fill this lack, we would like to propose the number and functions of leukocyte subsets as indicators for the evaluation of alternative therapies.

## 2. HISTORICAL ANALYSIS OF HYDROTHERAPY

Throughout the world, the phenomenon of the natural hot spring has spawned places of healing and treatment. Mongolia[1] has a long history of so-called immersion

---

N. Yamaguchi, S. Shimizu, and H. Izumi • Department of Serology, Kanazawa Meidical University, Daigaku 1-1, Uchinada, Kahoku-gun Ishikawa, 920-0293, Japan. Ishikawa Natural Medicinal Products Research Center, Fukubatake, Kanazawa, 920-0167, Japan

*Complementary and Alternative Approaches to Biomedicine*, edited by Edwin L. Cooper and Nobuo Yamaguchi. Kluwer Academic/Plenum Publishers, 2004.

therapy recorded in Chinese or Mongolian medical works. In Gallo-Roman France, a most important number of documents, coins, various china, and votive stones survive from the province of Aquitaine. Even after two thousand years the therapeutic qualities of several mineral hot waters springs of that era are still recognized.[2] This impressive history expands up into Germany, the Netherlands[3] and through the whole of middle and eastern Europe, extending as far as the well known bathing centers and customs of Turkey.

Japan's geological position on the edge of the Pacific plate has given it volatile underpinnings, always changing and bursting forth in volcanic and earthquake activity. The legacy of this activity is the plethora of hot springs scattered all through its many islands. For centuries, these hot springs have been therapeutic destinations for the islands' inhabitants. Traditionally, each hot spring has its own character and efficacy for various complaints. Through the years, each water source was evaluated for its specific properties and with the advent of better transportation in our mountainous land, even remote springs in the mountains were visited for their specific medicinal effect. For example, the hot spring of Fukatani, which is located just below our research facility was known all over the area to cure hemorrhoids. The proprietors of the hot spring inns say that until about 20 years ago, many of the guests came to cure that sort of ailment. Now in Japan, hot spring hydrotherapy is often used as a supplementary therapy for many.[4,5,6] It has been shown to reduce surgical complaints such as shoulder pain, amyloidosis and various rheumatic problems. It can also lighten the burden to the heart and improve the condition of patients who suffer from emphysema as well as other respiratory ailments.

Hot springs have also traditionally functioned as places to relax and enjoy oneself. In their most extreme form, some even retain a kind of Japanesque Las Vegas atmosphere offering various forms of entertainment to accompany long relaxing dips in the bath. Although resort-type hot springs always existed along with those for illnesses, in recent years, trends have transformed even remote hot springs, such as the one below our facility, into fashionable resorts for the healthy to visit for relaxation and stress release. It is interesting to note that this historical duplicity encompassed by hot springs has also entered the world of Western medicine as the release of mental stress and physical fatigue has been shown to be essential for good health. Put into other words: the relaxation side of the hot spring promotes prevention, including health enhancement, and the medicinal side, the treatment of illness.

Recently, hot spring hydrotherapy has sparked more and more interest and many objective studies have been established. What is the mechanism of hot spring hydrotherapy that affects humans? It is known that the function of the immune system is closely related with the nervous and endocrine systems and varies with physical conditions and changes in the surroundings.[7-14] It seems reasonable that hot spring hydrotherapy should also influence the immune system, but so far no objective studies have been made. Finding the answer to this query became the focus of our investigation.

In order to investigate the influence of hot spring hydrotherapy on the immune system, we began by examining 126 people during the spring and winter of 1997 and 1998. Our results showed that within 24 hours after hot spring hydrotherapy, the white blood cells in the peripheral blood had changed significantly, not only in cell count but also cell function as well. We hope that our work will attract more attention to the mechanisms by which hot spring hydrotherapy regulates the human immune system.

## 3. SUBJECTS AND METHODS

### 3.1. Subjects and Methods:

Twelve to 17 individuals per group were tested from the spring of 1997 to the winter of 1998. The total number of volunteers was 126, ranging from 18 to 81 years. We divided them into two groups, the hydrotherapy group and the control group. In addition, the hydrotherapy group was further divided into two groups according to the variation after hydrotherapy. After preliminary results showed that a critical change in up and down regulation occurred at around 35 years old, the participants were divided according to age: the younger group was 35 years old and under, and the older group was over 36. According to the percentage of lymphocytes or granulocytes in the total leukocytes, the hydrotherapy group was further divided into two types: subjects whose granulocyte number was over 70% belonged to the G type, those whose lymphocyte number was over 40% belonged to the L type (Tables 16.1, 16.2).

The controls were divided into two groups. The negative control group (no exercise, just daily life activities; 1–2 mets) and the light exercise group (walking for exercise). Table 1 shows the background of the subjects.

**Table 16.1. Description of the Volunteers:** The volunteers were divided into two groups, the hydrotherapy group and the control group. The hydrotherapy group was further divided into two types, the G type and the L type (See Materials and Methods).

| | age | number | sex | systemic disorder |
|---|---|---|---|---|
| Hot spring group (3 ~ 4 mets) | 18–81 | n = 126 total | female n = 48 male n = 78 | hypertension (n = 4) diabetes (n = 3) |
| Control group (1 ~ 2 mets) | 20–81 | n = 20 | female n = 7 male n = 13 | hypertension (n = 1) |
| Control group (3 ~ 4 mets) (7 ~ 8 mets) | 20–65 | n = 41 | female n = 16 male n = 25 | hypertension (n = 2) diabetes (n = 1) |

**Table 16.2. The regulation in number of WBC according to age:** The volunteers were divided into two groups, 35 years old and younger, and 36 and older, corresponding to the up-regulation or down-regulation of peripheral WBC in the text. The WBC subsets were then counted morphologically according to granulocyte and lymphocyte number.

| | | The number of cells (X ± SD/mm$^2$) | | |
|---|---|---|---|---|
| Group | type of cells | before bathing | after bathing | P value |
| Older | WBC | 5940.00 ± 1035.85 | 7400.00 ± 1017.34 | <0.01 |
| | granulocyte | 3247.60 ± 841.78 | 3769.24 ± 844.98 | <0.05 |
| | lymphocyte | 2127.61 ± 577.92 | 2935.22 ± 655.83 | <0.01 |
| Younger | WBC | 6631.24 ± 1413.30 | 6098.21 ± 1326.50 | <0.05 |
| | granulocyte | 4100.86 ± 1003.51 | 3251.44 ± 885.13 | <0.01 |
| | lymphocyte | 1788.00 ± 598.52 | 1670.86 ± 431.76 | >0.05 |

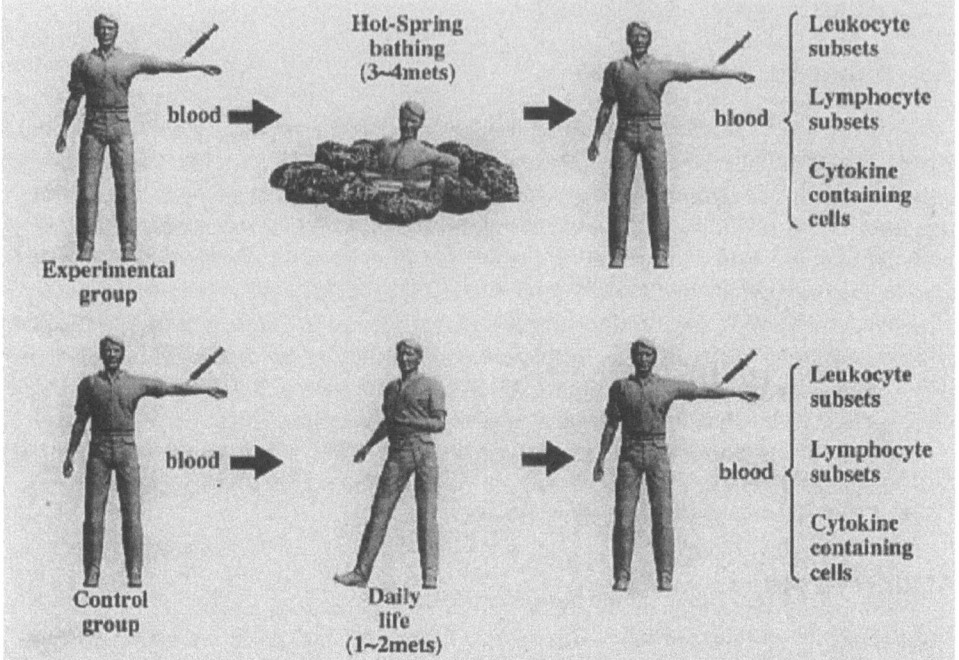

**Figure 16.1.** Experimental protocol: The age of the volunteers ranged from 18 to 87 (n = 126). They were separated into two groups, 35 years old and younger, and 36 and older, according to the results in this text. One group took a bath two to three times for 20 min, within a 24 hr period. Blood was collected from a peripheral venula before hydrotherapy and 24-hrs later. The blood was prepared as plasma and examined by a FACScan. Details of the procedures are within the text.

The hot springs used for these tests were Wakura Onsen Spa (a conc. sodium chloride with chlorinated soils, Nanao City, Ishikawa Pref.), Chugu Onsen Spa (a dil. sodium chloride with sodium carbohydrate, Oguchi Village, Ishikawa Pref.) and Gero Onsen Spa (a simple alkaline hot spring, Gero Town, Gifu Pref.) The water temperature was 41 degrees centigrade, fluctuating up or down by 1 degree centigrade.

On the first day of the trial, five-milliliters of blood were drawn from the forearm vein of all the subjects at 4 o'clock p.m. During that evening and the next morning, the hydrotherapy group bathed in the hot spring two to three times, for 20 minutes each time. Meanwhile, the light exercise control group took a 4 km walk in an hour (3–4 mets) or half an hour (7–8 mets) the same number of times that the hydrotherapy group took baths. Finally, at 4 o'clock p.m. on the second day, the blood of all the subjects was sampled again. Heparin was used as an anti-coagulating reagent (Fig. 16.1).

## 3.2. Leukocyte Count

The total number of leukocytes was recorded with a standard counter. In the differential counting, 200 cells were counted in each May-Grunwald-Giemsa stained smear. The numbers of granulocytes, monocytes and lymphocytes were determined, respectively.

## 3.3. Lymphocytes and Lymphocyte Subset Analysis

The whole blood obtained from the subjects was washed twice with PBS (phosphate buffered saline, pH 7.2). The suspensions were treated with fluorescent monoclonal antibodies (FITC-conjugated anti-human CD2, CD4, CD8, CD16, CD19 and CD56) separately. After 30 minutes of staining at 4 degrees centigrade, the cells were analyzed by a FACScan (Becton Dickinson Co. Ltd. U.S.A.).

## 3.4. Cytokine-Containing Cell Analysis

The blood cell suspensions were cultured with PMA (phorbol 12-myristate 13-acetate), Ionomycin and BSA (bovine serum albumin) for 4–5 hours at 37 degrees centigrade. After that, the cell suspensions were stained using the monoclonal antibodies of PE-IL-4, FITC-IFN-$\gamma$ and FITC-IL-1$\beta$, respectively. Then they were analyzed by the FACScan (Becton Dickinson Co. Ltd. U.S.A.). The antibodies and reagents used in the entire test were purchased from Becton Dickinson Immunocytometry system (U.S.A.).

## 3.5. Statistical Analysis

Statistical comparisons between the results obtained from the different groups were performed by the use of a Peason's correlation coefficiency test. $P < 0.05$ was considered significant.

## 4. RESULTS

*The effect of hot spring hydrotherapy on the leukocyte count correlated with the age and original basic count of the individuals.* Table 16.2 shows the total numbers of leukocytes, granulocytes and lymphocytes from peripheral blood collected before and after hot spring hydrotherapy in the older and younger groups. The quantitative variation of the two groups differed. In the younger group, the total number of WBC ($p < 0.05$) and granulocytes ($p < 0.01$) clearly decreased. On the other hand, the total number of WBC and lymphocytes significantly increased ($P < 0.01$) after hydrotherapy in the older group. Furthermore, the results show that changes in the leukocyte count and subset count had a negative relationship before and after hydrotherapy. In other words, subjects who had a higher cell count level before hydrotherapy tended to show a decrease in the number of WBC 24 hrs after hydrotherapy. There was a significant correlation between age and rate of change (Fig. 16.2, $r = 0.9$, $p < 0.001$). Since the turning point occurred at 35 years old, we separated the participants into a younger (under 35) and an older (over 36) group.

*The effect of hot spring hydrotherapy on the number of WBC and lymphocyte subsets also correlated with the age and theoriginal basic number of the individuals.* The results varied among the lymphocyte subsets before and after hot spring hydrotherapy. After hydrotherapy, the CD16+, CD8+ and CD19+ cells increased in the younger group. Meanwhile, the CD16+ cells increased in the older group while the CD19+ cells decreased remarkably (Fig. 16.3). This quantitative change in lymphocytes is shown in Fig. 16.4. Except for CD8+ cells, the CD2+, CD4+, CD16+, CD19+ and CD56+ cells all showed

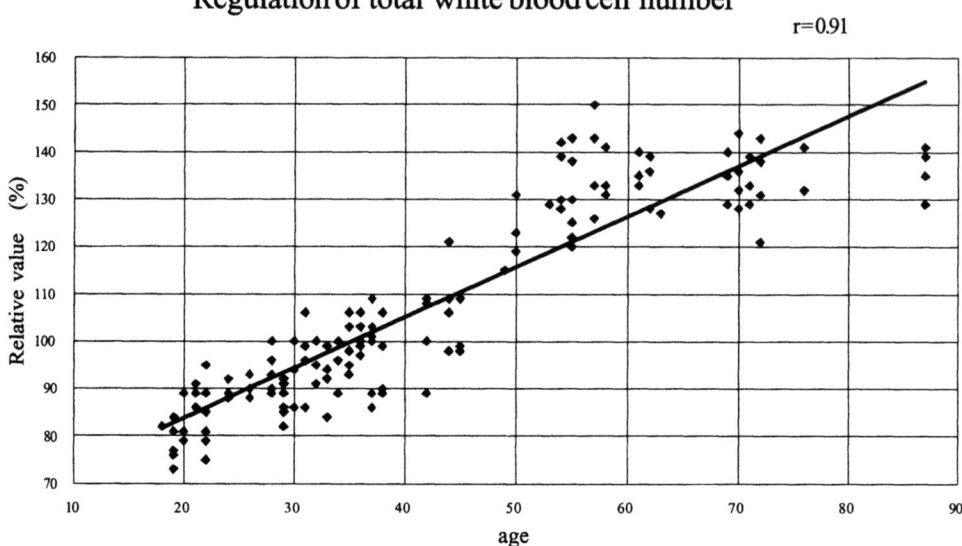

**Figure 16.2.** The variation in the number of leukocytes before and after hot spring hydrotherapy. The relative values (%) of WBC before and after hydrotherapy are shown according to age (n = 126). The significant correlation was obtained in the abscissa and compared according to age and variation (r = 0.9, p < 0.001).

negative variations. There was a strong correlation between the variable ratio and the value before hydrotherapy: high values tended to decrease and low values tended to increase (r = 0.692, p < 0.01).

A comparison of the CD4+/CD8+ cells and CD16+/CD56+ cells between the two groups is shown in Fig. 16.5. All the ratios of the CD16+/CD56+ cells increased in both groups. The changes in the CD4+/CD8+ cells, however, were different: the younger group showed a decrease while the older group showed an increase.

*Hot spring hydrotherapy regulated the leukocyte count differently in G and L type individuals.* In order to clarify the effect on individuals according to the type of regulation, we divided the volunteers into two groups: the G-type group had a granulocyte count over 70% and the L Type group had a lymphocyte count over 40%. Figs. 16.5 and 16.6 show that after hydrotherapy, the lymphocytes or lymphocyte subsets increased in the G type group, accompanied by a decrease in granulocytes. However, in the L type group, the granulocytes increased notably as did the number of lymphocyte populations.

*Hot spring hydrotherapy effected the functional changes in cytokine-containing cells.* To test whether hot spring hydrotherapy affected the functional maturation of immunocytes, we further investigated the number of cytokine containing cells by FACS analysis. Fig. 16.7 shows the effect of hot spring hydrotherapy on cytokine production. Even though the increase in IFN-$\gamma$ containing cells had no statistical significance, the increase in IL-4 was remarkable (r = 0.560, p < 0.01) while the IL-1$\beta$ containing cell counts showed a decrease. In addition, before and after hydrotherapy, the levels of IFN-$\gamma$ and IL4 or IL-1$\beta$ also had meaningful negative changes (See Fig. 16.7). After hydrotherapy, there was a decrease of cytokine-producing cells in the subjects who had previously had a higher level.

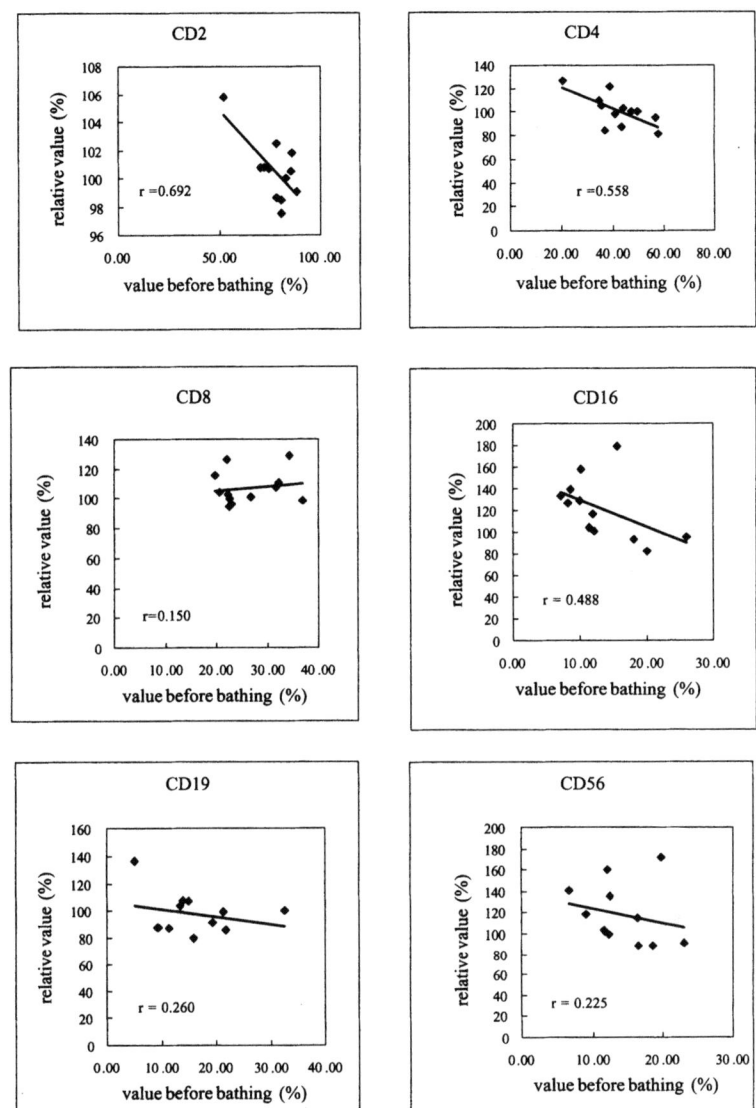

**Figure 16.3.** The variations in the lymphocyte subsets of the younger and older groups. The variation rate is shown for CD positive cells in the younger (black column) and older volunteer groups (gray column).

The changes depended on the age and the basic level of the white blood cells in the blood of the individuals.

We set up both positive and negative control experiments to show how the WBC were regulated during a 24 hr period. There were no clear changes in the peripheral blood of the general group that did only daily activities, indicating 1–2 mets (data not shown). In order to understand changes in the peripheral blood cells with other types of exercise such as walking, Figure 16.8 shows the changes in numbers and functions of white blood cells with 3 ~ 4 mets ($r = 0.689$, $p < 0.01$) and 7 ~ 8 mets ($r = 0.711$, $p < 0.01$, Fig. 16.8).

Regulation in cytokine containing cell

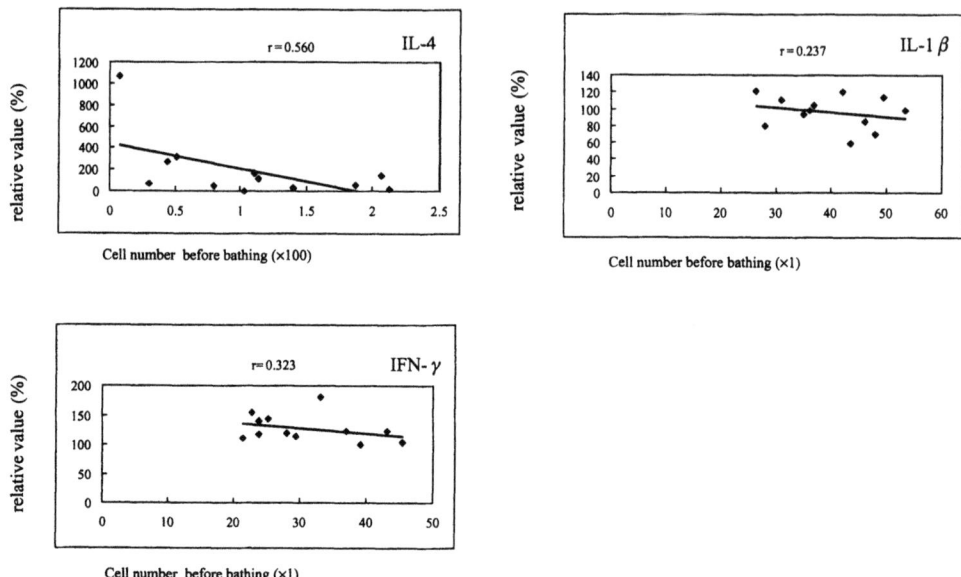

**Figure 16.4.** The variation in the number of lymphocyte subsets before and after hot spring hydrotherapy. Each value of the subsets exhibited by CD is indicated for the two different age groups.

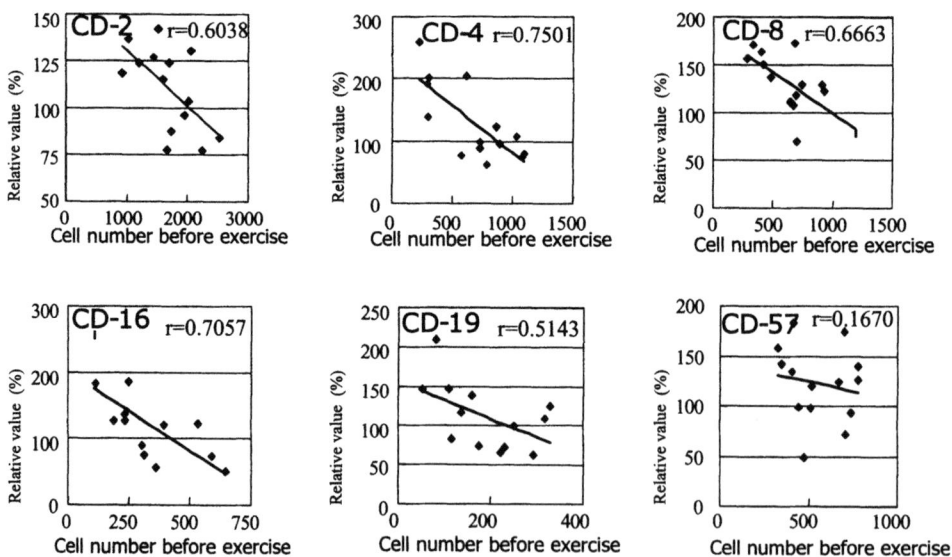

**Figure 16.5.** The relative values of CD positive cells such as CD 4/8 and 16/56. In order to estimate the qualitative status of the immune response as judged by CD markers, the ratios of CD 4/8 and 16/56 are shown for the two different age groups, which were further divided into the granulocyte rich (G) and lymphocyte rich (L) types. The details of the procedure are explained in Materials and Methods.

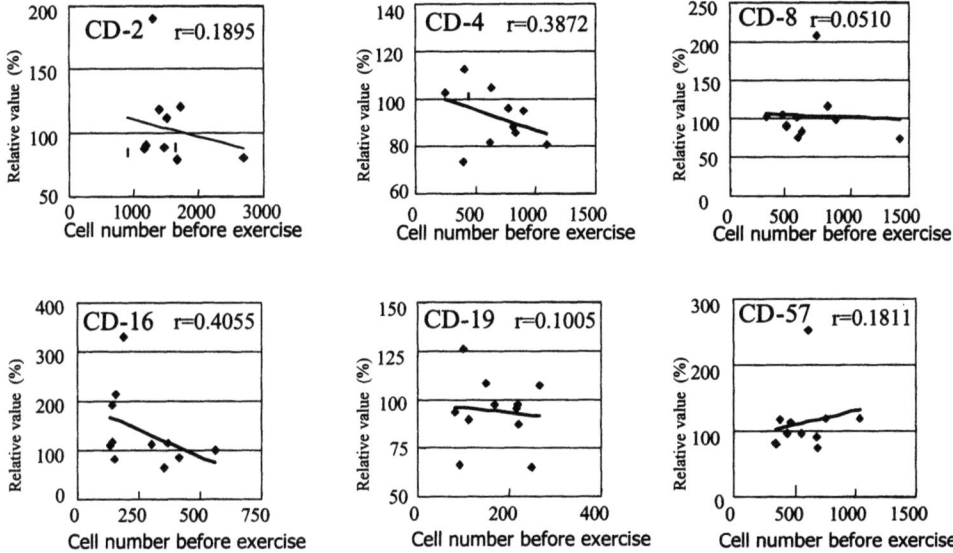

**Figure 16.6.** The relative values of subsets were sorted morphologically and by CD positive markers were remarked in the G and L types of the volunteers. Each subset was calculated and shown according to white blood cell type and CD marker. The volunteers were separated into G and L types with over 70% granulocytes and over 40% lymphocytes, respectively.

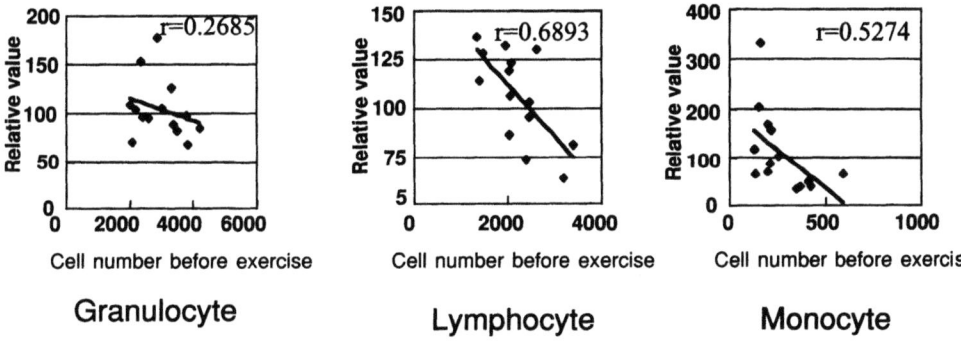

**Figure 16.7.** Cytokine containing cells: The relative numbers of cytokine containing cells in IL-1β, IL-4 and IFNγ were counted and shown by the vertical line. Each value on the first day is indicated at the horizon point. The correlation efficiency $r = 0.237$, $r = 0.560$ and $r = 0.323$ indicating, IL-4 containing cells were significantly regulated by hot-spring hydrotherapy.

The white blood cell changes in the light exercise group in Fig. 16.9, confirmed that the effects of the two types of exercise were the same as those experienced with hot spring bathing. As for regulation of the immune system in cell number and function: light exercise indicated 3 ~ 4 mets which was higher than hot spring bathing and strenuous walking showed even higher values (Fig. 16.10, 7–8 mets).

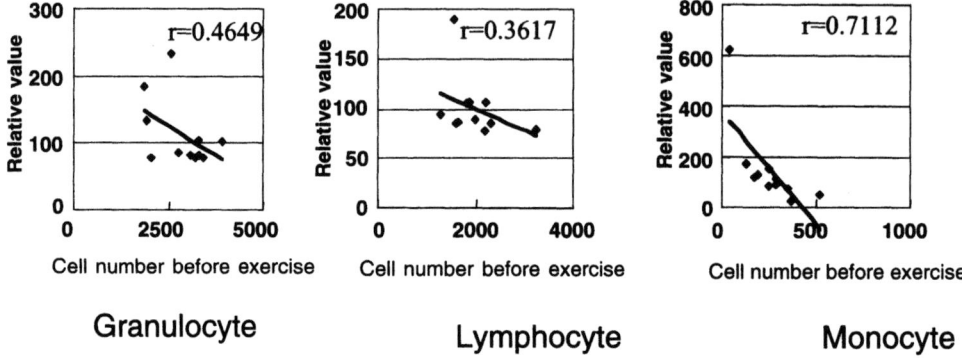

**Figure 16.8.** The relative values of WBC subsets after two different types of walking exercise: 4 km in 60 min (3 ~ 4 mets) and 4 km in 30 min (7 ~ 8 mets as judged by heart rate and breathing). As with the hot spring experiment, each WBC subset was counted.

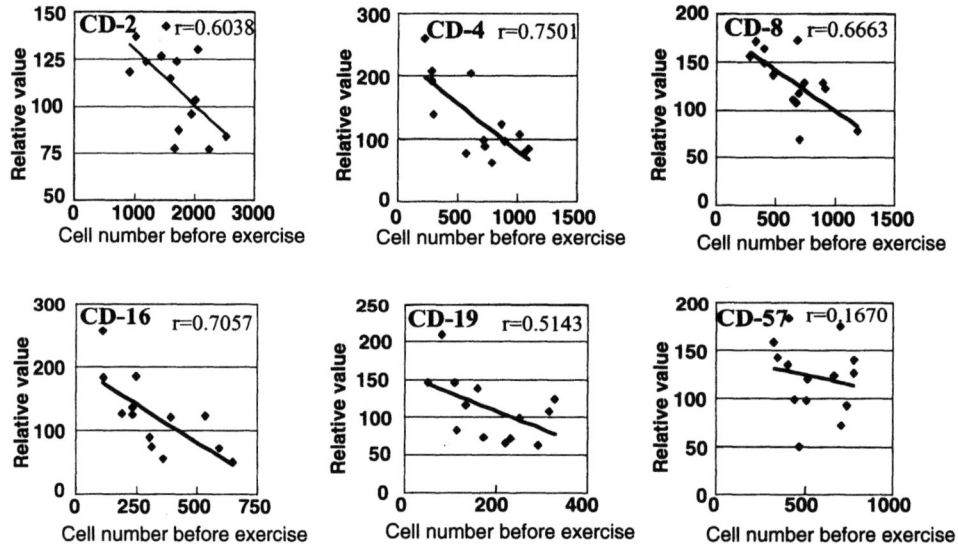

**Figure 16.9.** The regulation of lymphocyte subsets with 3 ~ 4 mets of exercise. The relative values of lymphocyte subsets with 3 ~ 4 mets of exercise are shown.

## 5. DISCUSSION

In Japan, it is traditionally accepted that hot spring hydrotherapy is as beneficial to the health as moderate exercise. Many studies have shown that moderate exercise can improve physical immune function, even though the details of the mechanism remain unclear. With our study and reproducible results, we can say that the effect of hot spring hydrotherapy is also connected with the function of the immune system. The cell count and function of granulocyte, monocyte, lymphocyte and lymphocyte subsets in the peripheral blood had changed after hydrotherapy. Our data also showed that the variations in peripheral blood

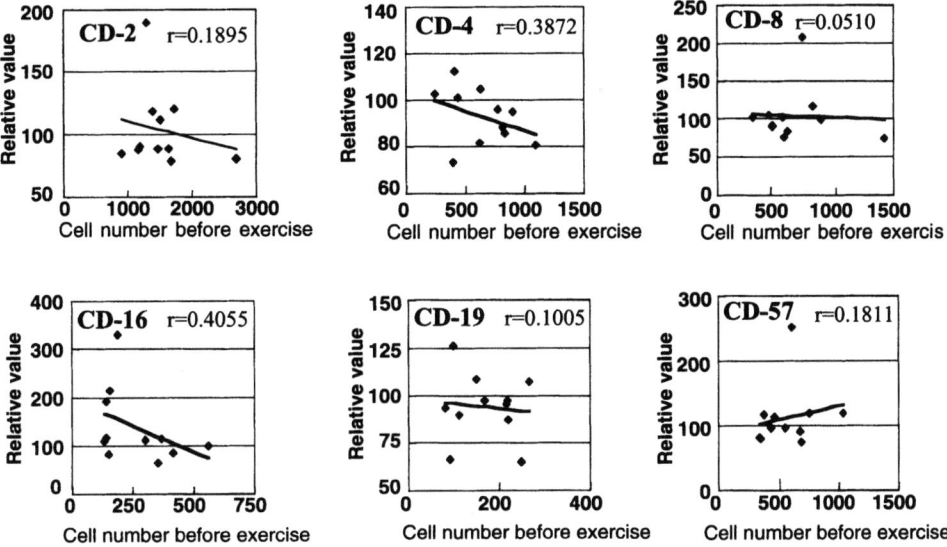

**Figure 16.10.** The regulation of lymphocyte subsets with 7 ~ 8 mets of exercise. The relative values of lymphocyte subsets are shown from the walking exercise with 7 ~ 8 mets.

occurred according to age. The younger group showed a clear decrease in the number of granulocytes while the total number of leukocytes increased in the older group. Other interesting results included the negative relationship in cell counts of the younger and older groups and differeing data between the G and L types. One possible reason for the cell count increase may be that the temperature of a hot spring is always higher than normal body temperature, causing perspiration that may result in hemoconcentration. In order to avoid this short term phenomenon, we did not measure the blood samples immediately, but 24 hours after hot spring hydrotherapy.[4] This gave the subjects enough time to drink and supplement the extracellular fluid that was lost during hot spring hydrotherapy. The other reason for the collection schedule is that the lymphocyte and granulocyte numbers change with the circadian rhythm.[7]

The leukocytosis in individuals may also be associated with the regulation of the nervous system and hormones.[9] It is well known that corticosteroids are involved in the maturation and differentiation of lymphocytes and monocytes. On the other hand, taking corticosteroids can result in transient lymphocytopenia. We thought that hot spring hydrotherapy would release stress and therefore reduce the levels of corticosteroids and other similar hormones.[11] The leukocytes were then redistributed from the body compartments such as central lymphoid organs from the circulation into other body compartments in the younger participants and into the circulation of the older individuals. In this way, the cell counts in the blood could be maintained at constant and average levels or so called homeostasis.[10] Such regulation may occur not only through hot spring bathing but also by the hot water hydrotherapy such as that of the Japanese public bath system. A critical assessment of this issue will be a major focus of future research.

After hot spring hydrotherapy, the subsets of the T lymphocytes changed more clearly in the younger group than in the older one. The percentages of CD4+ and CD8+ subsets were enhanced significantly, especially the percentage of CD4+ cells. This may have also been associated with hormones, which mobilized lymphocytes into the marginal pool.[10]

We found that the function of lymphocytes significantly increased after hot spring hydrotherapy. We know that cytokine IFN-γ, which is closely associated with the activity of cellular immunity, is secreted from Th1 cells.[9] The IL-4 associated with humoral immunity is regulated by Th2 cells. Simon et al. confirmed that the level of IL-4 in the circulation was enhanced after exercise, and since IL-1 promoted immune responsiveness, it may be an important link between fever and the function that enables the host to defend itself against infection.[9,12,13] The increase of cytokine-containing cells in IL-4 demonstrated that hot spring hydrotherapy could enhance the immune system's ability to fight off infection by augmenting humoral immunity. It has been reported that all kinds of environmental stimulation activate the autonomic nervous system and then influence the immune system through adrenergic receptors on granulocytes and cholinergic receptors on lymphocytes.[14-23] The variations in the hormone and β-adrenergic receptors of lymphocytes may cause the rise in lymphocyte activity. This would be an important and interesting point to investigate further.

It is well known that the ratio of CD4+/CD8+ reflects immune function. The function increases when the ratio is high, or is suppressed when it is low.[24,25] After hydrotherapy, the ratio increased clearly in the older group, but decreased in the younger group. We know that both cellular and humoral immune functions lower with aging. This lowered immune function is enhanced and raised to a normal level or higher by hot spring hydrotherapy. This result and others, such as that which occurred with moderate exercise, indicate that hot spring hydrotherapy can improve the function of the immune system, especially in older people.

The CD16+ cell and CD56+ cell counts also increased in the two groups, showing that hot spring hydrotherapy can promote the production of Natural killer (NK) cells which play a vital role in protecting against viral infections and tumors.[26,27] This alteration in NK cell count following hot spring hydrotherapy is being followed with intense interest.

There was much variation in the negative relationships found in this test, showing similar tendencies to those seen in the rate of breathing and heart rate with moderate exercise (3 ~ 4 mets). This means that hot spring hydrotherapy can enhance the activity that regulates and maintains the balance of human beings, leading to immune function regulation.

We hope that our work will attract further investigation into the mechanism of how hot spring hydrotherapy benefits the health. In further research, we plan to focus on the mutual interaction between the immune system function and the efficacious period of hydrotherapy, the influence of the hot spring's components and temperature, the physical effect of the spring water, as well as the effect on the individual's mental condition.

## 6. References

1. Zhonghua Yi, Shi Za Zhi, Development of immersion therapy in Mongolian medicine, 1995 Jan;25(1):43–5.
2. Dubarry J.J., Richard M., Spas in Gallo-roman Aquitaine, Hist Sci Med 1994;28(3):217–22.
3. van Tubergen A, van der Linden S., A brief history of spa therapy, Ann Rheum Dis 2002 Mar;61(3):273–5.
4. Tanizaki, Y., Kitani, H., Okazaki, M., et al.: Clinical effects of spa therapy of bronchial asthma. 2. Relationship to ventilatory function. J.J.A.Phys.M.Baln.Clim. 55:82–86, 1992.
5. Tanizaki, Y., Kitani, H., Okazaki, M., et al.: Clinical effects of spa therapy on bronchial asthma. 1. Relationship to clinical asthma type and patient age. J.J.A.Phys.M.Baln.Clim. 55:77–81, 1992.

6. Mitsunobu, F., Kitani, H., Okazaki, M., et al.: Clinical effects of spa therapy on bronchial asthma. 6. Comparison among three kinds of spa therapies. J.J.A.Phys.M.Baln.Clim. 55:185–190, 1992.
7. Abo, T., Kawate, T., Itoh, K., et al.: Studies on the bioperiodicity of the immune response:1. Circadian rhythms of human T,B and K cell traffic in the peripheral blood. J. Immunol. 126:1360–1363, 1981.
8. Tonnesen, E., Christensen, N.J., Brinklov, M.M.: Natural Killer cell activity during cortisol and adrenaline infusion in healthy volunteers. European J. Clin. Inv. 17:497–503, 1987.
9. Mandler, R.N., Biddison, W.E., Mandler, R., and et al.: β-Endorphin augments the cytolytic activity and interferon production of natural killer cells. J. Immunol. 136:934–939, 1986.
10. Wang, X-X., Kitada, Y., Kenichiro Matsui, K., Ohkawa, S., Sugiyama, T., Kohno, H., Shimizu, S., Lai, J.E. Matsuno, H., Yamaguchi, M., Yamaguchi, N.: Variation of Cell Populations Taking Charge of Immunity in Human Peripheral Blood Following Hot Spring Hydrotherapy Quantitative Discussion-. The Journal of Japanese Association of PhysicalMedicine, Balneology and Climatology. 62:129–134, 1999.
11. Matsuno, H., Wang, X., Wan, W., Matsui, K., Ohkawa, S., Sugiyama, T., Kohno, H., Shimizu, S., Lai, J.E. Yamaguchi, N.: Variation of Cell Populations Taking Charge of Immunity in Human Peripheral Blood Following Hot Spring Hydrotherapy Qualitative Discussion-. The Journal of Japanese Association of Physical Medicine, Balneology and Climatology. 62:135–140, 1999.
12. Yamaguchi, N., Hashimoto, H., Arai, M., Takada, S., Kawada, N., Taru, A., Li, A-L., Izumi, H., Sugiyama, K: Effect of Acupuncture on Leukocyte and Lymphocyte Subpopulation in Human Peripheral Blood-Quantitative discussion-. The Journal of Japanese Association of Physical Medicine, Balneology and Climatology. 65:199–206, 2002.
13. Wan, W., Li, A-L., Izumi, H., Kawada, N., Arai, M., Takada, A., Taru, A., Hashimoto, H., Yamaguchi, N: Effect of Acupuncture on Leukocyte and Lymphocyte Subpopulation in Human Peripheral Blood Qualitative discussion-. The Journal of Japanese Association of Physical Medicine, Balneology and Climatology. 65:207–211, 2002.
14. Wang, X-X., Katoh, S., Liu, B-X., et al.: Effect of physical exercise on leukocyte and lymphocyte subpopulations in human peripheral blood. Cyto. Res. 8:53–61, 1998.
15. Galun, E., Burstein, R., Assia, E., et al.: Changes of white blood cell count during prolonged exercise. Int. J. Sports Med. 8:253–255, 1987.
16. Oshida, Y., Yamanouchi, K., Hayamizu, S., et al. : Effect of acute physical exercise on lymphocyte subpopulation in trained and untrained subjects. Int. J. Sports Med. 9:137–140, 1988.
17. Soppi, E., Varjo, P., Eskola, J., et al.: Effect of strenuous physical stress on circulating lymphocyte number and function before and after training. J. Clin. Lab. Immunol. 8: 43–46, 1982.
18. Simpson, J.R., Hoffman-Goetz, L.: Exercise stress and murine natural killer cell function. Proc. Soc. Exp. Biol. Med. 195:129–135, 1990.
19. Landmann, R.M., Muller, F.B., Perini, C., et al.: Changes of immunoregulatory cells induced by psychological and physical stress: relationship to plasma catecholamines. Clin. Exp. Immunol. 58:127–135, 1984.
20. Hoffman-Goetz, L., Thorne, R., Simpson, J.A., et al.: Exercise stress alters murine lymphocytes subset distribution in spleen, lymph nodes and thymus. Clin. Exp. Immunol.76:307–310, 1989.
21. Gimenez, M., Mohan, K.T., Humbert, J.C., et al.: Leukocyte, lymphocyt and platelet response to dynamic exercise. Eur. J. Appl. Physiol. 55:465–470, 1986.
22. Espersen, G.T., Elbaek, A., Ernst, E., et al.: Effect of physical exercise on cytokines and lymphocyte subpopulations in human peripheral blood. APMIS 98:395–400, 1990.
23. Kappel, M., Tvede, N., Galbo, H., et al.: Evidence that the effect of physical exercise on NK cell activity is mediated by epinephprine. J. Appl. Physiol. 70:2530–2534, 1991.
24. Moriguchi, S., Kayashita, J., Okada, S., et al.: Effects of nutrition and exercise on the decreased cellular immune functions in the aged rats. Bulletin of the Physical Fitness Research Institute. 75:144–151, 1990.
25. Moriguchi, S.: Beneficial effects of nutrition and exercise on cellular immune function. J. Jap. Soc. Nut. Fo. Sci. 48:1–8, 1995.
26. Gatti, G., Cavallo, R., Satori, M.L., et al.: Inhibition by Cortisol of human natural killer cell activity. J. Steroid Biochem. 26: 49–58, 1987.
27. Cabriel, H., Kindemann, W.: The acute immune response to exercise: what does it mean? Int. J. Sports Med. 18:528–545, 1997.

# 17

# Music Therapy, Wellness, and Stress Reduction

## Andrea M. Scheve, MM, MT-BC[1]

### 1. What Is Music Therapy?

Music therapy (MT) is an established health profession using music and music activities or interventions to address physical, emotional, cognitive, social, psychological, and physiological needs of infants, children, adolescents, and adults including the elderly, with disabilities or illnesses. MT may also be used with healthy children and adults to maintain wellness throughout life. MT builds on the power of music, using music in a focused and concentrated way for healing and for change. It is a non-invasive medical treatment designed to prevent illness and disease, alleviate pain and stress, help people express feelings, promote physical rehabilitation, positively affect moods and emotional states, enhance memory recall, and provide unique opportunities for social interaction and emotional intimacy.[1]

### 2. A Brief History of Music as Medicine

One may only speculate how far back the use of music as medicine dates. Written history dating back to 6,000 B. C. from Iraq identifies music's role in magic, healing ceremonies, and medicine. Evidence of the use of music in other ancient cultures exists as well. Music was used in chant therapies in Egypt circa 5,000 B. C., and in Babylonian culture circa 1850 B. C. in religious ceremonies as treatment for sin, which caused illness. In Greece circa 600 B. C., music was prescribed for emotionally disturbed individuals.[2] In recent history, the first acknowledgement of the use of music in general hospital treatment came in 1914 by the American Medical Association. Dr. Kane reported using a phonograph in his operating room for "calming and distracting patients from the horror of the situation."[3]

The profession of music therapy was established in 1950 as a result of Red Cross nurses treating wounded soldiers of World War II. The nurses implemented music into the daily regiment of the wounded, frail, and sick for entertainment, diversion, recreation,

---

[1] Andrea M. Scheve, MM, MT-BC, Director of the UPMC Music Therapy Program. Children's Hospital of Pittsburgh, Suite 5725, Pittsburgh, PA 15226.

*Complementary and Alternative Approaches to Biomedicine*, edited by Edwin L. Cooper and Nobuo Yamaguchi. Kluwer Academic/Plenum Publishers, 2004.

and to raise morale.[4] This also targeted their physiological and psychological well-being. Today, over 5,000 music therapists are employed throughout the United States in over 50 different settings including hospitals, clinics, day care facilities, schools, community health centers, substance abuse facilities, nursing homes, hospices, rehabilitation centers, correctional facilities, and private practices.

## 3. Music Therapy in the Hospital Setting

Currently MT is utilized in the hospital setting across populations including premature infants, pediatrics, adolescents, adults, and geriatrics. Each of these populations present unique medical, physiological, social, emotional, physical, and psychological problems that music therapists may address. MT interventions are determined on a case by case basis and are unique to each patient and each individual situation.

### 3.1. Music Therapy with Premature Infants

Caine studied the effects of music listening on 52 pre-term and low birth weight newborns who were stable and in isolettes. The experimental group received 60 minutes of tape recorded vocal lullabies, children's songs and auditory stimulation three times daily. She found this significantly reduced initial weight loss, increased daily average weight, increased formula and caloric intake, decreased the length of hospitalization, and significantly reduced stress behaviors of these infants.[5] Standley studied the effects of music and multimodal stimulation on 40 newborns in a Newborn Intensive Care Unit (NICU) and found a significant decrease in length of hospitalization for female infants and increased weight gain per day for both genders.[6] Another study by Standley demonstrates music's effectiveness as a contingency to teach non-nutritive sucking in 12 infants in the NICU using the Pacifier Activated Lullaby. Pacifiers were fitted with pressure transducers so each time the infant sucks, 10 seconds of recorded female vocal music plays. Results of this study indicate the suck rate during the experimental condition was significantly higher than the no music condition. This transfers to nutritive feeding, allowing the infant to develop proper suck, swallow, and breath schemas necessary for bottle feeding.[7] Whipple conducted a study concerning training parents of NICU patients to use music and multimodal stimulation and the frequency of stress behaviors exhibited by the infants during parent-neonate interactions. She found significantly less stress behaviors exhibited during parent-neonate interactions for the experimental group compared to the untrained control group. She also found appropriateness of parent responses and actions was significantly greater for the experimental group.[8]

### 3.2. Music Therapy and Oncology Patients

Another patient population within the hospital with which music therapy is effective include cancer/oncology patients. Standley researched the effects of music listening on reducing nausea and emesis in chemotherapy patients. Fifteen participants were divided into two experimental groups and two comparable control groups. Both experimental groups reported less nausea and nausea was greatly delayed for these groups as well.[9] Boldt conducted research on 6 bone marrow transplant patients to determine music therapy's effectiveness on motivation, physical comfort, psychological well-being, and exercise endurance. Results indicate across 10 sessions, music therapy is effective in increasing subject's self report of relaxation and comfort levels, but was less effective in reducing pain and nausea. More cooperative behavior during exercise and an increase in endurance was noted for the long term patients. For short term patients, only lasting 2–3 sessions, the same results were noted but endurance did not increase.[10] Burns conducted a study on cancer survivors concerning elevating mood and increasing life quality of these patients using the Bonny Method Guided Imagery in Music. She found an increase in mood and quality of life scores for the experimental group that endured after the 10 sessions had ended.[11] Bailey examined the difference between live music listening and tape recorded music listening on mood and anxiety of 50 cancer patients ranging in age from 17–69. The author found the live music group reported significantly less tension and anxiety, and changes in mood for the better compared to the tape recorded music listening group. This suggests human interaction plays a role in reducing tension and elevating mood.[12] Brodsky demonstrates music therapy's effectiveness with isolated pediatric cancer patients in stimulating verbalization, resolving anxiety and fear provoking fantasies, and establishing a positive affect when compared to other therapies.[13] Smith et al. conducted a study on 42 men in a V. A. Hospital receiving radiation treatment for pelvic or abdominal malignancies. The experimental group listened to patient-chosen music through headphones before and during daily treatments for the duration of the planned course of therapy. The control group followed standard protocols. No significant difference was found in this study between the two groups but a post-hoc analysis revealed a decrease in anxiety scores for the experimental group, suggesting music is beneficial during radiotherapy.[14]

### 3.3. Music Therapy and Cardiovascular Patients

Barnason et al. studied 96 patients who underwent elective heart bypass surgery to examine the influence of music therapy and music-video therapy during the early postoperative period on mood and anxiety. The authors found a significant change over time for heart rate and systolic and diastolic blood pressure, which indicated a relaxation response for experimental groups.[15] Guzzetta conducted research in a coronary care unit on 80 patients diagnosed with myocardial infarction (heart attack). The patients were randomly assigned to a control, relaxation, or music group and evaluated by apical heart rate, peripheral temperatures, cardiac complications, and qualitative patient evaluative data. This author found both the relaxation group and the music group were effective in lowering apical heart rates and increasing peripheral temperatures when compared to the control group, and less cardiac complications were reported for the experimental groups.[16] Mandle et al. performed a study on 45 patients undergoing angiography procedures (an X-ray of the inside of the heart and blood vessels) divided into three groups: relaxation audiotape, music audiotape, and blank audiotape. Patients were instructed to listen to their audiotape throughout the procedure and radiologists were not told the group assignment or tape contents. Patients who received the relaxation tape experienced significantly less anxiety, pain, and received significantly less medication (fentanyl citrate and diazepam) than the other two groups.[17]

## 3.4. Music Therapy and Childbirth/Gynecological Procedures

Many studies concerning women in childbirth or women undergoing gynecological procedures have found music therapy to be beneficial to the patient. Walters researched psychological and physiological effects of vibrotactile stimulation of patients awaiting scheduled gynecological surgery. Thirty-nine participants ranging between 19–65 years of age were randomly assigned to one of three groups (control, music, and vibrotactile) and completed pre- and post-self reports of psychological data including tension, anxiety, relaxation, stress, mood, and apprehension. Physiological measurements such as systolic and diastolic blood pressure, pulse rate, and temperature were also measured. Results indicate that both the music and vibrotactile groups experienced reduced levels of tension, anxiety, and stress, and increases in relaxation and mood. Both groups also spent less time in surgery, less time in the post-op recovery room, and received significantly less post-operative medication.[18] Good et al. did a study on three nonpharmacological interventions: relaxation, music, and the combination of relaxation and music on 311 gynecological surgery patients. Results indicate a significant decrease in pain scores for all experimental groups, indicating a need for postoperative intervention with gynecological surgery patients.[19]

Clark, McCorkle, & Williams studied the effects of music therapy during the third trimester on successful experiences in childbirth in 20 expecting women. The experimental group consisted of 13 women who participated in 6 pre-delivery music therapy training sessions. Seven women in the control group followed standard procedures. Data were collected by a questionnaire reflecting subjective perceptions and reports of home practice. Results indicate the experimental group achieved significantly greater "success" scores and practiced more at home than the control group.[20] Hanser, Larson, & O'Connell did a study with 7 expecting women measuring overt pain responses during delivery with music therapy. Music was used for relaxation, to cue rhythmic breathing, as comfort through positive associations, and to mask anxiety-producing noise. All subjects emitted fewer pain responses with music than without music.[21] Liebman & MacLaren studied the effects of music and relaxation on expecting adolescents. This study found music therapy significantly reduces state and trait anxiety with these patients during third trimester pregnancy when compared to the control group who did not receive music therapy.[22]

## 3.5. Music Therapy and Comatose Patients

Other patients who benefit from music therapy include comatose and traumatic brain injury patients. Boyle & Greer studied the effects of using music as a contingency for performing simple motor tasks on three comatose patients. The contingent music affected all behaviors for patient #1 who had been comatose for 6 months, but fewer responses were observed for patients #2 and #3 who had been comatose for 10 and 38 months respectively.[23] Jones et al. published a case study on one comatose traumatic brain injury patient. The authors presented the patient with four different auditory stimuli: voices of family members and friends, classical music, popular music, and nature sounds to determine which auditory stimuli the patient would respond. Behavioral observations of body movements and measures of pulse rate were determined the best indicators of responsiveness, and the family's and friend's taped voices elicited the greatest response.[24] Scheve did a case study on a coma patient and found the relationship between the patient and music therapist to elicit overt responses. Over time, the patient responded to the initial greeting of the music therapist's voice.[25] Hurt, Rice, McIntosh, & Thaut tested rhythmic auditory stimulation (RAS), a MT technique, on 8 traumatically brain injured patients with persisting gait disorder. Rhythmic auditory stimulation consists of pairing rhythm with walking and slowly increasing the rhythm over time which will in theory increase walking speed, velocity, cadence, and stride length. Significant results for velocity, cadence, and stride length were reported for 5 patients. Two patients did not entrain to faster RAS tempi.[26]

## 3.6. Music Therapy and COPD, CVA, and Parkinson's Patients

Chronic obstructive pulmonary disease (COPD) patients, stroke (CVA) victims, and Parkinson's patients have all been shown to be positively affected by music therapy intervention. Pfister, Berrol, & Caplan did a study measuring the effects of music on exercise tolerance and perceived symptoms during treadmill walking. No significant differences were found in this study but the authors note 60% of patients spontaneously reported enjoying listening to music during exercise.[27] Thornby, Haas, & Axen studied the effects of music, grey noise, and silence on exercise time and external work of 36 COPD patients. This study found significant increases in exercise time and external work for the music group and a significantly decreased perceived exertion rate as compared to the grey noise and silence groups.[28] Thaut et al. studied the difference between conventional rehabilitation protocols and RAS with hemiparetic stroke victims. Pre- and post-tests revealed a significant increase in velocity and stride length of patients receiving RAS compared to conventional therapy.[29] Thaut et al. also found RAS to be effective with Parkinson's patients. Fifteen patients who trained with RAS significantly increased their gait velocity, stride length, and step cadence compared to the eleven people in the two control conditions: no intervention and an internally self-paced training program.[30]

## 3.7. Music Therapy and Operative Patients

Preoperative, perioperative, and postoperative patients have been found to benefit from music therapy. Haun, Mainous, & Looney found 20 minutes of music listening to be an effective anxiolytic for women undergoing breast biopsy surgery. Results of this study indicate a statistically significant decrease in anxiety levels for the music group ($n = 10$) compared to the non-music group ($n = 10$) and a statistically significant decrease in respiration rate for the music group.[31] In a similar study, Gaberson investigated the effects of both a 20 minute humorous tape and a 20 minute music tape on 46 preoperative patients awaiting elective surgery. This researcher found no significant difference between the group anxiety means.[32] Tusek, Church, & Fazio studied the effects of guided imagery on anxiety, pain, and narcotic medication usage of 130 patients undergoing elective surgical procedures. Guided imagery tapes were used by the experimental group three days prior to surgery, directly before surgery, perioperatively, and postoperatively for six days after the surgery. These authors found a significant decrease in pain, anxiety, and a significant decrease in the amount of pain mediations (almost 50% less) required.[33] Sanderson conducted research concerning reduction of anxiety and pain pre- and postoperatively in the recovery room. Similarly, she found a significant decrease in the music listening group as compared to the control group in preoperative anxiety ratings. The experimental group also made fewer pain and anxiety statements, behaved less anxiously, and used less pain and nausea medications. However, both groups reported the same amount of recovery time.[34] Miluk-Kolasa, Matejek, & Stupnicki conducted research on the effects of music therapy on physiological responses of 100 in-patients awaiting non-orthopedic surgery. The physiological measurements taken were: arterial pressure, heart rate, cardiac output, skin temperature, and glucose count. The day preceding surgery, all measurements were taken, then the patient was told about the surgery, and measurements were taken again every 20 minutes for an hour. The experimental music listening group listened to relaxing music following surgical instruction and their physiological measurements returned to baseline. The control group's measurements remained at the stressor-induced level for the duration of the one hour period.[35] Good et al. studied the effects of jaw relaxation, music, and a combination of the two on postoperative pain after major abdominal surgery on 500 subjects. The authors found a significant decrease in pain for all experimental groups compared to the control group. After ambulation, there was no significant difference between the pain scores of the experimental and control groups. This suggests a need for reminders of relaxation techniques during ambulation following surgery.[36]

## 3.8. Music Therapy and Pediatric Care

Music therapy is effectively utilized in pediatric care with many different diagnoses, although less research exists with this population. Fowler-Kerry & Lander researched the effects of suggestion and music distraction on 200 pediatric patients receiving routine immunization injections. The music distraction group reported significantly less pain than the suggestion or control group.[37] Malone researched the effects of live music on distress of 40 pediatric patients receiving needle-sticks. She found live music significantly reduced behavioral distress and it was especially effective for children 1 year old and younger.[38] Froelich examined the differences in verbalizations elicited between medical play therapy and music therapy for 40 school-aged pediatric patients. She found music therapy significantly increased verbalizations about the hospital experience compared to medical play therapy.[39] Verbalizations about the hospital experience enhances coping skills in children. Marley did a case study of 27 hospitalized infants and toddlers and found music therapy, plus the interaction with a music therapist, effectively reduces stress-related behaviors.[40] Music therapy is also used with the pediatric population to allow the child control over one aspect of their environment, allow for verbal and musical expression, and distract the child from the situation, which also promote coping skills. Physiologically, singing requires deep breathing, which increases oxygenation to the bloodstream and active music making increases respiration, and utilizes gross and fine motor movements.

## 4. Using Music to Reduce Stress

### 4.1. What Is Stress?

Stress is the result of a situation that occurs which exceeds the mind's capability to effectively respond. It occurs when you cannot cope with what you believe to be the perceived consequences or demands of the event. Stress-reactive areas of the brain are activated which increases the stress hormone concentration in the bloodstream, and in turn causes an increase in heart rate, sweat deposits, feelings of nervousness or depression, a decrease in the effectiveness of the immune system to protect against infection, an increase in cholesterol deposits in the blood vessels of the heart, and possible cell damage in the brain.[41]

### 4.2. What Kind of Music Should I Use?

The selection of music to use in stress reduction is a unique and personal choice. Some people may find themselves late to work in the midst of rush hour traffic, gripping the steering wheel and muttering obscenities and turn the radio to the classical station to help calm down. Others may find themselves in the same situation but choose to sing at top volume to their favorite pop song. Either way, the choice is your own. The first step in calming down and reducing stress is choosing the music, one aspect of the situation that can be controlled.

### 4.3. Singing

Singing is the next step. This is even more personal than the music selection, but also extremely important because it requires deep diaphragmatic breathing. Deep breathing, also known as abdominal or diaphragmatic breathing, rapidly increases the amount of oxygen in your blood. The brain detects the increase in oxygen and responds by decreasing the concentration of stress hormones in the blood. This can be accomplished without singing, but it must be practiced.

Deep breathing sounds easy enough, but if you take notice, most adults take shallow breaths that can be observed filling only the chest cavity. If someone is deep breathing, you can observe their entire abdomen expand with each inhale and retract with each exhale. More oxygen stimulates the longer, slower alpha brain waves associated with relaxation and calm mind states. Therefore, deep breathing and singing are helpful in eliciting a relaxation response and protecting the body from harmful side effects of stress.

Take a moment before you go to sleep while lying in bed and place one hand on your chest and the other on your abdomen. Take in a few deep breaths, and notice how the hand on your abdomen rises and falls with each inhale and exhale. The hand on your chest will only move slightly. Breathe in through the nose, and out through the mouth.

### 4.4. Musical Associations

Have you ever heard an old song come on the radio that you haven't heard in years, and suddenly you are transported back to another time? You remember every word and it brings back memories and associations. Maybe it's your wedding song or a song your sweetheart used to sing to attract your attention. This is a musical association. This same principle can be used when developing stress reduction techniques.

Again, choose the music. Some guidelines to follow for choosing music for this exercise is finding a CD or a tape that is relaxing to you. Maybe something with a slow tempo, or beat, with which you can practice deep breathing, and a nice flowing melody line. There can be lyrics or it may be entirely instrumental.

Next, play the same CD each night while falling asleep. Do some deep breathing to the rhythm of the music as you form an association with that particular music. The association you form when falling asleep to a piece of music is a relaxation association. As you become more and more relaxed, and fall asleep, you still hear the music.

Once the relaxation association is formed between you and the particular piece of music, and you find yourself in a stressful situation, just play the CD you chose, and you will have an automatic relaxation response. Block out any other stimuli, listen to the music, and take deep cleansing breaths.

## 4.5. Music and Visualization or Meditation

Another stress reduction technique that helps develop a musical relaxation association is using music during visualization exercises. Visualization is another term for guided imagery. This is different from meditation because with visualization, you focus on mentally escaping the present situation and taking a mini-vacation by imagining yourself somewhere pleasant and peaceful. With meditation, the goal is to clear your mind of negative thoughts and focus on something neutral. Using the same piece of music while practicing either of these techniques will form a musical association with the relaxation response in your body.

## 5. Music and Wellness

### 5.1. Getting Well

We all know the best way to be well is to prevent illness. Don't get sick in the first place. However, because many of us are ill, getting well must be achieved before maintaining wellness. In addition to the standard medical care, music therapy is being utilized in the hospital setting for many reasons and with myriad patients ranging from premature infants in the Neonatal Intensive Care Unit (NICU) to Palliative Care. Music has been found to be an effective anxiolytic (anxiety reducer) for patients undergoing anxiety-producing procedures and allows for less pain medications, less sedatives, and an increased recovery time.[35] Music has also been found to reduce stress hormones and enhance the immune system, which allows healing to occur more rapidly in sick patients and promotes long-term wellness in healthy adults.[42] Active music making while in the hospital promotes movement (you can't play a drum unless you move your arms and hands) which increases recovery time from surgery.

## 5.2. Maintaining Wellness

Active music making throughout life is a great addition to maintaining wellness as well as proper nutrition and exercise, regular visits to the doctor, having an optimistic outlook, and participation in religion or spirituality. Active music making has several benefits in addition to the anxiolytic and stress reducing properties. First of all, it's active. You must physically move to play an instrument or sing. Next, music can be made by one person or in a group. It can be a social activity in which people come together and support one another. Making music is a form of emotional expression, in a safe environment. Many people have trouble expressing their feelings verbally but expression occurs freely on an instrument or within a musical context.

## 6. Conclusion

Music therapy is a valid and efficacious treatment for many different diagnoses when combined with the standard of care. It is an effective anxiolytic and can also aid in stress reduction. Music is a powerful medium and it can be used in our everyday lives to enhance the quality of life as we grow older. It is easy to get involved in music. Join a choir or find out when and where group drumming occurs in your community. Simply use your favorite CD or tape for relaxation. Take piano lessons. Music is easy, fun, and accessible to almost anyone. It is never too late to learn how to play an instrument, sing, or dance. So enjoy music and enjoy life.

## 7. References

1. Excerpt from "Friends of Music Therapy" American Music Therapy Association (AMTA).
2. Davis, Gfeller, & Thaut (1992). Music therapy: A Historical perspective. In Morgan, M. M. (Ed.), *An Introduction to Music Therapy Theory and Practice* (pp. 16–30). Dubuque, IA: Wm. C. Brown, Publishers.
3. Taylor, D. B. (1981). Music in general hospital treatment from 1900–1950. *Journal of Music Therapy, 30*(4), 210–223.
4. Robb, S. L. (1999). Marian Erdman: Contributions of an American red cross hospital recreation worker. *Journal of Music Therapy 36*(4), 314–329.
5. Caine, J. (1991). The effects of music on the selected stress behaviors, weight, caloric and formula intake, and length of hospital stay of premature and low birth weight neonates in a newborn intensive care unit. *Journal of Music Therapy, 28*(4), 146–152.
6. Standley, J. M. (1998). The effect of music and multimodal stimulation on physiological and developmental responses of premature infants in neonatal intensive care. *Pediatric Nursing, 24*(6), 532–539.
7. Standley, J. M. (2000). The effect of contingent music to increase non-nutritive sucking of premature infants. *Pediatric Nursing, 26*(5), 493.
8. Whipple, J. (2000). The effect of parent training in music and multimodal stimulation on parent-neonate interactions in the neonatal intensive care unit. *Journal of Music Therapy, 37*(4), 250–268.
9. Standley, J. (1992). Clinical applications of music and chemotherapy: The effects on nausea and emesis. *Music Therapy Perspectives, 10*(1), 27–35.
10. Boldt, S. (1996). The effects of music therapy on motivation, psychological well-being, physical comfort, and exercise endurance of bone marrow transplant patients. *Journal of Music Therapy, 33*(3), 164–188.
11. Burns, D. S. (2001). The effect of the Bonny method of guided imagery and music on the mood and life quality of cancer patients. *Journal of Music Therapy, 38*(1), 51–65.
12. Bailey, L. M. (1983). The effects of live music versus tape-recorded music on hospitalized cancer patients. *Music Therapy, 3*(1) 17–28.

13. Brodsky, W. (1989). Music therapy as an intervention for children with cancer in isolation rooms. *Music Therapy 8,* 17–34.
14. Smith, M., Casey, L., Johnson, D., Gwede, C., Riggin, O. Z. (2002). Music as a therapeutic intervention for anxiety in patients receiving radiation therapy. *Kentucky Nurse, 50*(2), 7–12.
15. Barnason, S., Zimmerman, L., & Nieveen, J. (1995). The effects of music interventions on anxiety in the patient after coronary artery bypass grafting. *Heart & Lung, 24*(2), 124–132.
16. Guzetta, C. (1989). Effects of relaxation and music therapy on patients in a coronary care unit with presumptive acute myocardial infarction. *Heart & Lung, 18,* 609–618.
17. Mandle, C. L., Domar, A. D., Harrington, D. P., Leserman, J., Bozadjian, E. M., Friedman, R., Benson, H. (1990). Relaxation response and femoral angiography. *Radiology, 174*(3–1), 737–739.
18. Walters, C. L. (1996). The psychological and physiological effects of vibrotactile stimulation, via a somatron, on patients awaiting scheduled gynecological surgery. *Journal of Music Therapy, 33*(4), 261–287.
19. Good, M., Anderson, G. C., Stanton-Hicks, M., Grass, J. A., Makii, M. (1999). Relaxation and music reduce pain after gynecologic surgery. *Pain, 81*(1–2), 163–172.
20. Clark, M. E., McCorkle, R. R., & Williams, S. B. (1981). Music therapy assisted labor and delivery. *Journal of Music Therapy, 18*(2), 74–87.
21. Hanser, S. B., Larson, S. C., & O'Connell, A. S. (1983). The effect of music on relaxation of expectant mothers during labor. *Journal of Music Therapy, 20*(1), 50–58.
22. Liebman, S. S., & MacLaren, A. (1993). The effects of music and relaxation on third trimester anxiety in adolescent pregnancy. In F. J. Bejjani (Ed.), *Current research in arts medicine* (pp. 427–430).
23. Boyle, M., & Greer, R. (1983). Operant procedures and the comatose patient. *Journal of Applied Behavioral Analysis, 16*(1), 3–12.
24. Jones, R., Hux, K., Morton-Anderson, K. A., & Knepper, L. (1994). Auditory stimulation effect on a compatose survivor. *Archives of Physical Medicine and Rehabilitation, 75*(2), 164–171.
25. Scheve, A. M. (2001). Music therapy and a coma patient: A case study. Unpublished research paper.
26. Thaut, M. H. & Davis, W. B. (1993). The influence of subject selected versus experimenter chosen music on affect, anxiety, and relaxation. *Journal of Music Therapy, 30*(4), 210–223.
27. Pfister, T., Berrol, C., & Caplan, C. (1998). Effects of music on exercise and perceived symptoms in patients with chronic obstructive pulmonary disease. *Journal of Cardiopulmonary Rehabilitation, 18*(3), 228–232.
28. Thornby, M. A., Haas, F., & Axen, K. (1995). Effect of distractive auditory stimuli on exercise tolerance in patients with COPD. *Chest, 107*(5), 1213–1217.
29. Thaut, M. H., McIntosh, G. C., Prassas, S. G., & Rice, R. R. (1993). Effect of rhythmic cuing on temporal stride parameters and EMG patterns in hemiparetic gait of stroke patients. *Journal of Neurologic Rehabilitation, 6,* 185–190.
30. Thaut, M. H., McIntosh, G. C., Rice, R. R., Miller, R. A., Rathbun, J., & Braut, J. M. (1996). Rhythmic auditory stimulation in gait training for Parkinson's disease patients. *Movement Disorders, 11,* 1–8.
31. Haun, M., Mainous, R. O., Looney, S. W. (2001). Effect of music on anxiety of women awaiting breast biopsy. *Behavioral Medicine, 27*(3), 127–133.
32. Gaberson, K. B. (1995). The effect of humorous and musical distraction on preoperative anxiety. *AORN Journal, 62*(5), 784–791.
33. Tusek, D., Church, J. M., & Fazio, V. W. (1997). Guided imagery as a coping strategy for perioperative patients. *AORN Journal, 66*(4), 644–649.
34. Sanderson, S. K. (1986). *The effect of music on reducing preoperative anxiety and postoperative anxiety and pain in the recovery room.* Unpublished master's thesis. The Florida State University.
35. Miluk-Kolasa, B., Obminski, Z., Stupnicki, R., & Golec, L. (1994). Effects of music treatment on salivary cortisol in patients exposed to pre-surgical stress. *Experimental and Clinical Endocrinology, 102*(2), 118–120.
36. Good, M. (1995). A comparison of the effects of jaw relaxation and music on postoperative pain. *Nursing Research, 44*(1), 52–57.
37. Fowler-Kerry, S., & Lander, J. (1987). Management of injection pain in children. *Pain, 30,* 169–175.
38. Malone, A. B. (1996). The effects of live music on the distress of pediatric patients receiving intravenous starts, venipunctures, injections, and heel sticks. *Journal of Music Therapy, 33*(1), 19–33.
39. Froelich, M. (1984). A comparison of the effect of music therapy and medical play therapy on the verbalization behavior of pediatric patients. *Journal of Music Therapy 21*(1), 2–15.

40. Marley, L. S. (1984). The use of music with hospitalized infants and toddlers: A descriptive study. *Journal of Music Therapy, 21*(3), 126–132.
41. Rabin, B. S. (1999). Stress, Immune Function, and Health: The Connection. New York, NY: John Wiley & Sons, Inc., Publishers.
42. Knight, W. E., & Rickard, N. S. (2001). Relaxing music prevents stress-induced increases in subjective anxiety, systolic blood pressure, and heart rate in healthy males and females. *Journal of Music Therapy, 38*(4), 254–272.

#  18

# Music Therapy, a Future Alternative Intervention Against Diseases

Haruhisa Wago and Shinji Kasahara

## 1. Introduction

It is a serious problem that a lot of stresses caused by external stimuli have malignant influences on both physiological and psychological functions in humans. Particularly, those stresses compromise bioactivities by adversely affecting the nervous and endocrine systems as well as the immune system, which plays a key role in prevention of infectious diseases and cancer[1-4]. Music, a universal language with many purposes, has been used in the health care setting to reduce stress and anxiety[5]. However, its biomedical significance is not fully understood. In this chapter we review effects of music therapy on human bio-functions such as nervous, endocrine and immune systems and then report our investigation on how music can enhance the immune system.

In general, stresses can be classified into three types: 1) physical/chemical, such as heat or cold; 2) physiological, such as fatigue or infection; 3) psychological, such as dissatisfaction, anxiety, and disappointment. In modern society, however, it is challenging to figure out the best way to remove those latter two types[6]. Thus, it would be important to establish effective methods to prevent the growing decline of human health caused by stresses and to augment the defense system of immuno-compromised host. Recently, music therapy has been used in order to alleviate various psychological disorders. Although it is still controversial, a number of studies suggest that music may facilitate a reduction in the stress response. Decreases in heart rate, greater tolerance of pain and suffering, skin conductance, muscle activity, subjective reports of relaxation, lessened anxiety and depression have been described in stressed patients exposed to classical music[7-9]. It has also been reported that music therapy is effective in dissolution of insomnia, dementia, antenatal training, culture of sentiments, and autism[10].

---

**Haruhisa Wago** • Department of Medical Technology, Saitama Medical School Junior College, Saitama 350-0495, Japan  **Shinji Kasahara** • Department of Neurobiology, David Geffen School of Medicine at UCLA, Los Angeles, California 90095, USA

*Complementary and Alternative Approaches to Biomedicine*, edited by Edwin L. Cooper and Nobuo Yamaguchi. Kluwer Academic/Plenum Publishers, 2004.

## 2. Music on Neuro-Endocrine-Immunology

*Music modulates neuroendocrine molecules*

The hypothalamic-pituitary-adrenal (HPA) axis, melatonin, prolactin, and biogenic amines (serotonin and catecholamines) are involved in the alteration in functions of neuroendocrines in the brain in response to stressful input. These neurohormones and neurotransmitters have also been implicated in regulating behavioral responses to various stimuli and in modulating aggression and depression[11-14]. Hormones of the HPA axis as well as melatonin and serotonin modulate various physiological functions and behaviors including circadian rhythms, sleep, sexual behaviors, mood, and tumor growth[15-18].

Neuroendocrine also modulates the immune system as a consequence of stress or depression. A molecular basis for reciprocal communication between the immune and neuroendocrine systems has been described[19,20]. It is now commonly known that neuromolecules, such as ACTH, dynorphin, dopamine, growth hormone (GH), opioid peptides, prolactin (PRL), thyrotropin (TSH), and others have the ability to modulate immune functions. Stress responses of neuroendocrine, such as an increase of plasma corticosterone concentrations through ACTH secretion, could have deteriorating effects on specific cells and tissues that are required for optimum immunological defense[21].

Music has been found to palliate stress induced hyperactivity of the HPA axis, involving ACTH and corticosteroid secretion[8,1-2] as well as to modulate norepinephrine, epinephrine, GH, PRL, β-endorphin secretion[24-26]. With regard to hormonal changes and behavioral modifications, it has been reported that the anxiolytic effects of music are accompanied by reduced levels of plasma cortisol among night shift nurses[27]. Similar results have been reported also for obstetric patients[28].

Thus, recent investigations on music therapy intervention for stress and behavioral modification is directed toward measurements of changes in stress-related neuroendocrine molecules such as cortisol. These studies suggest that the relaxing effects of music are critically involved in stimulation of neuroendocrine systems.

*Music, as an alternative intervention against immunological diseases*

Recently, investigations on stress and physiological responses to it have been directed to interdisciplinary characters of the stress concept. This concept has brought about the overlapping of certain scientific areas and medical disciplines such as neuroimmunology and psychoneuroimmunology[29]. This move to a unified approach suggests the fact that stress is involved in the immunological disease process. It is occasionally involved in the onset of inflammation, infection, autoimmune processes, and even the development of malignant tumors. It is known that stress disrupts the efficacy of the immune system against infection and malignant tumor development[30,31]. For example, chronic stress suppresses the immune system and may impair induction of apoptosis of tumor cells, presumably involving nitric oxide pathways[32-34]. Stress-induced anxiety and depression are also associated with increased mortality and with an augmented development of neoplastic diseases[35,36].

Noise can be a detrimental stress, which has been known to produce adverse effects on immune parameters both in human and animal models[37,38]. This modulation of the immune system has been considered to play a major role in the adverse effects on cancer[36]. When loss or inactivation of T cells or other innate defense elements are induced, the host surveillance system could fail to terminate transformed malignant cells during their immunologically vulnerable stages[39]. However, some of these immune suppressions are reversible with pharmacological interventions[40-43]. With regard to clinical aspects, many patients with cancer are highly stressed. In order to alleviate patients' stress and to improve their quality of life, music programs have been developed in several countries[44]. However, there is still little knowledge about the effects of non-pharmacological treatments such as music on immunological responses and evolution of diseases.

Nunez et al. has evaluated the effects of music on the immune system and cancer development in rodents subjected to sound stress[45]. Music reduced stress-induced immune suppression and development of lung metastases provoked by carcinosarcoma cells in mice. Music activated immune parameters [T cell populations, proliferation of spleen cells by concanavalin A (Con A), and natural killer (NK) cell activity] and anti-tumor responses in unstressed rodents. Music can positively soften adverse effects of stress on the number and capacities of lymphocytes that are required for an optimal immunological response against cancer in rodents.

Thus, music appeared to be as effective as pharmacological interventions (i.e. benzodiazepines, 5 HT agonists) to reverse stress induced immunosuppression[46-48].

## 3. Experiments: Active and Passive Music Therapy on Immune Functions

### Purpose and overview of our investigations

Music therapy can be characterized into two types: 1) active music therapy which means playing musical instruments or singing pleasant songs; 2) passive music therapy such as music appreciation or listening.

Detailed effects of music on the human immune system remain unclear. We have studied both active and passive music therapies on immune responses focusing on salivary IgA and neutrophil functions since these are responsible for the prevention of microbial infections. We examined the effectiveness of singing (as active music therapy) on immune parameters by measuring levels of IgA secretion and the IL-1 production by peripheral neutrophils[49]. We also analyzed how music appreciation (as passive music therapy) alters: 1) $\alpha$-wave in electroencephalogram; 2) neutrophil functions such as the proportion in total leukocytes, production of active oxygen, phagocytosis, and levels of interleukin-1 (IL-1); 3) salivary IgA secretion[50].

### Experimental models and preparations

Two groups of three normal healthy adults (6 adults in total) at the age of 20–21 were examined for the effect of singing and music appreciation on immune responses. Since all of these normal volunteers differ from each other in life style including diet, they were requested to have almost the same foods and sleeping time of 6 hours for 2 days. Additionally, they refrained from visual and acoustic stimuli such as watching TV and listening to the radio for 2 days prior to the experiments.

For active music therapy, each of the volunteers selected a popular song to sing, with which they felt comfortable. They sang in a relaxing condition for 30 min in an isolated quiet room. Before and after singing, 10 ml of the peripheral blood was collected into small glass tubes with anti-coagulant (3.8% sodium citrate), and diluted 1:2 with phosphate-buffered saline (PBS, pH 7.2) to obtain the peripheral neutrophils. Saliva was collected into 1.5 ml microtubes before and after singing, and stored at $-20°C$ until the experiments.

For passive music therapy, healing music such as Mr. Lonely, Moon River, When You Wish Upon A Star, and A Whiter Shade Of Pale (Ark Co.) was selected for 60 min listening. These songs are known to induce 20–70 μV of $\alpha$-waves in electroencephalogram, which provides humans with ease and relaxed feelings. Our preliminary examination using elecroencephalogram revealed that $\alpha$-waves were detected in the volunteers by listening to these songs. For the immunological analysis, volunteers were bled before and after listening to music. Preparation of blood and saliva was done as described for active music therapy (see above).

Results were expressed as mean + standard deviation (SD) of 3 or 4 independent experiments, and the data before and after listening was compared. Based on the Student's t test, the data were analyzed and differences were considered to be significant with $p < 0.05$.

## *Music enhanced salivary IgA secretion*

Secretory IgA plays an essential role in the local host defense system in humans by eliminating invading pathogens. Particularly, salivary IgA is an important immunoglobulin molecule, which responds quickly to antigens present in the mouth. We examined the effect of music on saliva IgA levels using the Single Radial Immuno Diffusion (SRID) method. Briefly, 3 μl of sample saliva and IgA standards was poured into each well of the SRID plates with a microsyringe, and then the plate was incubated for 48 hours at room temperature. Then, the secretory IgA content in saliva was estimated according to the standard curve.

Results revealed that IgA secretion in saliva was significantly increased by 30-min singing as shown in the upper histogram of Fig. 18.1 (A: 1.5; B: 1.4; C: 1.3 times increased each; $p < 0.05$). Comparison before and after singing revealed that those who sang experienced an increase in IgA. As shown in the lower histogram, there were significant increases of IgA secretion after a 60-min music appreciation in A and B: 1.2 times higher after listening ($p < 0.05$).

From our results, it was shown that levels of secreted salivary IgA were enhanced by music appreciation, which suggests that the local immunity is positively influenced by music listening. Our result coincided with the report that the salivary IgA levels were enhanced when the music therapy was given to children for 30 min[51].

Thus, there is a possibility that singing and listening to pleasant songs can be useful in prevention of infections: the more IgA we have in saliva by music, the better protected we are against pathogens.

## *Listening to music augmented neutrophil proportion but not total leukocytes number*

We performed preliminary experiments to examine the proportion of neutrophils occupied in the total number of leukocytes by Wright's Giemsa stain to elucidate the effect of singing on the number of neutrophils. Our result revealed that the number of neutrophils tended to increase after singing in all of the volunteers (data not shown). Also, in order to verify the effects of music appreciation on the numbers of human peripheral leukocytes and neutrophils, we examined: 1) the total number of leukocytes; 2) the ratio of neutrophils occupied in total leukocytes.

The peripheral blood was directly diluted with Turk solution in a gradated manner. A Melangeur-diluted pipette was used for leukocyte count. The number of total leukocytes/μl was measured in a hemocytometer under a light microscope. Then, a small amount of

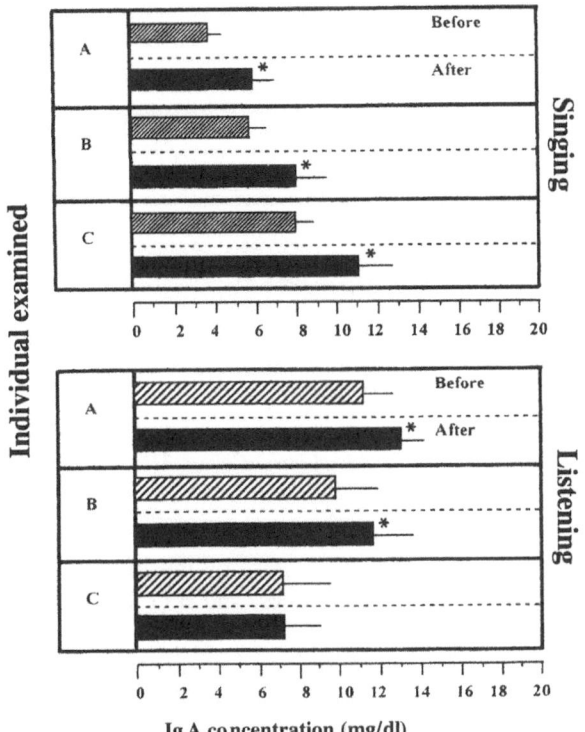

**Figure 18.1.** Effect of music singing and listening on salivary IgA levels [Mean + S.D. in 3 (singing) or 4 (listening) experiments, p < 0.05] [From Wago et al. *Bull. Saitama Med. School Junior College.* **11**, 11 (2000); Wago et al. *Bull. Saitama Med. School Junior College.* **11**, 19 (2000), with permission.]

blood was smeared on glass slides, and stained with Wright-Giemsa to count neutrophils. Three hundred stained cells were counted, and the ratio of peripheral neutrophils in total leukocytes was determined. As shown in Fig. 18.2, comparison of results before and after listening to music revealed that the total number of peripheral leukocytes was a little affected

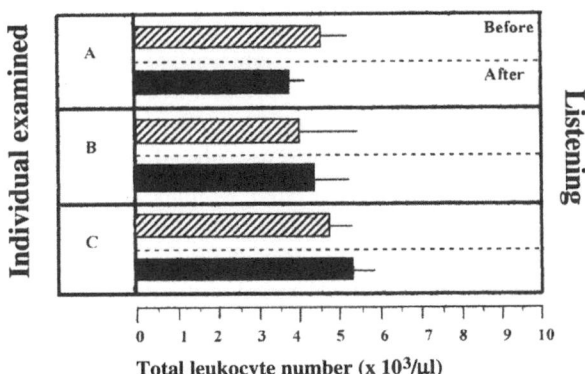

**Figure 18.2.** Effect of music listening on the number of leukocytes in the peripheral blood (Mean + S.D. in 4 experiments) [From Wago et al. *Bull. Saitama Med. School Junior College.* **11**, 19 (2000), with permission.]

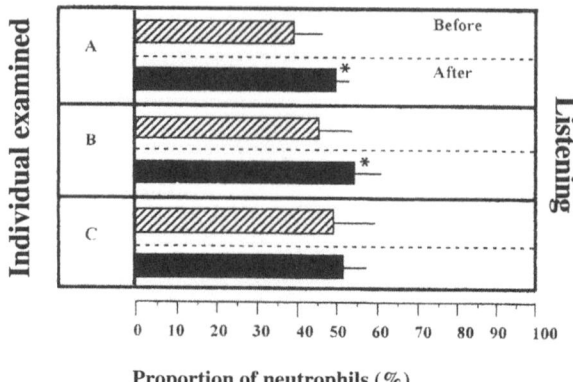

**Figure 18.3.** Effect of music listening on proportions of neutrophils in leukocytes (Mean + S.D. in 4 experiments) [From Wago et al. *Bull. Saitama Med. School Junior College.* **11**, 11 (2000), with permission.]

by music-listening (no significant change) and the ratio of neutrophils in total leukocytes of three adults tested had a tendency to increase by music appreciation. Particularly, the proportion of neurtrophils in both A (1.3 times) and B (1.2 times) significantly increased after listening ($p < 0.05$) (Fig. 18.3).

Our results revealed that the proportion of neutrophils in the total leukocytes was increased by music-listening. Thus, music appreciation had some effect on the neutrophil number circulating in the blood, but not on the total leukocyte number. Also, singing a pleasant song affected the neutrophil number: the proportion of neutrophils in the total leukocytes had a tendency to increase. These results suggest that music singing and listening have positive influences on the mobilization of stored neutrophils in lymph nodes or other possible secondary immuno competent organs.

### *Improved killing activity of neutrophils by music listening*

To perform a functional analysis of peripheral neutrophils, we examined the production of active oxygen by neutrophils before and after music listening, depending on microscopic observations of the formation of blue formazan pigments via NBT reduction reaction.

**First**, peripheral neutrophils from the volunteers were separated using density gradient medium (Dainihon Seiyaku, Tokoyo, density = 1.114). After centrifugation, neutrophil cell bands were collected in small tubes at 4°C, and washed once with PBS. Cells were then resuspended in Eagle's MEM with 0.1% gelatin, and the cell concentration was adjusted to $1 \times 10^6$ cells/ml using Tatai's hemocytometer.

**Second**, 1 ml of neutrophil suspension was placed on a glass slide, and incubated for 30 min at 37°C for the cells to attach. After washing with PBS, one ml of nitroblue tetrazolium (NBT) solution was added to the neutrophil monolayer and incubated for 20 min at 37°C. Washed with PBS again, cells were dried in room temperature, fixed with methanol for one min, dried and post-stained with safranin-O for 5 min.

**Finally**, cells were observed under a light microscope, and the formation of blue formazan pigments in the cytoplasm was assessed. This formation indicates sites where active oxygen has been produced. In this experiment, 300 neutrophils were counted and the ratio of formazan-positive cells per total neutrophils were estimated. Fig. 18.4 shows that

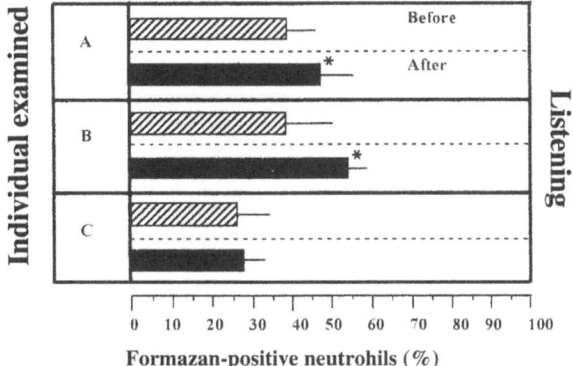

**Figure 18.4.** Effect of music listening on the number of formazan-positive neutrophils (Mean + S.D. in 4 experiments) [From Wago et al. *Bull. Saitama Med. School Junior College.* **11**, 11 (2000), with permission.]

the numbers of formazan-positive neutrophils in both A (1.2 times) and B (1.4 times) were significantly increased by music listening ($p < 0.05$). For the volunteer C, there were no significant differences between before and after listening.

### Activation of neutrophil phagocytosis after music listening

Phagocytic activity of neutrophils plays a critical role as an initial cellular immune response. Therefore it would be reasonable to examine the effect of music on this innate immune function. We conducted experiments to assess phagocytosis using FITC-labeled latex particles as foreign objects to be phagocytosed.

One ml of neutrophil suspension in Eagle's MEM was incubated on a glass slide for 30 min at 37°C for attachment and washed with PBS. Then, 0.1 ml of 0.1 % FITC-labeled latex particle suspension was added to these neutrophil monolayer slides. Then they were incubated for 40 min at 37°C in moist petri dishes. Following incubation, the cells were washed three times with cold PBS to remove latex particles which had not been uptaken.

Phagocytic activity was evaluated under a fluorescent microscope from the number of neutrophils which phagocytosed more than one FITC-latex particle in their cytoplasm. Three hundred neutrophils were observed for measurement. Observations revealed that phagocytic activity of peripheral neutrophils in A (1.1 times; $p < 0.05$) and C (1.3 times; $p < 0.01$) were significantly enhanced by a 60-min music appreciation (Fig. 18.5). For the volunteer B, the phagocytic activity was almost the same even after listening.

Thus, together with the data for active oxygen production (see above), our results suggest that listening to pleasant music affects innate immune functions of neutrophils such as killing activity and phagocytosis.

### IL-1 levels were increased by music singing and listening

Since IL-1 is well known to be produced by neutrophils and be functioning as an essential monokine in regulating T cell response, we also examined the effect of singing on its production. Neutrophils collected before and after music singing and listening were co-cultured with LPS for 48 hours and their supernatants were evaluated for the IL-1 content with ELISA assay.

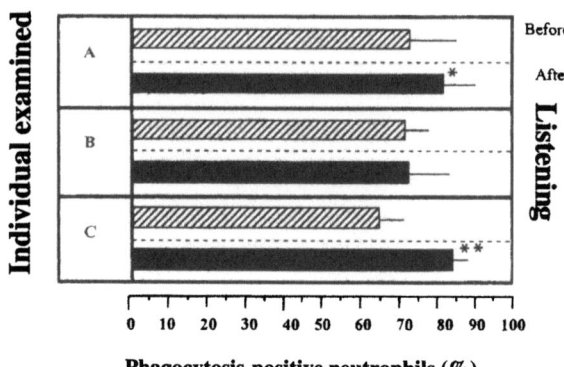

**Figure 18.5.** Effect of music listening on the number of phagocytosis-positive neutrophils (Mean + S.D. in 4 experiments) [From Wago et al. *Bull. Saitama Med. School Junior College.* **11**, 11 (2000), with permission.]

One ml of neutrophil suspension at the concentration of $1 \times 10^6$ cells/ml in Eagle's MEM was incubated in 12-well plates for 30 min at 37°C for adhesion. Then, after washing twice with PBS, the attached cells were cultured in MEM with 10% fetal bovine serum (FBS) and E. coli LPS (10 µg/ml) for 48 hours at 37°C. IL-1 levels in the supernatant were measured by specific ELISA assay kit (solid-phase sandwich ELISA for IL-1; Genzyme Co.) according to the manufacturer's protocol.

Results showed that music-singing significantly enhanced the secretion of IL-1 in two volunteers [A: 1.2 ($p < 0.05$); B: 4.7 ($p < 0.01$) times] (Fig. 18.6, upper histogram). Only a little increase was seen after singing in C. Music listening also significantly augmented the IL-1 production [A: 1.9; B: 2.7; C: 2.2 ($p < 0.01$) times] (Fig. 18.6, lower histogram). Particularly, volunteer B showed a drastic increase in levels of IL-1 after music-listening.

The increase of IL-1 production may also broadly promote the human defense reactions. It has been documented that IL-1 secreted by neutrophils or macrophages elicits the activation of T lymphocytes and NK cells and the induction of acute phase proteins such as C-reactive protein in the liver[52].

## 4. Conclusion and Future Directions

It has been reported that a variety of stressors affect human bio-functions leading to immune suppressions. Early studies have revealed, for example, 1) uneasiness of space aviators when they depart or return decrease the total number of peripheral lymphocytes and other immune responsiveness[53]; 2) sorrow during bereavement impairs natural killer activities and suppresses blastogenesis of T lymphocytes[54,55]; 3) stress of taking examinations suppresses NK cells[56]. These results suggest relationships between physical or psychological stresses and immune suppression. Therefore, stress exerts a strong influence on human defensive functions, presumably through neuro-endocrine system. Regarding medical aspects, it would be essential to soften these stresses by efficient mental or psychological ways to improve states of diseases which are involved in psychological, behavioral, or other neuro-endocrine-immune dysfunctions.

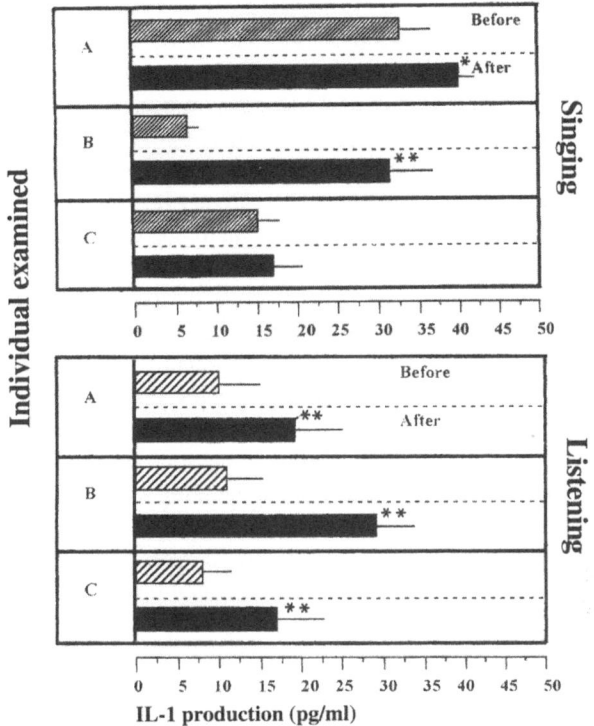

**Figure 18.6.** Effect of music listening on the number of phagocytosis-positive neutrophils [Mean + S.D. in 3 (singing) or 4 (listening) experiments) [From Wago et al. *Bull. Saitama Med. School Junior College.* **11**, 11 (2000); Wago et al. *Bull. Saitama Med. School Junior College.* **11**, 19 (2000), with permission.]

It has been understood that mental therapies are useful to alleviate stresses. Music applications, one of the effective psychological therapies, has been recently paid attention by medical facilities because it is effective in softening stresses and also in dissolution of insomnia, dementia, and autism[57]. As found in these reports, early studies on music therapy have been mainly focusing on modifying neuro-endocrine systems by music stimuli. Various neurohormones and neurotransmitters including melatonin, prolactin, serotonin, and catecholamines play systematic regulatory roles in behavioral and physiological functions[17,18,58–60]. Interaction among these neurohormones and neurotransmitters suggest possible simultaneous changes in these neuronal factors as a result of stimuli which induce a relaxing condition including music therapy. We expect that monitoring changes of these neuronal factors may elucidate potential mechanisms which are involved in the reversal of behaviorally and physiologically stressed conditions. This would provide biological markers for behavioral changes in response to relaxing effects such as given by music therapy. However, studies involving the stimulatory effects of music on the neuro-endocrine systems remain largely uninvestigated. In addition to that, little is known about how music applications affect immune reactions in relation to neuro-endocrine functions.

Humans possess innate (nonspecific) and adaptive (specific) defense elements: phagocytic cellular responses shown by neutrophils and macrophages are major factors to eliminate invading pathogens as the initial defense system, and the antigen-specific responses by immunoglobulin molecules are also induced to discriminate and remove antigens[61]. We therefore focused on peripheral leukocyte function and salivary IgA levels to assess effects of music singing and listening on both innate and adaptive immune functions. Our experiments demonstrated that music singing and listening activate: 1) neutrophil responses such as migration (as suggested by increased cell ratio), active oxygen induction, phagocytosis, and IL-1 production; 2) salivary IgA secretions. These data suggest that music singing and listening can be: 1) beneficial to prevent diseases which are involved in immune systems; 2) interventions against stress-induced immune suppressions.

It is also an important point that type of music, personality traits and temperament may influence the wide inter individual variability in response to music. McCraty et al. examined the effects of rock, new age, and designer music alone and found that each music induced different autonomic activity and serum IgA levels[1]. This suggests that music can be designed to enhance the beneficial effects of positive emotional states on immunity, and that this effect may be mediated by the autonomic nervous system (Fig. 18.7).

Particularly, there are surprising claims that listening to Mozart's music enhances spatial reasoning and IQ by 8-9 points, lowers blood pressure, improves cognitive performance of patients of Alzheimer's dementia (AD), and ameliorates auditory disturbance[62-64]. Although data is not shown in this section, we investigated the efficacy of *Mozart's music* in music therapy[65]. **First**, changes of heart rate and blood pressure were examined with and without listening to Mozart after 15 min of light exercise. Results showed faster decreases of heart rate and blood pressure in all volunteers with *Mozart's music*. **Second**, secretory IgA in saliva was measured before and after listening to Mozart using single radial immunodiffusion (SRID). The result clearly showed that salivary IgA levels were elevated after 60 min of listening. **Third**, peripheral neutrophil functions such as phagocytosis and active oxygen production were investigated. Our results revealed that *Mozart's music* augmented phagocytosis and production of active oxygen by neutrophils after listening. Therefore, listening to Mozart appears to be effective in decreasing blood pressure and heart rate leading to their quick stability and also in enhancing immunity by IgA and neutrophils.

With regard to deviations of results, our data differed among volunteers and experiments although there were significant differences in each volunteer before/after music applications. This reason is possibly due to differences of life-style or to psychological/physiological conditions of volunteers on each experimental day. Furthermore, it is known that circadian or daily rhythm affects the number of human immunocytes including neutrophils and lymphocytes. Neutrophils increase in the daytime in response to adrenalin, while lymphocytes are augmented in the nighttime by secretions of acetylcholine[66]. It is most likely that these environmental, psychological, and physiological rhythms are different among volunteers, and therefore these rhythms may affect experimental data. A possible solution would be to examine volunteers who are living together in common life-style environments such as public facilities for elderly people, because their environmental conditions are assumed to be almost the same. In addition, it would also be necessary to look at different generations in order to verify general effects of music listening on immunity. Lastly, it is also essential to establish methods to assess stress levels. Recently, more and more scales, scores, and experimental procedures have been accepted[29,67,68].

In conclusion, it is evident that music is associated with the human immune system[51]. Pleasant music can be used as an intervention against diseases which are involved in

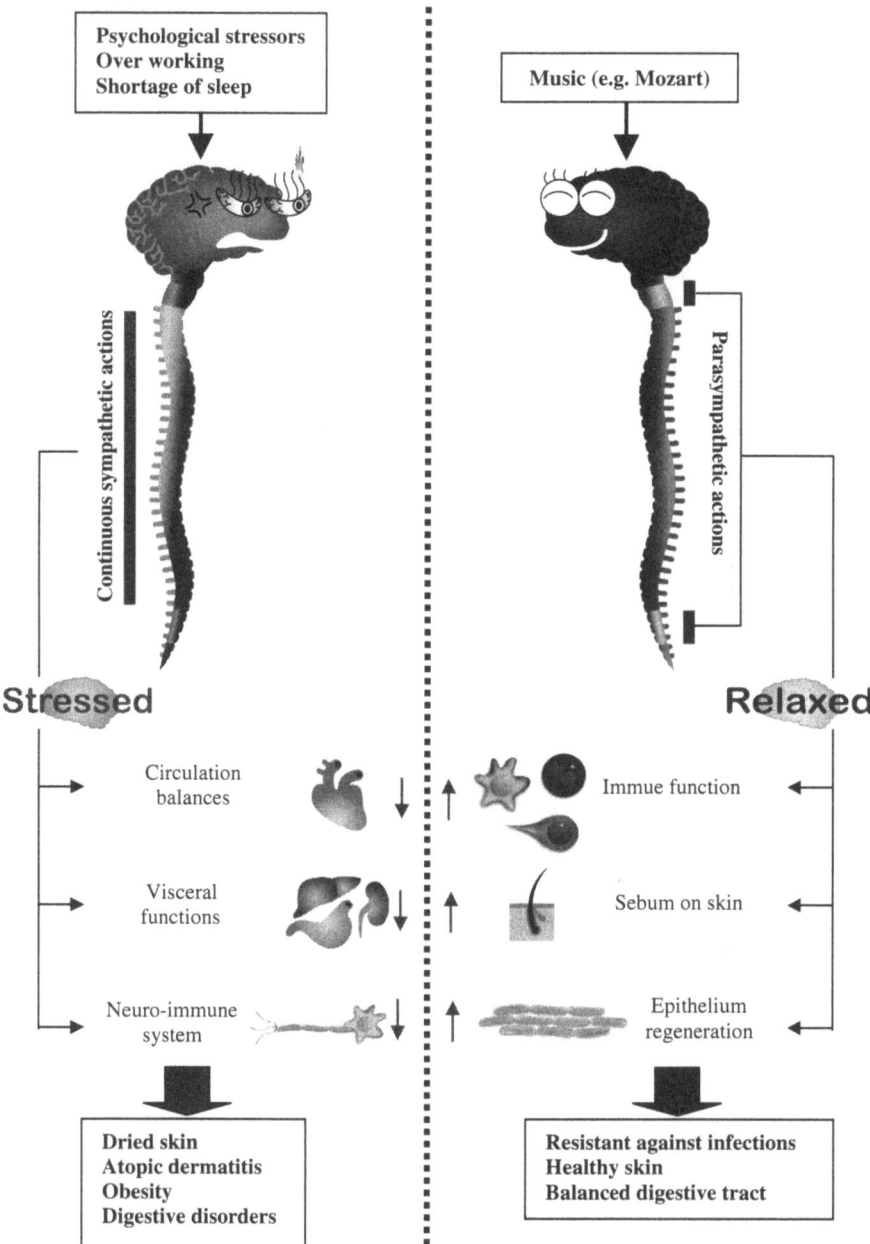

**Figure 18.7.** A scheme representing contrasting physical consequences by stress and music through autonomic nervous system. Stress-induced hyper-activation of sympathetic pathways can provoke impairments of circulation regulations, general visceral functions, and neuro-immune responses. Relaxed conditions by music stimuli leads to parasympathetic actions which enhances immune responses, sebum on skin, and epithelium regenerations. These opposite physical alterations are associated with incidences and preventions of diseases

immunological defects. It would become more important to develop music therapy for impaired defensive elements in humans in hospitals or other medical facilities. In the light of the results presented here, we would like to propose that it is also necessary to further elucidate how music affects functions of NK cells and T or B lymphocyte or other immunocytes. In addition, our society has serious problems that immunological or psycho-behavioral disorders such as cancer or AD patients suffer from depression, agitation, aggression, violence, suicide, and sleep disturbances[69,70]. These behavioral problems compromise the quality of their lives. Moreover, there is an increasing concern that care for these patients is becoming a heavy and costly burden on caregivers and society[69,71]. Integrative use of music therapy in society, medical facilities, and the patient's family would: 1) improve abnormal and disruptive behaviors of these patients and minimize the need for tranquilizing medication[72]; 2) improve quality of life by providing social interactions with others[73]; 3) help to generate a resource for long-term economic benefit for society and patients.

## 5. Acknowledgement

We appreciate the help of Ms. Atsue Kamata, a graphic designer, for the illustrations.

## 6. References

1. R. McCraty, M. Atkinson, G. Rein and A.D. Watkins, *Stress Med.* **12**, 167 (1996).
2. S.D. VanderArk and D. Ely, *Percept. Mot. Skills.* **77**, 227 (1993).
3. J.F. Byers and K.A. Smith, *American J. Critical Care*, **6**, 183 (1997).
4. F. Kimura and H. Negoro, in: *Simple Physiology*, Nankoudou, Tokyo, 1996. p. 71.
5. H. Covington and C. Crosby, *J. Psychosoc. Nurs. Ment. Health Serv.* **35**, 34 (1997).
6. K. Hoshi, *Stress and Immunity*, Koudansya, Tokyo, 1993.
7. H.L. Bonny, *Music Therapy*, **3**, 4 (1983).
8. P.A. Brauchili. *Zeitschrift fur Experimentellle und Angewandte Psychologie*, **40**, 179 (1993).
9. K. Allen and J. Blascovich, *JAMA*, **272**, 1724 (1994).
10. A. Komatsu and H. Sasaki, in: *The Front of Music Therapy*, edited by A. Komatsu and H. Sasaki (Ningen to Rekishisya, Tokoyo, 1996).
11. L. Valzelli, *Prog. Neuropsuchopharmacol. Biol. Psychiatry*, **8**, 311 (1984).
12. A. Coppen, A.J. Prange Jr, P.C. Whybrow and R. Noguera, *Arch. Gen. Psychiatry.* **26**, 474 (1972).
13. H.Y. Meltzer, *Ann. N. Y. Acad. Sci.* **600**, 486 (1990).
14. R.M. Salomon, P.L. Delgado, J. Licinio, J.H. Krystal, G.R. Heninger and D.S. Charney. *Biol. Psychiatry.* **36**, 840 (1994).
15. K. Uchida, N. Okamoto, K. Ohara and Y. Morita. *Brain Res.* **717**, 154 (1996).
16. J.R. Reiter, *Endocr. Rev.* **12**, 151 (1991).
17. L. Wetterberg and B. Aperia, *Psychoneuroendocrinology* **8**, 75 (1983).
18. A. Brzezinski A, *N. Engl. J. Med.* **336**, 186 (1997).
19. J.E. Blalock, *Int. J. Neurosci.* **51**, 363 (1990).
20. E.L. Cooper, *Animal Biol.* **1**, 169 (1992).
21. V. Riley, *Science* **212**, 1100 (1981).
22. L. Montello, *Int. J. Arts. Med.* **4**, 14 (1995).
23. D.S. Pope, *Image. J. Nurs. Sch.* **27**, 291 (1995).
24. C.H. McKinney, F.C. Tims, A.M. Kumar and M. Kumar. *J. Behav. Med.* **20**, 85 (1997).
25. C.H. McKinney, M.N. Antoni, A.M. Kumar and M Kumar, *J. Assoc. Music Imagery* **4**, 67 (1995).
26. C.H. McKinney, M.N. Antoni, M. Kumar, F.C. Tims and P.M. McCabe, *Health Psychol.* **16**, 390 (1997).
27. M. Rider, J. Floyd and J. Kirkpatrick, *J. Music Ther.* **22**, 46 (1985).

28. B. Halpaap, R. Spingte, R. Droth, W. Kummert and W. Koegel, in: *Musik In Der Medizen/Music In Medicien*, edited by R. Spingte and R. Droth (Springer-Verlag, New York, 1987), p. 232.
29. T. Esch, *Gesundheitswesem*, **64**, 73 (2002).
30. R. McCarty and P.E. Gold, *Psychosom. Med.* **58**, 590 (1996).
31. K.H. Coker, *Semin. Urol. Oncol.* **17**, 111 (1999).
32. B.S. McEwen, *Brain Res.* **886**, 172 (2000).
33. P. Secchiero, A. Gonelli, C. Celeghini, P. Mirandola, L. Guidotti, G. Visani, S. Capitani and G. Zauli. *Blood* **98**, 2220 (2001).
34. J.P. Kolb, V. Roman, F. Mentz, H. Zhao, D. Rouillard, N. Dugas, B. Dugas and F. Sigaux, *Leuk. Lymphoma* **40**, 243 (2001).
35. J.M. Murphy, R.R. Monson, D.C. Olivier, A.M. Sobol, and A.H. Leighton, *Arch. Gen. Psychiatry* **44**, 473 (1987).
36. R.B. Shekelle, W.J. Raynor Jr, A.M. Ostfeld, D.C. Garron, L.A. Bieliauskas, S.C. Liu, C. Maliza and O. Paul. *Psychosom. Med.* **43**, 117 (1981).
37. K.W. Kelley and R. Dantzer, in: *Stress And Immunity*, edited by N. Plotnikoff, A. Murgo, R. Faith and J. Wybran (CRE Press, Florida, 1991), p. 433.
38. D.O. McCarthy, M.E. Ouimet and J.M. Daun, *Res. Nurs. Health* **15**, 131 (1992).
39. H. Zimel and A. Zimel, *Rev. Roum. Endocrn.* **11**, 213 (1974).
40. M. Freire-Garabal, A. Belmonte, F. Orallo, J. Couceiro and M.J. Nunez, *Cancer Lett.* **58**, 183 (1991).
41. M. Freire-Garabal, M.J. Nu nez-Iglesias, J.L. Balboa, J.C. Fernandez-Rial and M. Rey-Mendez, *Pharmacol. Biochem. Behav.* **51**, 821 (1995).
42. M. Freire-Garabal, M.J. Nunez, D. Pereiro, P. Riveiro, C. Losada, J.C. Fernandez-Rial, E. Garcia-Iglesias, J. Prizmic, J.M. Mayan and M. Rey-Mendez, *Life Sci.* **63**, 31 (1998).
43. M. Faisal, F. Chiappelli, I.I. Ahmed, E.L. Cooper and H. Weiner, *Brain Behav. Immun.* **3**, 223 (1990).
44. M.F. Cunningham, B. Monson and M. Bookbinder, *AORN J.* 66, 674 (1997).
45. M.J. Nunez, P. Mana, D. Linares, M.P. Riveiro, J. Balboa, J. Suarez-Quintanilla, M. Maracchi, M.R. Mendez, J.M. Lopez, M. Freire-Garabal, *Life Sci.* **71**, 1047 (2002).
46. T. Hattori, Y. Hamai, H. Ikeda, T. Harada and T. Ikeda, *Gann.* **69**, 401 (1978).
47. M. Freire-Garabal, M.J. Nunez, J.L. Balboa, J.A. Suarez and A. Belmonte. *Cancer Lett.* **62**:185 (1992).
48. M. Freire-Garabal, M.J. Nunez-Iglesias, J.L. Balboa, J.C. Fernandez-Rial, L. Garcia-Vallejo and M. Rey-Mendez, *Cancer Detect. Prev.* **20**, 160 (1996).
49. H. Wago, C. Orikasa, C. Torisawa and Y Suto, *Bull. Saitama Med. School Junior College.* **11**, 19 (2000).
50. H. Wago, K. Goto, N. Sakamoto and M Sakurada, *Bull. Saitama Med. School Junior College.* **11**, 11 (2000).
51. D. Lane, *Oncol. Nurs. Forum* **19**, 863 (1992).
52. S. Kasahara, in: *Cytokine*, edited by S. Kasahara (Nihon Igakukan, Tokyo, 1991), p. 13.
53. C.L. Fischer, J.C. Daniels, W.C. Levin, S.L. Kimzey, E.K. Cobb and S.E. Ritzmann, *Aerosp. Med.* **43**, 1122 (1972).
54. M. Irwin, M. Daniels, T.L. Smith, E. Bloom and H. Weiner. *Brain Behav. Immun.* **1**, 98 (1987).
55. S.J. Schleifer, S.E. Keller, M. Camerino, J.C. Thornton and M. Stein, *JAMA.* **250**, 374 (1983).
56. J.K. Kiecolt-Glaser, W. Garner, C. Speicher, G.M. Penn, J. Holliday and R. Glaser, *Psychosom. Med.* **46**, 7 (1984).
57. A. Komatsu and H. Sasaki, in: *The Front of Music Therapy*, edited by H. Sasaki, Ningen to Rekishisha, Tokyo, (1996).
58. L.D. Van de Kar and C.L. Bethea. *Neuroendocrinology* **35**, 225 (1982).
59. M.L. Rao and T. Mager. *Psychoneuroendocrinology* **12**, 141 (1987).
60. Preslock JP, *Endocr. Rev.* 5, 282 (1984).
61. H. Wago, in: *Introduction To Animal Immunology*, edited by H. Wago, (Asakura Shoten, Tokyo, 1994).
62. F.H. Rauscher, G.L. Shaw, and K.N. Ky, *Nature* **365**, 611 (1993).
63. J.S. Jenkins, *J. Roy. Soc. Med.* **94**, 170 (2001).
64. H. Wago, *The Healing Power Of Amadeus Sound*, Makino Publishing, Tokyo, 2003.
65. H. Wago, M. Kimura, J. Inoue, R. Kobayashi and S Nakamura, *Bull. Saitama Med. School Junior College*, **13**, 45 (2002).
66. T. Abo, *Future Immunology*, Intermedical, Tokyo, 1997.
67. J. Briere, K. Johnson, A. Bissada, L. Damon, J. Crouch, E. Gil, R. Hanson and V. Ernst. *Child Abuse Negl.* **25**, 1001 (2001).

68. M. Maes, J. Mylle, L. Delmeire and A. Janca, *Psychiatry Res.* **105**, 1 (2001).
69. B. Reisberg, J. Borenstein, S.P. Salob, S.H. Ferris, E. Franssen and A. Georgotas, *J. Clin. Psychiatry* **48**, 9 (1987).
70. F. Fernandez, J.K. Levy, *Psychiatr. Med.* **9**, 377 (1991).
71. L. Teri, E.B. Larson and B.V. Reifler, *J. Am. Geriatr. Soc.* **36**, 1 (1988).
72. D.S. Knopman and S. Sawyer-DeMaris, *Geriatrics.* **45**, 27 (1990).
73. D. Aldridge, *Biomed. Pharmacother.* **48**, 275 (1994).

# IV

# Dietary Intervention in Specific Diseases

# 19

# Diet Instruction for Japanese Traditional Food in Therapy for Atopic Dermatitis

Hiromi Kobayashi, Nobuyuki Mizuno, Hiroyuki Teramae, Haruo Kutsuna, Mika Nanatsue, Kazuko Hirai, and Masamitsu Ishii

## 1. Introduction

Atopic dermatitis is a disease that has chronic recurrent eczematous lesions as the main symptoms,[1] and its progression to intractable cases has become a severe problem in recent years.[2] Detection and elimination of allergens and aggravating factors, external treatment mainly through the use of topical steroids, oral administration of antipurtic drugs, and proper skin care are the first choices of treatment. For cases that do not respond to these efforts, we have been performing concomitant Japanese traditional herbal treatment for more than 20 years.[3]

Based on practices that were introduced from China via Korea in the 7th century, Japanese traditional medicine developed independently and was the mainstream of public healthcare until the latter half of the 19th century when Western medicine became the official medical system. After passing through an era in which a limited number of physicians performed Japanese traditional medicine, it was re-incorporated into the medical care system supported by national health insurance in the 1970's. Since then, Japanese traditional medicine has been used as a treatment concomitantly applicable with modern Western medicine in all fields.[4] While modern Western medicine tends to analytically understand diseases, Japanese traditional medicine observes diseases and takes into consideration the living environment of the patients.[5] Accordingly, instruction of life-style, diet instruction in particular, has been the focus of Japanese traditional medicine.

The effect of diet on skin symptoms has been reported in Western medicine, such as Lind's discovery in 1742 regarding prevention of scurvy by the ingestion of fruit.[6] Later, skin lesions caused by deficiency of vitamin C,[7] vitamin B complex,[8] vitamin B$_6$,[9] pantothenic acid,[10] and linolenic acid[11] were reported. In the first half of the 1900's, the cause of pellagra was experimentally demonstrated to be in the diet.[12]

---

**Hiromi Kobayashi, Nobuyuki Mizuno, Hiroyuki Teramae, Haruo Kutsuna, and Masamitsu Ishii** • Department of Dermatology, Osaka City University Graduate School of Medicine   **Mika Nanatsue and Kazuko Hirai** • Department of Health and Nutrition, Osaka City University Graduate School of Human Life Science, ...

*Complementary and Alternative Approaches to Biomedicine*, edited by Edwin L. Cooper and Nobuo Yamaguchi. Kluwer Academic/Plenum Publishers, 2004.

Diet therapy has a broad meaning and has already been practiced not only in complementary alternative therapy but also in the field of modern medicine. In this article, reports of dietary therapy in dermatology are summarized, focusing on atopic dermatitis. Next, diets considered important in Japanese traditional medicine for atopic dermatitis are presented with examples, and results of studies that provide scientific evidence are described.

## 2. Reported Cases of Diet Therapy for Atopic Dermatitis

Diet therapy for atopic dermatitis can be roughly divided into allergic and non-allergic approaches. Elimination diet therapy (limited diet therapy) searches for and eliminates dietary factors as allergens. Other complementary alternative diet therapy, namely, impairments due to excessive, lack, and imbalance of food compositions are eliminated, and ingestion of food considered to promote healing is recommended. However, the overwhelming majority of reports are of elimination diet therapy (limited diet therapy), and diet therapy tends to be recognized as elimination diet therapy. However, evaluation of therapy is not consistent and physicians use their own methods.[12] Some researchers insist that limiting consumption of egg whites and milk, known as typical allergens, in pregnant women and lactating mothers exhibits a preventive effect,[14] while others oppose it, observing no preventive effect,[15] and no clear guidelines have been established. The frequency of food-induced aggravation of atopic dermatitis was previously reported to be 0.6%,[16] but it has recently been reported that chocolate and coffee occupied the upper positions of aggravating factors, suggesting a non-allergic mechanism.[17] Although some cases require elimination of food involved in developing symptoms as allergens, attention should also be paid to malnutrition, common antigenicity, and specific antigenicity in elimination diet therapy. This therapy should be performed based on accurate diagnosis using an appropriate method, with sufficient attention paid to adverse effects such as growth disorders.

As a complementary alternative diet therapy,[18] low-calorie diet therapy and hunger cures limiting calorie intake, a method using $\omega 3$ (n-3) fatty acid food as a supplement, vitamin therapy, herb therapy, low-salt diet, and vegetable juice methods have been introduced. Recently, the usefulness of beverages such as oolong tea[19] and persimmon leaf tea[20] has been reported. Defect of $\delta 6$-desaturase that converts linoleic acid to linolenic acid in atopic dermatitis has been reported,[21] and the efficacy of evening primrose oil[22] containing linolenic acid and sausage containing borage oil has been reported,[23] although these are n-6 lipids. However, other reports describe that the possibility of decreased 6-desaturase was low[24] and evening primrose oil was ineffective,[25] pointing out that the relationship between symptoms and ingestion of specific fatty acids should be carefully discussed.[26]

Another diet therapy that differs from the use of these supplemental foods is diet instruction focusing on balance. Several types of diet instruction have been reported to be useful for intractable cases of atopic dermatitis: restriction of fats and fatty oils[27], prohibition of coffee, cocoa, potato chips, and snacks, recommendation of a balanced diet[28] consisting of fish and vegetables with restrictions on juice, meats, and oily food such as tempura, and a gastrointestinal Candida-inhibiting diet[29] in which ingestion of sugar, fruits, and alcohol is restricted. In a survey performed in England, wheezing and hypersensitivity were less frequently observed in children who had Asian type eating habits, while the risks of hypersensitivity and atopic dermatitis were significantly higher in children who had Western type eating habits,[30] suggesting that Asian type eating habits with rice as the principle food may have preventive benefits. There are several reports that recommend Japanese traditional diets,[31–40] and Japanese traditional diets can be recommended as a therapeutic method for a life-style disease such as atopic dermatitis.

## 3. Problems in Present Diets

The Japanese diet rapidly changed after World War II from one mainly consisting of grains, vegetables, seaweed and fish to a high-protein, high-lipid, high-carbohydrate diet, in which large amounts of meat, fat and fatty oil, sugar, and dairy products are ingested.[41,42,43] It has been pointed out that excess ingestion of carbohydrates may induce abnormal growth of intestinal Candida and cause injury of intestinal mucosa and abnormal immunity.[44] A tendency of excess ingestion of n-6 lipids has also been reported,[45] suggesting that present diets may increase inflammatory diseases.

Daily consumption of sweets, stir-fried and deep fried dishes is regularly seen in recent diaries of dietary patterns. Table 19.1 outlines problems associated with these habits.

| 14-year-old female | | | | | 21-year-old male | | | | |
|---|---|---|---|---|---|---|---|---|---|
| Date | Breakfast | Lunch | Supper | Between meals | Date | Breakfast | Lunch | Supper | Between meals |
| April 16 | - | Spaghetti | Tirashi sushi, miso soup | Ice cream | June 19 | - | Chiken curry, miso soup | Zarusoba, mabotofu, roast pork, boiled rice | |
| April 17 | Rice ball | Boild rice, stif fried vegetables, salad, deep fried globefish, konnyaku, canned orange | Roll cabbage, salad, burdock- Ten, | Chocolate, bread, ice cream, yogurt | June 20 | Hamburger, hashed potato, coffee | Grilled mackerek with salt, spinach, kimuchi, boiled rice | Stewed meat and potato, tempura, coiled rice | |
| April 18 | Boiled rice, laver, salmon, | Hamburger steake, boiled rice, sausage, broccoli, canned orange | Boiled rice, teriyaki yellowtail, spinach, miso soup | Chocolate | June 21 | - | Steamed cheese bun | Suchi, baked udon | |
| April 19 | Boiled rice, miso soup, sausage | Chiken rice, deep fried chicken, brocolli, grapefruit | Salad (ham, lettuce, broccoli), pork chop, boiled rice, miso soup (tofu, wakame) | Popcorn | June 22 | - | Bread (cheese, mayonnaise) | Grilled meat, kimuchi, boiled rice | Shark fin Chinese noodles, deep fried chicken |
| April 20 | Sweet bun | Fried pork dutlet, boiled rice, bamboo sprout, wakame | Fried rice, Chimese meat dumpling | Ice cream | June 23 | - | Udon | Fried shrimp, chicken, potato, boiled rice, miso sopu, pikles | Cola, soft cream |
| April 21 | Pie | Fried chicken, juice | Croqeutte, boiled epinach seasoned with soy sauce, boiled rice with ingredients, miso sopu (tofu, wakame) | Ice cream wheat flour sweet | June 24 | - | Beef bowl, egg | Beaf stake, onion, lettuce, tomato, potato salad, crab-taste boiled fish paste, cheese, French bread, boiled rice | Cheese cake, coffee |
| April 22 | Boiled rice with ingredients, miso soup | Fried pork cutlet, bamboo sprout, wakame, chicken, quin gin cai, natsumikan, bhamkuhen | Roast pork, chopped skipjack, tibular boiled fish paste, cucumber, lettuce, boiled rice | Cake, ice cream | June 25 | Rice ball | Rice ball, instant Chinese noodles | Boiled fish, Chopped beef, spinach, boiled rice | Grapefruit juice |

**Figure 19.1.** Example of a diet diary.

**Table 19.1.** Problems in present diets

1. Environmental medical problem: Contamination.
2. Nutritional problem: Excess ingestion of sweets, fat and fatty oil, and protein. Insufficient ingestion of vitamins and minerals. Deviation from eating habits suitable for weather and climate.
3. Psychosomatic problem: Fewer meals with family members.
4. Oral surgical problem: Increase in soft food and decrease in mastication.

## 4. Effect of Japanese Traditional Food on Diseases other than Atopic Dermatitis

In 1999, Professor Yoshiyuki Ohno et al., Graduate School of Nagoya University, Department of Medical Research, Laboratory of Preventive Medicine/Medical Statistics-Judgment, reported the results of an epoch-making epidemiological survey, a multi center study by the Ministry of Health and Welfare Study Group, showing the relationship between diseases and life-styles.[46] In this survey, risk factors and preventive factors of 11 intractable diseases including von Recklinghausen disease, scleroderma, mixed type connective-tissue diseases, and Behcet's disease were investigated in various items of lifestyle such as smoking, exercise, and diet.

According to this survey, dietary factors that increase risks of many diseases include excess ingestion of beef rich in fat (3–4 times or more per week), yogurt, cheese, butter, margarine (1–2 times or more per week), and tomatoes. Inversely, factors that prevent and inhibit diseases had the Japanese traditional diet pattern such as ingestion of fish and shellfish, pickles, and green and yellow vegetables. These findings suggested that eating habits should be considered more than they were previously in their relationship to disease. It was also suggested that measures toward the improvement of intractable diseases are present, although the improvement is slight. Furthermore, the survey clarified that Japanese traditional food inhibits many diseases.

It has been reported that similar diet instruction is effective for treatment of psoriasis,[47,48] and diet instruction is also useful for other skin diseases that cannot be easily healed by conventional therapy alone.

## 5. Diet Instruction that Complements Therapy for Atopic Dermatitis

In the Japanese traditional medicine concomitantly administered for 20 years in our department, we have focused on not only medication with herbs, but also dietary advice. Diet instruction is essentially recommendation of balanced diets, mainly Japanese food, and we have designed practical and useful diets for individual patients. Particularly, use of a diet diary is useful for increasing patient awareness. Our diet instruction corrects eating habits that may have aggravated diseases, and proposes acceptable eating habits. We basically focus on items presented in Table 19.2, check the diet diary, and individually advise patients in consideration of their lifestyles. Instruction is continuously or intermittently given, and the state of compliance is evaluated.

Patient 1: 36-year-old male with no particular past medical history or familial medical history. The patient had had atopic dermatitis since childhood, and had been treated mainly

**Table 19.2.** Diet instruction

---
1. The principle food is low-polished rice from which the influence of contamination is eliminated as much as possible. Boiled rice is recommended for breakfast.
2. For dishes other than the principle food, vegetables in season, fish and shellfish/seaweed, soybean products.
3. Care not to ingest excess sweets, alcohol, coffee, or juice.
4. Care not to ingest excess meats, fats, or fatty oils (n-6 fatty acids in particular).
5. Avoid processed food and additives.
6. Bland and homemade dishes are recommended.
7. Moderation in eating and pleasant meals with sufficient chewing.

---

with external steroids by a physician, but the effect had decreased since about five years before, and oral steroids were additionally administered. Since dermatitis recurred repeatedly due to dose reduction of oral steroids, steroid medication was discontinued for more than one month, and oral antihistaminic agents and external steroids were administered. However, systemic erythroderma developed, and the patient consulted the Department of Dermatology, Hospital of Osaka City University School of Medicine. Upon first examination, systemic desquamation of erythema and edema were noted. In addition to the previous treatment, traditional herbal remedies (Hochu-ekki-to, Goreisan, and Yokuinin) were administered, and a diet mainly consisting of Japanese traditional food was recommended. The symptoms moderately improved after two weeks, and the condition without exanthema persisted for six months. When therapeutic drug administration was completed, only diet recommendations were continued, and exanthema did not appear for more than six months.

Since the above diet therapy is a complex treatment, evaluation by the current techniques of modern medicine such as the double blind method is difficult. Moreover, dietary improvement as a non-allergic factor requires a relatively long period in many cases. Thus, we investigated the implication of the usefulness of each characteristic factor of Japanese traditional food.

## 6. Usefulness of Pre-Germinated Brown Rice in Therapy for Intractable Atopic Dermatitis

### 6.1. Background

We have recommended low-polished rice and brown rice in diet instruction because they are rich in active components such as vitamins and minerals. Since pre-germinated brown rice (Table 19.3), which is more easily boiled than brown rice and rich in aminobutyric acid, has become commercially available in recent years, we investigated its usefulness.[49,50]

### 6.2. Subjects and Methods

Patients with intractable atopic dermatitis, in whom standard treatment alone did not resolve exanthema during one-year or a longer course of observation, were explained the content of the study protocol based on the Declaration of Helsinki, and 15 consenting patients were selected. The patients were from 18 to 33 years of age and consisted of nine males and six females.

**Table 19.3.** Comparison of components of pre-germinated brown rice, brown rice, polished rice, and wheat flour

|  | Pre-germinated brown rice | Brown rice | Rice | Wheat flour |
|---|---|---|---|---|
| GABA (mg) | 15 | 4 | 1.5 | — |
| Fibar (g) | 3.2 | 2.3 | 0.8 | 2.7 |
| vitamin B1 (mg) | 0.38 | 0.4 | 0.09 | 0.1 |
| vitamin E (g) | 1.2 | 1.3 | 0.3 | 0.3 |
| inositol (mg) | 265 | 257 | 44 | — |
| calcium (mg) | 10 | 9.8 | 5 | 20 |
| magnesium (mg) | 111 | 128 | 37 | 23 |
| IP6 (mg) | 474 | 874 | 94.8 | — |
| ferulic acid (mg) | 26 | 24 | 3.7 | — |
| oryzanol (mg) | 30 | 35.5 | 4.3 | — |

In the method, as a rule, the previous treatments were not changed, and the only change was replacement of polished rice and bread in the previous diet with boiled pre-germinated brown rice (Fancl Co.), 100–200 g per day. Clinical symptoms were graded by six steps from the absence of exanthema to severe symptoms (no exanthema = 0, very mild = 1, mild = 2, moderate = 3, slightly severe = 4, severe = 5). Observation over the course of six months or longer was considered important, and improvement by three steps or more was rated as markedly effective, two steps as effective, and one step as slightly effective. General blood test, blood chemistry including AST, ALT, and LDH, and measurement of blood aminobutyric acid (GABA) and blood total IgE level by radioimmunosolvent test were performed monthly as often as possible. In addition, acceptance of pre-germinated brown rice ingestion by patients was investigated by a questionnaire.

### 6.3. Results

Eight out of 15 patients continued eating pre-germinated brown rice for six months or longer, and the diet was effective in three and slightly effective in five patients. Three patients continued the diet for three months or longer but less than six months. There were four dropouts who continued the diet for less than three months: one patient did not want to continue and the other three had difficulty visiting the hospital. No aggravation was noted after ingestion of pre-germinated brown rice. No hematological abnormality or liver function disorder was detected in the blood chemistry of any patients. The IgE level was abnormally high before ingestion in eight patients, and it tended to decrease after ingestion in five patients. The LDH level was high in five patients before ingestion and it tended to decrease in all five patients after ingestion. The GABA level changed in relation to symptoms in some patients.

Patient 2: 31-year-old male. Familial medical history: None in particular. Past medical history: Asthma. Present illness: The patient had atopic dermatitis complicating asthma since infancy and had been treated by a physician. The patient received various treatments such as external steroids, external moisturizing agents, counseling, and traditional herb remedies, but exanthema did not disappear. It was suddenly aggravated three months before, and the patient visited the Department of Dermatology, Hospital of Osaka City University School

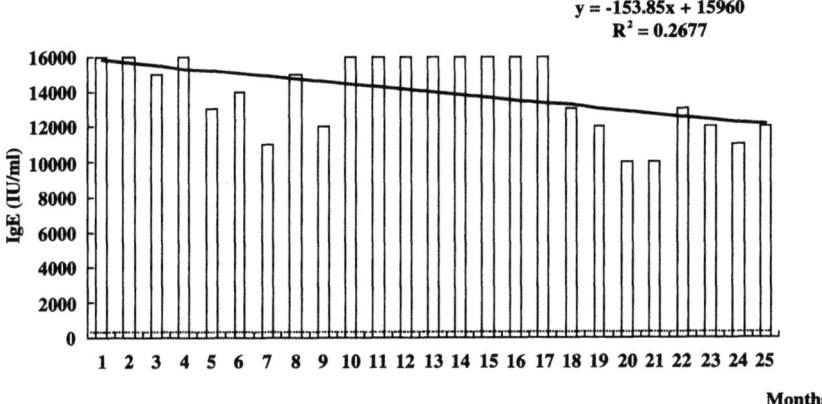

**Figure 19.2.** Patient 2: IgE level.

of Medicine. At the first examination, the skin of the whole body exhibited erythema accompanied by desquamation and itching. In addition to external steroids and oral antiallergic agents, traditional herbal remedies were given along with instruction for a Japanese diet. Dermatitis slightly subsided within a year, but the effect was insufficient. Thus, after the study was explained to the patient and consent was obtained, polished rice was replaced with pre-germinated brown rice. The area of erythema narrowed and the itching was also reduced after ingestion for one year, and moist skin resumed after ingestion for two years. The levels of IgE (Figure 19.2) and LDH (Figure 19.3) decreased slowly with reduction of the skin symptoms, and the GABA level (Figure 19.4) tended to change with these changes.

Patient 3: A 21-year-old female had suffered from systemic exanthema accompanied by itching since four years of age and was treated with external steroids by a physician, but the exanthema had not healed. The patient was treated by another pediatrician who administered traditional herbal medicine, and the treatment was effective to some extent.

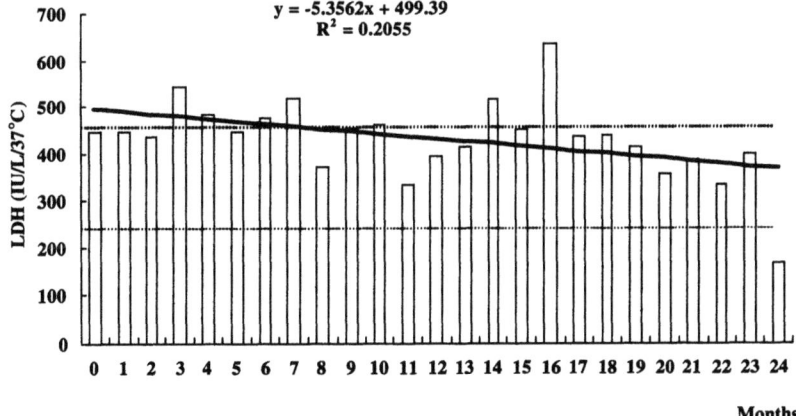

**Figure 19.3.** Patient 2: LDH level.

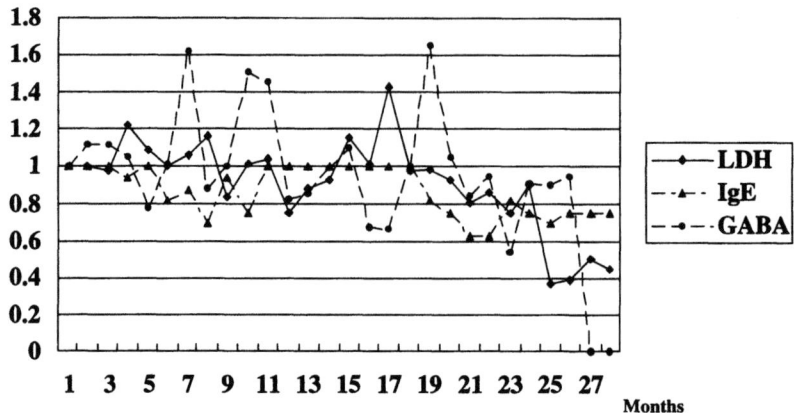

**Figure 19.4.** Patient 2: Relative changes in GABA level from the value before ingestion of pre-germinated brown rice.

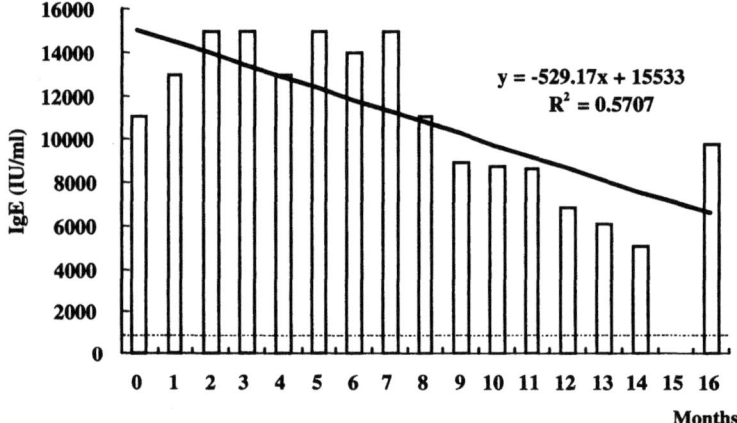

**Figure 19.5.** Patient 3: IgE level.

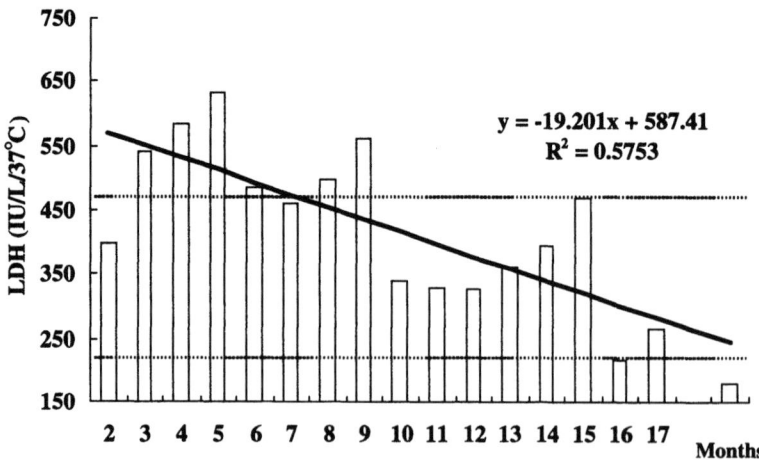

**Figure 19.6.** Patient 3: LDH level.

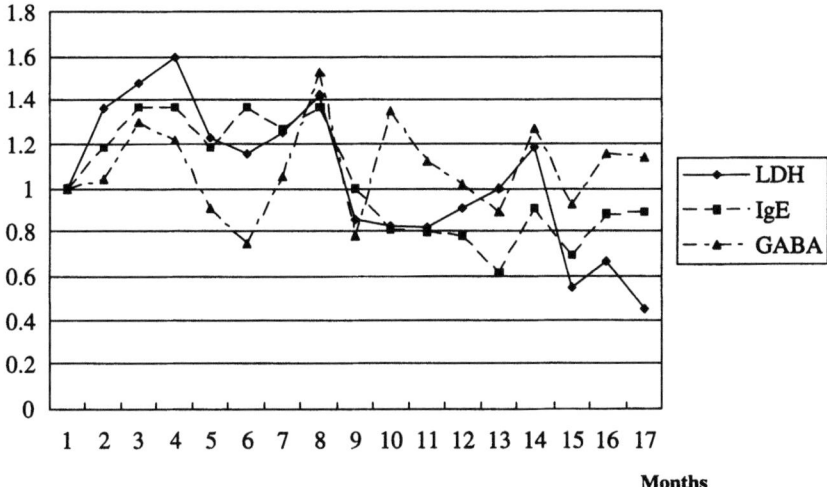

**Figure 19.7.** Patient 3: Relative changes in GABA level from the value before ingestion of pre-germinated brown rice.

However, the patient became unable to receive the treatment by the pediatrician, and visited the Department of Dermatology, Hospital of Osaka City University School of Medicine, requesting treatment with traditional herbal medicine. At the first examination, erythema accompanied by desquamation was noted on the face and neck and erythematous regions with scratch scars noted on the chest and four limbs. Several oral herbal remedies and external steroids were concurrently administered, and diet instruction was also given. The symptoms improved, but skin dryness and mild erythema remained even after two years. Thus, after we explained this study and obtained the patient's consent, ingestion of pre-germinated brown rice was initiated. Almost no exanthema remained one year later, and the skin recovered its moistness. Both the IgE and LDH levels decreased with remission of skin symptoms.

### 6.4. Discussion

Symptoms gradually improved in patients with intractable atopic dermatitis who ate pre-germinated brown rice in diet therapy, and acceptance of the diet by the patients was good. There was no abnormal laboratory test value or adverse effect attributable to ingestion of pre-germinated brown rice. The study is being continued.

## 7. Significance of Traditional Medicinal Diet Instruction Based on Blood Fatty Acid Balance in Patients with Atopic Dermatitis

### 7.1. Background

We investigated the significance of this therapy, focusing on unbalanced unsaturated fatty acids from the aspect of fat and oil science.

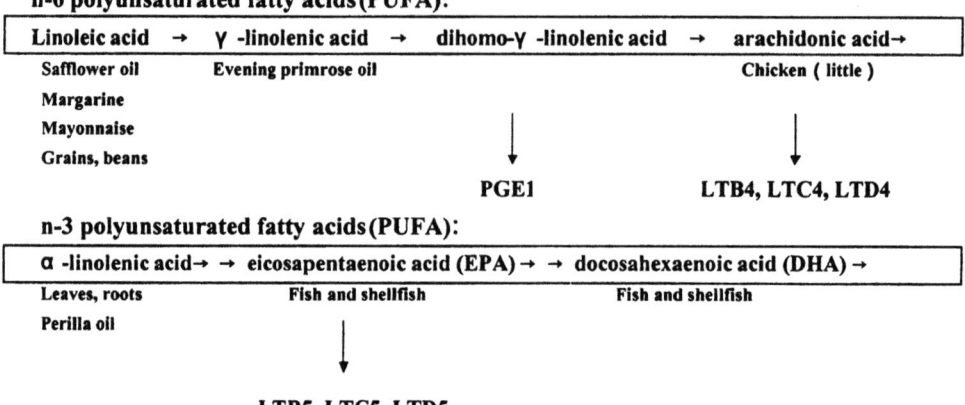

**Figure 19.8.** n-3/n-6 polyunsaturated fatty acids and chemical mediators (partially modified figure reported by Okuyama[57]).

In recent Japanese eating habits, ingestion of $\omega 6$ (n-6) polyunsaturated fatty acids represented by linoleic acid has increased, while ingestion of n-3 fats and oils contained in fish oil and green and yellow vegetables has decreased.

$\omega 6$(n-6) polyunsaturated fatty acids change to a potent prophlogistic substance such as leukotriene B4 via the arachidonic acid cascade. In contrast, n-3 fats and oils have been clarified to exhibit an inhibitory action on inflammation[51] and inhibit allergic inflammation[24,52,53] (Figure 19.8[57]). Based on this information, n-3 polyunsaturated fatty acids including eicosapentaenoic acid (EPA) have been used for treatment of atopic dermatitis, and Bjorneboe A. et al.[54] and others reported its usefulness.

Our diet instruction is to decrease the intake of meats, stir-fried dishes, and fried food, and we recommend eating fish, shellfish, and green and yellow vegetables. Following the instruction may lead to an increase in $\omega 3$ (n-3) and decrease in ingestion of $\omega 6$ (n-6). The state of serum multi-unsaturated acid balance in patients with atopic dermatitis and the effects of Japanese traditional medicinal diet instruction on symptoms and blood fatty acids were investigated.[55,56]

## 7.2. Subjects and Methods

The objective and methods of this study, based on the Declaration of Helsinki, were explained to patients with atopic dermatitis being treated at the Department of Dermatology, Hospital of Osaka City University School of Medicine between 1999 and 2000. 112 consenting patients were selected for the study (male: 37, female: 75).

For the treatment, in addition to the standard treatment described in the guidelines of the Japanese Dermatological Association (treatment mainly by a balance of external steroids and oral antiallergic agents), diet instruction was performed as a part of Japanese traditional medicine. The content of diets was continuously recorded for a specified period, and meats, fat, oil, and sweets were reduced from the menus. Patients were instructed to ingest diets mainly consisting of Japanese food made of rice, vegetables, fish, and soybean products. Intractable cases received concomitant oral herbal remedies.

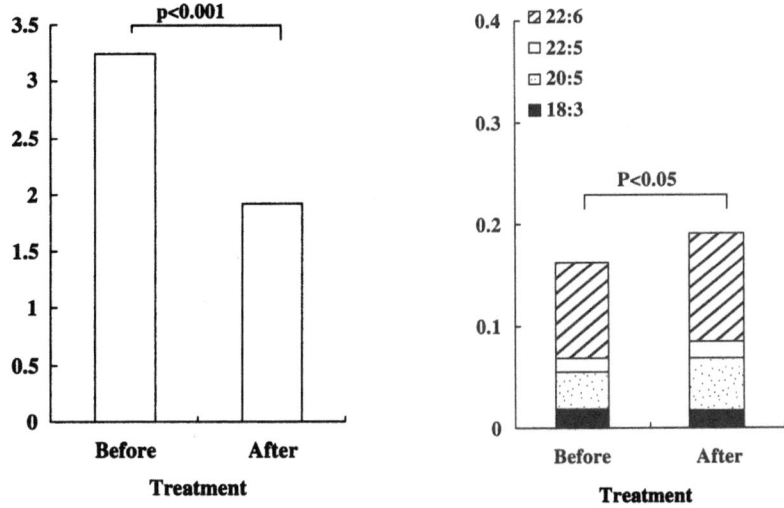

**Figure 19.9.** Changes in severity after therapy and n-3 PUFA.

Serum fatty acids in fasting sera collected before and after therapy were measured by capillary gas chromatography after methylation of the serum samples. The severity of clinical symptoms was evaluated by six-step grading (no exanthema = 0, very mild = 1, mild = 2, moderate = 3, severe = 4, severest = 5), and compared between before and after therapy.

### 7.3. Statistical Analysis

Regarding the value before therapy as the baseline, measured values and scores were presented as the mean values. Differences between the two items were analyzed by student T test, and $p < 0.05$ was regarded as significant. Correlation was analyzed using Spearman's correlation coefficient test.

### 7.4. Results

The mean measured values of n-3 and n-6 fatty acids for all patients before therapy were 0.16 and 1.04, respectively, and the n-3/n-6 ratio was 0.16.

Compared to the score before therapy, the severity score significantly decreased from 3.25 to 1.91 after therapy, and the total n-3 polyunsaturated fatty acid level significantly increased (Figure 19.9). There was hardly any difference between the n-6 polyunsaturated fatty acid levels before and after therapy. Both the n-3/n-6 ratio and the n-3 eicosapentaenoic acid/n-6 arachidonic acid ratio increased after therapy (Figure 19.10).

In a comparison of the severity and n-3/n-6 ratio, no correlation was observed for the total fatty acid, but in patients with an EPA/AA ratio of less than 0.3 before therapy, who comprised 70% of all patients, the severity and n-3/n-6 ratio were negatively correlated. Regarding the improvement rate of the symptoms, they improved as n-3 lipids increased (Figure 19.11).

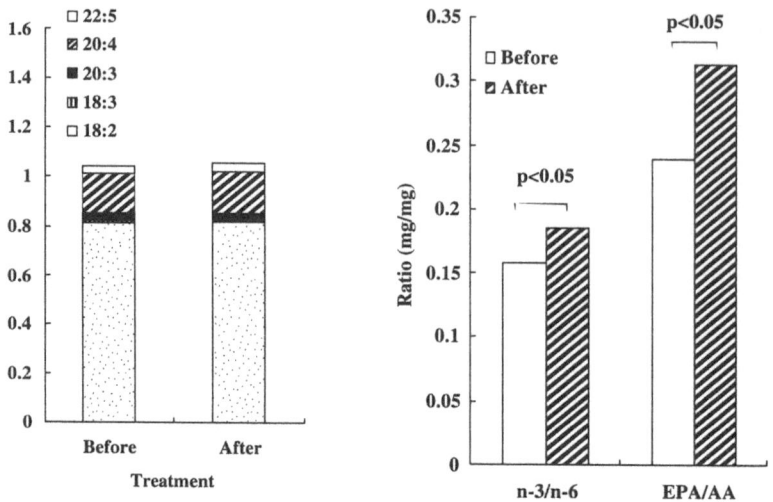

**Figure 19.10.** n-6 PUFA and n-3/n-6 ratio before and after therapy.

## 7.5. Discussion

In patients with atopic dermatitis in this study, the mean n-3/n-6 ratio was 0.158 before therapy, which was lower than the expected value for disease prevention, 0.3,[42] and slightly lower than the value recently measured in healthy individuals in urban areas, 0.2,[58] and the measured value in patients with atopic dermatitis, 0.18.[39]

The increase in an n-3 polyunsaturated fatty acid, eicosapentaenoic acid (EPA), after therapy may have resulted from increased ingestion of fish, shellfish, and green and yellow vegetables after diet instruction. In contrast, the n-6 polyunsaturated fatty acid level did not

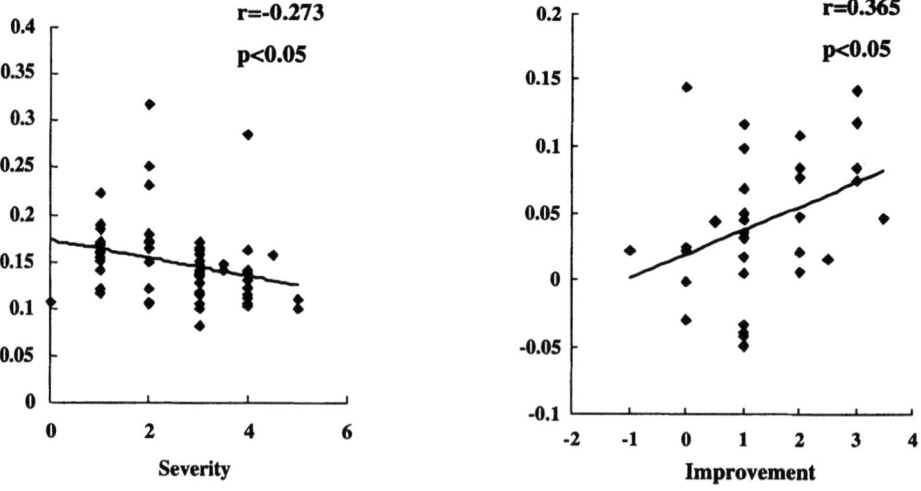

**Figure 19.11.** Correlation between severity and n-3/n-6.

change after therapy. Since n-6 fatty acids are contained in many food products including plant fat and oil, instruction to consciously limit n-6 fat and oil may be necessary.

There have been many reports of the effects of n-3 fat and oil supplements in the therapy of atopic dermatitis.[54] Our diet instruction recommends Japanese traditional food without the use of supplements. This instruction increased n-3 fat and oil and improved symptoms in patients in whom the ratio of representative n-3 to representative n-6, EPA/AA, was low. Such patients comprised 70% of all patients. However, an increase in n-3 fats and oils does not necessarily lead to improvement in some patients. It is of course necessary to consider various aggravating factors other than lipids, that is, dietary factors such as excess ingestion of sweets and the involvement of stress.

The influence of n-3/n-6 varies among races. The frequency of asthma induced by fried food is higher in Asian children than in non-Asian children.[59] It has been reported that deficiency of eicosapentaenoic acid (EPA) and prostaglandin E1 (PGE1) is likely to induce food allergy in Native Americans living on the coast of British Columbia,[60] suggesting that deviation from traditional life-style may cause an imbalance of the immune system.

This study suggests that a decrease in the n-3/n-6 ratio in serum fatty acid composition is an aggravating factor for atopic dermatitis in Japan. Although there are various aggravating factors, this study clarified that a medicinal diet recommending traditional Japanese food is useful for improving symptoms and the n-3/n-6 ratio for patients whose present eating habits are a cause of atopic dermatitis.

## 8. Conclusions

Problems with food are broad and deep. When growing food from seeds, it is necessary to pay attention to all processes from soil management, growing methods, harvesting, cooking, and ways of eating. It is also important to maintain useful factors that exist originally in the soils such as microorganisms and insects and begin growing crops with an effort to decrease contamination in the air and water. Food problems should be considered from a broad outlook, fully considering the possibility of the presence of important factors non-analyzable by analytical techniques currently available. There are foods that increase health and prevent diseases, and the effect of Japanese traditional food in atopic dermatitis therapy may be significant.

## 9. References

1. J. M. Hanifin and G. Rajka, Diagnostic features of atopic dermatitis, *Acta Derm. Venereol.* **92**(Suppl.), 44 (1980).
2. The International Study of Asthma and Allergies in Childhood (ISSAC) Steening Committee, Worldwide variation in prevalence of symptoms of asthma, allergic rhinoconjunctivitis, and atopic eczema: ISAAC, *Lancet* **351**, 1225–1232 (1998).
3. K. Takahashi, M. Ishii, Y. Asai, and T. Hamada, Experiences of therapy for skin disease by Japanese herbal medicine, *Kampo medicine* **5**(2), 9–11 (1981).
4. K. Terasawa, *KAMPO, Japanese-oriental medicine, Insights from clinical cases* (K.K.Standard McIntyre, Tokyo, 1993).
5. K. Takahashi, General herb treatments for skin disease: On basis of Chinese medical theory, *Skin* **39**(1), 1–23 (1997).

6. W. B. Shelly, Experimental disease in the skin of man. A review, *Acta Derm. Venereol. Suppl.* **108**, 5–38 (1982).
7. Brit Med Res Council, Vitamin-C requirement of human adults, *Lancet* i: **254**, 853–858 (1948).
8. W. H. Sebrell and R. E. Butler, Riboflavin deficiency in man (ariboflavionsis). *Publ. Health Report*, Washington, **54**, 2121–2131 (1939).
9. R. W. Vilter, J. F. Mueller, H. S. Glazer, T. Jarrold, J. Abraham, C. Thomson, and V. R. Hawkins, Effect of vitamin B6 deficiency induced by desoxypyridoxine in human beings, *J. Lab. Clin. Med.* **42**, 335–387 (1953).
10. R. E. Hodges, M. A. Ohlson, and W. B. Bean, Pantothenic acid deficiency in man, *J. Clin. Invest.* **37**, 1642–1657 (1958).
11. A. E. Hansen, H. F. Wiese, A. N. Boelsche, M. E. Haggard, D. J. D. Adam, and H. Davis, Role of linoleic acid in infant nutrition, clinical and chemical study of 428 infants fed on milk mixtures varying in kind and amounts of fat, *Pediatrics* **31**, 171–192 (1963).
12. M. G. Shultz, Joseph Goldberger and pellagra, *Am. J. Trop. Med. Hyg.* **26**, 1088–1092 (1977).
13. T. Yoshida, 4. Points of therapy for skin disease. Diet therapy for atopic dermatitis, *Clinical Dermatology* **53**(5, suppl.), 113–116 (1999).
14. S. Halken, H. P. Jacobsen, A. H. $\phi$ st, D. Holmenlund, The effect of hypo-allergenic formulas in infants at risk of allergic disease, *Eur. J. Clin. Nutr.* **49**(Suppl. 1), S77–83 (1995).
15. M. E. Herrmann ME, A. Dannemann, A. Gruters, B. Radisch, J. W. Dodenhausen, R. Bergman, A. Coumbos, H.-K. Weitzel, U. Wahn, Prospective study of the atopy preventive effect of maternal avoidance of milk and eggs during pregnancy and lactation, *Eur. J. Pediatr.* **155**, 770–774 (1996).
16. M. Uehara et al., Atopic dermatitis and food. Elimination/challenge test and clinical course, *Skin clinic* **32**(1), 87–90 (1990).
17. T. Uenishi, H.Sugiura, M. Uehara, Role of food in irregular aggravation of atopic dermatitis, *J. Dermatol.*, **30**, 91–97 (2003).
18. H. Hashizume, Special issue/Latest manual of treatment of atopic dermatitis. Complementary alternative therapy, *Monthly Derma.* **54**, 113–118 (2001).
19. M. Uehara, H. Sugiura, and K. Sakurai, A trial of oolong tea in the management of recalcitrant atopic dermatitis, *Arch. Dermatol.* **137**(1), 42–43 (2001).
20. M. Matsumoto, M. Kitani, A. Fujita, and T. Tanaka, Inhibitory effect of persimmon leaf extract on atopic dermatitis in NC/Nga mice, *Journal of Japan Society for Nutrition/Food* **54**(1), 3–7 (2001).
21. M. S. Manku, D. F. Horrobin, N. Morse, V. Kyte, K. Jenkins, Reduced levels of prostaglandin precursors in the blood of atopic dermatitis: Defective delta-6-desaturase function as a biochemical basis for atopy, *Prostaglandins. Leukot. Med.* **9**, 615–628, (1982).
22. C. R. Lovell et al., Treatment of atopic eczema with evening primrose oil, *Lancet* I, 278 (1981).
23. F. Furukawa, S. Tokura, M. Takigawa, and F. Morimatsu, Effect of substitute food containing $\gamma$-linoleic acid for allergy on atopic dermatitis, *Skin* **40**(suppl. 20), 73–78 (1998).
24. K. Sakai, H. Okuyama, H. Shimazaki, M. Katagiri, S. Torii, T. Matsushita, and S. Baba, Fatty acid compositions of plasma lipids in topic dermatitis/asthma patients, *Jpn. J. Allergol.* **43**(1), 37–43 (1994).
25. J. T. M. Bamford et al., Atopic eczema unresponsive primrose oil, *J. Am. Acad. Dermatol.* **13**, 959–965 (1985).
26. H. Okuyama, Improvement of allergic constitution by control of dietary fatty acids, *Adv. Clin. Pharmacol.* **19**, 51–62 (1998).
27. H. Kimata, Infantile atopic dermatitis and fatty liver: Improvement by oxatomide and diet instruction, *Therapeutic Research* **20**(2), 531–539 (1999).
28. T. Uesugi, Effect of diet instruction on patients with atopic dermatitis, *Clinical Dermatology* **52**(7), 495–497 (1998).
29. M. Matsuda, A case of severe atopic dermatitis markedly improved by intravenous antifungal agent and diet therapy.
30. O. J. Carey et al., The effect of lifestyle on wheeze, atopy, and bronchial hyperreactivity in Asian and white children, *Am. J. Respir. Crit. Care. Med.* **154**, 537–540 (1996).
31. Y. Nagata, *Journal of Japanese Society of Allergology* **34**(8), 763–764 (1985).
32. Y. Ikezawa, Recent progress and topics in diet therapy and nutrition therapy. 19. Allergic disease: Focused on atopic dermatitis, *Clin. Nutr.* **83**(4), 515–526 (1993).
33. O. Mizukami, Diet therapy for atopic dermatitis, *Therapy* **79**(1), 151–153 (1997).

34. M. Ishii, H. Kobayashi, and N. Mizuno, Japanese herb treatments of adult atopic dermatitis by diet and Japanese herb remedy—Evaluation of disappearance of disease phases—State of herb treatments in dermatology 9 (Sogoigaku Co., Tokyo, 1997), pp. 63–77.
35. H. Kobayashi and M. Ishii, Atopic dermatitis. Seize of complex system from a macroscopic viewpoint. Disappearance of disease phases by dietary improvement and Japanese herb treatments, *Medical View points* **18**(6), 1–2 (1997).
36. M. Ishii, West-East harmonized therapy—Summary (10) Dermatological disease, Radio Tanpa Hoso, Kotaro Kampo Salon no.778, 1997.
37. M. Ishii et al., Focus of therapy for allergic disease/Special edition. Atopic dermatitis: the disappearance of disease phases by diet and Japanese herb treatments, *Medical Review Bermedico* **13**(3), 21–22 (1998).
38. M. Ishii, in: *Combination therapy with diet and traditional Japanese medicine for intractable adult atopic dermatitis—Interpretation of dietary influence. State of herb treatments in dermatology 10*, edited by Dermatological Oriental Medicine Study Group (Kyowa Kikaku Tsushin, Tokyo, 1999), pp. 35–42.
39. Y. Nagata, Therapy for atopic dermatitis by Japanese traditional food (minus N-6, plus N-3)—Content and results of therapy—Lipid Nutrition series 3. Fat and oil and allergy, pp. 31–50, 1999.
40. Y. Nagata, Therapy for atopic dermatitis by Japanese traditional food (minus N-6, plus N-3)—Discussion—Lipid Nutrition series 3. Fat and oil and allergy, pp. 51–60, 1999.
41. K. Hirai, C. Shimazu, R. Takazoe, and Y. Ozeki, Cholesterol. Phytosterol and polyunsaturated fatty acid levels in 1982 and 1957 Japanese diets, *J. Nutr. Sci. Vitaminol. (Tokyo)* **32**(4), 363–372 (1986).
42. W. E. Lands, T. Hamazaki, K. Yamazaki, H. Okuyama, K. Sakai, Y. Goto, and V. S. Hubbard, Changing dietary patterns, *Am. J. Clin. Nutr.* **51**(6), 991–993 (1990).
43. K. Katanoda, Y. Matsumura. National Nutrition Survey in Japan—Its Methodological transition and current findings—, *J Nutr Sci Vitaminol*, **48**, 423–432 (2002).
44. W. G. Crook, *The yeast connection* (Vintage Books, Division of Random House, New York, 1984).
45. H. Okuyama, T. Kobayashi, and S. Watanabe, Dietary fatty acids—the N-6/N-3 balance and chronic elderly diseases. Excess linoleic acid and relative N-3 deficiency syndrome seen in Japan, *Prog. Lipid Res.* **35**(4), 409–457 (1996).
46. Y. Oono and T. Hashimoto, Life-style and development of intractable disease, *Igaku no ayumi* **190**(11), 1031–1033 (1999).
47. K. Takahashi, Therapy for psoriasis using Japanese herbal medicine, *Skin* **26**, 1166–1173 (1984).
48. E. Soyland, J. Funk, G. Rajka, M. Sandberg, P. Thune, L. Rustad, S. Helland, K. Middelfart, S. Odu, E. S. Falk, K. Solvoll, G.-E. A. Bjorneboe, C. A. Drevon, Effect of dietary supplementation with very-long-chain n-3 fatty acids in patients with psoriasis, *N. Engl. J. Med.* **328**(25), 1812–1816 (1993).
49. H. Kobayashi, N. Mizuno, M. Ishii, Y. Someya, M. Kise, and K. Ishiwata, Investigation of use of pre-germinated brown rice in diet instruction for intractable atopic dermatitis—Preliminary report. 4th meeting of the Japan Society for complementary alternative medicine, 2001.
50. H. Kobayashi, N. Mizuno, H. Teramae, H. Kutsuna, M. Ishii, Y. Ito, A. Mizuguchi, M. Kise, and H. Aoto, Investigation of use of pre-germinated brown rice in diet instruction for atopic dermatitis (2nd report). 5th meeting of the Japan Society for complementary alternative medicine, 2002.
51. B. R. Lokesh, J. M. Black, J. B. German, J. E. Kinsella, Docosahexaenoic acid and other dietary polyunsaturated fatty acids suppress leukotriene syntheses by mouse peritoneal macrophages, *Lipids* **23**, 968–972 (1988).
52. H. Okuyama, Minimum requirement of n-3 and n-6 essential fatty acids for the function of the central nervous system and for the prevention of chronic disease, *Proc. Soc. Exp. Biol. Med.* **200**(2), 174–176 (1992).
53. S. Watanabe, N. Sakai, Y. Yasui, Y, Kimura, T. Kobayashi, T. Mizutani, and H. Okuyama, A high alpha-linolenate diet suppresses antigen-induced immunoglobulin E response and anaphylactic shock in mice, *J. Nutr.* **124**(9), 1566–1573 (1994).
54. A. Bjorneboe, E. Soyland, G. E. A. Bjorneboe et al., Effect of n-3 fatty acid supplement to paients with atopic dermatitis, *J. Int. Med.* **225**(Suppl.), 233–236 (1989).
55. M. Nanatsue, I. Ichi, K. Hirai, H. Kobayashi, and M. Ishii, Serum fatty acid composition in patients with atopic dermatitis. Japan Society for Nutrition/Food 21C Memorial joint meeting of Kinki, Chugoku, and Shikoku branches and open symposium (40th memorial meeting of the Kinki branch, 34th meeting of the Chugoku/Shikoku branch), 2001.
56. H. Kobayashi, M. Nanatsue, N. Mizuno, H. Teramae, H. Kutsuna, K. Hirai, and M. Ishii, Serum fatty acid composition in patients with atopic dermatitis and recommendation of Japanese traditional

food.—n-3/n-6 balance—53rd Chubu branch general meeting and academic meeting of Japanese Dermatological Association (Gifu), 2002.
57. H. Okuyama, Environmental allergens and allergic reactivity of body—Possibility of improvement of allergic constitution by selection of food fatty acid—Lipid Nutrition series 3. Fat and oil and allergy, pp. 1–13, 1999.
58. H. Iso, S. Sato, A. R. Folsom, T. Shimamoto, A. Terao, R. G. Munger, A. Kitamura, M. Konishi, M. Iida, Y. Komachi, Serum fatty acids and fish intake in rural Japanese, urban Japanese, Japanese American and Caucasian American men., *Int. J. Epidemiol.* **18**, 374–381 (1989).
59. N. M. Wilson, Food related asthma: a difference between two ethnic groups, *Arch. Dis. Child.* **60**(9), 861–865 (1985).
60. C. E. Bates, Racially determined abnormal essential fatty acid and prostaglandin metabolism and food allergies linked to autoimmune, inflammatory, and psychiatric disorders among Costal British Columbia Indians, *Med. Hypotheses* **25**(2), 103–109 (1988).

# 20

# Kampo Therapy for Adult Atopic Dermatitis by Dieting and Herbal Medications: Evaluating the Disappearance of Disease Phases

DIETING AND HERBS FOR ATOPIC DERMATITIS

Masamitsu Ishii, Hiromi Kobayashi, Nobuyuki Mizuno, Hiroyuki Teramae, Takeshi Nakanishi, Haruo Kutsuna and Iwao Yamamoto

## 1. Introduction

Considerable time has passed since the gradual increase of refractory atopic dermatitis and its emergence as a social issue in Japan. Compared with the past, the disease already appears to be more refractory in children and most refractory in adults. We have treated atopic dermatitis by Kampo (Japanese Herbal Medicine)[1] therapies including dieting[2-4] and achieved better results than by Western treatments alone, but there were difficulties in demonstrating the genuine effectiveness of Kampo therapies. We, therefore, evaluated the effectiveness of Kampo therapies including dieting in the most refractory cases, i.e. severe cases of adult type atopic dermatitis or cases having persistent rash, according to prolonged control of symptoms and the disappearance of disease phases (alternating waves of remission and exacerbation) as indices. Since the treatment was markedly effective, we present our understanding of adult type atopic dermatitis, reasons for inclusion of dieting in the therapy, and therapeutic results.

## 2. Patterns of the Course of Adult Type Atopic Dermatitis

In Japan, many physicians have noticed new patterns in the course of adult type atopic dermatitis with careful observation of patients over a long period. The course of adult type

---

Masamitsu Ishii, Hiromi Kobayashi, and Nobuyuki Mizuno, Hiroyuki Teramae, Takeshi Nakanishi and Haruo Kutsuna • Department of Dermatology, Osaka City University Post Graduate School of Medicine, Japan. Iwao Yamamoto • Yamamoto Internal Clinic, Kyobasahi, Japan.
*Complementary and Alternative Approaches to Biomedicine*, edited by Edwin L. Cooper and Nobuo Yamaguchi. Kluwer Academic/Plenum Publishers, 2004.

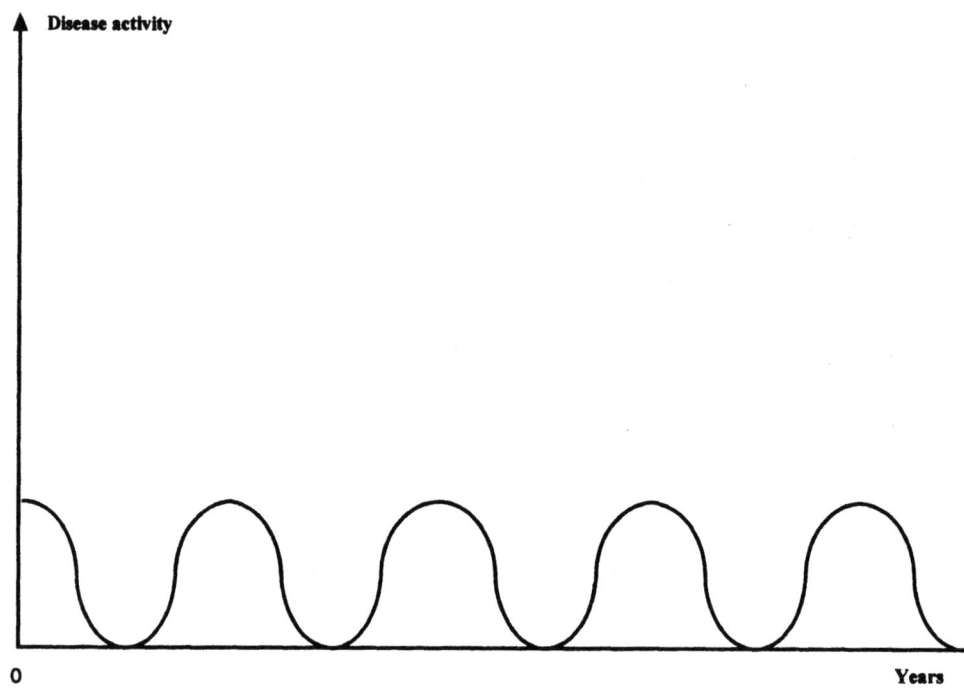

**Figure 20.1.** Conventional diagram of the therapeutic course of atopic dermatitis.

atopic dermatitis used to be described as a single pattern, or alternating waves of remission and exacerbation (Figure 20.1). Physicians generally agree that the treatment for this type of disease is controlling it with an external steroid during exacerbation, and with a moisture-retainer during remission to protect the skin and promote the repair of its barrier function.

Certainly, there are many patients with this type, and they are considered to represent a group in which the disease can be controlled by treatment with an external steroid and moisture retainer alone. However, it cannot be denied that the disease gradually becomes uncontrollable with conventional steroid therapy in many patients. Many clinicians would admit that dermatitis, mild at the onset, gradually deteriorates over years, and develops to systemic erythroderma or refractory erythema of the face with a tendency of effusion in a considerable number of patients despite exhaustive treatments. This condition has been suggested to be rosacea due to the use of steroids or, conversely, a rebound, but it is not alleviated even many years after withdrawal of steroids or their replacement with milder steroids in many patients. In view of these cases, it may be useful to understand such a course as a characteristic pattern of refractory adult type atopic dermatitis. Some authors emphasize exacerbating factors such as contact dermatitis, folk therapies, and discontinuation of steroid by the patient's judgment, which are certainly observed in many cases. However, if we frankly admit the occurrence of formerly unknown refractory adult type atopic dermatitis in a large number of patients, we should interpret it as a new disease type.

On the basis of clinical courses of adult atopic dermatitis, we present the following model patterns of the course of the disease as a new pathologic entity, its changes with steroid therapy and its discontinuation, and its healing process.

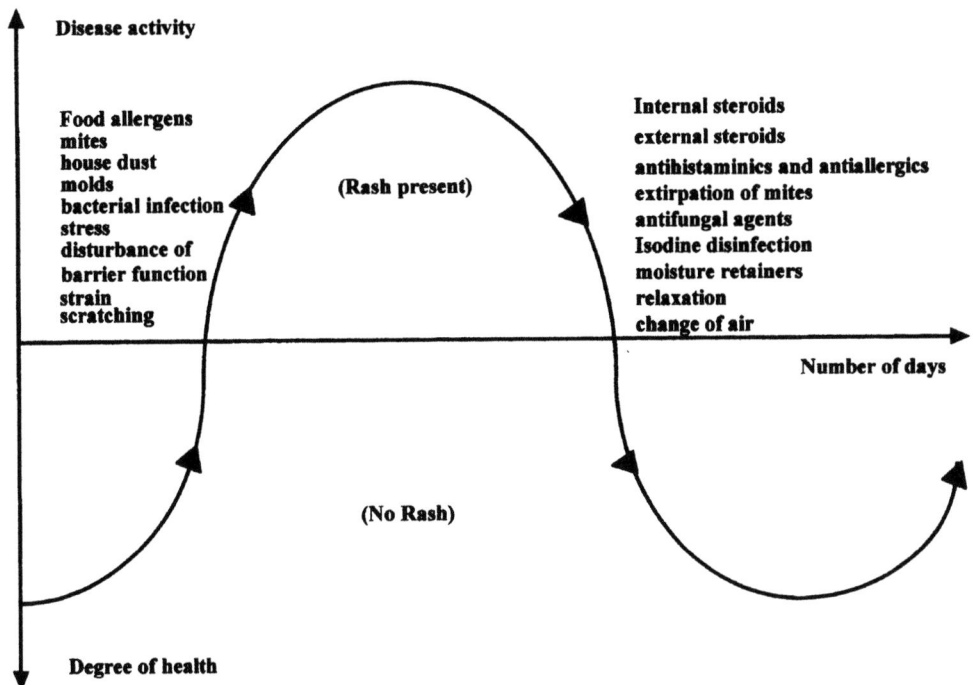

**Figure 20.2.** Diagram of one short cycle of atopic dermatitis in which the degree of health is incorporated.

The disease pattern can be expressed by a graph in which the activity of the disease is shown upward, and the degree of health downward, along a vertical scale. Only the activity of the disease used to be shown along the vertical scale, and conditions below the origin were disregarded. However, by showing also the degree of health downward along the vertical axis, the interrelation between the patient and disease can be better represented. The origin, at which both the value of the disease activity and the value of the degree of health are zero, is regarded as the threshold of the appearance of rash, the rash become severe higher along the scale, but it is not only absent but also becomes less likely to occur as the state of health improves further down along the scale. The horizontal axis represents the lapse of time so that alternation of remission and exacerbation with the lapse of time can be clearly represented.

Figure 20.2 shows a cycle of remission and exacerbation in days or weeks rendered in such coordinates. The graph also shows exacerbating factors such as mite antigen, food antigen, bacteria, fungi, and stress and remitting factors (treatments). Figure 20.3 shows such remission-exacerbation cycles in a longer time span of years. In this pattern, the average disease activity is considered to be horizontal to the right, and this type of disease can be controllable with steroids and moisture-retainers. Figure 20.4, on the other hand, shows a pattern in which the disease activity gradually surpasses treatments with repetition of remission and exacerbation and becomes intractable. In this pattern, the mean disease activity increases with each reactivation. For example, mild rash that used to appear several times a year comes to exist continuously and deteriorates to erythroderma. In Figure 20.4,

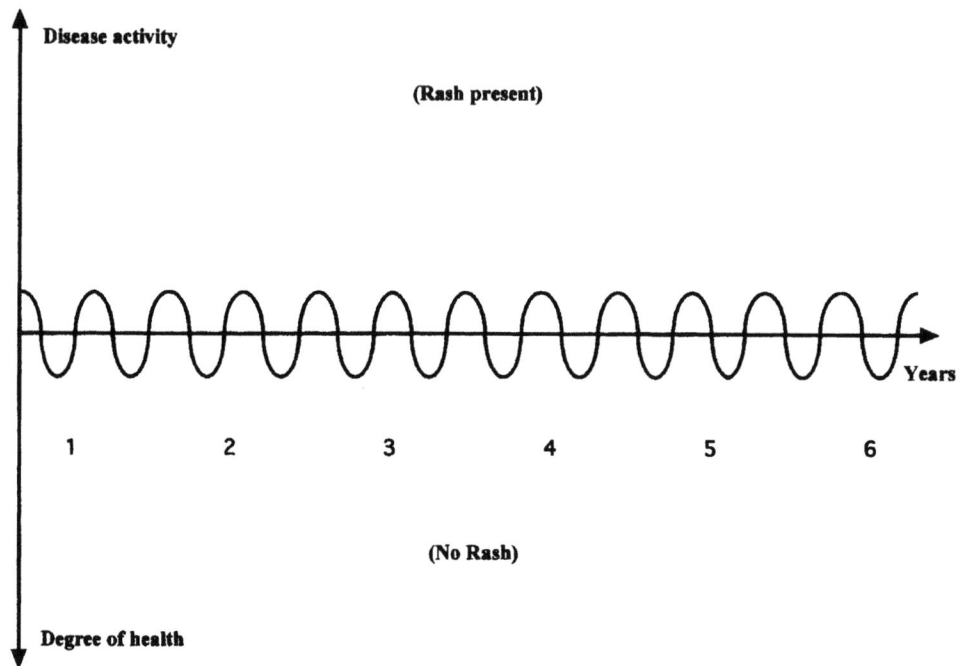

**Figure 20.3.** Conventional diagram of long-term course supplemented with the degree of health. Note the horizontal vector to the right.

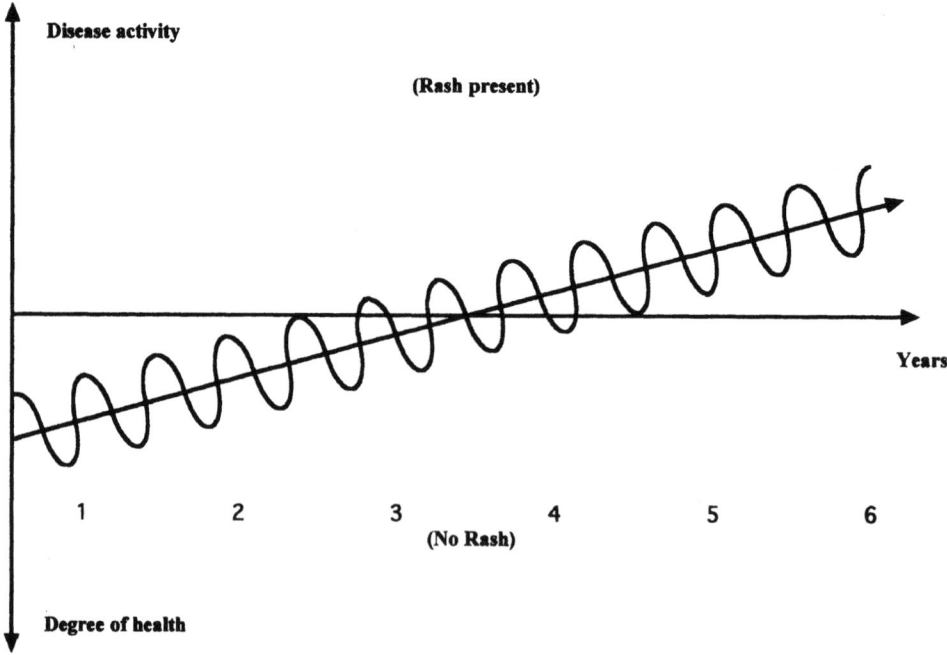

**Figure 20.4.** Diagram of a course progressively exacerbated with repetition of remission and exacerbation. An upward vector of exacerbation is supposed.

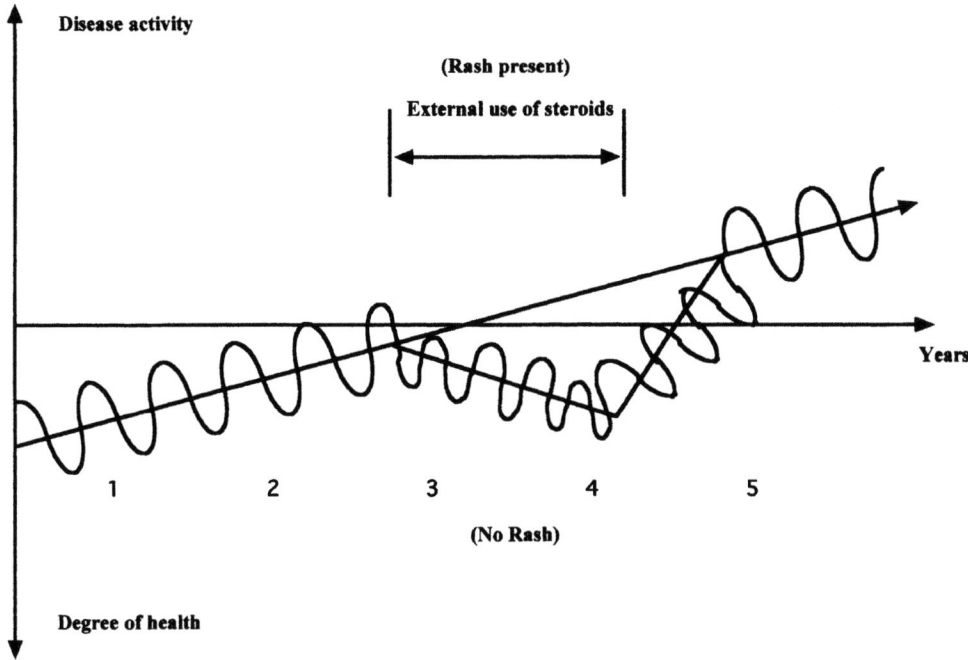

**Figure 20.5.** Problem of the use of external steroids in the presence of a vector of exacerbation.

an upward vector indicating progressive exacerbation appears unlike Figure 20.3 when the mean disease activities are connected. If external steroid therapy is carried out in the presence of such an exacerbating vector of disease activity, a period without rash may be maintained for months to years, but, if the therapy is discontinued, rash that corresponds to the disease activity at that point appears, giving an impression that steroid has considerably exacerbated the disease (Figure 20.5). Naturally, rebounds may also occur. However, as long as there is an exacerbating vector of disease activity, the true activity of the disease is bound to appear no matter how carefully steroids may be weakened. Moreover, if there are horizontal and upward vectors as seen above, a disease pattern having downward vectors as shown in Figure 20.6 may also be supposed. In such a pattern, the disease activity is gradually subsided despite recurrences, and the degree of health may eventually reach a level where no rash occurs even when the patient is exposed to some exacerbating factors.

Also, in a patient having such a downward vector of disease activity, if severe symptoms are controlled by an aggressive approach such as internal steroid therapy, symptoms that appear after discontinuation of the therapy are expected to be milder, corresponding to the decrease in the disease activity with the lapse of time (Figure 20.7).

## 3. Dietary Therapy: Returning to the Diet of the Days When the Patient Had No Refractory Atopic Dermatitis

On the basis of our clinical experience, we introduced the idea of representing the direction of the disease activity over a long period as a vector and assumed the presence of

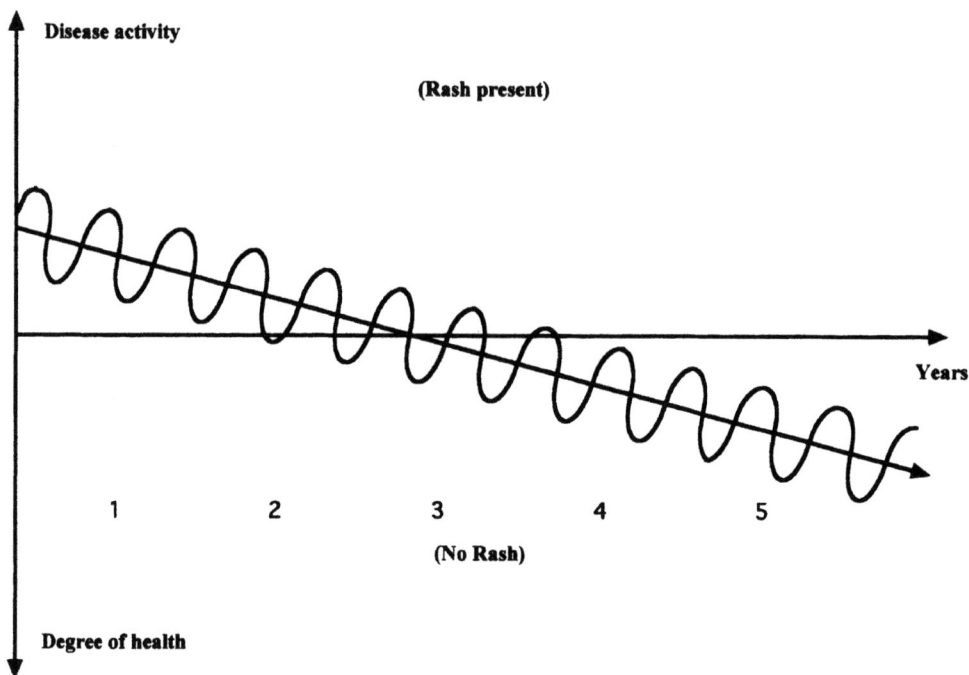

**Figure 20.6.** Diagram of a downward course with increasing degree of health.

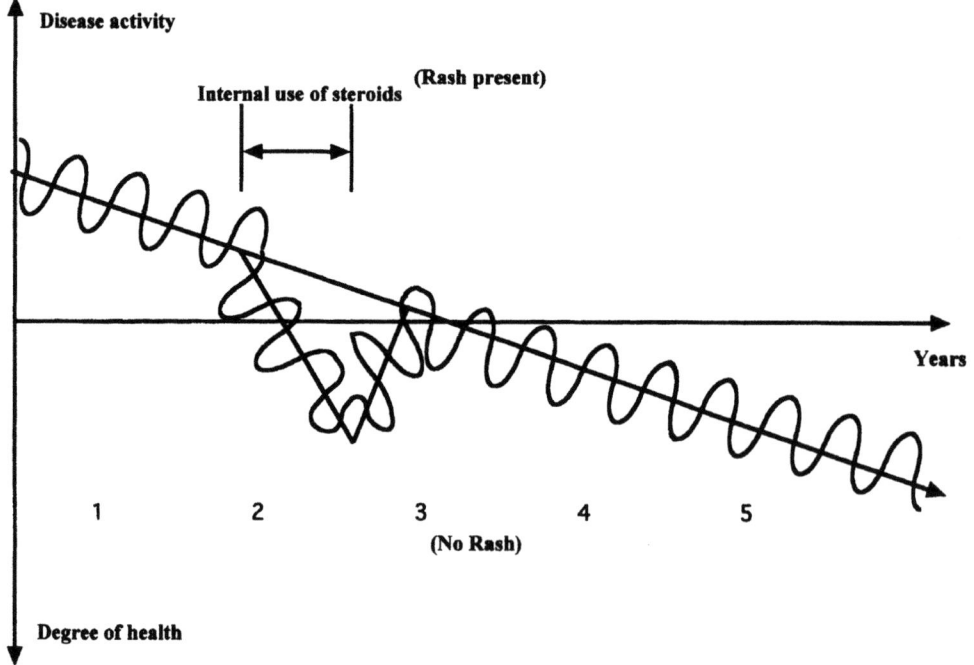

**Figure 20.7.** Validity of the internal use of steroids in the presence of a downward vector.

an exacerbating vector. The vector does not indicate individual exacerbating factors such as positive allergens, stressors, or infections, but it is considered to represent progressive changes in the vulnerability to adult atopic dermatitis acquired by the patients over a long period as part of their constitution, or serial changes in homeostasis of the body such as the hypersensitivity that makes the patient reactive to many allergens, inability to acquire tolerance, reduced defense against common bacteria and fungi, and reduced resistance to various stressors. What, then, may be basic exacerbating factors that progressively reduce the degree of health with time and advance the disease activity toward the direction of exacerbation? If refractory adult atopic dermatitis is assumed to have appeared about 20 years ago, it is necessary to review the drastic changes in clothing, diet, housing, and social life after World War II.

Environmental contamination and overeating or dietary changes are major problems that are clear to anyone, but no one has completely elucidated their effects. We have great interest in the dietary changes that have occurred since we started Kampo therapy for the treatment of atopic dermatitis 25 years ago. At that time, atopic dermatitis was not yet so troublesome and under the guidance of Yamamoto, a co-author and an instructor in Kampo therapy and dieting, we easily succeeded in treating many patients by advising them to reduce the intake of sweets and, to avoid cooling the body (disturbance of microcirculation). Yamamoto also contended that *oketsu* (a sign from the Kampo theory; i.e. disturbed circulation and unscavenged blood contents)[5-7] is a strong etiological factor for adult atopic dermatitis.[2,3]

However, with the appearance of refractory cases (from 20 years ago), we began to indicate additional restriction of the intake of lipids, which are a factor of *oketsu*, with other therapies. In the mean time, the concept of Yeast Connection was proposed from other fields of science, and an excessive intake of excessive sweets was shown to cause abnormal proliferation of intestinal *Candida* with associated damage of the intestinal mucosa and immunological abnormalities.[8,9] Studies of edible oils and fats have also shown that n-6 lipids such as linoleic acid, a polyunsaturated fatty acid the intake of which was highly recommended after World War II, are converted to markedly prophlogistic agents such as leucotriene (LT) $B_4$ through the arachidonic acid cascade and that lipids contained in fish and green and yellow vegetables consist mainly of n-3 lipids such as $\alpha$-linolenic acid, eicosapentaenoic acid, and docosahexaenoic acid, which have a suppressive effect on inflammation by inhibiting the production of prostaglandin $E_2$ and LT $B_4$, which are potent prophlogistic agents derived from arachidonic acid, through the eicosanoid synthesis channel.[10,11]

As for diet, there was a marked transition after World War II from a diet consisting mainly of grains, vegetables, and fish to one rich in protein, lipids, and carbohydrates, and as the intakes of meat, oil and fat, sugar, and dairy products increased, those of fish, grains, vegetables, and seaweed decreased. Twenty years have passed since the advent of the refractory adult type, and it takes 20 years for a baby to grow into adulthood. It seems certain that the above dietary changes have advanced considerably during these 40 years.

Based on these ideas, we suspected drastic changes in diet to be a central factor in the deteriorating vector of adult atopic dermatitis and decided to return to the diet in the late '50's to early '60's before the advent of the disease. That is, we increased fish, grains, vegetables, and seaweed and reduced meat, lipids, and sweets as much as possible as dietary modifications and prescribed them with treatment using herbal medicines. Unavoidable problems in planning dietary therapy include effects of environmental contaminants and additives contained in food and water and marked changes in the nutrient contents of farm or marine products that are apparently the same. In returning to the diet of the late '50's, these factors must also be returned to the state of those days which is difficult to realize now. All we can do is to use organic farm products, natural (not farmed) fish, and water from a good water purifier or good natural water. Since these organic or natural items are more expensive than commonly available items, they raise the cost of living, making it difficult to strongly recommend their use in this trial.

## 4. Results of Kampo Therapies (Including Dieting): Evaluating the Disappearance of Disease Phases

Demonstration of the effectiveness of Kampo medications is not easy. Some investigators strongly recommend the double-blind design, which has actually been employed. However, although there are Kampo prescriptions that act quickly, the effects of many Kampo medicines are seen by improvement of the constitution, often making it impossible to test their effects by the same short-term clinical trials as those used for Western drugs such as antibiotics and anti-cancer agents.

Also, as Kampo medicines' primary action is to gently restore homeostasis, they must often be used in combination with Western drugs for a period, adding to the difficulty of evaluating the effectiveness of Kampo medicines per se. In this study, also, the effectiveness of Kampo therapy was evaluated by the simultaneous use of ordinary Western medical treatments.

To evaluate the effects of long-term administration of Kampo medicines, the same doctor must continuously follow-up the same patients and record their courses over a long period. Unfortunately, however, many patients shop around for doctors, visiting one clinic after another. In addition, there is no attending physician system at many university hospitals, with doctors frequently transferred to other hospitals. All of these factors make the present medical system unsuitable for the long-term follow-up of patients and contributes to the difficulty of evaluating the effectiveness of Kampo medicines.

We, therefore, carefully designed the tests, paying attention to the following points, in order to understand the efficacy of Kampo medicines for atopic dermatitis. Presently, the goal of treatment for recurrent and refractory adult atopic dermatitis is the control of its course, with recurrence being regarded as unavoidable. While mild rash may be remitted over a considerable period by external steroid therapy, severe rash recurs soon. In patients with marked erythema, rash disappears temporarily with powerful external steroid therapy plus an internal steroid or cyclosporine regimen, but early recurrence is common in such patients. Therefore, adult atopic dermatitis is considered to usually recur despite temporary remissions with Western treatments. In consideration of the above situation, we assumed that the Kampo treatment was effective if disease phases, or waves of remission and exacerbation, disappeared with stabilization of the disease activity on the Kampo treatment performed in combination with Western treatments and designed the following testing procedure.

### 4.1. Patients and Modified Method for Efficacy Evaluation

1. Patients with a long history of adult atopic dermatitis who had severe or mild but persistent rash were selected as the subjects.
2. All patients who visited the Ishii Clinic during the 3 years from April, 1993 to March, 1996 and fulfilled the first criterion, were photographed so that the course of their

illnesses could also be traced with photographs. The therapeutic effect was evaluated 2 months after the end of this period.
3. In consideration of the time needed for Kampo medicines to show effects and to eliminate mild cases and cases of poor compliance, 113 (41 with severe, 52 with moderate, and 20 with mild dermatitis) who regularly visited the clinic and were compliant to Kampo medical therapy for 2 months or longer were selected from the patients described in 2. Patients who stopped visiting the clinic were followed up by telephone, and efforts were made to clarify the courses of all patients until final evaluation. As a result, 59 patients were still visiting the clinic, 36 were contacted by telephone, and 18 were lost due to moving or other reasons.
4. External steroid therapy was started with a preparation that was considered mildest for the condition. The preparation was changed to a milder one with alleviation of the condition, and the therapy was discontinued with disappearance of rash. Oral steroid was avoided as much as possible, and the therapy was regarded as ineffective if a patient took even a single tablet of internal steroid within the 6 months prior to judging that the disease phases had disappeared.
5. The effectiveness of the therapy was evaluated according to whether the rash had been alleviated and whether disease phases (waves of remission and exacerbation) had been absent for 6 months or longer. Thus, we attempted to evaluate the effectiveness of Kampo therapy on the basis of whether the results of the therapy were sustainable rather than temporary.

## 4.2. Method for Evaluation of Effectiveness

Symptoms were scored using a 5-point scale: 4 for severe, 3 for moderate, 2 for mild, 1 for slight, and 0 for no rash. The treatment was judged to be "markedly effective" when an improvement of 3 points or more continued for 6 months or longer, "effective" when an improvement of 2 points continued for 6 months or longer, "slightly effective" when an improvement of 1 point continued for 6 months or longer, "ineffective" when an improvement was observed with disease phases within 6 months or when no improvement was observed, and "exacerbated" when the score deteriorated. However, an improvement of 2 points from mild rash to no rash sustained for 6 months or longer was regarded as "markedly effective" by attaching importance to the long-term disappearance of refractory rash.

## 5. Contents of Kampo Therapy and Kampo Preparations Used

The principle of the diagnosis and treatment of Kampo medicine that we, particularly Yamamoto, have long maintained is that *sho* diagnosis based on the Kampo theory[1] must be determined on the basis of all information obtained by the use of all modern diagnostic techniques rather than by relying on *bo*(inspection), *mon*(talk), *bun*(listen), *setsu*(palpation) which are the 4 conventional diagnostic processes of Kampo medicine.[1] According to this view, *sho* changes constantly, and the prescription that matches *sho* must be prepared by closely evaluating the pharmacological action of each herbal drug and combining drugs for changing aspects of the condition. To live up to this idea, prescriptions must be adjusted delicately, but we mix 2 or 3 available extract preparations for convenience at our university clinic. In Japan, the Ministry of Welfare approves almost 200 kinds of heat packed average quality prescriptions for Kampo therapy. Outlines of the prescriptions used are described below.

True treatment (*honchi*; treatment for generalized, essential, and constitutional improvements) was often performed to improve 1. *ki*-deficiency, 2. *ketsu*-deficiency, and 3. *oketsu* (*KI* means elemental energy of body and also functions of every body system, *KETSU* means materials constituting the body, *SUI* means body fluid and *OKETSU* means blood stasis).[1,12]

1-3 type drugs were used concomitantly, i.e. 1) *Hochu-ekki-to*, was used for *ki* deficiency, 2) *Unsei-in* for *KETSU* deficiency and 3) *Keishi-bukuryo-gan*. for anti-*oketsu*. For *hyochi* (treatment for manifest symptoms and rash), 1-3 drugs were used concomitantly for inflammatory erythema such as 4) *Oren-gedoku-to* and for *SUI* retention (cutaneous edema) such as 5) *Gorei-san*, for inflammatory edema, such as 6) Eppi-ka-jutsu-to. *Unsei-in* is used for drying, cracking, and scaling, and *Keishibukuryougan* for thickening of the skin and lichenification.

A recent paper by Takahashi[13] describes the details of the Kampo prescriptions mentioned above and principles of their use since they are omitted here due to space limitations.

The constituents of each drug (extract from a prescribed mixture of herbs for one day) and their working nature is summarized as follows.[1]

## 1. *HOCHU-EKKI-TO* [14-15]

Components; ASTRAGALI RADIX 4.0 g, ATRACTYLODIS LANCEAE RHIZOMA 4.0 g, GINSENG RADIX 4.0 g, ANGELICAE RADIX 3.0 g, BUPLEURI RADIX 2.0 g, ZIZYPHI FRUCTUS 2.0 g, AURANTII NOBILIS PERICARPIUM 2.0 g, GLYCYRRHIZAE RADIX 1.5 g, CIMICIFUGAE RHIZOMA 1.0 g, ZINGIBERIS RHIZOMA 0.5 g

Target group; Weak patients, KI-deficiency due to chronic diseases, tuberculosis, anemia, surgery, loss of appetite, cough, mild fever, night sweats, palpitation, fear, weak feeble voice, slurred speech, disturbance of vision.

Indications: Dysfunction of the digestive system, weakness of muscles, weak physical condition, fatigue due to summer heat, recovery from an illness, tuberculosis, loss of appetite, gastric ptosis, common cold, hemorrhoids, prolapsus ani, ptosis of the uterus, impotence, hemiplegia, hyperhidorosis.

## 2. *UNSEI-IN* [16]

Components; REHMANNIAE RADIX 3.0 g, PAEONIAE RADIX 3.0 g, CNIDII RHIZOMA 3.0 g, ANGELICAE RADIX 3.0 g, SCUTELLARIAE RADIX 1.5 g, PHELLODENDRI CORTEX 1.5 g, COPTIDIS RHIZOMA 1.5 g, GARDENIAE FRUCTUS 1.5 g

Target group: Relatively strong patients, trophic disturbance of the skin and desquamation, dry skin, dotty accumulation of pigment, feeling of uprising heat, hot hands and feet, hypersensitivity, tendency for hemorrhage, reduced gland activity, red skin, eczema with severe itching, hypertonic rect.abd.muscle, resistance tender on pressure of the subcostal region.

Indication; Eczema, pruritus, dermatitis, stomatitis, menopause-syndrome, dysmenorrhea, bleeding from genitals and hemorrhoids, neurologic symptoms, psoriasis, urticaria, Behcet's disease.

3. *KEISHI-BUKURYO-GAN* [17]

Components; CINNAMOMI CORTEX 3.0 g, PAEONIAE RADIX 3.0 g, PERSICAE SEMEN 3.0 g, HOELEN 3.0 g, MOUTAN RADIX 3.0 g

Target group: Considerably strong patients, feeling of uprising heat, red face, resistance tender on pressure of the lower abdomen, OKETSU-syndrome, headache and painful tension of shoulder and cervical muscles, cold feet, amenorrhea, oligomenorrhea, dysmenorrhea.

Indications; Marked redness of the face, increased abdominal tension and resistance tender on pressure of the lower abdomen, inflammation of pelvic organs, endometoritis, menopause-syndrome, amenorrhea, oligomenorrhea, dysmenorrhea, colpitis, vertigo, painful tension of shoulder and cervical muscles, feeling of uprising heat, headache, sensitivity to cold, peritonitis, orchitis, hemorrhoids, hematomas after trauma.

4. *OREN-GEDOKU-TO* [18]

Components; SCUTELLARIAE RADIX 3.0 g, COPTIDIS RHIZOMA 2.0 g, GARDINIAE FRUCTUS 2.0 g, PHELLODENDRI CORTEX 1.5 g

Target group: Considerably strong patients, feeling of uprising heat, red face, restlessness, insomnia, irritability, nausea and fullness of the epigastric region epistaxis, pulmonary hemorrhage, bleeding hemorrhoids, anemia, tendency for hemorrhage, eruptive eczema, pruritus.

Indications; Choleric type, feeling of uprising heat, tendency for hemorrhage, pulmonary hemorrhage, hemorrhage of the brain, epitaxis, hypertension, tachycardia, efflorescences of the skin and pruritus, insomnia, neurosis, vertigo, feeling of cold gastritis, dysmenorrhea.

5. *GOREI-SAN* [19]

Components: ALISMATIS RHIZOMA 4.0 g, POLYPORUS 3.0 g, HOELEN 3.0 g CINNAMOMI CORTEX 1.5 g, ATRACTYLODIS LANCEAE RHIZOMA 3.0 g

Target group; Particular symptoms are a marked feeling of thirst and oliguria, headache, vertigo, nausea, emesis, sound of fluctuating liquid in the epigastric region.

Indications; Thirst, oliguria, nephrosis, symptoms due to excessive intake of food and alcohol, acute gastroenteritis, diarrhea, nausea, emesis, vertigo, headache, uremia, diabetes mellitus, condition due to the influence of extreme exogenous heat.

6. *EPPI-KA-JUTSU-TO*

Components; GYPSUM FIBROSUM 8.0 g, EPHEDRAE HERBA 6.0 g, ATRACTYLODIS LANCEAE RADIX 4.0 g, GLYCYRRHIZAE RADIX 3.0 g, ZIZYPHI FRUCTUS 2.0 g, ZINGIBERIS RHIZOMA 1.0 g

Target group; Considerably strong patients, picnic-athletic habitus, not sensitive to cold, tendency for developing edema and sweating, thirst, oliguria, swollen joints, arthralgia.

Indications; Edema, fits of sweating, oliguria, nephritis, nephrosis, rheumatoid polyarthritis, eczema

## 5.1. Results of Administration of Kampo Medicines (Figure 20.8)

The prescriptions mentioned above were used in various combinations in the 113 patients over a period of 2 months to 3 years.

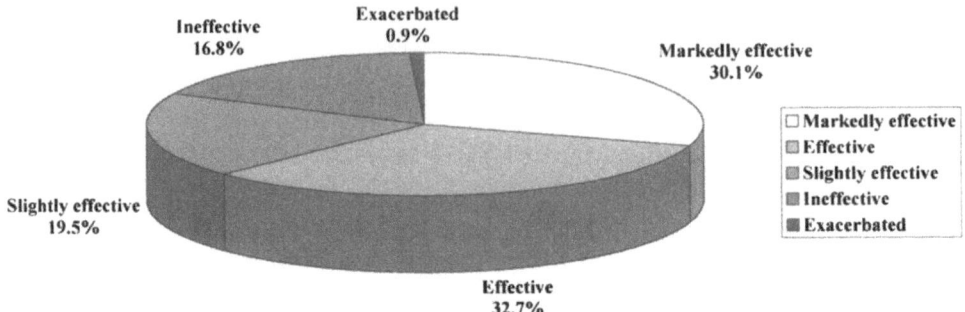

**Figure 20.8.** Results in all adult cases. Evaluation of treatment in 113 patients with refractory adult atopic dermatitis according to the disappearance of disease phases.

Markedly effective cases: An improvement in rash of 3 or more points continued for 6 months or longer or an improvement of 2 points with disappearance of rash continued for 6 months or longer in 34 patients (10 severe, 12 moderate, and 12 mild cases), or 30.1% of all patients. Of these patients, rash was absent for 6 months or longer (without external steroid) in 28, or 25% of all patients, and for 1 year or longer in 15. The Western treatments used concomitantly in these patients were mainly oral antihistaminic preparations and antiallergic preparations, and oral Celestamine was used in an early stage of the disease in 4 patients. Concerning external medications, 5 patients were treated without steroids from the beginning. In the other patients, mild-class or milder steroids were used at decreasing doses and were withdrawn with disappearance of rash. For the prurigo type, very-strong-class or milder steroids were used at decreasing doses and were eventually withdrawn in some patients.

Effective cases: An improvement of 2 or more points continued for 6 months or longer in 37 patients (9 severe, 26 moderate, and 2 mild cases), or 32.7% of all patients. The 2 mild cases were enrolled in the clinical trial late and had been without rash for 5 months at evaluation of the effect, but they were regarded as "effective" cases, because the rash was absent for less than 6 months.

Slightly effective cases: An improvement of 1 point continued for 6 months or longer in 22 patients (8 severe, 9 moderate, and 5 mild cases), or 19.5% of all patients.

Ineffective cases: No improvement was observed, or improvement was observed but disease phases recurred within 6 months in 19 (14 severe, 4 moderate, and 1 mild cases), or 16.8% of all patients.

Exacerbated cases: Exacerbation was observed in 1 case, or 0.9% of all patients.

## 6. Summary of the Results of Kampo Therapy (Including Dieting) and Discussion

The above results were not obtained after treating all 113 patients for 3 years. The patients were treated for 2 months to 3 years after their first visit to Ishii Clinic during the 3 years until evaluation so that evaluation was made after a shorter treatment period if the entry was later.

Under these conditions, the treatment was markedly effective (an improvement of 3 or more points sustained for 6 months or longer) in 34 patients, of whom 28 had no rash without external steroid, and 15 had no rash for 1 year or longer. These results suggest that the combination of Kampo and Western treatments is markedly more effective than Western treatments alone.

Also, the therapy was markedly effective or effective, meaning disappearance of disease phases for 6 months or longer rather than temporary improvements, in 63.8% of all patients. Therefore, the Kampo therapy is considered to have been extremely effective for the treatment of refractory adult atopic dermatitis as a whole.

A remaining problem is the concomitant dietary therapy. Dietary guidance has been a routine practice for physicians who perform Kampo therapies. However, we realized that the treatment for adult atopic dermatitis would be greatly facilitated by returning to the diet of the time before the advent of the disease. We talked to patients who had shown poor responses to the therapy and had them promise to be more compliant to our dietary instructions. We often observed better responses to Kampo medications in these patients after they made the promise. Compliance to dietary therapy is difficult to ensure, because it is left to the willingness of each patient, who may eat large quantities of lipids or sugar contrary to our advice. We attempt to check the patients' diet by having them keep an eating diary. Dieting is restriction of food and water that the patient eats or drinks. No one would disagree that food and water are the basis of the body and are essential for improving the degree of health. It must still be clarified whether the negative vector of disease activity is due to the marked changes in the diet including contamination of food and water or whether other factors are involved. The development of the body by good dietary habits is a slow process by nature. However, if a poor diet exerts an adverse effect on the vector of disease or health, its correction will change the direction of the vector for the better. Our results indicate this improvement in constitution by dietary modification.

We speculate that Kampo medications not only alleviate the existing symptoms but also reinforce, promote, and ensure improvement in the state of health, or recovery of homeostasis, initiated by an appropriate diet. If so, both dietary therapy and Kampo medications are considered to be extremely important as essential and complementary elements of the treatment for diseases.

## 7. References

1. K. Terasawa, Kampo;Japanese-Oriental medicine,Insights from clinical cases (Standard Mcintire, Tokyo,1993).
2. K. Takahashi, in: *Group of eczema and dermatitis (II) Kampo orthodox therapy for atopic dermatitis. Modern Kampo therapy—General discussion, cases, list of references*, editorial supervision by A. Kumagai (Toyo Gakujutsu Shuppansha, Ichikawa, 1985), pp. 526–527.
3. I. Yamamoto, Group of eczema and dermatitis II, Atopic dermatitis, THE KAMPO, 4(4), 1986, pp. 162–177.
4. H. Kobayashi, M. Ishii, T. Tanii, et al., Kampo therapies for atopic dermatitis: The effectiveness of *hochu-ekki-to*, *Nishinihon Hifu* **51**, 1003–1013 (1989).
5. K. Terasawa, K. Torizuka, H. Tosa, et al., Rheological studies on "oketsu" syndrome I. The blood viscosity and diagnostic criteria, *J.Med.Pharm.Soc.WAKAN-YAKU* **3**, 98–104 (1986).
6. K. Terasawa, T. Itou, Y. Morimoto, et al., The characteristics of the microcirculation of bulbar conjunctiva in "oketsu" syndrome, *J.Med.Pharm.Soc.WAKAN-YAKU* **5**, 200–205 (1988).

7. K. Terasawa, H. Shinoda, A. Imadaya, et al., The presentation of diagnostic criteria for "Yu-xie" (stagnated blood) conformation, *Int.J.Oriental Medicine* **14**, 194–213 (1989).
8. W. G. Crook, The yeast connection, in: *Vintage Books* (Division of Random House, New York, 1984).
9. G. F. Krocker, Food allergy and intolerance, (*Bailliere Tindall*, 1987).
10. B. R. Lokesh, J. M. Black, J. B. German, et al., Docosahexaenoic acid and other dietary polyunsaturated fatty acids suppress leukotriene synthesis by mouse peritoneal macrophages, *Lipids* **23**, 968–972 (1988).
11. S. Torii, Allergy and the n-6/n-3 ratio: Its clinical significance, *Lipid Nutrition* **4**, 56–63 (1995).
12. K. Kikuchi, N. Shibahara, Y. Shimada, et al., Correlation between "oketsu" syndrome and autonomic nervous activity, *J.Trad.Med.* **15**, 127–134 (1998).
13. K. Takahashi, General discussion of Kanpo treatments for skin disorders based on the theories of Kampo medicine, *Skin Research* **39**, 1–23 (1997).
14. M. Tohda, A. F. Mohamed, S. Nakamura, et al., Effects of Hochu-ekki-to a Kampo medicine, on serotonin 2C subtype receptor-evoked current response and the receptor mRNA expression, *J.Trad.Med.* **17**, 34–40(2000).
15. Y. Mseda, K. Tanaka, S. Sawamura, et al., Therapeutic effect of a traditional Chinese medicine, Bu-zhong-yi-tang (Japanese name: Hochu-ekki-to)through controlling Th1/Th2 balance, *J.Trad.Med.* **18**, 20–26 (2001).
16. E. Tahara, T. Satoh, C. Watanabe, et al., Effect of Kampo medicines on IgE-mediated biphasic cutaneous reaction in mice, *J.Trad.Med.* **15**, 100–108(1998).
17. T. Nagai, Recovering effect of Keishi-bukuryou-gan (Gui-Zhi-Fu-Ling-Wan) on erythrocyte membrane sialidase abnormality in glucocorticoid-induced "oketsu" model mouse, *J.Trad.Med.* **17**, 221–228 (2000).
18. T. Hong, G.-B. Jin, J.-C. Cyong, Effect of components of Oren-gedoku-to (Huang-Lian-Jie-Du-Tang) on murine colitis induced by dextran sulfate sodium, *J.Trad.Med.* **17**, 173–179 (2000).
19. M. Orita, K. Maeda, H. Higashino, et al., The control effectd of body water metabolism by Gorei-san analogue (Wu-Ling-San-analpgue) or Sairei-to (Chai-Ling-Tang) on stroke-prone spontaneously hypertensive rats (SHRSP), *J.Trad.Med.* **17**, 157–164(2000).

# 21

# Management of Nutritional and Health Needs of Malnourished and Vegetarian People in India

**H.D. Kumar**

## 1. Introduction

The maintenance of good health rests directly or indirectly on a strong nutritional foundation of plant foods. People everywhere require adequate amounts of essential macronutrients (proteins, carbohydrates, fats) and micronutrients such as vitamins and minerals. Both deficiency and excess of nutrients can be detrimental to health. Whereas many health-related problems of affluent or wealthy people arise from overconsumption, deficiencies of proteins and other nutrients afflict millions of poor people in developing countries.

Protein energy malnutrition, anaemia, vitamin A deficiency and iodine deficiency disorders are the four major forms of malnutrition in the developing world. Protein energy malnutrition manifests itself in growth retardation, wasting and stunting. The protein picture is dismal. The majority of Indian people are vegetarians and do not eat meat, fish or eggs. Their proteins come from pulses, lentils, legumes, and dairy products. The production of pulses has not kept pace with the rising population, further widening the protein-calorie imbalance. In poor people, it contributes substantially to weak health and predisposes them to infectious diseases and nutritional deficiency disorders. Malnutrition is most prevalent in the vulnerable young child when its synergistic relationship with disease can result in mortality.

Other forms of malnutrition are caused by deficiency of micronutrients like iron, vitamin A and iodine (Singh, 1992). The main nutritional cause of anaemia is iron and iron/folate deficiency. Nutritional blindness, which affects many million children in India annually results mainly from the deficiency of vitamin A.

Iodine deficiency disorders are associated with impairment of mental and intellectual functions in children and adults, and in severe cases with deafness and mutism, neuromuscular disorders, and perinatal and infant mortality.

The emergence of new viruses and drug-resistant bacteria has prompted scientists to employ the ability of certain foods and/or food additives to enhance the immune system and address problems associated with early aging. The microalga (cyanobacterium) *Spir-*

---

**H.D. Kumar** • Formerly Co-ordinator, Biotechnology Program, Banaras Hindu University, Varanasi, India
*Complementary and Alternative Approaches to Biomedicine*, edited by Edwin L. Cooper and Nobuo Yamaguchi. Kluwer Academic/Plenum Publishers, 2004.

*ulina* has attracted interest as one of the disease preventing, anti-aging and nutritious foods (Skulacher, 1994; Brown, 1998).

While *Spirulina* improves nutrition and general health, it does not prevent or cure the variety of infectious diseases (mostly water-borne) that afflict many people in developing countries. For guarding against these, neem (*Azadirachta indica*) and turmeric (*Curcuma longa*) have proved highly effective. A combination of *Spirulina*, neem and turmeric as the core package to which certain other specific plant products are added, can ward off many infectious and non-infectious diseases including cancer, gastrointestinal disorders, diabetes, skin troubles, dental problems and cardiovascular disorders. This chapter suggests some promising combinations for managing the health needs of malnourished people, especially vegetarians.

In the Indian sub-continent, medicinal plants have been traditionally used most widely in remote and inaccessible areas where qualified physicians trained in modern medicine are reluctant to practise and modern health facilities are lacking. Many tribal people rely on folk remedies for curing common ailments. Even in more accessible villages and urban areas, medicinal plants have become crucial to local livelihoods.

## 2. *Spirulina* for Nutritional Enhancement

Although a large number of edible algae are known, *Spirulina* has emerged as the most potent and nutritious food for humans. Indeed, spirulina may well be the most concentrated and nutritious whole food known. **Spirulina** is a filamentous, alkalophilic cyanobacterium that grows in many soda-rich lakes all over the world (Ciferri, 1983). It is widely distributed in India and occurs in those lakes, ponds and water-logged soils that are rich in soluble salts of sodium. When the pH exceeds 9.5, the alga virtually forms pure, unialgal blooms in the absence of competing organisms (Kumar, 1988).

### 2.1. Chemical Composition

Dried *Spirulina* contains (approximately) 60–70% protein, 6 to 8% lipid, 4% nucleic acids, 0.8% chlorophyll, 0.23% b-carotene, 0.12–0.15% xanthophylls and 12–15% phycobilins (Table 21.1). The amino acid spectrum of *Spirulina* protein generally resembles that of other microorganisms but is rather low in cysteine, methionine, and lysine (Becker, 1986). In terms of the protein standards specified by Food and Agriculture Organization (FAO) for egg albumin, milk and meat protein, *Spirulina* protein is somewhat inferior; but it is superior to other vegetable sources e.g. wheat, rice and legumes. The total protein content of *Spirulina* is much higher than vegetable protein, milk, meat and egg (Belitz and Grosch, 1999).

The percentage of usable protein in a food is determined by measuring the protein content, its digestibility, and its biological value. Table 21.1 compares some common foods for their usable protein. The only food with more usable protein than spirulina is eggs which most vegetarians do not eat. Even for non-vegetarians, not more than two eggs per week are recommended in view of their high cholesterol. This makes *Spirulina* spp. the first choice protein source for all people. Their useful proteins, vitamins, phycocyanin,

**Table 21.1.** Chemical constituents of *Spirulina* and their beneficial effects (after Behera and Kaur, 2003)

| Chemical | Percentage/ dry weight | Benefits | References |
|---|---|---|---|
| Protein | 65–68 | Relief of protein energy malnutrition; Treatment of Kwashiorkor and Marasmus disease | Narsimha et al., 1982; Ripley, 1996 |
| Polysaccharides | 10–18 | Enhancement of endonuclease activity and repair of DNA | Becker and Venkatraman, 1984 |
| Lipids | 6–7 | Lowering of serum cholesterol, triglyceride and LDL Prevention of AIDS | Nayaka, 1988; Begin and Das, 1986 |
| •-Linolenic acid | 1–1.5 | Cholesterol control | Mahajan and Kamat, 1995 |
|  |  | Pain relief of tropical pancreatitis and rheumatoid arthritis | Chaturvedi and Habib, 1999 |
| Ash | 4–5 | Provision of micronutrients | Narsimha et al., 1982 |
| Crude fibre | 3.0–8.0 | Obesity control | Becker, 1986 |
| •-carotene | 0.23 | Source of vitamin A, Prevention of blindness | Annapurna, 1991 |
| Vitamin B-12 | 0.001 | Promotes body's defense; improvement of nerves and tissues | www.spirulina.com |

carotenes, and long chain polyunsaturated fatty acids have contributed greatly to food, feed (Becker, 1994), medicine (Vonshak, 1990) and value-added biochemicals. *Spirulina* has been patented in Russia as a medical food for improving the immunity of the "Children of Chernobyl" suffering from radiation illness (Behera and Kaur, 2003).

Significant variations have been noticed in the fatty acid content of different species. In *Spirulina platensis* and *S. maxima*, free fatty acids account for 70 to 80% of the total lipids, the remaining being chiefly mono- and digalactosyl glycerides and phosphatidyl glycerol (Hudson and Karis, 1974). *S. platensis* synthesizes high concentration of •-linolenic acid (Nichols and Wood, 1968)—this is nutritionally and therapeutically beneficial for the treatment of hypertension, various atopic disorders, and several other conditions (see Behera and Kaur, 2003).

The high value biochemical constituents of *Spirulina* have found applications in health foods, therapeutics and specialized feeds for which *Spirulina* is cultured on a large scale (both in open and in closed bioreactors or tubular bioreactors; see Torzillo, 1997; Fox, 1998) to supply the growing market for products like beta carotene, phycocyanin and high value lipids and fatty acids.

Cereals like rice and wheat are rather deficient in lysine and threonine. The quality of these cereals can be upgraded by supplementation of *Spirulina* powder. Currently gamma linolenic acid (GLA) is extracted commercially from higher plant oils, especially from *Oenothera* spp. (evening primrose). Also the fungus (mold) *Mucor rouxii* produces about fifteen times more GLA than *Spirulina*. The chief merit of *Spirulina* is its suitability as a direct dietary supplement rather than as a source for extraction and purification of GLA.

## 2.2. Beta Carotene and Antioxidants

Beta carotene is an extremely important antioxidant. People who consume carotene-rich foods lower their risks of developing cancer. It is only beta carotene from natural foods that lowers the incidence of cancer; the synthetic form is not very effective.

Over 500 different carotenes (or carotenoids) have been detected in different plants but very little is known about them; the research so far has focussed on beta carotene. Beta carotene is used in food colouring, as animal feed, and in the health food industry. It has been used for long as a food colouring agent of natural origin. It has also been widely used as a source of vitamin A in animal feed. People who maintain above-average b-carotene levels usually have a lower incidence of several types of cancer (Peto et al., 1981). Inhibition of oral carcinogenesis by beta carotene has been reported by several workers (see Suda et al., 1986; Mathew et al., 1995; Sankaranarayanan et al., 1997). As it seems advisable to consume an extended range of carotenes, which are present in different plants, it is always prudent to eat a diversity of fruits and vegetables of different colours. This is because quite conceivably the other carotenes are just as valuable for our health. In our body, carotenoids are used and stored in the adrenal glands, the reproductive system, the pancreas and spleen, the skin, and the retina. Carotenoid depletion in these parts can disturb health despite adequate intake of beta carotene.

In several states of northern India, many people habitually chew betel ("paan") which contains arecanut, tobacco, cardamom, fennel, lime (calcium hydroxide) and other ingredients. While many of these ingredients are harmless, tobacco can induce oral cancer, and beta carotene from *Spirulina* or other sources (such as carrots) prevents such cancer (Garewal, 1995; Liede et al., 1998). A combination of turmeric (*Curcuma longa*), carrots, other seasonal vegetables and fruits, and *Spirulina* may be a treatment of choice for early stages of oral and other cancers.

Significant amounts of two important products, which are not 'health foods', have been produced from *Spirulina*. One is 'linablue', a phycobiliprotein concentrate sold in Japan as a colouring agent for ice creams and yoghurts, etc. The second is beta carotene, (pro-vitamin A and food colouring agent) that is effective as an anti-cancer agent.

Some important natural sources of beta carotene are spinach and other dark green leafy vegetables, drumstick fruit (*Moringa oleifera*), broccoli, carrots, strawberry, various other similar berries, papaya and other yellow and orange fruits and vegetables. The richest sources are *Spirulina* and drumstick. Vitamin A produced through beta-carotene helps prevent night blindness. Prolonged deficiency of this vitamin (or its precursor beta carotene) can result in xerophthalmia or even total blindness. This vitamin is needed for general body maintenance and repair. It prevents against cancer (Shukla, 1982).

Although oxygen is essential for metabolism, it is very reactive. Sometimes it combines with the complex metabolic macromolecules to make reactive, dangerous, intermediate compounds called free radicals. Free radical induced damage manifests itself outwardly in the form of dry skin, loss of muscle tone, and even skin cancer; internally, free radicals impair immune function, damage tissues and cellular DNA and generally weaken or destroy cells. An antioxidant reacts with a free radical and renders it harmless—in the same way that a sponge clears a spilled liquid. Some natural foods and herbs are rich in antioxidants, which is why diet is so important in a cancer-preventive lifestyle (see Garewal, 1995). Besides avoiding smoking, modifying our diet is the single most important factor in cancer avoidance. Beta carotene and vitamins C and E are powerful antioxidants.

## 2.3. Cultivation

Some factors that influence the growth of *Spirulina* include light, temperature, nutrients (including carbon availability), salinity, and pH. These parameters determine the operational protocol for pond management in order to establish a continuous culture for sustained production and to avoid contamination by grazers, predators or other algae (Vonshak and Richmond, 1988).

Microalgae can be cultivated either extensively or intensively either in open systems or in closed, non axenic or axenic, systems. The choice of culture system depends largely on the species and the product. Low grade *Spirulina* is produced in an extensive mode by using large, outdoor unlined ponds (about a meter deep) with little or only occasional stirring to induce turbulent flow (see Vonshak, 1990; Venkataraman and Becker, 1985).

Most commercial reactors used in large scale production of algal biomass are based on shallow raceways in which algal cultures are mixed in a turbulent flow by means of a paddle or wheel. Another approach involves the use of naturally occurring lagoons or lakes. The size of commercial ponds varies from 0.1 to 0.5 ha. in the raceway systems and up to 10 ha. In the lagoon system (Vonshak, 1990), *Spirulina* is usually cultivated outdoors in open raceway ponds. The growth medium used for its cultivation is highly alkaline (0.2M $NaH_2CO_3$) and pH is maintained at 9.5–10.5. These conditions permit the maintenance of a monoalgal culture under outdoor conditions.

The major production costs of high grade *Spirulina* are incurred in the energy-intensive harvesting and processing steps. Efficient harvesting requires an intensive step of water removal. In general, water removal may be achieved by filtration, centrifugation or flocculation and sedimentation. The best approach is centrifugation which, unfortunately, is very costly. But this difficulty is not serious for the filamentous *Spirulina* which can be harvested by filtration through polyester or muslin cloth.

The final processing step is determined by the kind and intended use of the product and its marketing. When intended for the health food market, the harvested biomass is spray dried. If specific chemicals or pigments are to be extracted, the biomass is usually used in wet form for the extraction process.

A great deal of work has been done at the Central Food Technology Research Institute Mysore on the cultivation of *Spirulina* and its use as human food (see Venkataraman and Becker, 1985). The A.M.M. Murugappa Chettiar Research Centre Madras has popularised the use and acceptability of *Spirulina* among several villagers in Tamil Nadu State (see Seshadri and Seshagiri, 1985). The National Botanical Research Institute in Lucknow has grown *Spirulina* biomass on city sewage and used it for feeding pigs and other animals. The spent medium is discharged into fish ponds where it exerts beneficial effects on the growth and yield of fish.

The Murugappa Chettiar Research Centre in Madras and the Central Food Technological Research Institute in Mysore, evaluated the usefulness and nutritional value of food items containing *Spirulina* (Seshadri and Seshagiri,1985; Venkataraman, 1983). It was shown that *Spirulina* is highly effective in relieving malnutrition in children and is even better than soybeans in this respect. Further, the alga is completely safe and non-toxic. In fact, for vegetarians, it appears to be the ideal solution to the chronic problem of protein-deficiency. A judicious mixture of wheat, soybeans, groundnuts, *Spirulina* and a little yeast can be the panacea for many of the ills that afflict the Indian population. In this mixture, *Spirulina* should not constitute more than 10% of the weight, and yeast not more than 5%.

The future prospects of *Spirulina* use rest largely on two factors: (a) The ability to reduce costs of production and thus make its biomass a commodity that can be traded in large quantities, not just limited to the health food market; and (b) The development of suitable reactors. In closed systems, cultures can be better protected from contaminants and thus the maintenance of monoalgal cultures is easier. Water loss and the ensuing increase in salinization of the medium are also inhibited. Optimal temperatures can be established and maintained better in closed systems, thus ensuring higher output rates.

## 2.4. Rural Development

Out of the total Indian population of about one billion, over 600 million live in villages; many of them are susceptible to malnutrition, protein-deficiency, and poor health. Many urban dwellers are also poor and malnourished.

A prominent feature of the Indian rural scene is the existence of vast stretches of derelict, barren wastelands. Being too alkaline or saline, these wastelands are unfit for cultivating conventional crops. However, these lands seem quite suitable for growing *Spirulina* which thrives well in alkaline media rich in sodium salts (Kumar, 1983).

At Banaras Hindu University, the approach mostly involved the use of *Spirulina* for reclamation of alkaline-saline wastelands, and devising a simple, low-cost method for raising inoculum cultures of *Spirulina* for distribution among farmers (Singh and Kumar, 1994). *S. major* was isolated from wasteland soil, grown in unialgal laboratory culture and then inoculated on the surface of mud pots containing wasteland soil moistened with water from village ponds. Before inoculation, the moistened pots are steam sterilized in pressure cooker; on cooling, they are inoculated. In some cases, a pinch of sodium carbonate is added to the soil to raise its pH so as to avoid contamination by competing organisms (Kumar et al., 1985). The inoculated pots are kept in a north window and moistened periodically for a few days by which time the alga would have formed a surface mat. All the above steps can be easily carried out by even illiterate farmers in their own houses.

The pots were used to inoculate wasteland fields during the rainy season (July through September). The fields were at first enclosed all round by constructing 1 foot earthen embankments (bunds) so as to allow water-logging and retention of rain water. A few of the *Spirulina* pots are then placed in the water-logged fields. Depending on environmental conditions, the alga forms a surface bloom within a week or two. It is harvested easily by pouring bucketfuls of the bloom suspension into baskets lined with muslin cloth. The algal filaments are retained on the cloth. The harvested alga may be used fresh or may be sun-dried.

Chapatis, poories, and idlis can be made from cereal doughs containing 10% *Spirulina*. At this level, the chapatis and idlis are quite palatable and acceptable in respect of smell and taste; the poories are acceptable even when *Spirulina* content is increased to 15%.

Many underdeveloped countries suffer from the vicious circle of poverty, malnutrition and infectious diseases (see Fig. 21.1). Till recently, the two major strategies to break this circle were increasing agricultural production and improving health and general hygiene. A third approach that may prove even more effective is through use of suitable strains of *Spirulina* as dietary supplements for malnourished people or malnourished animals (Fig. 21.1).

For the above purpose, we produced cobalt- and iodide-enriched (adapted, tolerant) strains of S. platensis by repeated subculturing in increasing concentrations of the respective

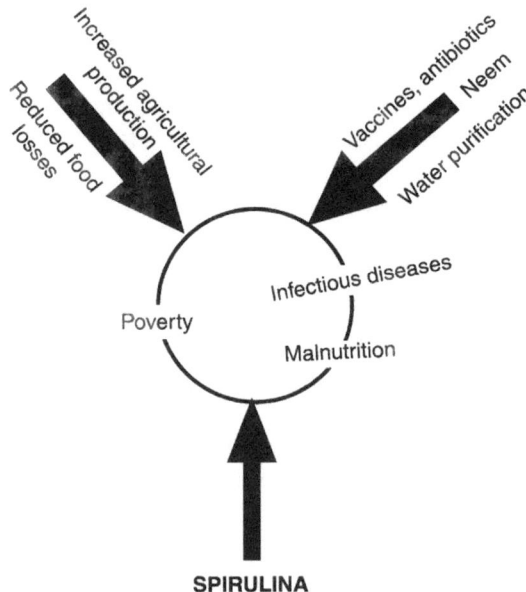

**Figure 21.1.** Use of Spirulina to break the vicious circle of poverty, malnutrition and disease in a developing country such as India.

micronutrients. The strains so produced showed greater uptake of the elements as compared to controls. Later, by growing the cells in media containing higher concentrations of both cobalt and iodine, a composite strain enriched in both these nutrients was successfully produced (Singh, 1992; Singh and Kumar, 1994). This enriched strain relieves the deficiencies of protein, vitamin A and $B_{12}$, iron and iodine. It is ideal for use by the poorer people in several states such as Uttar Pradesh, Bihar, Himachal Pradesh and others.

## 3. Balanced Nutrition for Good Health

Several of the problems resulting from improper nutrition can be avoided by following certain simple principles discussed by DellaPenna (1999), which may be outlined as follows:

1. Variety in food is the key to good health. A balanced diet includes moderate amounts of cereals, vegetables and fruits.
2. The food should include all the six natural tastes, viz., pungent, bitter, sweet, astringent, sour, and salty.
3. There is no single complete food. We should take food items of several different types—the items may be rotated on different days or times.
4. Avoid anything in excess. Moderation is best.
5. The daily diet should include one red or orange item (eg. carrot, tomato, or papaya which contain lycopene), two greens (e.g. spinach, cabbage, capsicum), five dry fruits in small amounts; 2–3 teaspoonfuls of oil; some garlic, soya and, if possible, oats; one small cup of pulses or lentils, and wheat or rice.

6. Seasonal and natural foods are preferable to preserved or unseasonal foods. Thoroughly washed raw vegetables are better than over-cooked vegetables. Properly peeled root vegetables are good in small amounts.
7. Alkaline foods (e.g. most fruits and vegetables) should constitute about 75% of the diet. Acidic foods (wheat bread, rice, fried items, coffee, tea and pickle) should make up no more than 25%).
8. Many plants contain nonessential but specific and useful secondary metabolites. Fruits and vegetables have antioxidants (e.g. vitamins C and E, beta carotene and selenium). Antioxidants destroy free radicals in the body—the free radicals, if not neutralized, damage cells and tissues. Four broad categories of phytochemicals account for plant-rich diets being associated with lower morbidity and mortality in adult life.

These four classes are (i) carotenoids that help ameliorate or prevent some cancers, cardiovascular disease and macular disease; (ii) glucosinolates (e.g. in broccoli) which ameliorate or prevent several cancers; (iii) phytoestrogens (e.g. in soybean) that ameliorate or prevent osteoporosis, cardiovascular diseases, and cancers of prostate, breast and colon; and (iv) phenolics (e.g. in red grapes) that ameliorate or prevent cancers and cardiovascular diseases (DellaPenna, 1999).

## 4. Ethnobotanical Management of Health Needs

### 4.1. Neem (Azadirachta indica, fam. Meliaceae)

Much influx of ethnobotanic knowledge came with the colonization of the New World—it was the empirical folk knowledge of the properties of various plants that slowly transformed itself into modern, allopathic medicine. In many countries multipurpose trees and other useful plants still provide fodder, firewood, food, and fibre to meet the basic needs of rural people. One extremely useful plant that has attracted universal attention is the neem tree which has been known since antiquity for its beneficial healing effects (Suri and Mehrotra, 1994; Jain, 2000; see Table 21.2). Its leaves, fruits, oil, and bark are

**Table 21.2.** Diverse uses of neem in India (after Jain, 2000).

| Culture/Region | Parts used | For prevention or control of |
|---|---|---|
| Nishi (Arunchal) | Bark, leaf | Skin diseases |
| Bodo (Assam) | Leaves plus root bark of *Moringa* | Caries |
| Folk (Manipur) | Leaves | Bowel complaints; small pox |
| Khasi and Jaintia tribe (Meghalaya) | Leaves plus bark of bael (*Aegle marmelos*) | Diarrhoea, dysentery |
| | Leaves and fruits tuberculosis | Cardiovascular disorders |
| Folk (Sikkim) | Leaves | Fever |
| | Poultice (of crushed leaves) | Abdominal pain |
| Bhil (Rajasthan) | Crushed leaves plus leaves of tamarind and *Capparis sepiaria* | Night blindness |

routinely used in Ayurveda and Unani (Greek) systems of medicine. It is widely used as medicine, pesticide, mosquito repellent, food for diabetics, and for cleaning teeth. Oil, extracted from neem fruits, is used against peptic ulcer, worm infections and rheumatism (Suri and Mehrotra, 1994).

An infusion or decoction of the fresh leaves is effective against malaria. Externally, the crushed leaves are applied over skin diseases such as boils, chronic ulcer, syphilitic sores, glandular swellings, and wounds. Fresh juice of leaves with honey is prescribed for jaundice and skin diseases. The tender leaves chewed along with a few black peppers (*Piper nigrum*) are effective in intestinal worms. In fact, neem leaves show effective antibacterial, anti-fungal, anti-protozoal and antiseptic activities against a wide variety of pathogenic organisms.

### 4.2. Turmeric (*Curcuma longa*, fam. Zingiberaceae)

The Indian indigenous (ayurvedic) system of medicine has long recognized the detoxifying action of turmeric (*Curcuma longa*). The raw (uncooked) rhizome of turmeric contains various sesquiterpenes and curcuminoids which show antioxidant, anti-inflammatory, wound healing, anticancer, antiproliferative, antifungal and antibacterial activity (Singh et al., 2002).

## 5. Prescription for Good Health

A core package for general health improvement of malnourished people and vegetarians includes the following components.

1. *Spirulina*-fortified chapatis made from whole wheat flour supplemented with 10% spirulina (flakes, powder or fresh). Daily intake of spirulina should not exceed 40 gm per person.
2. Five fresh leaflets of neem plus 1 gm of peeled, fresh, raw rhizome of turmeric to be chewed first thing in the morning, followed by brushing the teeth to clean the yellow stain. Fresh turmeric is available from July to October; it can also be planted in home gardens or in large pots. Neem leaves are generally available throughout the year, especially from March to December.

While the precise dosage, frequency and duration of consuming neem and turmeric will depend on several factors such as individual lifestyle, occupation, food habits, etc., the quantities suggested here are generally applicable to the majority of Indian people and can be taken by people of all age groups. Neem, turmeric and other similar remedies show slow but sustained benefits. They are mildly curative but strongly preventive in nature.

For people prone to respiratory disorders, 1 gm of peeled rhizome of ginger (*Zingiber officinale*) plus four whole black peppers (*Piper nigrum*) should be chewed daily, along with neem and turmeric. For those fond of fried and fat-rich foods, two cloves of garlic (*Allium sativum*) are added to the above prescription.

In India, as in many other countries, gastrointestinal disorders are the root cause of a variety of metabolic diseases. Patients suffering from gastrointestinal and liver diseases benefit greatly by eating the fruit pulp of *Aegle marmelos* (fam. Rutaceae), 1 tablespoonful

2 or 3 times a day; this fruit is available in north India from April to August. In Indian mythology, this plant is a favourite of Lord Shiva, and it possesses miraculous powers in healing and toning up the digestive and hepatic systems.

Diabetes, long regarded as a disease of the wealthy and affluent, is now spreading fast even among poor people. For them, the best remedies are prepared from seeds of fenugreek (*Trigonella foenum-graecum*), seeds of *Eugenia jambolina*, and whole fruit of the bitter gourd (*Momordica charantia*). All these need to be taken under proper medical advice in personalized dosage and duration.

Faithfully following the principles discussed by DellaPenna (1999) and adopting the prescriptions outlined above can cover a large spectrum of early stage mild illnesses such as skin diseases, cancer, diabetes, respiratory, digestive, hepatic, cardiovascular, dental and ophthalmic disorders, allergies, immune dysfunction, and various types of infections. Habitual consumption of the suggested remedies under expert advice and in proper doses (which must not be exceeded) can confer lasting benefits and ensure good health for all.

## 6. Acknowledgements

Sincere thanks are due to Professors Edwin L. Cooper, Nobuo Yamaguchi and Ms. Patty C. Willis for funding my participation in the Kanazawa Symposium and for their continuing friendship, guidance and cooperation since then. I thank my computer operator Mangala Pd. Dubey for patiently composing several drafts of this chapter.

## 7. References

Annapurna, V., 1991, Bioavailability of *Spirulina* carotene in pre school children, *J. Clin. Biochem. Nut.* **10**: 145–151.
Becker, E.W., Venkataraman, L.V., 1984, Production and utilization of the blue-green alga *Spirulina* in India, *Biomass* **4**: 105–125.
Becker, E.W., 1986, Nutritional properties of microalgae: potentials and constraints, In Richmond, A. (Ed.). *Handbook of Microalgal Mass Culture*. CRC Press. Boca Raton, FL.
Becker, E.W., 1994, *Microalgae: Biotechnology and Microbiology*, Cambridge University Press, Cambridge.
Begin, M.E., Das, U.N., 1986, A deficiency in dietary gamma linolenic acid or eicosapentaenoic acid may determine individual susceptibility to AIDS, *Med Hypoth.* **20**: 1–8.
Behera, B.K., Kaur, M., 2003, *Spirulina* in modern industries for manufacturing value added dietary packages, pp. 401–415. In Ahluwalia, A.S. (Ed.). *Phycology*. Daya Publishers, New Delhi.
Belitz, H.D., Grosch, W., 1999, *Food Chemistry*, pp. 473–693. Springer, Berlin.
Brown, I.I., 1998, The alternative bioenergetic patterns in cyanobacteria as factors of biochemical adaptation, In Subramanian, G., Kaushik, B.D., Venkataraman, G.S. (eds.). *Cyanobacterial Biotechnology*. pp. 21–34. Oxford & IBH, New Delhi.
Chaturvedi, U.K., Habib, I., 1999, Cyanobacteria as a source of food, *Proc. Acad. Environ. Biol.* **8**: 241–245.
Ciferri, O., 1983, *Spirulina*, the edible microorganism, *Microbiol. Rev.* **47**: 551–578.
DellaPenna, D., 1999, Nutritional genomics: manipulating plant micronutrients to improve human health, *Science* **285**: 376–379.
Fox, R.D., 1998, *Spirulina* farms—micro to macro, In Subramanian, G., Kaushik, B.D., Venkataraman, G.S. (eds.). *Cyanobacterial Biotechnology*. pp. 259–265. Oxford & IBH, New Delhi.
Garewal, H., 1995, Antioxidants in oral cancer prevention, *Am. J. Clin. Nutr.* **62 (6 Suppl)**: 1410S–1416S.
Hudson, J.F., Karris, G.I., 1974, The lipids of the alga *Spirulina*, *J. Sci. Food Agric.* **25**: 759–763.
Jain, S.K., 2000, Human aspects of plant diversity, *Economic Botany* **54**: 459–470.

Kumar, H.D., (March, 1983), Algae and rural development, pp. 1–11 (Presidential address) All India Phycological Congress, Kanpur.

Kumar, H.D. Phycotechnology of *Spirulina.* In Agrawal, S.K., Garg, R.K. (eds.), 1988, *Environmental Issues and Researches in India*, pp. 131–146. Himanshu Publications, Udaipur.

Kumar, H.D., Singh, D.V., Singh, Y., 1985, Phycotechnology for wasteland reclamation: growth of protein-rich *Spirulina major* on usar soil, Proc. Seminar on *Prospects and Problems of Green Vegetation Research in India.* pp. 45–51, 12–13 Dec., Calcutta.

Liede, K.E., Alfthan, G., Hietanen, J.H., Haukka, J.K., Saxen, L.M., Heinonen, O.P., 1998, Beta-carotene concentration in buccal mucosal cells with and without dysplastic oral leukoplakia after long-term beta-carotene supplementation in male smokers, *Eur. J. Clin. Nutr.* **52**: 872–876.

Mahajan, G., Kamat, M., 1995, Linolenic acid production from *Spirulina platensis, Appl. Microbiol. Biotechnol.* **43**: 466–469.

Mathew, B., Sankaranarayanan Nair, P.P., Varghese, C., Somanathan, T., Amma Amma, N.S., Nair, M.K., 1995, Evaluation of chemoprevention of oral cancer with *Spirulina fusiformis, Nutr. Cancer.* **24**: 197–202.

Narsimha, D.L.R., Venkataraman, G.S., Duggal, S.K., Bjorn, O.E., 1982, Nutritional quality of the blue-green alga *Spirulina platensis* Geiter, *J. Sci Food Agric.* **33**: 456–460.

Nayaka, N., 1988, Cholesterol lowering effect of *Spirulina, Nutr. Rep. Int.* **37**: 1329–1337.

Nichols, B.W., Wood, B.J.B., 1968, New glycolipid specific to nitrogen fixing blue-green algae, *Nature* **217**: 767–768.

Peto, R., Doll, R., Buchley, J.D., Sporn, M.B., 1981, Can dietary beta-carotene materially reduce human cancer rats? *Nature* **290**: 201–207.

Ripley, D.F., 1996, *Spirulina Production and Potential*, Edisud, La Calade R.N. 7, Aix-en-Province. France.

Sankaranarayanan, R., Mathew, B., Varghese, S.P.R., Menon, V., Jayadeep, A., Nair, M.K., Mathews, C., Mahalingam, T.R., Balaram, P., Nair, P.P., 1997, Chemoprevention of oral leukoplakia with vitamin A and beta carotene: an assessment, *Oral Oncol.* **33**: 231–236.

Seshadri, C.V., Seshagiri, S., 1985, *Spirulina—The Wonder Gift of Nature*, Sri AMM. Murugappa Chettiar Res. Centre, Madras.

Shukla, P.K., 1982, *Nutritional Problems of India*, Prentice Hall of India, New Delhi.

Singh, R., Chandra, R., Bose, M., Luthra, P.M., 2002, Antibacterial activity of *Curcuma longa* rhizome extract on pathogenic bacteria, *Current Sci.* **83**: 737–740.

Singh, Y., 1992, Mutagenesis and growth of the economically valuable cyanobacterium, *Spirulina*, PhD. Thesis, Banaras Hindu University, Varanasi.

Singh, Y., Kumar, H.D., 1994, Adaptation of a strain of *Spirulina platensis* to grow in cobalt and iodine-enriched media, *Jour. Appl. Bacteriol.* **76**: 149–154.

Skulacher, V.P., 1994, Bioenergetics: the evolution of molecular mechanisms and the development of bioenergetic concepts, *Antonie van Leeuwenhoek* **65**: 271–284.

Suda, D., Schwartz, J., Shklar, G., 1986, Inhibition of experimental oral carcinogenesis by topical beta-carotene, *Carcinogenesis* **7**: 711–715.

Suri, R.K., Mehrotra, A., 1994, *Neem (Azadirachta indica* A. Juss.). *A. Wonder Tree*, Soc. Forest Environ. Managers, Dehra Dun, India.

Torzillo, G., 1997, Tubular biroeactors, In Vonshak, A. (Ed.). *Spirulina platensis (Arthrospira)*, pp. 101–105. Taylor & Francis, London.

Venkataraman, L.V., 1983, Blue-green Alga *Spirulina*, Central Food Technol. Res. Inst., Mysore.

Venkataraman, L.V., Becker, E.W., 1985, *Biotechnology and Utilization of Algae: the Indian Experience*, Department of Science & Technology, New Delhi.

Vonshak, A., 1990, Recent advances in microalgal biotechnology, *Biotech. Adv.* **8**: 709–727.

Vonshak, A., Richmond, A., 1988, Mass production of *Spirulina*—an overview, *Biomass* **15**: 233–248.

# V

# Basic Science: Future Approaches to Novel Molecules for CAM

# 22

# Cultural Heritage: Porifera (Sponges), A Taxon Successfully Progressing Paleontology, Biology, Biochemistry, Biotechnology and Biomedicine

Werner E.G. Müller, Renato Batel, Isabel M. Müller and
Heinz C. Schröder

## 1. Introduction

In 1876, Campbell (Campbell, 1876 [p. 446]) wrote "those beautiful 'glass-rope sponges', *Hyalonema* etc., have been found by our researchers to be 'the most characteristic inhabitants of the great depths all over the world, and with them ordinary siliceous sponges, some of which rival Hyalospongiae in beauty'". The admiration for the beauty of sponges is documented since Aristotle (cited in Camus 1783), however the nature of these organisms and their phylogenetic position remained enigmatic until less than 10 years ago. E.g., in 1988 Loomis (Loomis, 1988 [p. 186]) wrote "the sponge cells are unspecialized flagellates held together by a glycoprotein extracellular matrix... they are multicellular, but just barely so". This view changed drastically since the introduction of modern molecular biological techniques; the informational genes investigated in detail now group the sponges to the Metazoa leaving any doubt on their evolutionary origin behind. The first breakthrough came with the study of the galectin molecule from *Geodia cydonium*, when it was discovered that the deduced polypeptide shared high sequence similarity only with metazoan proteins and comprised all characteristic amino acid moieties required for the binding to the sugar (Pfeifer et al., 1993).

The ever lasting interest in sponges originates also from their use in human therapy. The first notifications date back to Aristotle and Hippocrates (Arndt, 1925). Sponge extracts were used as hemostatic agents and the organic skeleton of sponges was used as material to extend blood vessels/body cavities or to replace excised tissue.

---

**Prof. Dr. W.E.G. Müller** • Institut für Physiologische Chemie, Abteilung Angewandte Molekularbiologie, Universität, Duesbergweg 6, 55099 Mainz; GERMANY. tel.: +6131-392-5910; fax.: +6131-392-5243; e-mail: wmueller@mail.uni-mainz.de
German "Center of Excellence *BIOTEC*marin"
*Complementary and Alternative Approaches to Biomedicine*, edited by Edwin L. Cooper and Nobuo Yamaguchi. Kluwer Academic/Plenum Publishers, 2004.

Focusing on the impact of sponges in biology, they were always one key taxon to be studied in order to clarify the borders of the kingdoms. 300 years ago the systematic position of sponges was not possible to define. Therefore, separate independent entities in systematics were created, e.g. plant-animals (Esper, 1794), animal-plants (Pallas, 1787) or Parazoa (Sollas, 1888 [p. xcv]). The present day used phylum name Porifera goes back to 1836 (Grant, 1836) and was originally termed "Poriphera/Poriphora". Today it should be accepted that this phylum must be grouped to the Metazoa, with only quantitatively different characters as other metazoan phyla, but qualitatively being identical to the "higher" Metazoa (Müller, 2001, 2003a).

During the last few years the interest in sponges strongly increased since basic technologies were elaborated to use these animals in a sustainable manner in molecular biotechnology (see: Müller, 2003b). It can be expected that in the coming years sponges will become one major source for novel bioactive drugs useful for the application in human therapy. Sponges will become key animals in conservation biology; their taxonomic/genetic diversity in relation to environmental changes will be actively studied in the complex and challenging discipline of molecular biodiversity (Müller et al., 2003c). As one prerequisite the creation of marine protection zones, e.g. the Evo-Devo Bay "Limski Canal" in Rovinj (Croatia), has been initiated.

In this chapter a historical survey of the different aspects in sponge research is given, highlighting the importance of these activities in their significant contribution to the present understanding ranging from biology to molecular biology and biotechnology. A series of aspects will not be discussed here, as the use of sponges in the daily and clerical use (reviewed in: Arndt, 1937). Needless to mention are the chapters in the Bible describing the use of a sponge to torture Christ (Mark 15/36; Matthew 27/48; St. John 19/29); perhaps this fact was the origin for the believe in sponge as a sacred relic.

## 2. Paleontology

It might be argued that in spite of the increased knowledge of Early Palaeozoic fossils these data have influenced our views about early evolution of Metazoa only to a minor extent (Dzik, 1991). Also some doubts on early fossil records have become overt; e.g. the controversy between Xiao (Xiao et al., 1998) and Xue (Xue et al., 1999) on "sponge/animal embryos". However, solid data strongly suggest that the evolutionary transition to Metazoa occurred very likely several times; e.g. to the Porifera and also to the Archaeocyatha (Okulitch, 1955). The latter taxon became extinct while the Porifera successfully survived the different ice periods until today. In the 1990's it was debated whether the multi-cellular animals also evolved through macro-evolutionary events and received their typical features immediately (Bergström, 1991). This would imply a polyphyletic origin of the Metazoa; a view which was strengthened also by early molecular biological findings using rDNA sequencing data (Field et al., 1988).

It was Mehl (Mehl et al., 1998) who, based on fossil records, impressively documented that the different sponge classes (Hexactinellida, Demospongiae and Calcarea) originated from one common ancestor approximately 600 million years ago. This finding was independently confirmed by application of molecular biological techniques (Müller et al., 1994). It appears that – based on the wider application of the latter methods and the more causal analytic explanations which can be drawn from their results over the more empiric paleontological observations – in the near future other "problematic taxa", e.g. the position of the Ctenophora, may be resolved by molecular biological analyses. Nonetheless, the confirmation of data through fossil records, as documented for Porifera, strengthens the validity of the conclusions from sequence comparisons.

## 3. Biology

Aristotle (in Camus, 1783) described the first sponge species. Among those are the three commercial sponges, the μανον (fluffy) species likely to be *Spongia equina* (see: Schmidt, 1862), the πυχνoζ (solid) species [*Spongia mollissima*] and the τραγοι (blocklike) species [perhaps *Ircinia*]. In addition, he mentioned the απλυσια species (aplysia [a name which was adopted later as *Aplysia* by Nardo {1833}; presently named *Aplysina*]; the "unwashable") which is, according to Schmidt (1862), a species from the genus *Cacospongia*.

### 3.1. Systematics

Very likely the oldest descriptions of sponge species, according to modern morphological and skeletal characters, are *Tetia sphaerica* (at present attributed to the genus *Tethya*) (Fig. 22.1A-a) and *Alcyonivm dioscoridis* (Fig. 22.1A-a) (Donati, 1753). Their descriptions were published in the period during which modern systematics were established by Linné (1788), a fruitful period which provided the solid basis for a classification of animals. However, the species descriptions provided by Linné are only sketchy (Linné, 1788 [vol. I/VI, p. 3812]). With regard to *Tethya* he named this species *Alcyonium lyncurium* and characterized it by the color golden-yellow [flavum]. Already Donati (1753) described two morpho-types of *Tetia sphaerica* based on the surface structure, the degree of being warty (Fig. 22.1A-a). Later Pallas (Pallas, 1787 [p. 192]) highlighted again these differences and termed both morpho-types *Alcyonium aurantium*. Therefore, two species names have to be considered today for this sponge species, namely *Tethya lyncurium* (Müller and Zahn, 1968) and *Tethya aurantium* (Baer, 1906). Distinct from *T. lyncurium* (Müller and Zahn, 1968) and *T. aurantium* is the endemic species *Tethya limski* (Fig. 22.1A-b) which is distinguished from *T. lyncurium/T. aurantium* both by morphology/spicule shape and the habitat where it is found (Müller and Zahn, 1968).

Molecular biological analyses to be performed in the future will surely help to define the species borders. A clear definition of species borders is crucial for the next step in taxonomy in order to be useful for studies devoted to molecular biodiversity, including also the economical section of it, the chemo-ecology. The experiences obtained with endemic sponge species from the Lake Baikal suggest that sequence data provide powerful characters to distinguish between (endemic) morpho-types (Schröder et al., 2003a).

### 3.2. Histology, Embryology

With the beginning of the 19$^{th}$ century studies emerged which explained morphogenesis on the basis of cell-cell and cell-matrix interactions; among the propagating scientists were T. Schwann, J. Müller and M.J. Schleiden. It must be stressed that it was only in 1831 that

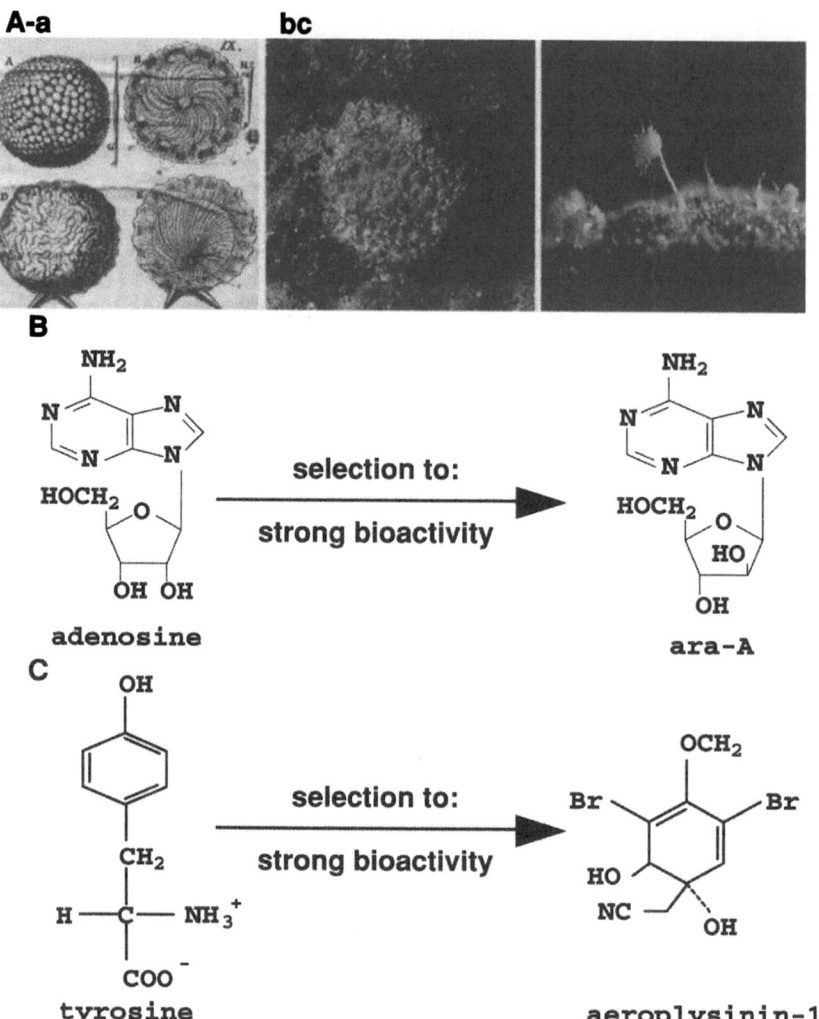

**Figure 22.1.** (A) Sponges of the genus Tethya (Demospongiae), views from the past and the presence. (A-a) Picture of *Tetia sphaerica*, outlining the two morpho-types (Donati 1753) [magnification x0.4]. (A-b) *Tethya limski*, an endemic species from the "Limski Canal" (Rovinj; Croatia) [x0.5]. (A-c) *Tethya lyncurium* or *Tethya aurantium* with its protruding buds [x3]. (B and C) Evolutionary shaping (chemo-evolution) of normal metabolites to bioactive secondary metabolites in sponges during their over 500 million years of biochemical selection for highest potency in action and selectivity in function. (B) Proposed origin of the bioactive compound arabinofuranosyladenine (ara-A) from the building block of DNA, adenosine; (C) structural relationship of the inhibitor aeroplysinin-1 from the amino acid tyrosine.

the cell nucleus was discovered by Brown (see: Mabberley, 1985). In 1856 (Lieberkühn, 1856 [Tab. 15]) already the first publication with sponges, the freshwater sponge *Spongilla fluviatilis* as a model, appeared in which detailed analyses on the development and differentiation of fertilized eggs are given. Even the differentiation stages of sponge cells, e.g. from the "Schwärmsporen" to the spicule-forming sclerocytes were already described in detail

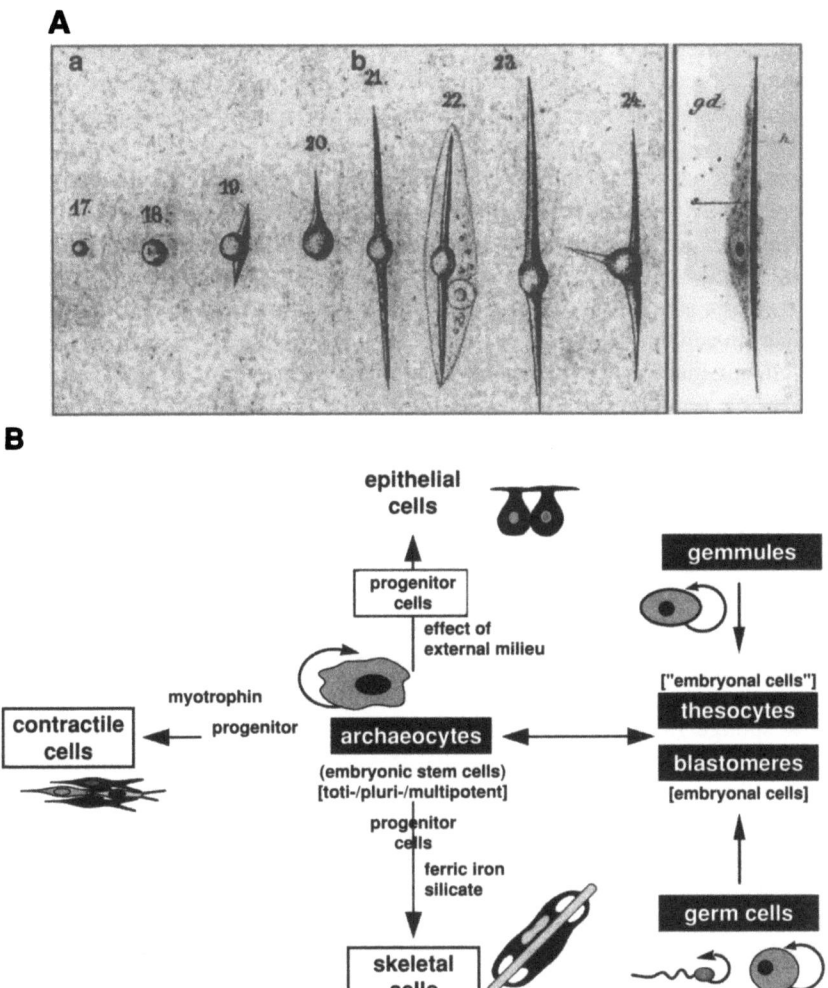

**Figure 22.2.** Histological analysis in sponge. (A) (A-a) Differentiation of embryonic "Schwärmsporen" to differentiated spicule-forming sclerocytes (in *Spongilla fluviatilis*) (Lieberkühn, 1856). (A-b) The intracellular formation of the spicules was also precisely observed by DeLage (1892). (B) Stem cell system in sponges, *S. domuncula* (modified according to Müller et al., 2003e). It is outlined that the toti-/pluri-/multipotent sponge embryonic stem cells, the archaeocytes, give rise to the germ cells on one side and to the three major differentiated cells, the epithelial-, the contractile- and the skeletal cells. It is indicated that during these transitions progenitor cells characteristic for these lineages have to be passed. In addition, it is outlined that under certain developmental phases committed progenitor cells are formed which respond to the silicate/Fe(+++) stimulus under differentiation to skeletal cells, the sclerocytes.

(Fig. 22.2A-a). A few years later, DeLage (1892 [Fig. 16/9d]) observed likewise precisely that the spicules are formed intracellularly (Fig. 22.2A-b). Very notably he mentioned that the spicule-containing cell "est de même nature que les amoeboïdes". This remark can be taken also as an indication that he had been aware of the new stem cells concept that originates from this time (e.g. Weismann, 1892 [p. 135]).

Based on the discovery that sponges comprise genes, and express them, that are characteristic of stem cells in higher metazoan phyla, e.g. *noggin* or the gene encoding the mesenchymal stem cell-like protein, it is now firmly established that this original metazoan phylum, the sponges, possesses stem cells from which differentiated cells originate (Müller et al., 2003d). It is agreed that the archaeocytes are totipotent. However, since for a series of sponge species it has been demonstrated that oocytes originate from archaeocytes, so far a clear distinction between totipotent and pluripotent stem cells in sponges cannot be given. Therefore, in sponges only a grouping of cells into embryonic stem cells, which are toti- to multipotent, and (terminally differentiated) somatic cells might be made. This classification could imply the existence of progenitor cells which have a lower stem cell propensity than the embryonic stem cells from which they arise. This view is strongly supported by the elegant findings showing that tissue explants can give rise to the formation of embryos (Maldonado and Uriz, 1999). In addition, we have described for the demosponge *S. domuncula* that morphogenic factors, e.g. silicon and Fe(III) are involved in the distinct differentiation paths from the archaeocytes to the formation of the somatic epithelial-, contractile- and skeletal cells on the molecular level (Fig. 22.2B).

In continuation of his early elegant work Lieberkühn (1856 [p. 413]) approached the question of whether sponges are colonies or integrated organisms. Moreover, Lieberkühn (1857 [p. 401]) also succeeded in distinguishing between sexual and asexual reproduction systems in freshwater sponges and compared these data to observations with marine sponges, e.g. *Suberites (Alcyonium) domuncula*. He described two states of propagation, the germ cells which give rise to embryos and the "gemmulis (gemmules)". The process of hatching a new sponge from gemmules was already described on cellular level by Laurent (1842), but misinterpreted in assuming that this process originates from fertilized eggs/ embryos.

Three reproduction bodies occur in sponges, primarily described in Demospongiae. In a series, but far from all sponge species (Graeffe, 1882), sexual reproduction cells are seen. Two examples are shown here, from *Geodia cydonium* and from *Suberites domuncula/Suberites ficus*. Semi-thin sections (Fig. 22.3) through *G. cydonium* tissue reveal especially during the season between July and October, in rare cases oocytes (A-d and A-e). They have a diameter of 100–130 µm and are hence much larger than the somatic cells which usually have a size of 10–20 µm. Staining with Carazzi-eosin/hemalum allows the discrimination between the nucleus and the nucleolus (A-e).

Secondly, *Suberites domuncula*. This demosponge has been intensively used for cellular as well as for molecular biological studies (see: Müller, 2001; Müller et al., 2001); Fig. 22.4A-a and A-b. The first description of this species under this name was given by Olivi (1792); however, a series of synonyms have been introduced later. Among those is the potential species *Suberites ficus* (Fig. 22.4A-c). The prominent difference between these two "species" is the fact that *S. domuncula* is growing comensally with a hermit crab (*Pagurites oculatus* [Decapoda: Paguridea], which resides predominantly in shells of the mollusk *Trunculariopsis trunculus* [Gastropoda: Muricidae]), while *S. ficus* is found as a crust on a bivalve *Chlamys* sp. Burton considered the two assumed species as the same taxon (Burton, 1953). Sequence data, presently assessed in our laboratory, should help to clarify this discrepancy. In *S. ficus* oocytes are found in the mesohyl compartment (Fig. 22.4A-d); the 100 µm large oocytes are surrounded by pinacocytes and are again relatively large with respect to archaeocytes, which measure 15 µm.

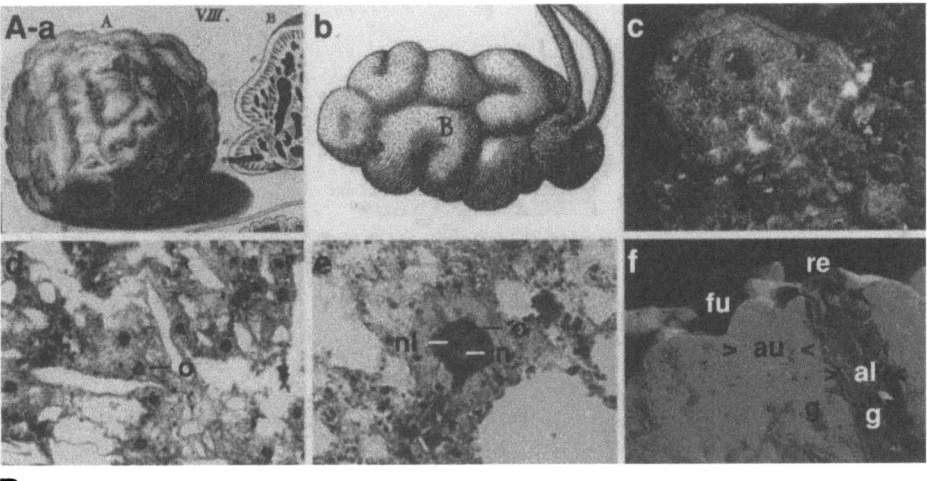

**Figure 22.3.** (A) Sponges of the genus Geodia (Demospongiae), views from the past and the present. (A-a) Picture of *Alcyonivm dioscoridis* (Donati, 1753) [x0.3]. (A-b) View of *Alcyonium durum*, given by Bianchi (1760) [x0.3]. (A-c) *Geodia rovinjensis* from the "Limski Canal" [x0.3]. (A-d and A-e) Semi-thin sections (1 μm) through *G. cydonium* tissue is shown, which comprises oocytes. In A-d: an overall aspect is shown; the oocytes (o) with a diameter of 100-130 μm are seen as prominent cells, much larger than the 15 μm sized somatic cells [x20]. A-e: a higher magnification of one oocyte, in which the nucleus (n) and the nucleolus (nl) can be distinguished [x100]. (A-f) Autograft fusion (> au <) and allograft (> al <) rejection in the sponge *G. cydonium*. The technique of insertion has been applied and the graft is marked (g). Autograft fusion (fu) and allograft rejection after 5 days (re) is shown. In the rejection experiment the graft undergoes apoptotic disintegration; the tissue turns from yellow to brownish [x0.5]. (B) Ig-like domains in sponges. Sequence comparison of the deduced aa from the Ig-like domains [V (variable) domains], present in mammalian Ig heavy-chain, Ig light-chain and murine T-cell receptor α sequences. The two Ig-like domains from the *G. cydonium* receptor tyrosine kinase (RTKIG1_GC and RTKIG2_GC) and from the sponge long form sponge adhesion molecule (SAMLIG1_GC and SAMLIG2_GC) are aligned with the Ig-like domain from the bovine Ig variable region (VIGH_BOS), the human Ig light-chain λ (VLAMB_HUM) and the variable murine T-cell receptor α (VTCRAS_MUS) The consensus sequence is given above; those aa which are present in at least four sequences are given as small letters. The vertical lines above the alignments indicate to the β-strands of the Ig domain A to G; the approximate location of the three CDR stretches are indicated. The bovine Ig variable sequence is given as a continuous line of white letters on black; the human Ig light-chain λ is marked white on grey.

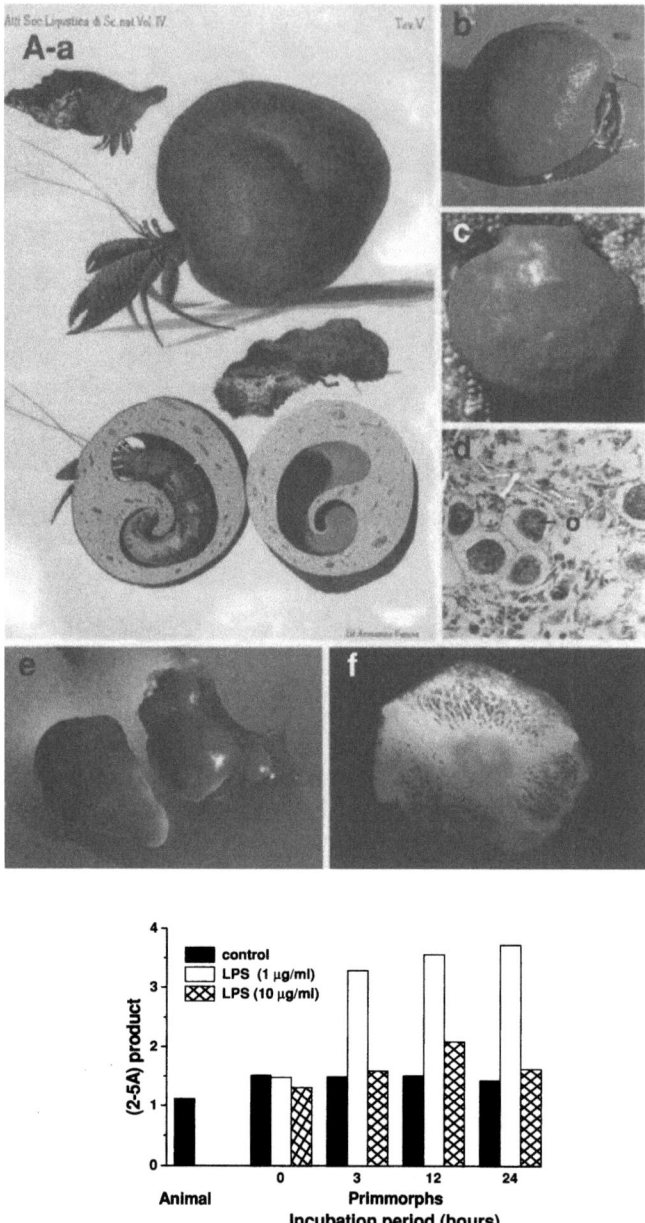

**Figure 22.4.** (A) Sponge of the genus Suberites (Demospongiae), views from the past and the present. (A-a) Picture of *S. domuncula* given by Celesia (1893) [x2]. (A-b) Photo of a *S. domuncula* specimen from the Adriatic Sea [x1]. (A-c) Photo of a *Suberites ficus* (*S. domuncula* ?) specimen from Scotland [x0.5]. (A-d) Semi-thin sections through *S. ficus* tissue; several oocytes (o) are seen [x40]. (A-e) Primmorphs formed in the absence of any organic matrix in the culture dish [x4]. The 3D-aggregates remained round and without canals; x3. (A-f) Formation of canals by incubation of cells on a galectin matrix in the culture dish (incubation: 5 days on non-coated plates followed by 10 days on coated plates); [x4]. (**B**) Production of (2–5)A in primmorphs from *S. domuncula*. Five days old primmorphs were incubated in seawater/0.2% with RPMI1640–medium and incubated in the absence of LPS (control; filled bars), or presence (1 [open bars] or 10 µg/ml [crosshatched bars]) of LPS for 0 to 24 hrs. The products were quantitated as described (Grebenjuk et al., 2002); the conversion of ATP to (2–5)A was calculated and is given in percent to the sum of ATP, ADP and AMP. In parallel, the level of (2–5)A in an animal kept for six months in the aquarium was determined.

Besides sexual reproduction, sponges very frequently propagate asexually through gemmulation, e.g. in *S. domuncula* (Wagner et al., 1998), or by bud formation. These two processes are distinguished by the way these asexual reproduction bodies are developing. Gemmules are bodies comprising very frequently special spicules, while buds are the results of tissue fragmentation (see: Simpson, 1984). The species *T. lyncurium* is especially known for the formation of buds (Fig. 22.1A-c); the buds are extruded on a thin stalk from which they can drop off and become free.

Based on the profound knowledge of cell biology in sponges important contributions on embryogenesis had already appeared by the 19$^{th}$ century. Most distinguished is the report of DeLage (1892) describing in detail the different types of larvae and the morphogenetic events proceeding in these larvae. Very impressive are the differentiation lineages of the cells which are formed from the archaeocytes and reach the choanocyte stage. It should be added that Heider (1886 [p. 7–29]) had already described shortly prior to DeLage very thoroughly the formation of the blastula from the blastosphaera and also the differentiation stages of some characteristic sponge cells, e.g. the choanocytes.

The reason for the frequent use of sponges in cell biological/immunological studies must be seen in the circumstance that these animals comprise a simple organization, or as outlined by Metchnikoff (1892 [p. 55]) "les éponges ont une organisation tellement peu developée,...". Hence the pathways that underlay this morphogenetic process during tissue development, or the response of the sponges against non-self factors, are of lower complexity than those seen in higher metazoan phyla.

## 3.3. Immunology

Sponges are characterized by a remarkably high regeneration capacity, qualifying them also for studies on the mode of tissue recognition (reviewed in: Müller et al., 1999; Müller and Müller, 2003). Since 1907 when Wilson (Wilson, 1907) described the species-specific reaggregation of cells from marine sponges (Porifera), they became a classical model system for the study of self/self- and self/non-self recognition of metazoan cells. Almost simultaneously, the same phenomenon was described for freshwater sponges as well (Müller, 1911). The potency of sponges to fuse with autografts and to reject allografts was discovered by Paris (1961) and Moscona (1968). Based on the occurrence of an accelerated rate of rejection of second-set grafts, a primordial immune memory was postulated for the sponge *Callyspongia diffusa* (Hildemann et al., 1979). Following the identification of the first cell-cell adhesion molecules in sponges, the intercellular aggregation factor (AF) and the cell surface- associated aggregation receptor (AR) to which the AF binds, studies about molecules potentially involved in immune response became possible on cellular level (Pancer et al., 1996). The AFs have been found simultaneously in two marine sponges, *Microciona prolifera* (Henkart et al., 1973; Weinbaum and Burger, 1973) and *G. cydonium* (Müller and Zahn, 1973).

Besides self-self fusion and self-nonself rejection an important observation was documented by Metchnikoff (1892) who found that specific cells in sponges are able to phagocytose and by that process eliminate bacteria. These observations contributed considerably to our present day understanding of cellular immunity.

The first molecular biological studies on the immune system in Porifera have been performed with the species *S. domuncula* and *G. cydonium*. Again, the history of the description and the nomenclature of the species *G. cydonium* is complex. A first drawing was given by Donati (1753) (Fig. 22.3A-a), and the animal was named *Alcyonivm dioscoridis*. It was again Linné (Linné, 1788 [vol. I/VI, p. 3813) who named the sponge *Alcyonium cydonium*. In between, Bianchi (1760 [p. 127]) published a description of the same species calling it *Alcyonium durum*. The name *G. cydonium* was first introduced by Jameson (1811) but later several synonyms were used, e.g. *Geodia mülleri* (Babi, 1922). Based on histo(in)compatibility reactions, as well as on differences in the habitat and the morphology a new species has been described, *Geodia rovinjensis* (Müller et al., 1983); Fig. 22.3A-c. Future molecular biological investigations must clarify also in this case the species borders in order to extend the studies from the species to the individual level.

*Defense against microbes/parasites:* Almost all marine demosponges contain bacteria. First data are now available which help to gain insight into the molecular mechanism by which the host (sponge) might discriminate between symbiotic or commensal and parasitic bacteria. First, it was demonstrated that defined bacterial strains can be engulfed by specific sponge cells, the bacteriocytes (Böhm et al., 2001). Furthermore, it was shown that protein synthesis in tissue from *S. domuncula* is inhibited after incubation with the bacterial endotoxin lipopolysaccharide (LPS) Böhm et al., 2001). Since serine-threonine directed mitogen-activated protein (MAP) kinases are essential components of the LPS-mediated pathway, evidence of activation of these kinases in response to LPS was sought (Böhm et al., 2001).

*Molecules involved:* One powerful mechanism to eliminate microbes is intracellular digestion. This cellular defense mechanism against foreign invaders is well developed from sponges to insects and humans. Mammalian macrophages are the first cells to encounter non-self material. They express several receptors, termed scavenger receptors, that bind to bacteria or their constituents, and hence act as key molecules in innate immunity. Among them is the type I macrophage scavenger receptor which comprises highly conserved SRCR domains. With regard to sponges, molecules comprising SRCR domains have been shown first in *G. cydonium* (Pancer et al., 1997; Blumbach et al., 1998). A phylogenetic analysis revealed that the sponge SRCR domain present in the "multiadhesive protein" displays high similarity to the mammalian WC1 surface antigens, e.g. from bovine, the human CD6 antigen, the human CD5 surface glycoprotein, as well as the human M130 antigen. These data strongly suggest that sponges comprise SRCR- domain(s) containing cell-surface molecules which might be involved in the recognition of bacteria. In addition, it is likely that the ingested "non-self" bacteria are killed by an oxidative and a nonoxidative (enzymatic) mechanism. Several cDNAs coding for lysosomal enzymes, e. g. cathepsin which is abundant in *G. cydonium* (Krasko et al., 1997), have been isolated from sponges.

*The interferon-related system: (2–5)A synthetase:* Very recently, a further (putative) defense system against invading bacteria and/or viruses has been detected in Demospongiae: the (2–5)A (2'–5')oligoadenylate synthetase [(2–5)A synthetase] system. The first sponge species studied that was found to display higher levels of (2–5)A oligoadenylate synthetase and its products other than vertebrate cells (Kuusksalu et al., 1995) was *G. cydonium*. The sponge (2–5)A synthetase was cloned (Wiens et al., 1999). Recently, functional assays were performed to elucidate the role of the (2–5)A synthetase in sponges, especially with respect to a potential infection with foreign, pathogenic microorganisms. The sponge cellular system, which proved to be suitable for this approach are the sponge primmorphs – special cultured aggregates of cells (Custodio et al., 1998; Müller et al., 1999b). If the primmorphs were cultivated in non-coated petri dishes they remained round-shaped and comprised no canals (Fig. 22.4A-e). In contrast, if they were incubated on a galectin matrix in the culture dish (incubation: 5 days on non- coated plates followed by 10 days on coated plates); canal-like structures were observed (Fig. 22.4A-f).

Using primmorphs from *S. domuncula* as a model system it could be demonstrated that they synthesize (2-5)A in larger amounts when they were incubated with LPS, suggesting an activation of the synthetase through a LPS-initiated pathway (Fig. 22.4B). To clarify if LPS also caused increased expression of the gene on the transcriptional level, the cDNA encoding the (2-5)A synthetase was cloned from *S. domuncula* (Grebenjuk et al., 2002). Again it was found that the expression of the gene encoding the (2-5)A synthetase became upregulated.

*Histo(in)compatibility responses in sponges:* Studies of histo(in)compatibility response in sponges have been performed for 30 years. Initially it was reported that sponges have only a low capacity for allorecognition (Moscona, 1968). However, after defining the system, it became apparent that sponges have a very high degree of precision when discriminating between self/self and self/non-self (see: Hildemann et al., 1979). The use of histology and light microscopy to observe the detection of autograft fusion and allograft rejection in sponges was superceded by the introduction of molecular biological techniques. Again the two demosponges *S. domuncula* and *G. cydonium* have been used for those studies. This new approach led to the discovery that there are immune molecules in the Porifera which share high sequence similarity to those of higher metazoan phyla, and especially to deuterostomes (see: Müller et al., 1999). It was established for both sponge species that, under controlled conditions, practically all autografts/syngrafts fused, while the allografts were rejected.

*Molecules involved in histocompatibility response of sponges:* Using these transplantation models it was established that macrophage-derived cytokine-like molecules are activated during allograft rejection. Among those sponge cytokines activated is the allograft inflammatory factor 1. In parallel with this change in expression, a second characteristic molecule was identified which resulted in increased expression of the Tcf-like transcription factor after transplantation in *S. domuncula* (Müller et al., 2002). Further molecules/factors very likely involved in histo(in)compatibility reactions are the glutathione peroxidase and endothelial-monocyte-activating polypeptide.

*Molecules in sponges comprising polymorphic Ig-like domains:* The most striking similarity between molecules involved in the human adaptive immunity and sequences isolated from *G. cydonium* are among those which contain immunoglobulin (Ig)-like domains, the receptor tyrosine kinase (RTK) and the sponge adhesion molecules (SAM). The *G. cydonium* RTK molecule possesses in the deduced polypeptide structure two complete Ig-like domains (Müller and Schäcke, 1996). Two other SAM species have been isolated from this sponge, which do not encode a tyrosine kinase but also contain in the extracellular part two Ig-like domains (GC-SAM) (Blumbach et al., 1999). The longer form of the SAM, GC-SAML, and the short form of SAM, GC-SAMS. The Ig-like domains found in GC-SAML and GC-SAMS as well as in the RTK display high sequence similarity to the V (variable) domain of mammalian immunoglobulin domains (Fig. 22.3B). Furthermore, the Ig-like domains of *G. cydonium* are polymorphic. In the mammalian Ig domains the $\beta$-pleated sheets, which provide the framework region for binding to the antigens, are connected by hypervariable regions, termed complementary- determining regions (CDRs). The conserved aa units of CDR1-3 are present in the *G. cydonium* molecules (Fig. 22.3B).

*Apoptosis:* Again, transplantation techniques were applied to determine which molecular mechanism, resulting ultimately in allograft rejection is induced after transplantation of allogeneic tissue. From *G. cydonium*, tissue pieces were removed with a cork borer from one specimen and were inserted into holes in the recipients (insertion technique). All autografts fused (Fig. 22.3A-f [fu]), while allografts initially fused together but after approximately 3 to 5 days the rejected graft tissue formed a pronounced demarcation boundary and underwent apoptotic degeneration (Fig. 22.3A-f [re]) and finally resorption.

Until recently it was proposed that physiological cell death is restricted to multicellular organisms, which have separate germ and somatic cells (Vaux et al., 1994). Two lines of evidence led us to assume that sponges are also provided with complex apoptotic pathways. In 1992 Pfeifer (Pfeifer et al., 1992) and others found that a factor could be identified in xenografts from *G. cydonium* that cross-reacted immunologically with an antibody raised against a mammalian tumor necrosis factor (TNF). Furthermore, it was shown that sponge cells have a high level of telomerase activity, when they are present in the state of cell-cell contact [both in intact organism and in primmorphs]. Consequently we postulated that, in order to maintain a defined "Bauplan", sponge cells in tissue organization must undergo apoptosis.

*Metazoan pro-apoptotic molecules in sponges:* As the most promising segment to screen for a pro-apoptotic molecule, we selected the death domain part which is found in the mammalian apoptosis controlling proteins Fas, tumor necrosis factor-$\alpha$ or its receptor, and FADD; it is absent in the nematode. This approach was successful; the molecule isolated from *G. cydonium* even comprises two death domains (Wiens et al., 2000a). Sequence comparisons revealed that the two domains found in the sponge molecule are to be grouped within the death domain family. Functional assays were performed with allografts from *G. cydonium* which revealed that in rejecting tissue a strong increase of the expression of the death domain-comprising gene (*GCDD2*) occurs (Wiens et al., 2001).

*Caspases:* In vertebrates, the death domain containing receptors/adapter molecules interact intracellularly with the caspase-8 proenzyme through the death-effector domain with a similar region in the caspase. Interestingly enough, until now, only one gene has been identified in *G. cydonium* which encodes two transcript forms, both for caspase-8 and for -3 equivalents (Wiens et al., 2003b). Functional studies indicate that the two forms of the sponge caspases act in *G. cydonium* in the apoptotic pathway.

*Metazoan anti-apoptotic/cell survival proteins:* In line with the biological evidence that in both *S. domuncula* and in *G. cydonium* apoptosis can be initiated by environmental stress factors, an intense screening for members of the Bcl-2 family was started. This effort resulted in the functional analysis of these anti-apoptotic/cell survival proteins, Bcl- 2, from these two sponge species (Wiens et al., 2000a and b; Wiens et al., 2001). The proof that the sponge gene product acts as a cell survival protein was performed by transfection studies using mammalian cells. It could be shown that mammalian cells transfected with the sponge *Bcl-2* related gene confer resistance against heat shock and growth factor deprivation (Wiens et al., 2001).

Taken together, the bulk of evidence shows that sponges have a complex apoptotic machinery, which allows the elimination of unwanted tissue (e.g. in allo-transplantation) and very likely also in the establishment of an organized body plan. In addition, these data show that the Porifera are provided with complex immune and apoptotic systems that allow the formation of an "integrated system". Considering the fact that the different sponge species are not "amorphous, asymmetrical creatures" as suggested earlier, but comprise a defined phenotype, a sponge might be defined as "integrated colony" or an individual, composed of functional units, allowing the formation of a defined body plan (reviewed in: Müller and Müller, 2003).

## 3.4. Evolution

The evolution to multicellular organisms occurred several times in different kingdoms, in plants, fungi and animals (Spencer, 1867 [vol. II, p. 10–98]). However, it remained enigmatic if these processes towards increased complexity have some elements in common or proceeded separately. It is conceivable that especially the most simple taxa of a respective kingdom, e.g. sponges (Porifera), have been in the center of intense research activities for centuries.

Aristotle (in Camus, 1783) suggested that sponges are animals, provided with some sensitive and contractile elements. However, later in the 17$^{th}$ century (Esper, 1794 [p. 167]) it was inclined to group the sponges ("Saugschwämme") to the plants ("würkliche Pflanzen", real plants); as an example, Jussieu (1789 [p. lxiii]) grouped the "Spongia" to the "Algae". This uncertainty remained until Haeckel, (1896 [vol. 2; p. 49]); the term plant-animals was created to find a place for the sponges within the multicellular organisms. Then with Saville Kent (Saville Kent, 1880–1881 [vol. 1, p. 37]) the sponges were intermittently subdivided to the Protozoa (Fig. 22.5A). Only with the discovery that sponges undergo the process of gastrulation they were included into the Metazoa, even though always strongly highlighting that they were only distantly related to them.

Since Donati (1753 [appendix p. 70]) it remained undoubted that links between the different plant/animal taxa exist, originating from inorganic units to the organic word and from there via the plants to the animals (Fig. 22.6A). Later, Lamarck (1797 [Tab. I; p. 314]) supplemented the process of evolution by concrete data and established a new systematics of animals (Fig. 22.6B), even though the cause of this process was seen by him as modification steps. But again also in his view, a steady increase in complexity was assumed to be the basis for his classification.

Finally with the application of the molecular biological techniques and the compilation of the sequence data for informative molecules, which are primarily involved in cell-cell- and cell-matrix recognition, proteins characteristic of Metazoa, it could clearly be demonstrated that sponges share one common hypothetical ancestor with the other metazoan phyla, which was termed Urmetazoa (Müller, 1995, 1997, 1998, 2001). The evolutionary novelties which appeared after the emergence of the sponges are depicted in Fig. 22.5B. It is outlined that for the transition to the Urmetazoa/sponges the basic elements of the regulatory system of apoptosis and of cell-cell adhesion were essential.

For reasons of comparison, and for a further validation of our data, also the calcareous sponge *Sycon raphanus* and the two Hexactinellida, *Rhabdocalyptus dawsoni* and *Aphrocallistes vastus*, were included into our sequencing program, based primarily on the demosponge *S. domuncula*. As a member of the calcareous sponges, *S. raphanus*, was chosen (Fig. 22.7A-c). The earliest description of a calcareous sponge, *Spongia coronata*, dates back to Ellis (1786 [p. 190; Tab. 68/8–9]); Fig. 22.7A-a. An artistic, imaginary species of a calcareous sponge, *Ascetta primordialis*, was published by Haeckel (1872 [Tab. 1/1]); Fig. 22.7A-b.

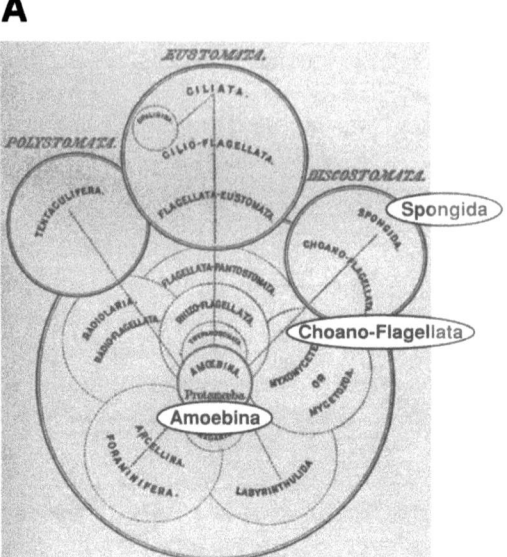
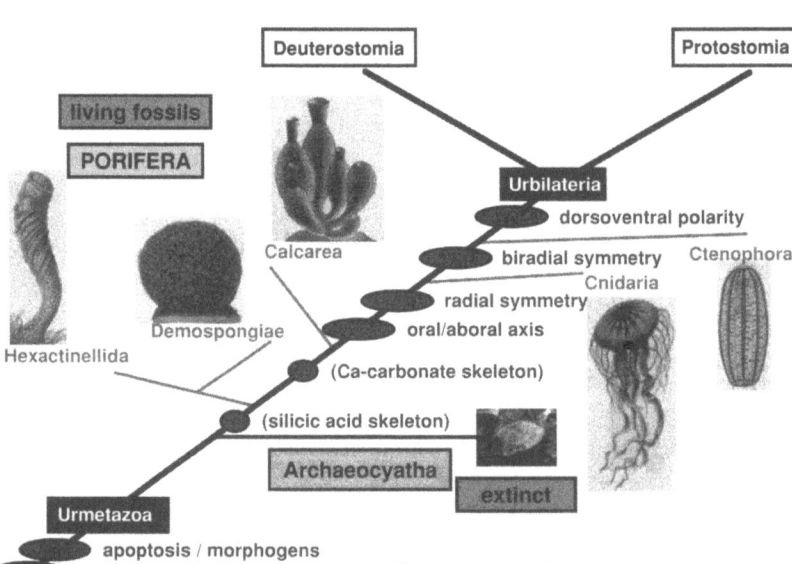

**Figure 22.5.** Relationship of sponges within the animals. (**A**) Saville Kent (1880–1881) grouped the sponges (Spongida) to the Protozoa, next to the Choanoflagellata, to indicate the empiric similar appearance of the sponge choanocytes with the flagellated Protozoa. (**B**) Present day metazoan phylogenetic relationships. The metazoan taxa evolved from a hypothetical ancestor of all metazoan phyla, the Urmetazoa. The major evolutionary novelties which had been attributed to the Urmetazoa were those molecules which mediate apoptosis, control morphogenesis, the immune molecules and primarily the cell adhesion molecules. Perhaps the first class/phylum which emerged had been the Archaeocyatha which however became extinct. Then the siliceous sponges with the two classes Hexactinellida and Demospongiae emerged and finally the Calcarea, which possess a calcareous skeleton. These three classes of Porifera are living fossils that provide a reservoir for molecular biological studies. The Calcarea are very likely a sister group of the Cnidaria. From the latter phylum the Ctenophora evolved which comprise not only an oral/aboral polarity but also a biradial symmetry. Finally the Urbilateria emerged from which the Protostomia and the Deuterostomia originated.

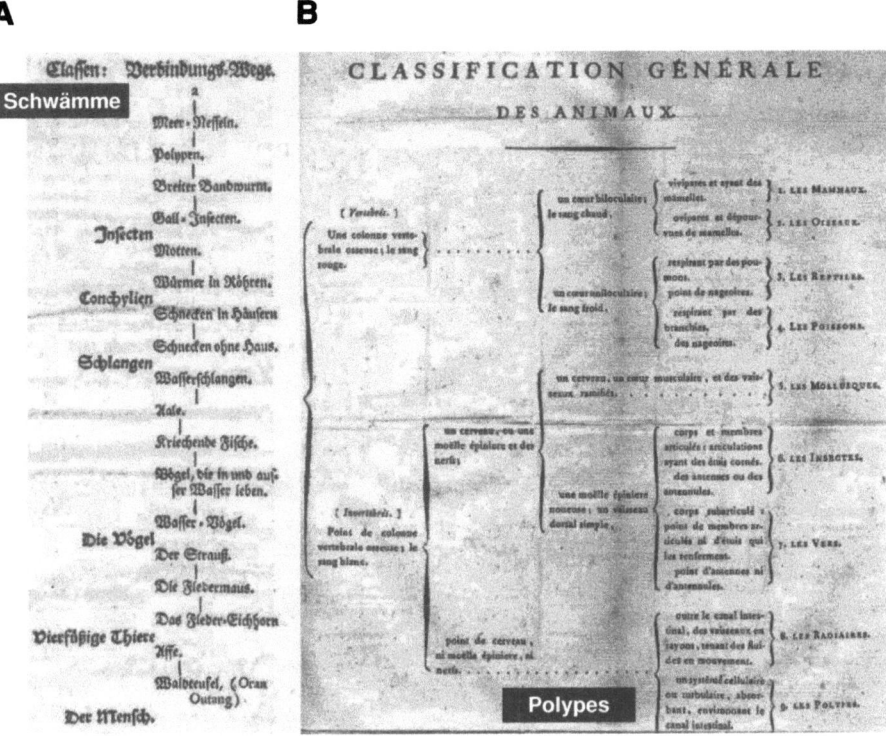

**Figure 22.6.** Systematic relationships of the organismic world in the literature from the 18$^{th}$ century. (A) "Verbindungswege [relationships]" between the different organisms as outlined by Donati (1753). He grouped the sponges together with the corals at the basis of the plants. (B) Lamarck (1797) in his systematics grouped the sponges together with the coelenterates and some protostoans to the "Polypes".

Only relatively late, the Hexactinellida (commonly termed also glass sponge) were separated from the siliceous sponges by Schulze (1887 [p. 25]). The first picture of a glass sponge was published by Thomson (1868), *Pheronema carpentari* (Fig. 22.8A) a species which this author grouped to the "vitreous" sponges. Schulze (1887) compiled a phylogenetic tree (Fig. 22.8B) which positioned the Hexactinellida together with the Calcarea at the basis of the taxon "Spongien", and hence closer to the other metazoans, than the Demospongiae including the Lithistida and the Tetractinellida/Monaxonida. Two photos are given here, *Farrea occa* and *Euplectella aspergillum* (Fig. 22.8C and D), showing the luxurious complexity of the hexactinellid body plan; their skeleton can be reduced to hexactinic/triaxonic spicules.

Major regulatory molecules, characteristic for metazoans, have been selected to obtain further insights into the phylogenetic relationships of the three different poriferan classes. First, the protein kinase C proteins, Ser/Thr kinases (Kruse et al., 1997 and 1998). These are molecules which are involved in signal transduction processes. Sequences identified from the three sponge classes and representative molecules from members of Protostomia, and Deuterostomia have been extracted from the data bases. One example is shown in Fig. 22.7B. The trees built after alignment of the sequences and after rooting with a distantly

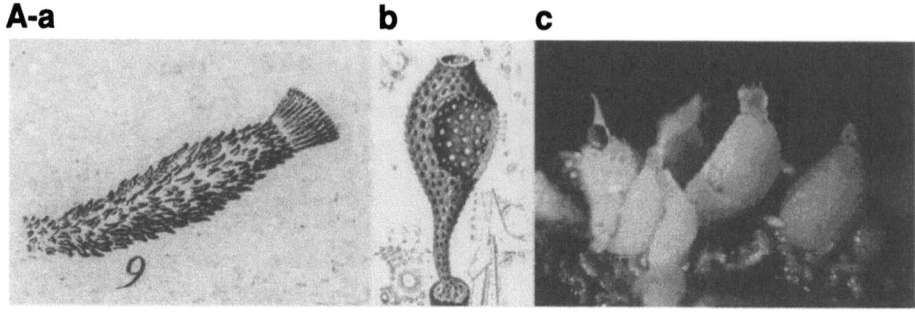

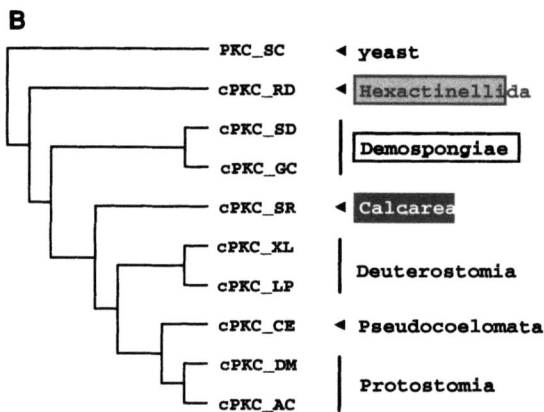

**Figure 22.7.** (**A**) Sponges of the class of Calcarea in the views from the past and the present. (A-a) Picture of *Spongia coronata* given by Ellis (1786) [x1]. (A-b) Picture of *Ascetta primordialis* published by Haeckel (1872) [x5]. (A-c) Photo of a *Sycon raphanus* specimen from the Adriatic Sea [x2]. (**B**) Resolution of the deep of phylogenic deep branches of the three classes of the Porifera, Hexactinellida, Demospongiae and Calcarea through the analyses of informative protein sequences. The phylogenetic tree was computed using the following protein kinase C (PKC) sequences (catalytic domain) from *Metazoa*, (*i*) the deuterostomes *Xenopus laevis* [frog—cPKC_XL] and *Lytechinus pictus* [sea urchin – cPKC_LP], (*ii*) the protostomes cPKC from *Drosophila melanogaster* [fruit fly – cPKC_DM] and *Aplysia californica* [mollusc, cPKC_AC], (*iii*) the pseudocoelomate *Caenorhabditis elegans* [cPKC_CE] and those from the sponges of the classes Demospongiae (*G. cydonium* [cPKC_GC] and *S. domuncula* [cPKC_SD]), Calcarea (*S. raphanus* [cPKC_SR]) and Hexactinellida (*R. dawsoni* [cPKC_RD]) as well as from Yeast *Saccharomyces cerevisiae* [PKC_SC], which was also used to root the tree. The sequences were analyzed by the "Neighbor" program; modified from Kruse et al. (1997, 1998).

related molecule from the yeast *Saccharomyces cerevisiae* showed that the hexactinellid sequence (*A. vastus* or *R. dawsoni*) branches off first from the common ancestor (Kruse et al., 1997 and 1998). Later the demosponge sequences (*S. domuncula* and *G. cydonium*) appeared and finally the sequence of the calcareous sponge *S. raphanus* could be resolved (Fig. 22.7B). It is important to note that the calcareous sponge groups together with the higher phyla and hence might be considered as a possible sister taxon to the Cnidaria. Based on these molecular data it could be established that the class Hexactinellida diverged first from a common ancestor the Urmetazoa, while the Calcarea and Demospongiae appeared

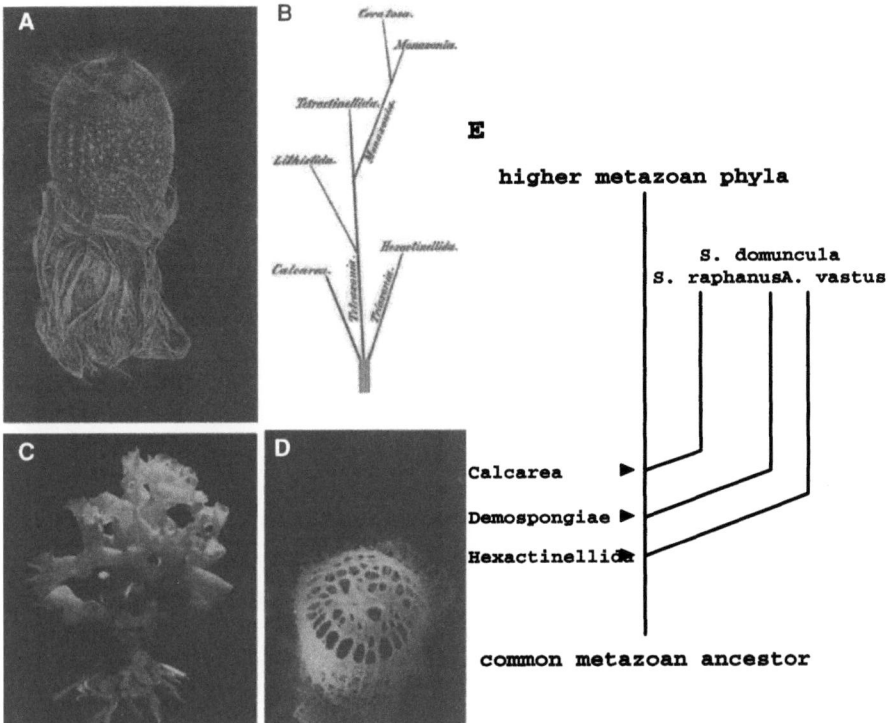

**Figure 22.8.** Sponges of the class of Hexactinellida in the views from the past and the present. (**A**) Picture of the "vitreous" sponges *Pheronema carpentari* (Thomson, 1868). (**B**) First phylogenetic tree including Hexactinellida on the basis of the sponge taxon (Schulze, 1887). (**C** and **D**) Photos of *Farrea occa* [x0.2] and *Euplectella aspergillum* [x0.4]. (**E**) Present day grouping of the three sponge groups, based on sequencing data of receptor tyrosine kinases (modified after Skorokhod et al., 1999). The catalytic domains of the insulin-like receptors as well as of the related sequences were selected from the data base and compared with the sequences obtained from Hexactinellida *A. vastus*, the Demospongiae *S. domuncula* and three sequences from the Calcarea *S. raphanus* type 1 (INR_SR1), *S. raphanus* type 2 (INR_SR2), *S. raphanus* type 3 (INR_SR3). After the alignment a rooted phylogenetic tree was constructed from which a proposed branching order of the three classes of the phylum Porifera, Hexactinellida, Demospongiae and Calcarea, from a common metazoan ancestor, is shown schematically.

later. This relationship could be confirmed by using the corresponding sequences from heat shock proteins (Koziol et al., 1998) and also by the receptor tyrosine kinase sequences (Skorokhod et al., 1999); Fig. 22.8E.

Started as a comparative study, only to confirm the sequence data gathered from *S. domuncula*, the sequence data of informative proteins from the three different Porifera classes revealed a surprising result. The phylogenetic trees obtained with the mentioned kinases and also with other molecules, e.g. tubulin or sequences from heat shock proteins (Koziol et al., 1998), revealed that the sponges groups are paraphyletic. The congruent results indicated that the Calcarea are monophyletic with the other metazoan phyla, while the Demospongiae and the phylogenetically oldest class Hexactinellida are separated from this cluster. It is interesting that these protein sequence data could also be confirmed recently by 18S rDNA sequence data (Borchiellini et al., 2001).

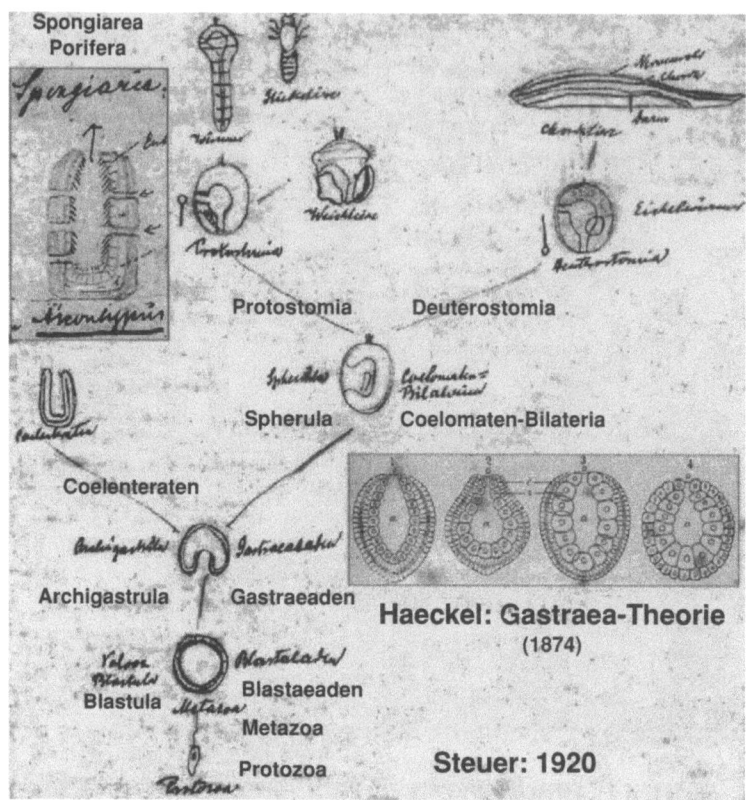

**Figure 22.9.** Animal radiation based on embryological characters. The scheme is taken from A. Steuer (1920). In this phylogenetic tree the evolution from the Protozoa to the Metazoa is driven by embryological novelties which appeared during the transition from the Blastaeden (the free-living form of the blastula larval stage) to the Gastraeden (gastrula). From there the Coelenterata branched while the evolution from the Gastraeden proceeded to the Coelomaten/Bilateria (spherula, an intermediate larvae to the pluteus), which also comprise the common ancestor of the Protostomia and Deuterostomia. Insert: Gastraea-hypothesis from Haeckel (1874). Four different forms of the gastrula are shown; (*1*) from the sponge *Olynthus*, (*2*) the coral *Actinia*, (*3*) an acoelomate, and (*4*) a tunicate *Ascidia*.

Focusing on the approaches presently used to clarify the early radiation of Metazoa it is necessary to go back to Haeckel (1874). His Gastraea-hypothesis is still the basis for the phylogenetic analysis, especially of the metazoan stem phyla (Nielsen, 2001). It is assumed that the free swimming gastrula/Gastraea, with its two germ layers is the hypothetical stem-species of all Metazoa. It might be stressed here, that Haeckel (1874 [p. 18]) already pointed out that sponges, especially the Calcarea, comprise both an ectoderm and an endoderm. As an example for this view, the scheme of A. Steuer (former director of the German-Italian Marine Station, Rovinj) from his manuscript for the student's lecture 1920 is given (Fig. 22.9). Looking at the embryonic characters it is at present assumed that the Porifera evolved from the Blastaea (the free-living form of the blastula), while the higher evolved Gastraea gave rise to the Cnidaria from which finally the Protostomia and the Deuterostomia derived (Nielsen, 2001).

In the last years these embryological characters as the basis for the phylogenetic relationship of metazoan phyla have been abandoned. It appears that the elucidation of the genetic innovations, especially those which underlay developmental processes, are more informative for the resolution of the deep branches in early animal evolution (Knoll and Carroll, 1999). These are especially the genes encoding the transcription factors which are involved in the pattern and axis formation of animals. It is well established that some homeobox genes are responsible for the establishment of anterior patterning; e.g. the Paired-class, the Antennapedia-class and the Lim-class (Galliot and Miller, 2000). Recently, we could establish with the *S. domuncula* model system that sponges contain a homeobox gene of the Lim-class whose expression is correlated with the canal formation in primmorphs (Fig. 22.4A–f). Based on the molecular data available it is proposed (Wiens et al., 2003a) that the hypothetical ancestor of Metazoa, the Urmetazoa, was provided with the basic regulatory repertoire, like cell adhesion molecules and cell differentiation factors as well as with a highly elaborated immune system (Müller et al., 1999) which allowed a pattern formation due to an expression of the already published morphogens as well as Pax-A like HD protein, Hox-like molecules or a Lim-class HD protein (reviewed in: Wiens et al., 2003a). In addition, recent studies in our laboratory demonstrated that *S. domuncula* expresses *Forkhead/T-box* genes and also a retinoic acid receptor (to be published). From these recent data it can be deduced that sponges have amazingly rich and diversified regulatory molecules allowing pattern formation. These data show the impact which sponges provided over the last 200 years in our understanding of the metazoan evolution.

## 3.5. Endemic Species: Conservation Biology

It is generally agreed that sponges evolve slowly compared to other metazoan phyla. However, this view may have to be revised since a series of endemic sponge species have been described, especially in the freshwater. As an example of the latter the Baikalian sponges must be addressed (Schröder et al., 2003a). In this lake, with an age of $\approx 24$ million years, as many as 1,500 endemic species have been described. Among those are also 10–20 sponges; they are morphologically very distinct from other freshwater sponges. By sequencing and analysis of intron stretches from tubulin and of a mitochondrial gene, the relationship among these sponges could be established as being monophyletic. More importantly it could be demonstrated that the evolution of the freshwater sponges in this lake was comparatively fast.

Data suggest that also in the marine environment a rapid evolution of sponges can take place. First data have been obtained from the "Limski Canal" which is located $\approx 15$ km to the North of the marine station in Rovinj (Croatia); the canal like bight is 10 km long and is distinguished from the outer sea by chemical and physical parameters. Hence, the canal represents an ecological isolate which is also separated from the open Adriatic Sea by a sediment barrier (reviewed in: Müller et al., 2003c). In this canal some demosponges have already been found which apparently are endemic, *Tethya limski* (Müller and Zahn, 1968) (Fig. 22.1A–b), *Geodia rovinjensis* (Müller et al., 1983) (Fig. 22.3A–c) and *Thoosa istriaca* (Müller et al., 1979). The Limski Canal (that has also been declared a natural reserve area by the Croatian government) was already used for biodiversity studies regarding immune molecules; the organizing of an Evo-Devo Bay "Limski Canal" is in progress.

## 4. Biochemistry – Biotechnology

Recently, sponges gained strong attention since it becomes increasingly evident that these animals are rich sources for bioactive compounds useful for human therapy. Since new techniques are available or are under development a sustainable use of sponges for biotechnological exploitation becomes, very likely, feasible. While comparatively little is known about the general biochemical pathways in sponges, especially on the molecular level, the application of sponges in biotechnology is already advanced.

### 4.1. (Bio)chemistry

Reports on (bio)chemical investigations of sponges date back to Geoffroy (1731). In a surprisingly quantitative manner the author described the separation of a sponge into inorganic as well as organic fractions. This work was devoted to an application of the "sel volatil" for human therapy, for the treatment of struma "tumeurs scrophuleuses" and goitre "goëtres".

Fragmentarily and parallel with biochemical studies in other lower metazoans, also in Porifera a series of metabolites and enzymes producing them have been reported (see: Arndt, 1930 and 1937). However, until recently, only those enzymes gained wider interest which are involved in the metabolism of silica, the inorganic material of the siliceous sponges. Two classes of enzymes have been identified; those involved in the anabolism of the spicules, the major enzyme is silicatein (reviewed in: Sumerel and Morse, 2003; Müller et al., 2003b), and those of the catabolism, the silicase (reviewed in: Schröder et al., 2003c). For further information these reviews should be consulted.

Very recently enzymes and their genes which are involved in the production of bioactive compounds could also be identified; e.g. the enzyme 3-hydroxyanthranilate 3,4-dioxygenase (Schröder et al., 2002d). This enzyme has been found to be rate-limiting in the formation of the neurotoxin quinolinic acid. It can be expected that the progress in the understanding of biochemical pathways will be accelerated by the rapid increase in our knowledge in the field of the gene repertoire of sponges.

### 4.2. Biotechnology

The approaches to use sponges for commercial purposes are old. They can be dated back to the Stone and Bronze Age and became very fashionable during the Crete/Mycenaen period (2000 BC) (reviewed in: Arndt, 1937). These efforts are primarily concerned with the bath sponges, *Spongia officinalis* and *Hippospongia equina*. Skin diving or harpooning had been the most favorite techniques to collect these species (Arndt, 1937). In the late 19th century the first diving equipment, scaphander-diving, was used for collection of sponges also to resist against the temperature during the winter season (Fig. 22.10A). The main areas of commercial sponge fisheries were in Europe (Spain/France, Adriatic Sea [Krapanj, Hvar], Aegean Sea [Calymnos], Levant Egypt, Tunis, Algeria) and America (Florida, Cuba, Bahamas); Moore (1908) and Arndt (1937).

The following avenues have been taken to exploit the commercial and/or bioactive value of sponges for human use, especially in medicine and biomaterials; mariculture, cell culture and cloning of gene clusters. Apparently mariculture was not used in America in

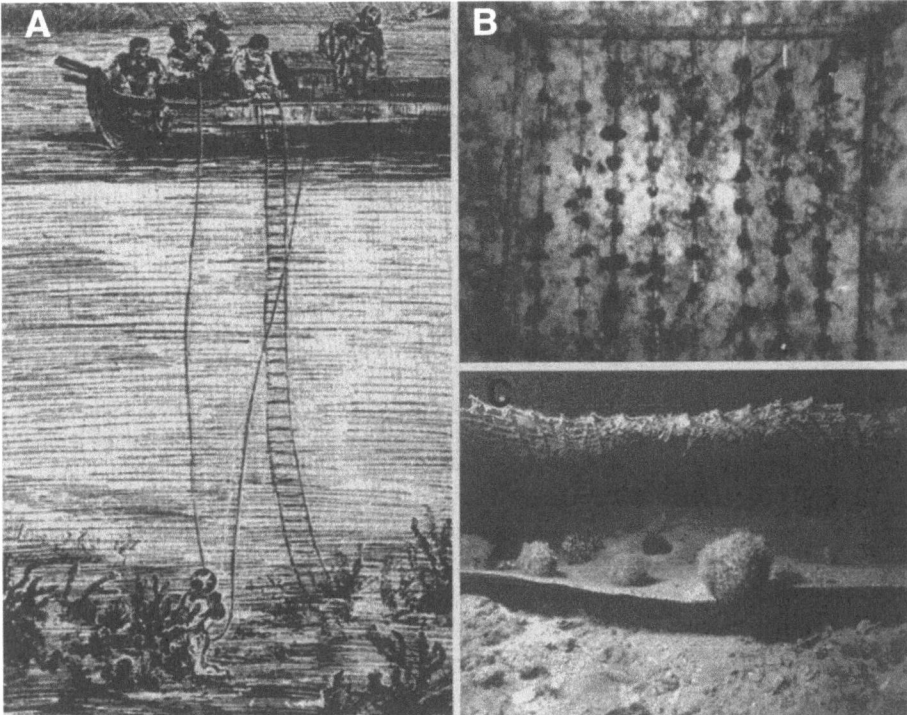

**Figure 22.10.** Mariculture of sponges. (**A**) Scaphander-diving as seen by Coupin (1902). (**B**) Traditional mariculture of commercial sponges as shown here in Calymnos, Greece. The fragments are lined on a rope and placed in about 10–15 m depth (x0.2). (**C**) Cultivation of sponges on a substrate, produced by electrochemical precipitation. Here the cultivation platform with several *Tethya lyncurium* specimens, growing at the mariculture platform in the "Limski Canal" (Rovinj), is shown (x0.2).

the 19th century; instead of this, regulation had been released to prohibit scaphander diving and dredging in some areas and some seasons (Moore, 1908 [p. 503]). In contrast, based on a highly developed conservation awareness in Europe (see: Haeckel, 1870 [p. 9 "diese Plünderung des Meeresgrundes"] or Lendenfeld, 1889 [p. 244 "At present sponge- fishing is carried out in a most barbarous and reckless manner"]), mariculture activities started already at the end of the 19th. The knowledge that sponges are comprised with a high regeneration capacity goes back to the Romans and was highlighted again in the 18th by Ellis (1787 [p. 213]). Based on these characteristics Schmidt (1864 [p. 25]) started the first rational approach to cultivate sponges which he named "Schwammfischerei". The term mariculture goes back to the term "Pisciculture" introduced by Phipson (1864 [p. 6]). Schmidt (1864) proposed two ways for the propagation of the commercial sponges. First by cultivation of hatched young specimens or by dissecting medium-sized specimens into 1 cm large fragments which were then attached to wooden pegs. The latter approach was performed by him together with Buccich from Hvar and was more than partially successful. This cultivation method, by fragmentation/transplantation is still used in Calymnos, Greece, and is steadily optimized (Pronzato, 1999) (Fig. 22.10B).

A modern approach was recently introduced. Fragments/recruits from corals or sponges were transplanted to a cathode mesh of a DC electrolytical system (e.g. applying the "Arcon" technology); the fragments are embedded into the "Arcon" substrate fed by solar energy during the precipitation phases (Fig. 22.10C). First experiments showed that the survival of the fragments is almost 100% and their growth rate is very encouraging (Schuhmacher and Schillak, 1994; Schillak et al., 2001).

The second technique for a sustainable use of sponges for applied purposes is cell culture. It appears that sponge cells in suspension do not proliferate (Ilan et al., 1996). To overcome the blockade of division of sponge cells in culture we elucidated some metabolic pathways. Based on previous findings, which indicated that single cells from *G. cydonium* and *S. domuncula* are telomerase-negative (Koziol et al., 1998) we concluded that suspension cultures of single sponge cells are more difficult to establish than those with tissue-like aggregates. This implies that cells from sponges will require stimuli resulting from cell-cell and/or cell-matrix contact in order to proliferate. This approach was successful (Custodio et al,. 1998; Müller et al., 1999b) and was applied for the production of the bioactive compound avarol (Müller et al., 2000 and reviewed in: Sipkema et al., 2003 and Schröder et al., 2003b). The 3D-cell aggregates, termed primmorphs, contain proliferating and differentiating cells (Fig. 22.4A–e). After transfer of these round-shaped aggregates onto a homologous galectin matrix, the primmorphs start to develop canal-like structures (Fig. 22.4A-f).

The third approach is the isolation and expression of *gene clusters*. These clusters may be defined as linearly arranged units of genes. When they are functionally linked and expressed in concert, transcripts are formed which encode for intermediary metabolic pathways, involved *(i)* in the catabolism of nutrients, *(ii)* the control of the initial and terminal differentiation and *(iii)* in the synthesis of secondary metabolites. Gene clusters usually interact through *cis*-regulatory elements, a system which may be initiated by a *trans*-regulatory intra- or extracellular stimulus (see: Breter et al., 2003). Until now gene clusters for the production of bioactive compounds have been successfully used for the synthesis of macrolide antibiotics, isolated from bacteria (Rawlings, 1997).

Sponges are known to produce polyketides, e.g. swinholide from the lithistid sponge *Theonella swinhoei* (Bewley and Faulkner, 1998), a class of compounds which display highly potent (especially) antibiotic activities. However, most of the bacteria present in sponges are considered to be uncultureable. To use and to utilize the bioactive potential of those microorganisms, molecular biotechnological approaches have to be applied. One promising way is to clone the polyketide synthases, which synthesize the core of the bioactive compound, and to express them in a heterologous system. While attempting this, polyketide synthases have been isolated from bacteria of the sponge *S. domuncula* (Müller et al., 2003a); the length is approximately 40 kb. The polyketide synthases are multienzyme complexes ($M_r$ 100 to 10,000 kDa) that use simple organic building blocks, e.g. acetyl-CoA, and their derivatives. As schematically outlined in Fig. 22.11, the polyketide synthases comprise at least the following enzymic activities; ß- keto acylthioester synthase, acyltransferase, and acyl carrier protein domain (reviewed in: Khosla et al., 1994). In addition, other enzymic activities can be included, e.g. ketoacyl- reductase, dehydratase, enoyl reductase, and thioesterase domains. Multiple copies of active site units reminiscent of those of the fatty acid synthases exist.

The overwhelming portion of bacteria existing in sponges usually cannot be cultivated (see: Thakur et al., 2003). Therefore, we have extended our sequencing attempts to the

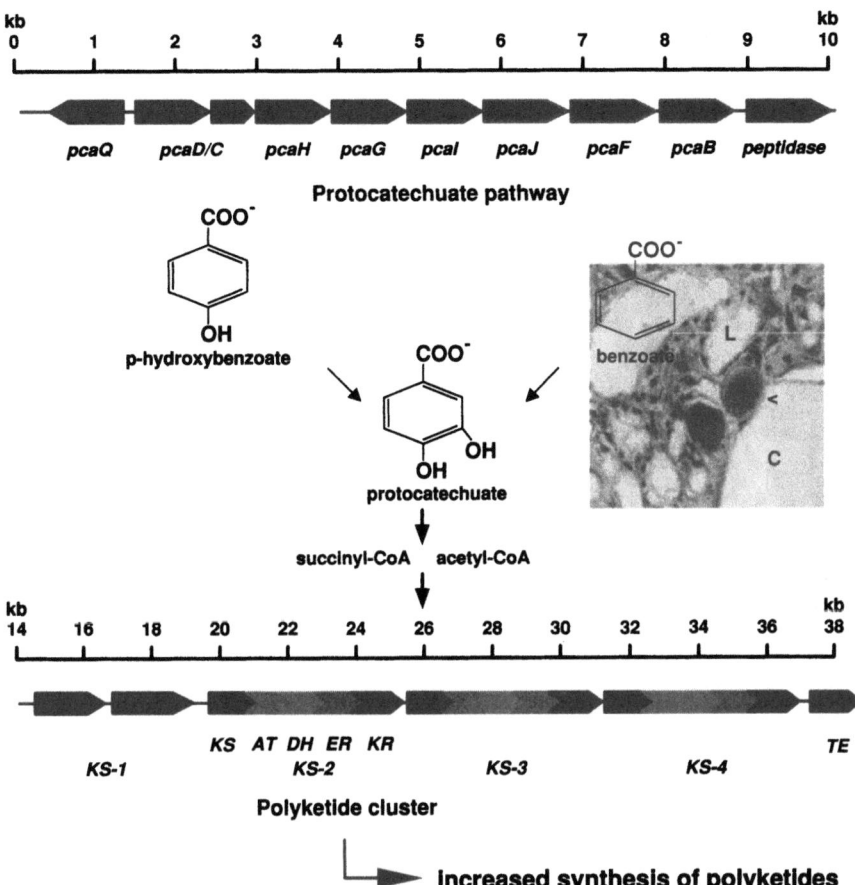

**Figure 22.11.** Schematic model of the domains encoding the polyketide synthase (KS), cloned from a bacterium that had been isolated from the sponge *S. domuncula* (the scales indicate the stretches on the DNA [given in kb]). *Lower scheme:* The modular polyketide synthetase is composed of at least the following enzymic activities; the ß-keto acylthioester synthase (KS), the acyltransferase (AT), the ketoacyl-reductase (KR), dehydratase (DH), enoyl reductase (ER), and thioesterase domain (TE). Multiple copies of active site units (here: KS-1 to KS-4) are usually present. The *photo in the center* of the figure shows a semi-thin section through the mesohyl compartment adjacent to lacunae, which are surrounded by an epithelium (formed by endopinacocytes). One bacteriocyte (arrow head) is shown which is embedded in the epithelium which surrounds a water canal (C); it is almost completely filled by bacteria (light micrograph; [x250]); modified according to Böhm et al. (2001). *Upper scheme:* Upstream of the polyketide synthetase cluster of this sponge-associated bacterium, genes encoding for enzymes of the protocatechuate/ß-ketoadipate pathway have been identified (abbreviated as *pcaB-pcaQ*). This pathway uses benzoate or *p*-hydroxybenzoate and forms in the cascade succinyl-CoA and acetyl-CoA, metabolites which also enter the polyketide synthase cycle.

region upstream the polyketide synthases from a bacteria of *S. domuncula* which only grows slowly on conventional medium. In the gene cluster of this bacterium we found, adjacent to the polyketide cluster, genes coding for enzymes of the protocatechuate/ß-ketoadipate pathway. The end-product of this pathway is succinyl-CoA and acetyl-CoA, metabolites which also enter the polyketide synthase cycle. As shown for a bacteria of the *Roseobacter* lineage, such a protocatechuate/ß-ketoadipate cluster, comprises a series of genes which encode for enzymes involved in the protocatechuate branch of the ß-ketoadipate pathway (Buchan et al., 2000). Exactly these genes are also present in the *S. domuncula* polyketide synthase-protocatechuate/ß-ketoadipate gene cluster (to be published). In turn, we also studied cultivation of the *S. domuncula* bacteria on medium composed of either benzoate or *p*-hydroxybenzoate. Indeed it was determined that after optimization of the medium with these two components both the gene expression of the polyketide cluster and the growth rate of the microorganisms increased. These data also give hope that many more bacteria, living in the sponge perhaps in a commensalic way, might be cultivated in future.

## 5. Biomedicine

Since ancient times, sponges have been intensively used as raw material for the isolation/preparation of apparently beneficial remedies (see: Arndt, 1925). It is however queer to mention that on the scale of the most nutritional and "tasty" items from the aquatic environment the sponges are at position five, beating the eel (Darwin–Brandis, (1799 [vol. 5; p. 39]).

### 5.1. Sponge Secondary Metabolites: Maximal Activity, Specificity and Selectivity Due to Evolutionary Shaping (Evolutionary Combinatorial Chemistry)

Being aware of the high biodiversity of sponge species and the phylogenetic age of these animals of more than 600 million years, one main question arises. Why had these metazoans been so successful during evolution and did not become extinct? Reasons might be seen in the fact that sponges, as sessile filter feeders, do not suffer from nutrient shortage and – in addition – have strong defense systems to defend against foreign invaders. Sponges have developed an amazingly efficient immune system which is very reminiscent of that found in vertebrates (Müller et al., 1999). In addition, the sponges have strategies to defend themselves against foreign organisms, prokaryotic, eukaryotic or viral attackers, by the production of secondary metabolites that repel them (Proksch, 1994; Sarma et al., 1993). It is known that the most potent anti-microbial compounds, especially the antibiotics, are produced by microorganisms, such as bacteria (e.g. *Streptomyces*) and fungi ("Fungi imperfecti", e.g. *Penicillium*). Consequently, and teleologically speaking, the sponges have utilized the capability of these micro-organisms as symbionts to enhance their potency to synthesize bioactive compounds. It can be stated that during evolution the bioactive compounds have been/are shaped in a three- dimensional structure for optimal bioactivity in order to fit and squeeze in an optimal way between a receptor-ligand system, "evolutionary combinatorial chemistry" [chemo-evolution].

The modern application of bioactive compounds/secondary metabolites from sponges started with the pioneering studies by Bergmann (Bergmann and Feeney, 1951) and Scheuer (1990) as well as by the yearly reports of Faulkner (e.g. Faulkner, 1995). The diversity of secondary metabolites produced in sponges has been highlighted in a large number of reviews (e.g. Faulkner, 1995; Sarma et al., 1993). They range from derivatives of amino acids and nucleosides to macrolides, porphyrins, terpenoids to aliphatic cyclic peroxides and sterols. This diversity reflects the efficient mechanisms of combinatorial biochemistry which the animals had acquired during their evolutionary history.

## 5.2. (Potential) use in therapy

Sponge specimens from the subclass of Ceractinomorpha comprise a well developed net of spongin fibers (Fig. 22.12A). This matrix of organic fibers, which can vary in size/diameter and also in their content in inorganic inclusions, is very resistant. Arndt (1937 [p. 1588]) writes that the sponge sample, exhibited in the "Stiftskirche" of Aachen originated from Charlemagne who bought it from Constantinople. Hence, these matrices have been widely used to soak them with fluid, e.g. vinegar (see Bible) or drugs. With respect to the latter use fluid extracts from narcotics, opium, had been used in the Middle Age to narcotize patients prior to cranial operations (1937); from this application the sponges received the name "sleeping sponge" or "spongia somnifera/spongium somniferum".

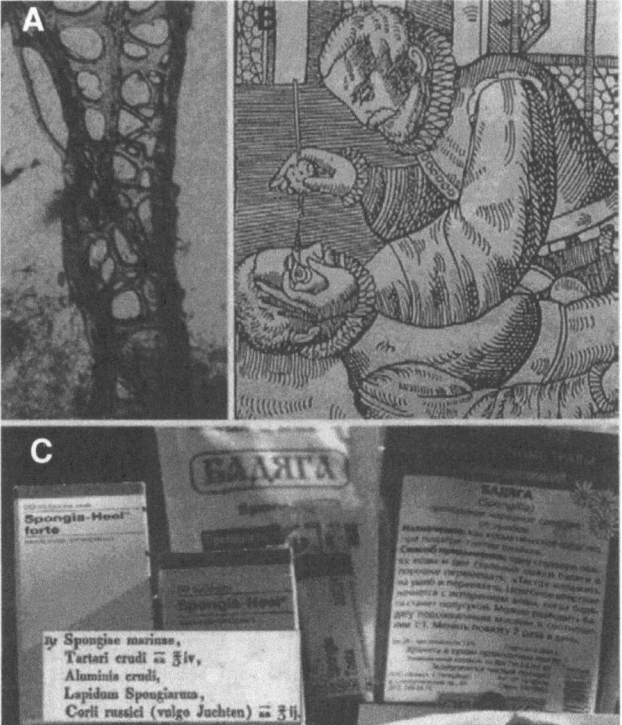

**Figure 22.12.** Use of sponges in traditional medicine. (**A**) Primary fibers of the sponge *Ircinia fascicularis* with some inclusion of foreign material; in the lower part the fine filament bundles are seen [x40]. (**B**) Sponge matrix used to drop medicine into the eye. (**C**) At present roasted commercial sponges, as powder, are used in cosmetics and for the treatment of wounds (badiaga). In addition, in some recipes the sponge powder is used as homeopathy to ameliorate migraine. The latter application originates from recipes, introduced at the end of the Middle Age, which was termed e.g. "Spongia marina". The recipe shown is taken from Rust (1835).

Until the 19th century the sponge matrix was frequently used to soak up blood and to stabilize organs during operations, to drop medicine into the eye (Fig. 22.12B) or to enlarge the blood vessels especially during cesareans (Gräfe, 1826 [p. 60]).

In traditional medicine the commercial sponges (*Spongia* sp, *Hippospongia* sp) were roasted on temperate fire and then used in cosmetics but also to treat sores and wounds. In Eastern European countries this powder can be bought in any drugstore (Fig. 22.12C), e.g. as БАДЯГА (badiaga). In Western European countries in another composition, the powder is used in cosmetics (Fig. 22.12C). The recipe is old and prepared as "Spongia tosta" (Hufeland, 1798 [p. 139]) or "Spongia marina" (Rust, 1835 [p. 484]). The composition of "Spongia marina" is shown in Fig. 22.12C; besides the ash from sponges, it contains tartaric acid, aluminum salt, willow and birch extract.

A target oriented use of sponges for human therapy started when the successful application of "Spongia tosta" was established for the treatment of scrofulae, thyroid gland disease (Muralt, 1692 [p. 368]). The sponge extract was prepared and applied as syrup; but also topical use was advised. Hufeland (1798 [p. 139]) considered "Spongia tosta" as the most effective treatment against scrofulae. In 1910 (Scott, 1910) it was discovered that iodine is the effective ingredient in the sponge preparation.

With modern marine biotechnology and marine natural product chemistry a rapid increase in knowledge in secondary metabolites from sponges and their potential application for human therapy occurred. A recent review surveys the progress (Proksch et al., 2002).

The following examples should be given here: sorbicillactone A, aeroplysinin and arabinofuranosyladenine.

*Sorbicillactone A:* In addition to bacteria, fungi have also been described as associated with sponges. One sponge-associated fungus is shown in Fig. 22.13A. In this line of research different fungal strains were isolated from Mediterranean sponges and tested for bioactivity. Among those was Sorbicillactone A (Fig. 22.13B), which was recently isolated, within the Center of Excellence "*BIOTEC*marin", from a strain of fungus *Penicillium chrysogenum* (Bringmann et al., 2003). *P. chrysogenum* had been isolated from a sample of the Mediterranean sponge *Ircinia fasciculata;* it possesses a unique bicyclic lactone structure, seemingly derived from sorbicillin. Among the numerous known sorbicillin-derived structures it is the first to contain nitrogen and thus the first representative of a novel type of "sorbicillin alkaloids". Remarkably, this compound exhibits promising activities in several mammalian and viral test systems. It is highly selective in its cytostatic activity against murine leukaemic lymphoblasts (L5178y) (Fig. 22.13C). These properties may qualify sorbicillactone A or one of its derivatives for animal and (hopefully) also future therapeutical human trials.

*Aeroplysinin:* It has been demonstrated that the spectrum of bioactive compounds within different sponge taxa varies. Examples are the brominated metabolites derived from tyrosine that are characteristic for the order Verongia, like aeroplysinin, or the isonitriles for the order Halichondria (Van Soest, 1994). Consequently, marker compounds have been described for different sponge groups and successfully used to support taxonomy (Van Soest and Braekman, 1999). Evidence has been presented that the sponge *Aplysina aerophoba*, which produces the bioactive compound aeroplysinin in certain biotopes (Kreuter et al., 1992), is a highly bioactive compound which inhibits proliferation of leukemia cells and prevents the formation of vessels in the chick chorio- allantoic membrane assay (Rodrigues-Nieto et al., 2001). Very likely, aeroplysinin is formed from a (brominated)

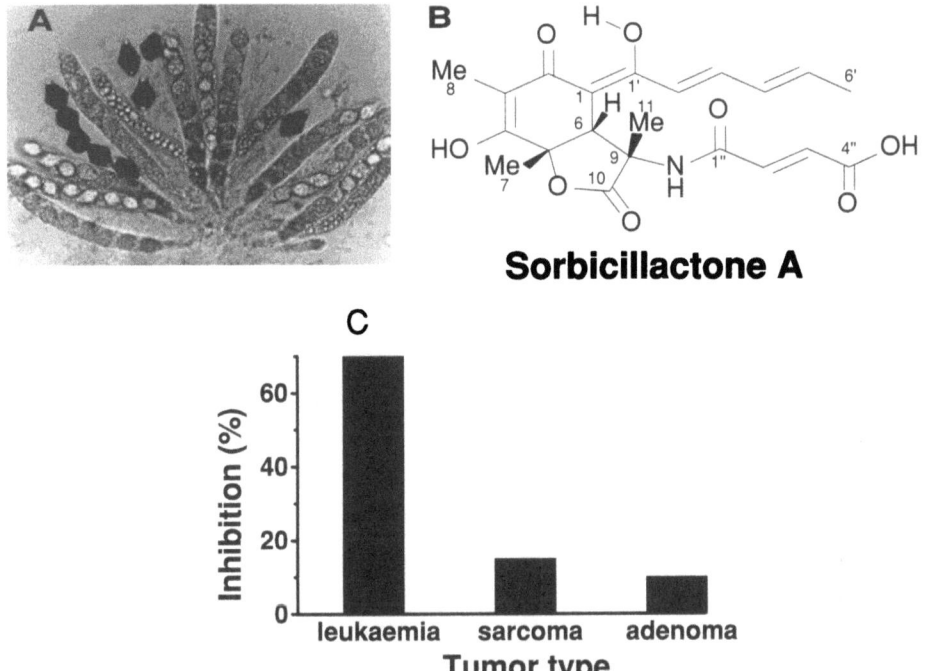

**Figure 22.13.** Bioactive compound from a sponge-associated fungus. (**A**) A fungus, *Biconiosporella corniculata* Schaumann (Ascomycota), had been isolated from a sponge. (**B**) Structures of sorbicillactone A. (**C**) Inhibitory activity of 3 µg/ml of sorbicillactone A on murine leukaemic lymphoblasts L5178y (leukaemia), human cervix HeLa S3 cells (sarcoma) and rat adrenal pheochromocytoma cells PC-12 (adenoma). The inhibitory activity of the compound on the respective tumor cell lines are given in percentages.

tyrosine (Fig. 22.1C) during wound formation (Ebel et al., 1997). Hence, this bioactive compound is a striking example for the process of evolutionary combinatorial chemistry during which a highly bioactive compound is formed from an "inactive" metabolite during evolution.

*Arabinofuranosyladenine:* However, despite the large number and the high variety of structurally different natural products only very few of these marine secondary metabolites have been tested in clinical trials. So far, arabinofuranosyladenine [ara-A; isolated from the gorgonian *Eunicella cavolini* (Cimino et al., 1984)], a derivative of ara- U, isolated from the sponge *Cryptotethya crypta* (Bergmann and Feeney, 1951) is the only secondary metabolite (Cohen, 1963) which has been approved for human application in clinics; it displays potent anti-herpes virus activity (Müller et al., 1975; Müller et al., 1977). Again, ara-A/ara-U are striking examples for the acting principle in lower metazoan phyla, especially those which are sessile, to form potent bioactive secondary metabolites from bioactively "inert" metabolites; this strategy is called evolutionary combinatorial chemistry. It is most likely that ara-A/ara-U originate from the nucleosides adenine, uridine via an isomerase (Fig. 22.1B). The next task of the future is to identify those enzymes which are acquired as evolutionary novelties in those species.

## 6. Conclusion

Sponges (phylum Porifera) are distinguished from other metazoan phyla by a series of characteristics. They were – as commercially important products – in the center of many ancient reports from Greek times. Besides the description of some species also remarks on physiological aspects are found in those documents. The emphasized importance of the sponges is also reflected in Christian mythology. During the end of the Middle Ages, sponges gained target oriented importance as therapeutically valuable raw materials against defined diseases. In the early Modern Age Porifera were put in the center of evolutionary studies to understand the origin and derivation of the animals in general and to define the borders between the multicellular kingdoms. Recently, the interest in sponge research is accelerated by the fact that they had been "discovered" as the starting material for the isolation of a series of highly active, amazingly selective, bioactive secondary metabolites which might become lead compounds for drugs to be used in human therapy. This direction is flanked by recent advances in marine biotechnology, especially in the field of bio-processing. Finally, molecular biology significantly contributed to the different disciplines in that their results gave the basis for a deeper understanding of molecular evolution and provided the rational platform for the progress in our understanding to imitate and mimic evolutionary combinatorial chemistry (chemo-evolution). The aim will be to produce the highly active secondary metabolites in a recombinant manner through the application of gene clusters which have to be isolated both from the sponge itself and from its associated microorganisms, especially bacteria and fungi. Hence, sponges retained their central importance from Ancient to Modern Times and can therefore be considered as one animal taxon which deserves the attribute of a cultural driving element, especially in our western civilized world.

## 7. Acknowledgements

We thank Dr. C. Moss (Argyll, Scotland [*Suberites ficus*]), Dr. Karsten Schaumann (AWI, Bremerhaven, Germany [fungus *Biconiosporella corniculata*]), Prof. M. Sciscioli (Bari, Italy [eggs from *Geodia cydonium*]) and Prof. R. Pronzato (Genova, Italy [*Suberites domuncula* from Celesia]) for the permission to use the indicated figures. Supported by the International Human Frontier Science Program [RG-333/96-M] and the Bundesministerium für Bildung und Forschung (project: Center of Excellence BIOTECmarin).

## 8. References

Arndt, W., 1925, Die Verwendung der Spongien in der Medizin, *Arch. Naturgesch. (Abt. A)* **90**:149–174.
Arndt, W., 1930, Porifera (Schwämme, Spongien), in: *Tabulae Biologicae*, W. Junk ed., W. Junk, Berlin; Suppl. II (vol. 6): 38–120.
Arndt, W., 1937, Schwämme. In: *Die Rohstoffe des Tierreichs*, F. Pax. and W. Arndt, eds., Bornträger, Berlin; vol. I/2:1577–2000.
Babi, K., 1922, Monactinellida und Tetractinellida des Adriatischen Meeres, *Zool. Jahrb Abt. Syst.* **46**:217–302.
Baer, L., 1906, Silicispongien von Sansibar, Kapstadt und Papeete. Inaugural Dissertation: Berlin.
Bengtson, S., 1998, Animal embryos in deep time. *Nature* **391**:529–530.
Bergmann, W., and Feeney, R.J., 1951, Contribution to the study of marine sponges. 32. The nucleosides of sponges, *J. Org. Chem.* **16**:981–987.

Bergström, J., 1991, Metazoan evolution around the Precambrian-Cambriantransition, in: *The Early Evolution of Metazoa and the Significance of Problematic Taxa,* A.M. Simonetta and S. Conway Morris, eds., Cambridge University Press, Cambridge; pp. 25–34.
Bewley, C.A., and Faulkner, D.J., 1998, Lithistid sponges: star performers or hosts to the stars? *Angew. Chem. (Int. Ed. Eng.)* **37**:2162–2178.
Bianchi (Planci), J., 1760, *De Conchis Minvs Notis Liber,* Palladis, Romae.
Blumbach, B., Pancer, Z., Diehl-Seifert, B., Steffen, R., Münkner, J., Müller, I., and Müller, W.E.G., 1998, The putative sponge aggregation receptor: isolation and characterization of a molecule composed of scavenger receptor cysteine-rich domains and short consensus repeats, *J. Cell. Sci.* **111**:2635–2644.
Blumbach, B., Diehl-Seifert, B., Seack, J., Steffen, R., Müller, I.M., and Müller, W.E.G., 1999, Cloning and expression of new receptors belonging to the immunoglobulin superfamily from the marine sponge *Geodia cydonium, Immunogenetics* **49**:751–763.
Böhm, M., Hentschel, U., Friedrich, A., Fieseler, L., Steffen, R., Gamulin, V., Müller, I.M., and Müller, W.E.G., 2001, Molecular response of the sponge *Suberites domuncula* to bacterial infection, *Marine Biology* **139**:1037–1045.
Borchiellini, C., Manuel, M., Alivon, E., Boury-Esnault, N., Vacelet, J., Le Parco, Y., 2001, Sponge paraphyly and the origin of Metazoa, *J. Evol. Biol.* **14**:171–179.
Breter, H.J., Grebenjuk, V.A., Skorokhod, A., and Müller, W.E.G., 2003, Approaches for a sustainable use of the bioactive potential in sponges: analysis of gene clusters, differential display of mRNA and DNA chips. in: *Sponge (Porifera),* W.E.G. Müller, ed., Marine Molecular Biotechnology – Springer, Berlin; pp. 199–230.
Bringmann, G., Lang, G, Mühlbacher, J., Schaumann, K., Steffens, S., Rytik, P.G., Hentschel, U., Morschhäuser, J., and Müller, W.E.G., 2003, Sorbicillactone A, a structurally unprecedented bioactive novel-type alkaloid From a sponge-derived fungus, in: in: *Sponge (Porifera),* W.E.G. Müller, ed., Marine Molecular Biotechnology – Springer, Berlin. pp. 231–253.
Buchan, A., Collier, L.S., Neidle, E.L., and Moran, M.A., 2000, Key aromatic-ring-cleaving enzyme, protocatechuate 3,4-dioxygenase, in the ecologically important marine *Roseobacter* lineage, *Appl. Environ. Microbiol.* **66**:4662–4672.
Burton, M., 1953, *Suberites domuncula* (Olivi): its synonymy, distribution, and ecology, *Bull. Brit. Mus. (Nat. Hist.)* **1**:353–378.
Campbell, L.G., 1876, *Log Letters from "The Challenger",* MacMillan,London.
Camus, M., 1783, *Histoire des Animaux d' Aristote.* Desaint, Paris.
Celesia, P., 1893, Della *Suberites domuncula* e della sua simbiosi coi Paguri, *Atti Sic. Ligustica di Sc. nat.* Vol. IV, Tav. V.
Cimino, G., and Ghiselin, M.T., 2001, Marine natural products chemistry as an evolutionary narrative, in: *Marine Chemical Ecology,* J.B. McClintock, B.J. Baker, eds., CRC Press, Boca Raton, pp. 115–154.
Coupin, H., 1902, *Les Animaux Excentrique.* Vuibert, Paris.
Custodio, M.R., Prokic, I., Steffen, R., Koziol, C., Borojevic, R., Brümmer, F., Nickel, M., and Müller, W.E.G., 1998, Primmorphs generated from dissociated cells of the sponge *Suberites domuncula*: A model system for studies of cell proliferation and cell death. *Mech. Ageing Develop.* **105**:45–59.
Darwin, E. – Brandis, J.D., 1799, *Zoonomie oder Gesetze des Organischen Lebens.* Hahn, Hannover.
DeLage, Y., 1892, Embryogénie des éponge, *Arch. De Zool. Exp. (sér. 2)* **10**:345–498.
Donati, V., 1753, *Auszug seiner Natur-Geschichte des Adriatischen Meers.* CP Franckens, Halle.
Dzik, J., 1991, The fossil evidence consistent with traditional views of the early metazoan phylogeny, in: *The Early Evolution of Metazoa and the Significance of Problematic Taxa,* A.M. Simonetta and S. Conway Morris, eds., Cambridge University Press, Cambridge; pp. 47–56.
Ebel, R., Brenzinger, M., Kunze, A., Gross, H., and Proksch, P., 1997, Wound activation of prototoxins in the marine sponge Aplysina *aerophoba. J. Chem. Ecol.* **23**:1451–1462.
Ellis, J., 1786, *The Natural History of Many Curious and Uncommon Zoophytes, Collected from Various Parts of the Globe.* Benjamin White, London.
Esper, E.J.C., 1794, *Die Pflanzenthiere,* Raspe, Nürnberg.
Faulkner, D.J., 1995, Marine natural products, *Nat. Prod. Rep.* **13**:259–302.
Faulkner, D.J., 2000, Marine natural products, *Nat. Prod. Rep.* **17**:7–55
Field, K.G., Olsen, G.J., Lane, D.J., Giovannoni, S.J., Ghiselin, M.T., Raff, E.C., Pace, N.R., and Raff, R.A., 1988, Molecular phylogeny of the animal kingdom. *Science* **239**:748–753.
Galliot, B., and Miller, D., 2000, Origin of anterior patterning – how old is our head? *Trends Genet,* **16**:1–5.

Geoffroy, M., 1731, Analyse chimique de l'eponge de la moyenne espece, *Histoire de l'Acad. Roy. Scie.* **1731**:507–508.
Graeffe, E., 1882, Übersicht der Seethierfauna des Golfes von Triest, *Arb. Zool. Inst. Wien-Triest* **4**:313–321.
Gräfe, C.F., 1826, Ueber Minderung der Gefahr beim Kaiserschnitte, nebst der Geschichte eines Falles, in welchem Mutter und Kind erhalten wurden, *J. Journal der Chirurgie und Augen-Heilkunde* **9/1**:1–170.
Grant, R.E., 1836, Animal Kingdom, in: *The Cyclopaedia of Anatomy and Physiology*, R.B. Todd, ed., vol 1. Sherwood-Gilbert-Piper, London.
Grebenjuk, V.A., Kuusksalu, A., Kelve, M., Schütze, J., Schröder, H.C., and Müller, W.E.G., 2002, Induction of (2′–5′)oligoadenylate synthetase in the marine sponges *Suberites domuncula* and *Geodia cydonium* by the bacterial endotoxin lipopolysaccharide, *Europ. J. Biochem.* **269**:1382–1392.
Haeckel, E., 1870, *Das Leben in den grössten Meerestiefen*, Lüderitz-Charisius, Berlin.
Haeckel, E., 1872, *Atlas der Kalkschwämme*, Reimer, Berlin.
Haeckel, E., 1874, Die Gastrae-Theorie, die phylogenetische Classification des Thierreichs und die Homologie der Keimblätter, *Jenaische Z. f. Naturwiss.* 8:1–55.
Haeckel, E., 1896, *Systematische Phylogenie der Wirbellosen Thiere*, Reimer, Berlin.
Heider, K., 1886, Zur Metamorphose der *Oscalella lobularis* O. Schm. *Arbeiten Zool. Inst. Univ. Wien* **6–2**:27–30.
Henkart, P., Humphreys, S., and Humphreys, T., 1973, Characterization of sponge aggregation factor. A unique proteoglycan complex. *Biochem.* **12**: 3045–3050.
Hildemann, W.H., Johnston, I.S., and Jokiel, P.L., 1979, Immunocompetence in the lowest metazoan phylum: Transplantation immunity in sponges, *Science* **204**:420–422.
Hirabayashi, J., and Kasai, K., 1993, The family of metazoan metal-independent ß-galactoside-binding lectins: structure, function and molecular evolution, *Glycobiol.* 3:297–304.
Hufeland, C.W., 1798, *Ueber die Natur, Erkenntnißmittel und Heilart der Skrofelkrankheit*, Ghelensche Schriften, Wien.
Jameson, R., 1811, Catalogues of animals, of the class Vermes, found in the Firth of Fourth, an other parts of Scotland, *Mem. Werner. Soc.* **I**:556–565.
Jussieu, A.L., 1789, *Genera Plantarum Secundum Ordines Naturales Disposita*, Herrissant, Paris.
Khosla, C., Gokhale, R.S., Jacobsen, J.R., and Cane D.E., 1994, Tolerance and specificity of polyketide synthetases. *Annu. Rev. Biochem.* **68**:219–253.
Kimura, M., 1983, *The Neutral Theory of Molecular Evolution*, Cambridge University Press: Cambridge.
Knoll, A.H., and Carroll, S.B., 1999, Early animal evolution: emerging views from comparative biology and geology, *Science* **284**:2129–2137.
Koziol, C., Kobayashi, N., Müller, I.M., and Müller, W.E.G., 1998, Cloning of sponge heat shock proteins: Evolutionary relationships between the major kingdoms, *J. Zool. Syst. Evol. Res.* **36**:101–109.
Krasko, A., Gamulin, V., Seack, J., Steffen, R., Schröder, H.C., and Müller, W.E.G., 1997, Cathepsin, a major protease of the marine sponge *Geodia cydonium*: purification of the enzyme and molecular cloning of cDNA, *Molec. Marine Biol. & Biotechnol.* **6**:296–307.
Kreuter, M.H., Robitzki, A., Chang, S., Steffen, R., Michaelis, M., Kljajic, Z., Bachmann, M., Schröder, H.C., and Müller, W.E.G., 1992, Production of the cytostatic agent, aeroplysinin by the sponge *Verongia aerophoba* in *in vitro* culture, *Comp. Biochem. Physiol.* **101C**:183–187.
Kruse, M., Müller, I.M., and Müller, W.E.G., 1997, Early evolution of Metazoan serine/threonine- and tyrosine kinases: Identification of selected kinases in marine sponges, *Mol. Biol. Evol.* **14**:1326–1334.
Kruse. M., Leys. S.P., Müller, I.M., and Müller, W.E.G., 1998, Phylogenetic position of Hexactinellida within the phylum Porifera based on amino acid sequence of the protein kinase C from *Rhabdocalyptus dawsoni*, *J. Mol. Evolution* **46**:721–728.
Kuusksalu, A., Pihlak, A., Müller, W.E.G., and Kelve, M., 1995, The (2′-5′) oligoadenylate synthetase is present in the lowest multicellular organisms, the marine sponges: demonstration of the existence and identification of its reaction products. *Eur. J. Biochem.* **232**:351–357.
Lamarck, J.B.,1797, *Mémoires de Physique et d'Histoire Naturelle*, Agasse-Maradan, Paris.
Laurent, L., 1842, *Recherches sur L'Hydre et L'Éponge D'Eau Douce*, Bertrand, Paris.
Lendenfeld, R. v., 1889, *A Monograph of the Horny Sponges*, Royal Society London.
Lieberkühn, N., 1856, Zur Entwicklungsgeschichte der Spongillen. *Arch. Anat. Physiol.* **1856**:399–414.
Lieberkühn, N., 1857, Beiträge zur Anatomie der Spongien, *Arch. Anat. Physiol.* **1857**:376–403.
Linné, C., 1788, *Systema Naturae*, 13th edition. Beer, Lipsiae.
Loomis, W.F., 1988, *Four Billion Years*, Sinauer, Sunderland.

Mabberley, D.J., 1985, *Jupiter Botanicus. Robert Brown of the British Museum*, J. Cramer, Braunschweig.
Maldonado, M., and Uriz, M.J., 1999, Sexual propagation by sponge fragments, *Nature* **398**:476.
Mehl, D., Müller, I., and Müller, W.E.G., 1998, Molecular biological and palaeontological evidence that Eumetazoa, including Porifera (sponges), are of monophyletic origin, in: *Sponge Science–Multidisciplinary Perspectives*, Y Watanabe, and N Fusetani, eds., Tokyo: Springer-Verlag, pp. 133–156.
Metchnikoff, É., 1892, *Leçons sur la Pathologie Comparée de l'Inflammation*. Masson, Paris.
Moore, H.F., 1908, The commercial sponges and the sponge fisheries, *Bull. Bureau Fisheries* **28**:399–511.
Moscona, A.A., 1968, Cell aggregation: properties of specific cell-ligands and their role in the formation of multicellular systems, *Devel. Biol.* **18**: 250–277.
Müller, K., 1911, Das Regenerationsvermögen der Süßwasserschwämme, *Archiv f Entwicklungsmechanik* **32**: 397–446.
Müller, W.E.G., 1995, Molecular phylogeny of Metazoa (animals): monophyletic origin, *Naturwiss.* **82**:321–329.
Müller, W.E.G., 1997, Origin of metazoan adhesion molecules and adhesion receptors as deduced from their cDNA analyses from the marine sponge *Geodia cydonium*, *Cell & Tissue Res.* **289**:383–395.
Müller, W.E.G., 1998, Origin of Metazoa: Sponges as living fossils, *Naturwiss.* **85**:11–25.
Müller, W.E.G., 2001, How was metazoan threshold crossed: the hypothetical Urmetazoa, *Comp. Biochem. Physiol.* [A] **129**:433–460.
Müller, W.E.G., 2003a, The origin of metazoan complexity: Porifera as integrated animals. Integ Comp Biol; in press.
Müller W.E.G (ed.) 2003b, *Sponge (Porifera)*, Marine Molecular Biotechnology – Springer, Berlin.
Müller, W., and Zahn, R.K., 1968, Tethya limski n.sp, eine Tethyide aus der Adria (Porifera: Homosclerophorida: Tethyidae). *Senckenbergiana Biol.* **49**:469–478.
Müller, W.E.G., and Zahn, R.K., 1973, Purification and characterization of a species-specific aggregation factor in sponges, *Exp. Cell Res.* **80**:95–104.
Müller, W.E.G., and Schäcke, H., 1996, Characterization of the receptor protein-tyrosine kinase gene from the marine sponge *Geodia cydonium*, *Prog. Molec. Subcell. Biol.* **17**:183–208.
Müller, W.E.G., and Müller, I.M., 2003, Origin of the metazoan immune system: identification of the molecules and their functions in sponges. *Integr. Comp. Biol.* **43**; in press.
Müller, W.E.G., Rohde, H.J., Beyer, R., Maidhof, A., Lachmann, M., Taschner, H., and Zahn, R.K., 1975, Mode of action of 9-ß-D-arabinofuranosyladenine on the synthesis of DNA, RNA and protein *in vivo* and *in vitro*. *Cancer Res.* **35**:2160–2168.
Müller, W.E.G, Zahn, R.K., Bittlingmeier, K., and Falke, D., 1977, Inhibition of herpesvirus DNA-synthesis by 9-ß-D-arabinofuranosyladenine *in vitro* and *in vivo*, *Ann. New York Acad. Sci.* **284**:34–48.
Müller, W.E.G., Zahn, R.K., Rijavec, M., Britvic, S., Kurelec, B., and Müller, I., 1979, Aggregation of sponge cells. The aggregation factor as a tool to establish species. *Biochem. Systematics and Ecology* **7**:49–55.
Müller, W.E.G., Conrad, J., Schröder, C., Zahn, R.K., Kurelec, B., Dreesbach, K., and Uhlenbruck, G., 1983, Characterization of the trimeric, self-recognizing *Geodia cydonium* lectin I. *Europ. J. Biochem.* **133**: 263–267.
Müller, W.E.G., Müller, I.M., and Gamulin, V., 1994, On the monophyletic evolution of the Metazoa, *Brazil. J. Med. Biol. Res.* **27**:2083–2096.
Müller, W.E.G., Blumbach, B., and Müller, IM., 1999a, Evolution of the innate and adaptive immune systems: relationships between potential immune molecules in the lowest metazoan phylum [Porifera] and those in vertebrates, *Transplantation* **68**:1215–1227.
Müller, W.E.G., Wiens, M., Batel, R., Steffen, R., Borojevic, R., and Custodio, M.R., 1999b, Establishment of a primary cell culture from a sponge: primmorphs from *Suberites domuncula*. *Marine Ecol. Progr. Ser.* **178**:205–219.
Müller, W.E.G., Böhm, M., Batel, R., De Rosa, S., Tommonaro, G., Müller, I.M., and Schröder, H.C., 2000, Application of cell culture for the production of bioactive compounds from sponges: synthesis of avarol by primmorphs from *Dysidea avara*, *J. Nat. Prod.* **63**:1077–1081.
Müller, W.E.G., Schröder, H.C., Skorokhod, A., Bünz, C., Müller, I.M., and Grebenjuk, V.A., 2001, Contribution of sponge genes to unravel the genome of the hypothetical ancestor of Metazoa (Urmetazoa), *Gene* **276**: 161–173.
Müller, W.E.G., Krasko A., Skorokhod A., Bünz C., Grebenjuk V.A., Steffen R., Batel R., Müller I.M., and Schröder, H.C., 2002, Histocompatibility reaction in the sponge *Suberites domuncula* on tissue and cellular level: central role of the allograft inflammatory factor 1, *Immunogenetics* **54**:48–58.

Müller, W.E.G., Grebenjuk, V.A., Le Pennec, G., Schröder, H.C., Brümmer, F., Hentschel, U., Müller, I.M., and Breter, H.J., 2003a, Sustainable production of bioactive compounds by sponges: cell culture and gene cluster approach. Marine Biotechnol; in press.

Müller, W.E.G., Krasko, A., Le Pennec, G., Steffen, R., Ammar, M.S.A., Müller, I.M., and Schröder, H.C., 2003b, Molecular mechanism of spicule formation in the demosponge *Suberites domuncula*: Silicatein - collagen – myotrophin, *Progr. Molec. Subcell. Biol.* **33**:195–221.

Müller, W.E.G., Brümmer, F., Batel, R., Müller, I.M., and Schröder, H.C., 2003c, Molecular biodiversity. Case study: Porifera (sponges). *Naturwissenschaften* **90**:103–120.

Müller, W.E.G., Korzhev, M., Le Pennec, G., Müller, I.M. and Schröder, H.C., 2003d, Origin of metazoan stem cell system in sponges: first approach to establish the model (*Suberites domuncula*). *Biomolecular Engineering*; in press.

Muralt, J. v., 1692, *Hippocrates Helveticus oder der getreu-sichere und wohl-bewährte Eydgnössische Stadt-Land-und Hauß Artzt*, König, Basel.

Nardo, G.D., 1834, Possibile applicazione alle arti degli aghi silicei costituenti il tessuto solido di alcuni Spongiali del Mare Adriatico, *Giorn. Tecno. e Belle Art.*, p 83.

Nielsen, C., 2001, *Animal Evolution*, Oxford University Press, Oxford.

Okulitch, V.J., 1955, Archaeocyatha and Porifera, in: *Treatise on Invertebrate Paleontology*, R.C. Moore, ed., University of Kansas Press, New York. pp. E1–E20.

Olivi, G., 1792, Zoologia Adriatica ossia catalogo regionato degli animali del golfo e delle lagune di Venezia. Bassano 1–32.

Pallas, P.S., 1787, *Charakteristik der Thierpflanzen*, Raspe, Nürnberg.

Pancer, Z., Kruse, M., Schäcke, H., Scheffer, U., Steffen., R., Kovács., P, and Müller, W.E.G., 1996, Polymorphism in the immunoglobulin-like domains of the receptor tyrosine kinase from the sponge *Geodia cydonium*. *Cell Adhesion and Commun.* **4**:327–339.

Pancer, Z., Kruse, M., Müller, I., and Müller W.E.G., 1997, On the origin of adhesion receptors of metazoa: cloning of the integrin α subunit cDNA from the sponge *Geodia cydonium*, *Molec. Biol. Evol.* **14**:391–398.

Paris, J., 1961, Contribution a la biologie des éponges siliceuses *Tethya lyncurium* Lmck. et *Suberites domuncula* O.: histologie des greffes et sérologie, *Vie et Milieu* **11** (Suppl.): 1–74.

Pfeifer, K., Haasemann, M., Gamulin, V., Bretting, H., Fahrenholz, F., and Müller, W.E.G., 1993, S-type lectins occur also in invertebrates: high conservation of the carbohydrate recognition domain in the lectin genes from the marine sponge *Geodia cydonium*. *Glycobiol.* **3**:179–184.

Pfeifer, K., Schröder, H.C., Rinkevich, B., Uhlenbruck, G., Hanisch, F.-G., Kurelec, B., Scholz, P., and Müller, W.E.G., 1992, Immunological and biological identification of tumor necrosis factor in sponges: role of this factor in the formation of necrosis in xenografts. *Cytokine* **4**:161–169.

Phipson, T.L., 1864, *The Utilization of Minute Life*. Groombridge, London.

Proksch, P., 1994, Defensive roles for secondary metabolites from marine sponges and sponge-feeding nudibranchs, *Toxicon* **32**:639–655.

Proksch, P., Edrada, R.A., and Ebel, R., 2002, Drugs from the seas – current status and microbiological implications, *Appl. Microbiol. Biotechnol.* **59**:125–134.

Pronzato, R., 1999, Sponge-fishing, disease and farming in the Mediterranean Sea, *Aquatic Conser. Mar. Freshw. Ecosys.* **9**:485–493.

Rawlings, B.J., 1997, Biosynthesis of polyketides, *Natural Product Rep.* **1997**:523–556.

Reiswig, H., 1971, *In situ* pumping activities of tropical demospongiae. *Mar. Biol.* **9**:38–50.

Rodrigues-Nieto, S., Gozáles-Iriarte, M., Carmona, R., Munoz-Chápuli, R., Medina, M.A., and Quesada, A.R., 2001, Anti-angiogenic activity of aeroplysinin-1, a brominated compound isolated from a marine sponge. *FASEB J.* (published online Dec. 28, 2001).

Rust, J.N., 1835, *Theoretisch-praktisches Handbuch der Chirurgie, mit Einschluss der syphilitischen und Augen-Krankheiten*. Vol. 15. Enslin, Berlin.

Sarma, A.S., Daum, T., and Müller, W.E.G., 1993, *Secondary metabolites from marine sponges*. Akademie gemeinnütziger Wissenschaften zu Erfurt, Ullstein-Mosby Verlag, Berlin.

Saville Kent, W., 1880–1881, *A Manual of the Infusoria: Including a Description of all Known Flagellate, Ciliate, and Tentaculiferous Protozoa, British and Foreign, and an Account of the Organization and Affinities of the Sponges*. David Bouge, London.

Scheuer, P.J., 1990, Some marine ecological phenomena: chemical basis and biomedical potential, *Science* **248**:173–177.

Schillak, L., Ammar, M.S.A., and Müller, W.E.G., 2001, Transplantation of coral species to electrochemical produced hard substrata: *Stylophora pistillata* (Esper, 1797) and *Acropora humilis* (Dana, 1846). Mombasa, Kenya, 19–22 June 2000, *Brussels, ACP-EU Fish. Res. Rep.*, (10): p. 68–84

Schmidt, O., 1862, *Die Spongien des Adriatisches Meeres*, Engelmann, Leipzig.

Schmidt, O., 1864, *Spongien des Adriatisches Meeres* – Supplement, Engelmann, Leipzig.

Schröder, H.C., Efremova, S.M., Itskovich, V.B., Krasko, A., Müller, I.M., and Müller, W.E.G., 2003a, Molecular phylogeny of the freshwater sponges in Lake Baikal. *J. Zool. Syst. Evol. Research* **41**:80–86.

Schröder, H.C., Brümmer, F., Fattorusso, E., Aiello, A., Menna, M., De Rosa, S., Batel, R., and Müller, W.E.G., 2003b, Sustainable production of bioactive compounds from sponges: primmorphs as bioreactors, in: *Marine Molecular Biotechnology*, W.E.G. Müller, ed., pp. 163–197.

Schröder, H.C., Krasko, A., Le Pennec, G., Adell, T., Hassanein, H., Müller, I.M., Müller, W.E.G., 2003c, Silicase, an enzyme which degrades biogenous amorphous silica: contribution to the metabolism of silica deposition in the demosponge *Suberites domuncula*, *Progr. Molec. Subcell. Biol.* **33**:250–268.

Schröder, H.C., Sudek, S., De Caro, S., De Rosa, S., Perovic, S., Steffen, R., Müller, .IM., and Müller, W.E.G., 2002d, Synthesis of the neurotoxin quinolinic acid in apoptotic tissue from *Suberites domuncula*: cell biological, molecular biological and chemical analyses, *Marine Biotechnol.* **4**:546–558.

Schuhmacher, H., and Schillak, L., 1994, Integrated electrochemical and biogenic deposition of hard material: a nature-like colonization substrate, *Bull. Marine Sci.* **55**: 672–679.

Schulze, F.E., 1887, *Zur Stammesgeschichte der Hexactinelliden*, Reimer, Berlin.

Scott, L., 1910, Über Spongin, *Biochem. Z.* **27**:266–269.

Simpson, T.L., 1984, *The Cell Biology of Sponges*, Springer-Verlag, New York

Sipkema, D., van Wielink, R., van Lammeren, A.A.M., Tramper, J., Osinga, R., and Wijffels, R.H., 2003, Primmorphs from seven marine sponges: formation and structure, *J. Biotechnol.* **100**:127–139.

Skorokhod, A., Gamulin, V., Gundacker, D., Kavsan, V., Müller, I.M., and Müller, W.E.G., 1999, Origin of insulin receptor tyrosine kinase: cloning of the cDNAs from marine sponges, *Biol. Bull.* **197**:198–206.

Sollas, W.J., 1888, Report on the Tetractinellida, in: *Report on the Scientific Results of the Voyage of H.M.S. Challenger*; vol. 25.

Spencer, H., 1867, *The Principles of Biology*, Williams and Nogate, London.

Steuer, A., 1933, Zur Fauna des Canal di Leme bei Rovigno, *Thalassia* **1**:1–43.

Sumerel, J.L., and Morse, D.E., 2003, Biotechnological advances in biosilicification, *Progr. Molec. Subcell. Biol.* **33**:225–247.

Thakur, N.L., Hentschel, U., Krasko, A., Anil, A.C., and Müller, W.E.G., 2003, Antibacterial activity of the sponge *Suberites domuncula* and its primmorphs: potential basis for chemical defense, *Aquatic Microbiol. Ecol.* **31**:77–83.

Thomson, C.W., 1868, On the "vitreous" sponges. *Ann. Mag. Natur. Hist.* **1**:114–115.

Van Soest, R.W.M., 1994, Demosponge distribution patterns. in: *Sponges in Time and Space*, R.W.M. van Soest, and A.A. Balkema, eds., Brookfield, Rotterdam. pp. 213–223.

Van Soest, R.W.M., and Braekman, J.C., 1999, Chemosystematics of Porifera: a review. *Memoire of the Queensland Museum* **44**:569–589.

Wagner, C., Steffen, R., Koziol, C., Batel, R., Lacorn, M., Steinhart, H., Simat, T.m and Müller, W.E.G., 1998, Apoptosis in marine sponges: a biomarker for environmental stress (cadmium and bacteria), *Marine Biol.* **131**:411–421.

Weinbaum, G., and Burger, M.M., 1973, A two-component system for surface guided reassociation of animal cells, *Nature* **244**:510–512.

Weismann, A., 1892, *Das Keimplasma: Eine Theorie der Vererbung*, Fischer, Jena.

Wiens, M., Kuuksalu, A., Kelve, M., and Müller, W.E.G., 1999, Origin of the interferon-inducible (2'–5')oligoadenylate synthetases: cloning of the (2'–5')oligoadenylate synthetase from the marine sponge *Geodia cydonium*, *FEBS Letters* **462**:12–18.

Wiens, M., Krasko, A., Müller, C.I., and Müller, W.E.G., 2000a, Molecular evolution of apoptotic pathways: cloning of key domains from sponges (Bcl-2 homology domains and death domains) and their phylogenetic relationships, J. *Mol. Evol.* **50**:520–531.

Wiens, M., Krasko, A., Müller, I.M., and Müller, W.E.G., 2000b, Increased expression of the potential proapoptotic molecule DD2 and increased synthesis of leukotriene B4 during allograft rejection in a marine sponge, *Cell Death Diff.* **7**:461–469.

Wiens, M., Diehl-Seifert, B., and Müller, W.E.G., 2001, Sponge Bcl-2 homologous protein (BHP2-GC) confers distinct stress resistance to human HEK-293 cells, *Cell Death Diff.* **8**:887–898.

Wiens, M., Mangoni, A., D'Esposito, M., Fattorusso, E., Korchagina, N., Schröder, H.C., Grebenjuk, V.A, Krasko, A., Batel, R. , Müller, I.M., and Müller, W.E.G., 2003a, The molecular basis for the evolution of the metazoan bodyplan: extracellular matrix-mediated morphogenesis in marine demosponges, *J. Molec. Evol.*; in press.

Wiens M., Krasko A., Perovic S., and Müller W.E.G., 2003b, Caspase-mediated apoptosis in sponges: cloning and function of the phylogenetic oldest apoptotic proteases from metazoa, *Biochim. Biophys. Acta* **1593**:179–189.

Wilson, H.V., 1907, On some phenomena of coalescence and regeneration in sponges, *J. Exptl. Zool.* **5**:245–258.

Xiao, S., Zhang, Y., and Knoll, A.H., 1998, The three-dimensional preservation of algae and animal embryos in a neoproterozoic phosphorite, *Nature* **391**:553–558.

Xue, Y., Zhou, C., and Tang, T., 1999, "Animal embryos", a misinterpretation of neoproterozoic microfossils, *Acta Micropalaeontol. Sinica* **16**:1–4.

# 23

# Earthworms: Sources of Antimicrobial and Anticancer Molecules

Edwin L. Cooper, Binggen Ru and Ning Weng

## 1. INTRODUCTION

### 1.1. Earthworms in Medicine: A Brief Historical Excursion

*1.1.1. 20th Century Accounts*

According to Reynolds and Reynolds, (1972) few people know of the earthworm's long association with medicine, yet documents recording its use in various remedies date back to 1340 A.D.[1,2] For instance, doctors practicing folk medicine in Burma and India use earthworms in treatment of various diseases. In Burma, the primary use for earthworms is in the treatment of a disease called *ye se kun byo*, which display the symptoms of pyorrhea. The worms are heated in a closed pot until reduced to ashes and these ashes are used either alone as a tooth powder, or for greater palatability, are combined with roasted tamarind seeds and betel nuts.[3] With another disease *meephwanoyeekhun thwaykhan*, women generally feel postpartal weakness and are unable to nurse their infant. Worms are boiled in water with salt and onions and the clear fluid is decanted and mixed with the patient's food.[3] It is interesting to note that the patient is kept ignorant of the nature of the medicine!

*1.1.2. Healing In Southeast and The Middle East*

The medicinal properties of earthworms are known and have been used in many different countries and cultures. For instance, earthworms have been used to treat smallpox in Burma and Laos. Worms are soaked in water and the patient is bathed in the liquid. The worms are then roasted, powdered, mixed with coconut water, and drunk. This treatment hastens the course of the disease and reduces mortality from 100 percent to 25 percent[4] In Iran the earthworm is also considered to be an all-round wonder drug. Earthworms are

---

**Edwin L. Cooper** • Laboratory of Comparative Immunology, Department of Neurobiology, David Geffen School of Medicine at UCLA, University of California, Los Angeles   **Binggen Ru** • National Laboratory of Protein Engineering, College of Life Science, Peking University, Beijing   **Ning Weng** • Department of Biochemistry and Molecular Biology, College of Life Science, Peking University, Beijing

*Complementary and Alternative Approaches to Biomedicine*, edited by Edwin L. Cooper and Nobuo Yamaguchi. Kluwer Academic/Plenum Publishers, 2004.

baked and eaten with bread to reduce the size of a stone in the bladder and bring about its expulsion. In addition, they are dried and eaten to cure the yellowness of a jaundiced patient. Furthermore, earthworm ashes are applied to the head with oil of roses to promote hair growth.[3]

### 1.1.3. Biochemistry and Reports from USA

Closer to the US, Carr described how the Cherokee Indians of the Great Smokey Mountains used earthworm poultices to draw out thorns. "Just make your poultices of chopped up worms. Draws so powerfully, remove soon."[5] The Nanticoke Indians of Delaware used earthworms in a well-known remedy for the pain of rheumatism [6,5] "Put fishing worms in a bottle let die and apply to stiff joints. Smells bad, but sure helps."[5] Biochemists have investigated the lipids of earthworms and discovered earthworm fatty acids that enter into therapeutics. Scientists have also isolated a bronchial dilating substance from earthworms. Thus we see that the Indians have made discoveries anciently known to them in their medical lore, but known only relatively recently to the medical world at large.

### 1.1.4. An Anthology of Chinese Herbs and Medicines

Recently we have noted the book entitled: The Eu Yan Sang Heritage: An Anthology of Chinese Herbs and Medicines. According to the front note: "The information in this book is extracted from various established medical classics and is to be used as reference material only. Please consult a physician or seek professional advice before taking any medication[7]. The book is divided into eight major sections: I. The Eu Yan Sang Heritage; II Superior Chinese Herbs and Fine Medicines III. Commonly used Chinese Herbs: Benefits, Action and Indications:(sources from seeds, pericarp, fruits, bulbs, roots, rhizomes, leaves, flowers, whole plants, barks, branches, minerals and resins, marine, animals, insects, cryptic); IV. Traditional Herbal Remedies from the Medical Classics; V. Simple and Effective Collection of Home Remedies; VI. Specialized Pills, Fine Powder and Capsules, and First Aid Series; VII. Family-style Soups and Tonic Soups, Teas, Wines and Nourishing Essence (Index of Chinese Herbs; Index of Eu Yan Sang products featured in the Anthology). The earthworm source is of particular interest for this chapter is found on page 146 accompanied by a picture 259. Dilong, Earthworm (*Pheretima aspergillum*). "Cold, Influences the bladder, liver, lung and spleen channels. Stops spasms, reduces heat-toxins, settles wheezing, promotes urination, and unblocks and activates the channels. Commonly used to treat high fever with convulsions, swollen and painful joints, long-term cough, difficult urination and high blood pressure."

As to whether the remedies mentioned previously actually work, we do not know. If we believe the reports of improved conditions, we still do not know whether these are truly cause and effect situations. Not so long ago, many would have questioned the practice of inserting needles into the body as a cure for various aches and pains. The fact remains that these remedies have been employed for centuries were carried down to future generations and were recorded for posterity. With respect to the current review, there is pertinent work as will be reviewed later. According to Rong Fan, 1996, lumbrokinase (LK) is a group of proteolytic enzymes, including plasminogen activator and plasmin, separated biochemically from a certain earthworm type.[8] **In the remainder of the review we will present evidence concerning products from earthworms as related to: 1) thrombosis in general; 2) basic science concerning lytic properties of the earthworm's cellular and humoral immune system; 3) possible role of lytic molecules that may be currently employed in certain clinical situations that may be akin to those more folkloric approaches.**

## 1.2. Earthworm Immune System: Cells and Molecules

As will be seen later, knowledge of the earthworm's immune system has played a vital role in understanding the nature of the cells and the lytic molecules that they produce. This is especially the case when the immune system of the earthworm is analyzed for the sake of the earthworm, without any immediate intention of assigning a clinical role.[9–29] Earthworms are as numerous as there is habitable soil and they range from small familiar species to the gargantuan *Megascolides australis*, an Australian species that reaches lengths of over 3 meters! Innate immunity in earthworms is a system that has received less attention than that of *Drosophila* but which has much interest from the dual perspective of both immune system function and evolution. In fact, the capacity to destroy bacteria and other microbes apparently exists in parallel with the capacity to lyse foreign, eukaryotic cells, two responses found in many invertebrates [Tables 23.1, 23.2, 23.3, 23.4]. The earthworm's body cavity contains coelomic fluid and leukocytes that are as varied as they are in other equally complex invertebrates and they resemble certain vertebrate leukocytes with respect to morphology, cytochemistry and function[30]

Both the leukocytes and the fluid that they synthesize and secrete, affect immunobiological responses [13,18,31–36]. They do so by various routes: 1) opsonization; 2) inflammation and phagocytosis; 3) agglutination; 4) mitogenesis; 5) lysis; 6) destruction of experimentally introduced allogeneic, xenogeneic but not autogeneic transplants *in vivo* and various target cell types *in vitro*. With respect to leukocytes, Roch proposed that cellular activities are based mainly upon phagocytosis and leukocyte cell-to-cell recognition.[36] The latter leads to cytotoxicity, mixed lymphocyte stimulation like-reactions and cellular cooperation.[37–39]

**Table 23.1.** General innate immune components in invertebrates

Phagocytic cells
Encapsulation (granuloma/nodule formation see ERP)
Transplant rejection (evidence of accelerated second-set rejection)
Proteases and protease inhibitors
Lysozyme
Lysins
Agglutinins
C-Type lectins [Ca++-dependent lectins]
S-Type lectins (Galectins)
Prophenoloxidase system (proPO)
Metal-binding proteins (cysteine-rich metallothioneins)
Cytokine-like (functional analogues; no sequence data)
Transferrin and lactoferrin
Bacterial Permeability Increasing Protein (BPI)

**Table 23.2.** Innate immune components in particular invertebrates

| Component/Characteristic | Metazoan model |
|---|---|
| Encapsulating-Relating Proteins (ERPs) | Insect (*Tenebrio molitor*) |
| Eicosanoids | Insects |
| Inducible inflammatory system (Toll, NFκB) | Insects/arthropods, Caenorhabditis elegans |
| Acute phase proteins: Pentraxins, Ig superfamily complement | Insects/arthropods |
| $\alpha_2$-macroblobulin family (protease inhibitor) | Nematodes, crustaceans, mollusks, echinoderms tunicates; horseshoe crab (*Limulus polyphemus*) |
| Lipoprotein-receptor related protein (LRP/$\alpha_2$ M-R) | *C. elegans* |
| Mannose Binding Lectin (MBL) | *C. elegans*, tunicates |
| Antimicrobial peptides (Defensins): Cecropins, Drosomycin, Drosocin, Attacin Metchnikowin Mytilins; Myticins | *Drosophila Mytilus* |
| *Galloprovincialis* | |
| Lytic and Antimicrobial Components: Lysins (fetidins, lysenin, eiseniapore, Eisenin) Antimicrobial peptide: non-lytic lumbricin I TNF analogue Pattern recognition molecule (CCF) | Earthworm |
| Reactive oxygen species (ROS) | Mollusks |
| Fibrinogen Related Proteins (FREPS) | Mollusks (planorbid snails) |
| NK-like cells associated with CD markers | Sipunculids, annelids, mollusks |
| Tachylectins | Horseshoe crab |
| Acute Phase Proteins: | |
| C-reactive protein, LPS-binding protein (LBP) | Crustaceans, horseshoe crab |
| C3 homologue | Echinoderms; tunicates |
| Collectin (collagenous, carbohydrate binding protein) | Tunicates, horseshoe crab |
| Lectin-mediated complement pathway | Tunicates, horseshoe crab |
| Mannose Binding Protein (MBP) | Tunicates |
| Mannose Associated Serine Protease (MASP) | Tunicates |
| Antimicrobial peptides: Clavanins, styelins | Tunicates |

The focus of this paper concerns the activities of humoral products. Humoral activities include lysozyme, synthesis and secretion of agglutinins, a phenoloxidase/peroxidase system, and synthesis and expression of the first to be discovered lytic components, the fetidins.[15,40–43] We will focus on earthworm immunity by: 1) identifying certain molecules that sequester experimental antigens (bacteria, erythrocytes, cancer cells: the lysins [primarily fetidins lysenin, CCF-1, eiseniapore]; 2) presenting evidence of their possible therapeutic applications that may be already employed in certain countries. Because of the lytic properties the review will focus on the properties of the lytic components and how they may be applied to clinical situations notably thrombosis. It will become clear that products of the earthworm's immune system, the lytic components may not be the same as those molecules that have been isolated for clinical purposes. At most these molecules may be shared in

**Table 23.3.** Proteins from *Eisenia foetida* that exert cytolytic functions (except antimicrobial lumbricin I from *Lumbricus rubellus*)

| Name | Accession. No. | Homology | Presence | Function |
|---|---|---|---|---|
| **Lysenin, 42 kDa** <br> *EfL1* (L1) <br> *EfL2* (L2) <br> *EfL3* (L3) <br> [Sekizawa, et al., 1997] | D85846 <br> D85847 <br> D85848 | U02710 <br> (Fetidin 1) | Coelomic fluid, chloragocytes, coelomocytes | Contracts rat smooth muscle |
| **Fetidin 1, 40 kDa** <br> Four isoforms <br> **Fetidin 2, 45 kDa** <br> monomorphic <br> [Lassegues, et al., 1997] | U02710 | D85848 <br> (Lysenin) | Coelomic fluid, chloragocytes, inducible, increases after injecting pathogenic bacteria | Hemolysis, bacteriolysis, agglutination, clotting, opsonization, heme-binding enzymes, peroxidase |
| **Coelomic Cytolytic Factor, CCF-1 42 kDa** <br> [Beschin, et al., 1998] | AF030028 | AF395805 (CCF) | Coelomic fluid, coelomocytes | Opsonization, Not hemolytic, pattern recognition |
| **Lumbricin I, 7.2 kDa** <br> [Cho, et al., 1998] | AF060552 | | Whole worm, not inducible, constitutive | Antimicrobial, not hemolytic |
| **Eisenin I, 28 kDa** <br> [Xie, et al, 2003] | AY172839 | | Whole worm, inducible | Need to be defined |

**Table 23.4.** Lytic factors found in *Eisenia foetida*

| Name | Assay System | Mol. mass | References |
|---|---|---|---|
| Lytic Factor | NK-cell assay Targets: NK-dependent K562 NK-independent U937, BSM, CEM | ND | [Cooper, et al., 1994; Cossarizza, et al., 1996; Quaglino, et al., 1996] |
| Fetidin | Binding to sphingomyelin on RBC membranes, bactericidal activity | 40, 45 kDa | [Lassegues, et al., 1997; Roch, et al., 1979; 1981; 1989 Milochau, et al., 1997; Valembois, et al., 1985] |
| Lysenin | Binding to sphingomyelin on RBC membranes, liposome | 41 kDa | [Yamagi, et al., 1998] |
| Eiseniapore | Relief of fluorescence quenching from liposomes | 38 kDa | [Lange, et al., 1997; 1999] |
| Hemolysin (H1, H2, H3) | Isolated by preparative PAGE | 46, 43, 40 kDa | [Eue, et al., 1998] |

function. As we will read, the two most popular earthworms in these analyses are from the family Lumbricidae and have figured prominently in analyses of their immune systems. Now we carry this work further and look at how information of their immune systems has played a role in the eventual use in clinical studies.

## 2. Fibrinolysis

### 2.1. Lumbrokinase from *Lumbricus rubellus*

#### 2.1.1. Basic Characteristics Related to Clinical Trials

The earthworm has been used as a drug for various diseases in China and the Far East for thousands of years, however, without modern scientific pharmacological studies. According to the record in the most famous Chinese ancient medical publication *"Ben Cao Gang Mu (Compendium of Materia Medica)"*, the traditional medico material "Di Long" (earthworm) was regarded as being effective in treating limb numbness and hemiplegia. It was also found to be antifebrilic and capable of sedation and able to improve blood circulation and relieve clots. According to Rong Fan, 1996, lumbrokinase (LK) is a group of proteolytic enzymes, including plasminogen activator and plasmin, separated biochemically from a certain earthworm type[8]. The plasminogen activator (e-PA) in LK is similar to the plasminogen activator (t-PA) from other tissues. This makes it possible to show thrombolytic activity only in the presence of fibrin. Therefore, LK has the advantage of not causing hemorrhage due to hyperfibrinolysis during medication, as compared with streptokinase and urokinase. As a result, according to acute and subacute toxicological experiments, no negative effects of LK on nervous, cardiovascular, respiratory, and blood systems of rats, rabbits and dogs have been recorded. In addition, long-term animal toxicological experiments show no damage to hepatic or renal functions. There has been no negative influence on embryonic development, nor have there been tetratogenic or mutagenic effects observed in embryonic rats. In clinical experiments there have been no undesired effects on blood levels of glucose and lipids.

In experimental acute pulmonary artery embolism of rabbits, the embolus was labeled with $^{125}$I and the radioactivity in blood was tested 0.5, 1, 2, 3 and 5 hours after duodenal administration of LK. The obvious dosage correlation was demonstrated by the marked increase in radioactivity in blood at 3 and 5 hours after administration. Inferior vena cava thrombosis test in rats showed that after rectal administration of LK, the reduction of thrombosis was evident. LK has the effect of decreasing fibrinogen, lowering blood viscosity, and reducing platelet aggregation. It can be used in treatment and prevention of ischemic cerebrovascular diseases as well as other embolic and thrombotic diseases, such as coronary heart disease, myocardial infarction, arterial sudden deafness, thrombosis of central vein of retina, embolism of peripheral vein and pulmonary infarction. The LK capsule is soluble and must be taken before eating meals. Oral administration of one or two capsules 30 minutes before eating, three times daily, for four weeks of treatment completes one course. Two or three courses are usually needed to result in improvement or recovery from symptoms. Neither obvious side effect, nor toxic or allergic reactions were found in clinical tests.

#### 2.1.2. First Attempts at Isolation

As an attempt to discover possible therapeutic effects previously ascribed to historically recorded traditions, the groups of Mihara have obtained strong fibrinolytic enzymes in saline extracts of the earthworm, *Lumbricus rubellus* [44,45]. These enzymes hydrolyze both plasminogen-rich and free platelets, are heat-stable (up to 60°C) and display a broad optimal pH range (1–11). Three partially purified enzyme fractions have been further subdivided. The first fraction (F-I) is divided into three fractions (F-I-0, F-I-1, and F-I-2), that exhibit similar biochemical characteristics, but the second fraction (F-II) is not subdivided. The third fraction (F-III) is divided into two more fractions (F-III-1 and F-III-2). Based on enzymatic activities against various substrates, fraction I enzymes are chymotrypsin-like and fraction III enzymes are trypsin-like. Fraction II is neither a trypsin- nor an elastase or chymotrypsin-like enzyme. Amino acid compositions of the six enzymes have been estimated and found to contain abundant asparagine or aspartic acid, sparse amounts of proline or lysine, and contain no sugar components. These enzymes are therefore regarded as novel fibrinolytic enzymes, and referred to collectively as *lumbrokinase* from the earthworm's generic name. Analyses of substrate specificity and inhibition indicate that these enzymes are trypsin-like serine proteases. The N-terminal amino acid sequences of the enzymes revealed similarities to those of trypsin-like enzymes such as elastase and coagulation factor IX. These six enzyme proteins were suggested to derive as isozyme(s) from at least four different genes.

This group then designed clinical trials for in vivo experiments on human volunteers. 120 mg of lyophilized earthworm powder were administered orally to 7 healthy volunteers (aged 28–52 years old) three times after meals every day for 17 days. Blood was withdrawn once a day before and at 1, 2, 3, 8, 11 and 17 days after beginning the administration. The fibrin degradation products (FDP) value, tissue plasminogen activator (t-PA) antigen level and t-PA activities were measured in the blood. Before administration, the t-PA antigen level was $5.6 \pm 0.38$ ng/ml, and it gradually increased until the 17th day. The FDP level was increased on the 1st and 2nd days after administration but decreased and normalized by day 17. Fibrinolytic activities also tended to increase. These results suggested that earthworm powder (enzyme) could be a possible oral thrombolytic agent, applicable for treating patients with thalassemia.

Another annelid, the leech, has been a model for similar investigations. According to Rester et al., 1999 the serine proteinase plasmin is, together with tissue-type plasminogen activator (tPA) andurokinase-type plasminogen activator (uPA), involved in the dissolution of blood clots in a fibrin-dependent manner[46]. Moreover, plasmin plays a key role in a variety of other activation cascades such as the activation of metalloproteinases, and has also been implicated in wound healing, pathogen invasion, cancer invasion and metastasis. The leech-derived (*Hirudo medicinalis*) antistasin-type inhibitor bdellastasin represents a specific inhibitor of trypsin and plasmin and thus offers a unique opportunity to evaluate the concept of plasmin inhibition. The complexes formed between bdellastasin and bovine as well as porcine beta-trypsin have been crystallized in a monoclinic and a tetragonal crystal form, containing six molecules and one molecule per asymmetric unit, respectively. Both structures have been solved and refined to 3.3 A and 2.8 A resolution. Bdellastasin turns out to have an antistasin-like fold exhibiting a bis-domainal structure like the tissue kallikrein inhibitor hirustasin. The interaction between bdellastasin and trypsin is restricted to the C-terminal subdomain of bdellastasin, particularly to its primary binding loop, comprising residues Asp30-Glu38. The reactive site of bdellastasin differs from other antistasin-type inhibitors of trypsin-like proteinases exhibiting a lysine residue instead of an arginine residue at P1. A model of the bdellastasin-microplasmin complex has been created based on the X-ray structures. These modeling studies indicate that both trypsin and microplasmin recognize bdellastasin by interactions that are characteristic for canonically binding proteinase inhibitors. On the basis of these three-dimensional structures, and in comparison with the tissue-kallikrein-bound and free hirustasin and the antistasin structures, it is possible to postulate that binding of inhibitors toward trypsin and a switch of the primary binding loop segment P5-P3 accompanies plasmin. Moreover, in the factor Xa inhibitor antistasin, the core of this molecule would prevent an equivalent rotation of the P3 residue, making exosite interactions of antistasin with factor Xa imperative. Furthermore, Arg32 of antistasin would clash with Arg175 of plasmin, thus impairing a favorable antistasin-plasmin interaction and explaining its specificity.[46]

### 2.1.3. Characterization of the Potent Fibrinolytic Enzyme

In 1993, Nakajima et al. confirmed and extended these previous works. After purification to homogeneity they too obtained stable and potent fibrinolytic enzymes (six homogeneous proteins) from extracts of the lyophilized powder. However, the molecular weight of each enzyme estimated by SDS-polyacrylamide gel electrophoresis was different from those that had been analyzed by gel filtration chromatography. The exact molecular weight of each enzyme (F-III-2, F-III-1, F-II, F-I-2, F-I-1, and F-I-0) measured by ion-spray MS analysis was 29,662, 29,667, 24,664, 24,220, 24,196, and 23,013. The isoelectric point (pI) of each enzyme was 3.40, 3.60, 4.20, 4.00, 4.30, and 4.85. The enzymes were single polypeptide chains with strong fibrinolytic activity with maximum reactivity for chromogenic substrates from pH 9-11.[47]

### 2.1.4. Chemical Modification of Enzyme with Human Serum Albumin Fragment and Characterization of the Protease as a Therapeutic Enzyme

The strongest fibrinolytic protease of the six enzyme proteins purified from *Lumbricus rubellus* (F-III-2) have been modified chemically with fragmented human serum albumin (10–30 kDa)[48]. This modified enzyme lost antigenicity characterized by the native enzyme and reacted with antisera against human serum albumin as well as their fragments, and a conjugate with the native enzyme to form precipitation lines that fused with each other. The conjugate was increasingly resistant to inactivation by protease inhibitors in rat plasma. The enzyme was a nonhemorrhagic protein and induced no platelet aggregation. It also retained potent proteolytic activity for fibrin and fibrinogen as that of human plasmin. This protein easily solubilized actual fibrin clots (thrombi) of whole blood induced by thrombin in a rat vena cava. The continuous fibrinolysis for fibrin suspension in an enzyme reactor system that uses the modified enzyme immobilized to oxirane-activated acrylic beads had been achieved without any inactivation of the activity at least for more than 1 month.

### 2.1.5. Effects of Crude Earthworm Extract on Promoting Blood Circulation and Halting Stasis

According to Zhang and Wang, 1992, the crude extract of earthworm, which has a thrombolytic effect, could significantly decrease the plasma fibrinogen content and euglobulin lysis time ($P < 0.01$). An enzymatic preparation, containing many fibrinolytic enzymes, was prepared from the crude extract by ammonium sulphate precipitation and DEAE-cellulose chromatography. When administered in rabbits, the enzymatic preparation had an effect in hemorheology improvement. The experiment shows the enzymatic preparation could obviously lower the aggregation of platelets; decrease the viscosity of whole blood and plasma as well as the index of erythrocyte rigidity significantly ($P < 0.001$). All these effects demonstrate its ability in promoting blood circulation to remove stasis.[49] In 1994, Woo et al purified an endoprotease in earthworm (*Lumbricus rubellus*) to apparent homogeneity using $^{125}$I-lactalbumin as a substrate. The protease, a chymotrypsin-like serine protease has a molecular mass of 27 kDa, and is markedly activated by poly-L-lysine or poly-L-arginine. Its activity is distributed to coelomic fluid but relatively little to coelomocytes.[50]

### 2.1.6. Experimental Approaches using Lumbrokanase Immobilized Polyurethane

Lumbrokinase is a potent fibrinolytic enzyme when purified from *Lumbricus rubellus*, is stable and shows greater antithrombotic activity than other fibrinolytic proteins.[51,52] A lumbrokinase fraction showing the most potent fibrinolytic activity has been immobilized onto a polyurethane surface to investigate its enzymatic and antithrombotic activity. The stability of immobilized lumbrokinase was determined by caseinolytic activity assay and the specificity of immobilized lumbrokinase on fibrinogen/fibrin was observed by SDS-PAGE. Immobilized lumbrokinase retained about 34% of its activity, when compared to its soluble state and it also demonstrated stability against thermal inactivation and degradation and within a pH range. The optimal pH of immobilized lumbrokinase shifted 1.0 pH unit upward when compared with the soluble enzyme. Immobilized lumbrokinase demonstrated stable proteolytic activity during various incubation periods in addition to proteolyzing fibrinogen and fibrin almost specifically, while hardly hydrolyzing other plasma proteins including plasminogen and albumin.

Upon exposure to the human whole blood, less amounts of $^{125}$I-fibrinogen were adsorbed to the lumbrokinase-immobilized surface than to the polyurethane control surface. The lumbrokinase-immobilized surface showed less platelet adhesion than did the MAMEC-grafted surface. Initially, the number of adhered platelets increased on the lumbrokinase-immobilized surface with time; yet, the platelet number drastically decreased on the lumbrokinase-immobilized surface after 80-min incubation. This suggests that lumbrokinase-immobilized polyurethane digested the adsorbed fibrinogen and inhibited platelet adhesion on the surface, probably by making fibrinogen adsorption is highly antithrombogenic. In the *ex vivo* A-A shunt experiment, the lumbrokinase-immobilized surface significantly prolonged occlusion time over control surfaces, which is primarily due to the high thrombolytic activity of immobilized lumbrokinase. Consequently, a highly efficient surface modification method on the polyurethane surface was developed, and this lumbrokinase immobilization technique could be useful in improving blood compatibility of blood-contacting devices. Moreover, these results lead to the suggestion that clinical applications of this material to artificial organs should be developed in the near future.

### 2.1.7. Dose Dependency of Earthworm Powder on Antithrombotic and Fibrinolytic Effects

In 1998, Kim et al. investigated the fibrinolytic and antithrombotic effects by orally administering to rats the freeze-dried powder of *Lumbricus rubellus* earthworm. The fibrinolytic activity of plasma was determined by measuring plasmin activity of the euglobulin fraction. It was increased to two-folds of controls at a dose of 0.5 g/kg/day and five times with 1 g/kg/day after 4-day administration. The antithrombotic effect was studied in an arterio-venous shunt rat model. The thrombus weight decreased significantly from 43.2 mg to 32.4 mg at a dose of 0.5 g/kg/day after 8-days of treatment. The level of fibrinogen/fibrin degradation product (FDP) in serum was elevated in a dose-dependent manner during the treatment period. On the 8th day after administration, the FDP value was increased to 7.7 micrograms/ml compared with the control value of 3.3 micrograms/ml.[53] These results suggest that earthworm powder is valuable for preventing and or treating thrombotic conditions.

## 2.2. Fibrinolytic Activities Associated with Factors Derived from *Eisenia Fetida*

### 2.2.1. *A Fibrinolytic Enzyme Derived from Eisenia Fetida*

According to Xiong et al., 1997, a fibrinolytic enzyme form the earthworm *Eisenia fetida* was purified by incubation, extraction, alcohol precipitation and chromatographies on DEAE-Sepharose Fast Flow, Lysine-Sepharose 4B. The enzyme, with a molecular weight of 33 000, and a pI value of 3.5, is a single chain protein. This enzyme shared both fibrinolysis and plasminogen activator activities. The optimum temperature and pH on the activity of the enzyme were 65°C and 8.5 respectively. The amino acid composition of the enzyme showed that it was rich in acidic amino acids. Its N-terminal amino acid sequence is I-V-G-G-I-E-A-R-P-Y-E-F-P.[54]

### 2.2.2. *A Fibrinolytic Enzyme (eFE-D)*

Xing et al., 1997, isolated and purified the fibrinolytic enzyme, eFE-D from the earthworm *Eisenia fetida*. Gel-filtration on Sephacryle S-200, ion-exchange chromatography on DEAE-Sepharose Fast Flow and hydrophobic chromatography on Phenyl-Sepharose Fast Flow as detected by the fibrinolytic activity with a standard fibrin plate method. The strongest fibrinolytic component eFE-D not only hydrolyzed fibrin directly, but also activated the plasminogen to plasmin. Its apparent fibrinolytic value was equal to 2800 UK IU per mg. Its molecular weight as estimated by SDS-PAGE and MS analysis was 29 kD and 24829 Da, respectively while its isoelectric point (pI) was 4.0. Fibrinolytic enzyme eFE-D was significantly thermo stable with a single polypeptide chain. Studies with protease inhibitors indicated that eFE-D was a kind of serine protease. Its N-terminal amino acid sequence is V-I-G-G-T-N-A-S-P-G-E-F-P-W-Q-L-S-Q-Q-R.[55] The result of amino acid composition analysis showed that the enzyme contained abundant amino acids of low molecular weight, but few aromatic and alkaline amino acids. This may be similar to the serine proteases discovered by Roch as described in a later section.

### 2.2.3. *Indicators of Fibrinolytic Activity*

In the same year the above studies were extended. According to Yang et al., 1997 a sodium-dodecyl-sulfate-activated fibrinolytic enzyme from *Eisenia fetida* (the *Eisenia fetida* enzyme) was purified by chromatography on DEAE-Sepharose, Sephadex G-75, and Phenyl Sepharose 4. It (Mr = 45 kDa) was composed of two subunits (Mr = 26 kDa and Mr = 18 kDa) held together by hydrophobic interactions. The enzyme displayed four activities when fibrin plates were used to detect the proteolytic activity. These were designated as CFPg (complete fibrinolysis in the plasminogen-free plate), uCFPg (uncompleted fibrinolysis in the plasminogen-rich plate), CF (complete fibrinolysis in the plasminogen-free plate), and uCF (uncompleted fibrinolysis in the plasminogen-free plate). SDS activated CFPg and rendered the enzyme more sensitive to some inhibitors. Leupeptin, chyomostatin, pepstatin, aprotinin, phenylmethylsulfonyl fluoride, and dithioithreitol had no effect on uCF. Pepstatin stimulated CFPg and uCFPg, while E-64, a thiol inhibitor, activated uCFPg and uCF. The N-terminal sequence of the large subunit was analyzed and compared with some know proteins. The large subunit alone had catalytic activity, while the small subunit did not. Using plasminogen as the substrate for defining peptide bond specificity, the *Eisenia fetida* enzyme was observed to cleave the carboxyl side of basic amino acids, small neutral amino acids, and Met residue.[56]

## 2.3. A Plasminogen Activator (e-PA)

### 2.3.1. Purification

In 1998a, Yang et al. purified a plasminogen activator from *Eisenia fetida* to obtain an effective drug for thrombolysis. The procedure included several techniques: DEAE Sepharose Fast Flow column, Sephadex G-75 column and Phenyl Sepharose 4 fast Flow (low substitute) column chromatography in order. The purified enzyme is composed of two kinds of subunits (L, large; S, small), which are held together by hydrophobic interactions. In gel filtration, the molecular weight of intact enzyme is 45,000. According to SDS-PAGE, the molecular weights of L and S are 26,000 and 18,000 respectively, while determined by MS (mass spectrum), the molecular weights of L and S are 24556.7 and 15546.6 respectively. There was no Cys in the S subunit and Lys in the L subunit. The N-terminal sequence of L was analyzed, producing VIGGTNASPGEI-PWQLSQQRQSGSW, and was compared with some known proteins. It was serine proteases that shared higher similarities between the 25 amino acids sequence. The enzyme showed three different kinds of fibrinolytic activities on fibrin plate, which were designated as CFPg (complete fibrinolysis with plasminogen), uCFPg (uncompleted fibrinolysis with plasminogen) and uCF (uncompleted fibrinolysis without plasminogen).[57]

### 2.3.2. Characterization

Yang et al., 1998b purified a plasminogen activator from *Eisenia fetida* (e-PA), which showed three kinds of fibrinolysis activities on fibrin plates. They are designated as CFPg (complete fibrinolysis in plasminogen-rich plate), uCFPg (uncompleted fibrinolysis in plasminogen-rich plate) and uCF (uncompleted fibrinolysis in plasminogen-free plate). To investigate these different activities, SDS and some inhibitors are used to study their influence on the activities. SDS can activate CFPg and render e-PA more sensitive to some inhibitors. Leupeprin, chymostatin, pepstatin, aprotinin, phenylmethylsulfonyl fluoride (PMSF) and dithiothreitol (DTT) have no effect on uCF. Pepstatin can stimulate CFPg and uCFPg while E-64, a kind of thiol inhibitor, can stimulate uCFPg and uCF. All of these results make it more difficult to sort out e-PA as a serine protease or a thioprotease. Furthermore, the specificity of peptide bond using plasminogen as substrate is that e-PA can cleave the carboxyl side of basic amino acids, small neutral amino acids and Met residue. Specifically, e-PA can cleave plasminogen to produce plasmin, which is the biochemical base for e-PA's use for thrombolysis.[58]

### 2.3.3. Degradation Components, Enzymology and CD Spectra

*2.3.3a. Benzoyl-L-arginine ethyl ester (BAEE)* In 1998c, Yang et al. used benzoyl-L-arginine ethyl ester (BAEE) as substrate to detect the enzymology kinetic of e-PA (a plasminogen activator from *Eisenia fetida*). The optimum pH of the degradation was about 8.4, and Km and Kcat were $1.24 \pm 0.16 \times 10^5$ mol/L and $13.80 \pm 4.02$ s$^{-1}$, respectively. The $K$m of the large subunit was almost the same as that of e-PA, but was about ten times smaller than that of the small subunit, showing that the holoenzyme and the large subunit have stronger affinity for the substrate than the small subunit. The Kcat of the two subunits were close, 1/6 and 1/3 of that of the holoenzyme, respectively. Eight kinds of inhibitors were used to examine the mechanism. Pepstatin and E-64 could stimulate the degradations while TPCK, TLCK, PMSF, chymostatin and leupeptin could inhibit it. EDTA was non-effective.[59]

*2.3.3b. N-acetyl-L-tyrosine ethyl ester (ATEE)* According to Yang et al., 1998d a plasminogen activator from *Eisenia fetida* (e-PA) could degrade the synthetic substrate, N-acetyl-L-tyrosine ethyl ester (ATEE). Although the optimum pH was about 8.5, the degradation was affected by different buffer systems. The activity was stronger in 0.2 mol/L Na$_2$HPO$_4$ than in 0.05 mol/L of pH 8.5, Tris-HCl. The $K$m and $K$cat of L subunit, S subunit, and the holoenzyme, e-PA, were detected in two different buffer systems. The results were that the $K$m of e-PA was much smaller than that of each subunit in 0.2 mol/L Na$_2$HPO$_4$, and the $K$m of the large subunit was about ten times smaller than that of the small subunit in both of the two buffer systems. The $K$cat of the small subunit was much larger than that of both the holoenzyme and the large subunit in two buffer systems. This phenomenon was explained in view of the enzyme structure. Pepstatin, E-64, and EDTA could inhibit the degradation of ATEE, which was not the case with the degradation of BAEE.[60]

*2.3.3c. Active centers of small subunit* Yang et al., 1998e found that a plasminogen activator from *Eisenia fetida* (e-PA) has multiple substrate specificities. Knowledge concerning the active center of its subunits is important for understanding relationships between function and structure of e-PA. Using TLCK, a specific trypsin inhibitor, the active centers were then analyzed. There appeared reversible inhibition with TLCK, although the V$_{max}$ with TLCK, 0. 445 µmol. L-1 s-1 was smaller than that without TLCK, 0. 65 µmol. L-1 s-1. There were two kinds of active centers with different sensitivities to TLCK, responsible for different substrate specificities as confirmed by the difference of CD spectra under different concentrations of TLCK.[61]

### 2.3.4. Fibrinolytic and Anticoagulative Activities

According to Hrzenjak et al., (1998), a biologically active glycolipoprotein complex, G-90 isolated from whole earthworm tissue extract shows anticoagulative and fibrinolytic activities. They isolated two tyrosine like serine peptidases with molecular masses of 34 kDa (P I) and 23 kDa (P II). Both peptidases exhibit fibrinolytic and anticoagulative activities. The activity of P I is much higher. P I in concentration of 10 (5) ng/ml of plasma shortened the physiological time of fibrin clot lysis by 54% and completely inhibited blood clotting at a concentration of 10(3) ng/ml of venous blood.[62] As we will see later, this component has been used to inhibit tumor growth. Thus here it is considered with respect to its properties in relation to fibrinolysis. Could it be that its capacity to inhibit tumor growth resides in its ability to interfere with blood circulation? After all the formation of new blood vessels is crucial in the development of cancer.

## 3. Serine Proteases May Regulate Lytic Activity?

Before launching into the lytic systems that have been isolated notably from *Eisenia fetida*, it is well to mention the situation with serine proteases since these have been reported in some instances as described in previous paragraphs. Three serine proteases have been isolated from coelomocyte lysates of *Eisenia foetida andrei*, one of which was a trypsin-type (Table 23.4).[28,63] Serine protease activity has been reported in *Lumbricus terrestris*.[25,64] According to interactions observed principally in insects, extracellular serine protease inhibitors regulate intracellular proteolytic enzymes.[65] Purified to homogeneity by affinity chromatography on trypsin, the serine protease inhibitor from *Eisenia foetida* is a monomer of 14 kDa. Its partial N-terminal amino acid sequence revealed a basic hydrophobic fragment that shared 68–75% homologies and 47–60% identities with several plant serine protease inhibitors. The cytotoxic activity of fetidins was stimulated *in vitro* by several serine proteases while incubation with soybean trypsin inhibitor variant a (STIa) resulted in less cytotoxicity.[28] The inhibitory effect occurred only when STIa was added before cell lysis. According to one interpretation, cytotoxicity involves the release of intracellular cytotoxic proteins, intracellular trypsin-like activator and extracellular serine protease inhibitor. One can imagine that this proposed chain of events provides a regulatory mechanism for both cellular/humoral immune responses of earthworms, linking them.

## 4. Antimicrobial Molecules

### 4.1. Lysins from *Eisenia Foetida*: Fetidins, Lysenin, Coelomic Cytolytic Factor (CCF-1)

#### 4.1.1. Fetidins

In *Eisenia foetida* earthworms, fetidins are polymorphic and multifunctional, including responses that involve cytolysis, those that are antibacterial and certain clotting capacity.[66–69] The clotting effect probably evolved and was deployed to eliminate nonpathogenic bacteria, and is mediated by a serine protease/serine protease inhibitor equilibrium. Several bits of information suggest that clotting of the fetidins is a normal response occurring at low but constant rates on the outer surface of earthworms.[70] Mixed with mucus, fetidins cover the body and they constitute an external nonspecific antimicrobial barrier (Fig. 23.1).

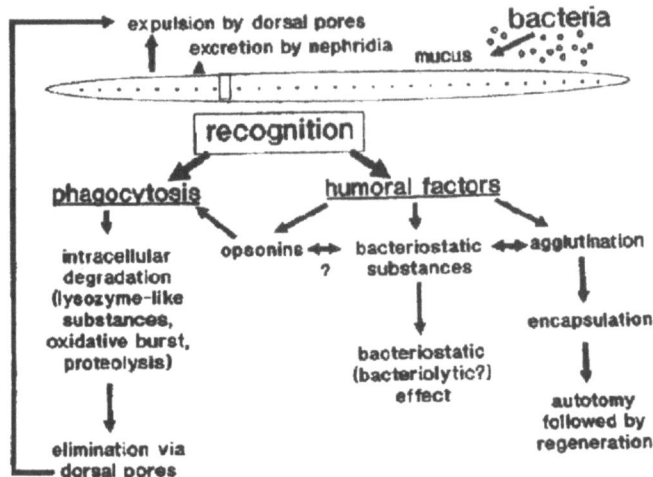

**Figure 23.1.** General armamentarium of natural resistance of the earthworm. The first protective barrier is represented by secreted mucus containing agglutinins. Bacteria invading the coelomic cavity are directly expelled via dorsal pores and excreted by nephridia or can be engulfed by coelomocytes. Moreover, humoral factors are involved in their elimination: agglutinated bacteria can be encapsulated, antibacterial substances prevent bacteria from multiplying, and opsonins facilitate phagocytosis. All humoral factors involved in antibacterial defense seem to be connected in complex protective system [from L. Tuckova and M. Bilej, Mechanisms of antigen processing in invertebrates: Are there receptors? Adv. Comp. Env. Physiol. 23 (1996) 41–72][from E.L. Cooper et al. Digging for innate immunity since Darwin and Metchnikoff. Bioessays, 24 (2002), 319–333.]

Mystery surrounding lytic components of coelomic fluid from *Eisenia* has been extended and partially clarified (Tables 23.3, 23.4). Fetidins appear to be controlled by two independent genes: one expresses and codes for a pI 6.1 hemolysin; the other is composed of four alleles that define ten genetic families. Significant similarities are characterized by large quantities of aspartic acid, glutamic acid, and glycine; polyclonal and monoclonal antibodies do not discriminate between purified hemolysin.[65] Lysis usually depends upon the presence of glycoprotein molecules and when erythrocytes are used as targets, lysis is mediated by two monomeric lipoproteins of 40 and 45 kDa termed fetidins. They are so named (from the earthworms, *Eisenia foetida*, because of their fetid odor) if they are disturbed, two related polymorphic immunodefense factors.[65] A cDNA containing an insert of 1.44-kb encoding a 34-kDa protein that corresponds to the size of deglycosylated fetidins has been cloned.[35] The cDNA sequence contains a peroxidase signature that confirms peroxidase activity of fetidins. The recombinant protein is antibacterial since it inhibits growth of *Bacillus megaterium*. Fetidin possesses varying degrees of significant homology with lysenins 1, 2, 3; there is no significant homology with CCF nor lumbricin I (Fig. 23.2).[71]

*4.1.2. Lysenin*

Lysenin, isolated from *Eisenia foetida*, is another protein of similar character, *i.e.* sharing biological roles and biochemical properties with fetidins, and eiseniapore, but it has been shown to affect different and perhaps more important functional properties (Tables 23.3, 23.4).[72,73] Lysenin, isolated as a 41-kDa protein causes contractions of smooth muscle

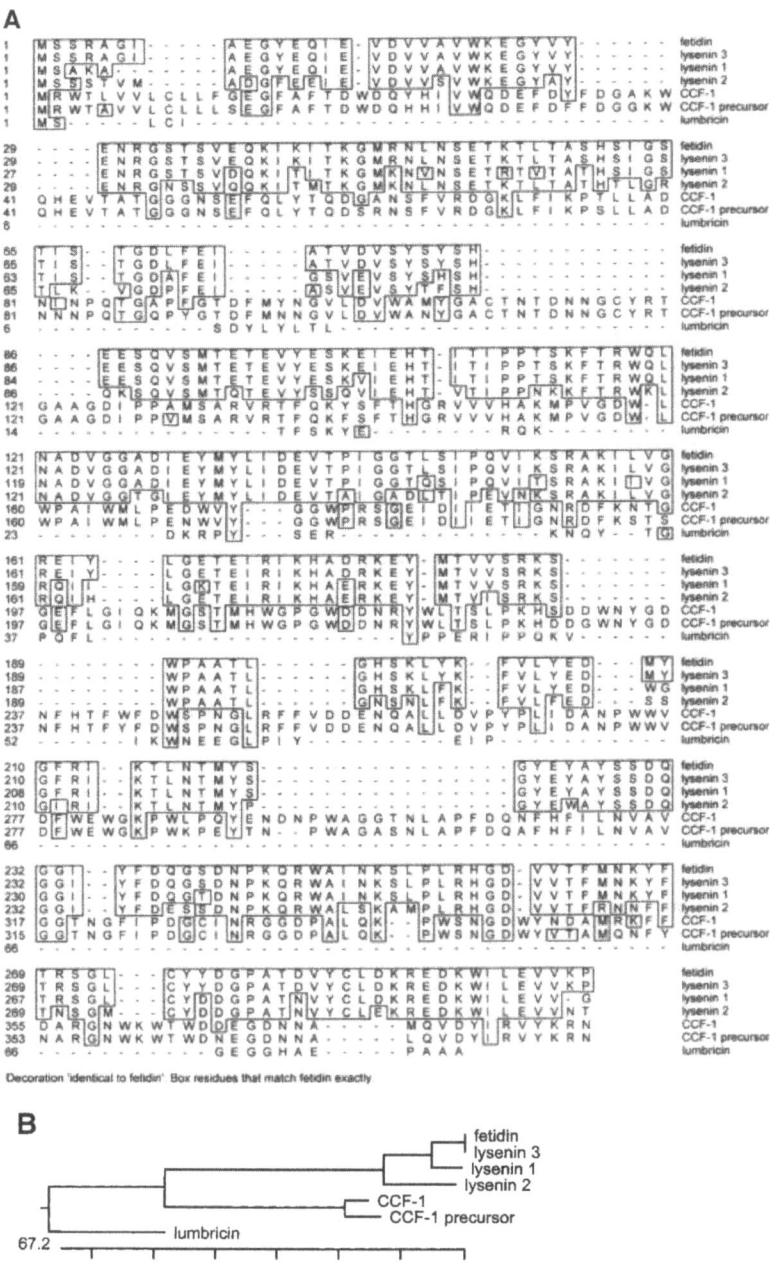

**Figure 23.2.** Molecular characteristics of the lysins from *Eisenia foetida*: fetidin, CCF-1, lysenin and the peptide lumbricin I from *Lumbricus rubellus* (See also Table 23.3). **A.** Clustal alignment MegAlign (DNA STAR Inc). Fetidin and lysenin 3 show 100% amino acid identity and 99% nucleotide homology, suggesting the same gene. **B.** Dendogram depicts the relationships among lytic molecules. The length of each pair of branches represents the distance between sequence pairs. The scale beneath measures the distance between sequences. Units may represent the number of substitution events or percent difference (Courtesy of Duane E. Keith, Jr.) [from E.L. Cooper et al. Digging for innate immunity since Darwin and Metchnikoff. Bioessays, 24 (2002), 319–333.]

from the rat-isolated aorta. Although recombinant and native lysenin had similar contractile activities when tested on rat aorta, the amino acid sequence of lysenin revealed no significant homology to previously characterized vasoactive substances. Like eiseniapore and other lytic components, lysenin induces hemolysis and among other various phospholipids binds specifically to sphingomyelin in cellular membranes. However, lysenin differs in amino acid sequence and certain biological activities from the other lytic proteins isolated from coelomic fluid of *Eisenia foetida*. There was no binding of lysenin to other phospholipids including sphingomyelin analogues such as sphingosine, ceramide, and sphingosylphosphorylcholine. This variance suggests that there is precise recognition of the molecular structure of sphingomyelin.

Further analyses revealed that if cholesterol incorporation changes the topological distribution of membrane, sphingomyelin accessibility to lysenin increases. Northern blot analysis of the RNA from earthworm tissues indicated that lysenin is produced by coelomocytes.[72,74] In particular, immunoreactive lysenin was detected in LC and in the free large chloragocytes present in the lumen of the typhlosole, a depression in the dorsal wall of the intestine. These coelomocytes and chloragocytes seemed to be mature and separate from the chloragogen tissue that lined the typhlosole. The free large chloragocytes in the typhlosole contained numerous vacuoles. The nuclei were small and irregular in shape, and glycogen granules and mitochondria were occasionally found between vacuoles. The chloragocytes of the chloragogen tissue that surrounded the coelomic side of the intestine and the dorsal blood vessel did not react with the lysenin antiserum and no expression of lysenin mRNA was detected in these cells. Furthermore, no evidence of the protein or of the mRNA was found in the cells of the pharyngeal gland. Their findings suggest that lysenin is produced in the free large chloragocytes in the lumen of the typhlosole.

Kobayashi *et al.* found that the coelomic fluid of *Eisenia foetida* was not toxic to 42 species, belonging to seven invertebrate phyla, almost all in aquatic adults and larvae. Eleven teleostean species tested died in 0.2-1% coelomic fluid mostly between 10 and 120 min and the effects were dose-dependent. Tadpoles of the toad *Bufo japonicus formosus* died in 0.4–2% coelomic fluid between 80 and 225 min depending upon size, with larger tadpoles surviving longer. Before dying, all experimental tadpoles developed curled and shrunken tails. The Okinawa tree lizard, soft-shelled turtle, Japanese quail, mouse and rat all died after i.v. injection of coelomic fluid (above 20 µl/kg). Thus, coelomic fluid was not toxic to invertebrates, but toxic to vertebrates. After heating, coelomic fluid lost its toxicity to fish, tadpoles and mice. Coelomic fluid incubated with sphingomyelin-liposomes was no longer toxic, suggesting the involvement of sphingomyelin in the toxicity. Since lysenin, which is a constituent of the coelomic fluid and known to bind specifically to sphingomyelin, exhibited toxicity similar to that of coelomic fluid, it was probably responsible for the toxic effects of coelomic fluid in vertebrate tissues. Accordingly, lysenin exerted lethal effects on spermatozoa in 5 of 33 invertebrate species tested and on spermatozoa taken from 30 of 39 vertebrates.[75,76]

It was postulated that plasma membranes of the spermatozoa of most invertebrates might not contain sphingomyelin whereas those of most vertebrate species might contain sphingomyelin. These possibilities were supported by the failure to detect sphingomyelin chemically in the testes of three species of invertebrates, none of which spermatozoa responded to lysenin. In contrast, sphingomyelin was detected in the testes of all 25 vertebrate species examined, irrespective of a negative or positive response of spermatozoa to lysenin. None of the six species of *Protista* examined was affected by lysenin. This work suggests that, in general, the spermatozoa of animals can be grouped into two categories, invertebrate and vertebrate, depending on the absence or presence of sphingomyelin in their plasma membrane. The incorporation of sphingomyelin into spermatozoa seems first to have occurred in protochordates during the course of evolution.

### 4.1.3. Coelomic Cytolytic Factor (CCF-1)

Bilej *et al.* in experiments aimed at identifying cytolytic molecules from the coelomic fluid of *Eisenia foetida* purified a 42-kDa protein that was named coelomic cytolytic factor (CCF). This protein is able to lyse TNF-sensitive tumor cell line in a protease-independent way but is not hemolytic. CCF displays high homology with invertebrate pattern recognition molecules, and participates in *Eisenia* defense mechanisms triggering the activation of the prophenoloxidase cascade upon binding to conserved microbial polysaccharides. In addition, *Trypanosoma cruzi* is lysed by coelomic cytolytic factor-1, an invertebrate analogue of TNF, and induces phenoloxidase activity in the coelomic fluid of *Eisenia foetida*. (Olivares et al, 2002) The lytic activity of CCF-1 observed on mammalian tumor cells might not be relevant for the immune defense of earthworms. It was shown that CCF displays functional analogies with the mammalian cytokine TNF, despite a lack of gene or amino acid sequence homologies.[77–85]

## 4.2. $H_1$, $H_2$, $H_3$: Mixed Effectors as Lysisns and Agglutinins

Because of different purification and assay procedures, we can often find two components, lysins and agglutinins in *Eisenia* and *Lumbricus*. However, another analysis reveals an apparent kinship between the lysins, fetidins and CCF-1 and three other factors called $H_1$, $H_2$, and H3 with molecular weights of 46, 43 and 40 kDa.[29] Isolated by preparative PAGE, $H_1$ and $H_2$ were stable in SDS and α-2-ME whereas $H_3$ splits into two fragments of 18 and 21 kDa. IEF indicates that each protein consists of different isoforms with pIs between 5.1 and 6.2. $H_3$ is a bifunctional protein that can lyse and agglutinate erythrocytes. At 56°C, hemolytic activity of all three proteins becomes inactivated, but agglutination activity of $H_3$ remains stable. Intracoelomic injection of erythrocytes reduced the number of hemolysins from three to two. Monospecific antisera raised against $H_1$, $H_2$ and $H_3$ and using these antibodies and carbohydrates as inhibitors of biological activity, revealed close structural relationship between the three components. Dual functional activity has also been observed in *Lumbricus rubellus* with the report of purification, characterization and cDNA cloning of a developmentally expressed peptide (Lumbricin I). This is particularly interesting and represents a departure from the three components $H_1$, $H_2$ and $H_3$ and perhaps the two fetidins and CCF-1. The humoral immune system (lysins and agglutinins) seems to possess contrasting effects on vertebrate immunocytes. This dichotomy suggests two points: 1) species variability although *Eisenia* and *Lumbricus* are members of the same family of segmented worms (Annelida-Lumbricidae); 2) protein functional variability in different species.

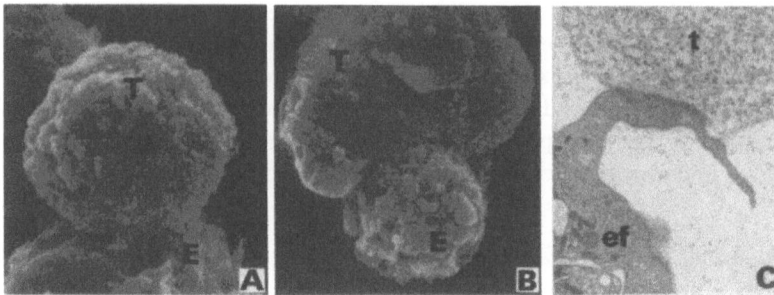

**Figure 23.3.** Interaction between an effector (E) and a target (T) after 5-min. Contact has been established, but lytic activity has not commenced (A). Interaction between an effector (E) and a target (T) after 15 min. The effector is embedding itself into the target (B) (x 6000)[from M.M. Suzuki and E.L. Cooper, Spontaneous cytotoxic earthworm leukocytes kill K562 tumor cells, Zool. Sci. 12 (1995) 443-451]. Small effector coelomocytes in contact with K562 human targets. Magnification higher that 4000 x showing tight binding (C) [from A. Cossarizza et al., Earthworm leukocytes that are not phagocytic and cross-react with several human epitopes can kill human tumor cell lines, Exp. Cell Res. 224 (1996) 174–182] [from E.L. Cooper et al. Digging for innate immunity since Darwin and Metchnikoff. Bioessays, 24 (2002), 319–333.]

### 4.3. Lumbricin I: Antimicrobial and Non-Lytic

Up to now we only considered large protein antimicrobial molecules that are also lytic. Lumbricin I isolated from *Lumbricus rubellus* is different, (Table 23.3, Fig. 23.3). This molecule exhibits antimicrobial activity against Gram negative (Gram-e), Gram positive (Gram+e) bacteria and fungi without hemolytic activity against human erythrocytes.[71] (Cho et al., 1998). It is interesting that Lumbricin I is rich in proline as in apidacins, drosocin, metchnikowin, bactenecins and PR39. For the first time we describe an antimicrobial peptide in earthworms that shares certain characteristics with those that have been isolated from arthropods (insects and *Drosophila* in particular).

### 5. Eiseniapore: Functionally Related, Isolated but not Sequenced

To approach the mechanism of synthesis and secretion of lytic components in *Eisenia foetida*, a novel new system has been defined and lytic activity ascribed to eiseniapore, a protein of 38-kDa.[26] (Tables 23.3, 23.4). Lipid vesicles of various compositions were used to determine whether specific lipids might serve as receptors to eiseniapore. The lysins bind to and disturb the lipid bilayer only when distinct sphingolipids consisting of a hydrophilic head group as phosphorylcholine or galactosyl and the ceramide backbone, (*e.g.* sphingomyelin), are present. Cholesterol enhances eiseniapore lytic activity toward sphingomyelin-containing vesicles probably due to interaction with sphingomyelin. Leakage of vesicles was most efficient when the lipid composition resembled that of the outer leaflet of human erythrocytes.

Further analysis revealed that the secondary structure of eiseniapore did not change upon binding to lipid membranes and lytic activity of eiseniapore was completely abolished after its denaturation or after preincubation with polyclonal antibodies. Absence of lysis suggests two mechanisms: 1) presence of specific sphingolipids is sufficient for

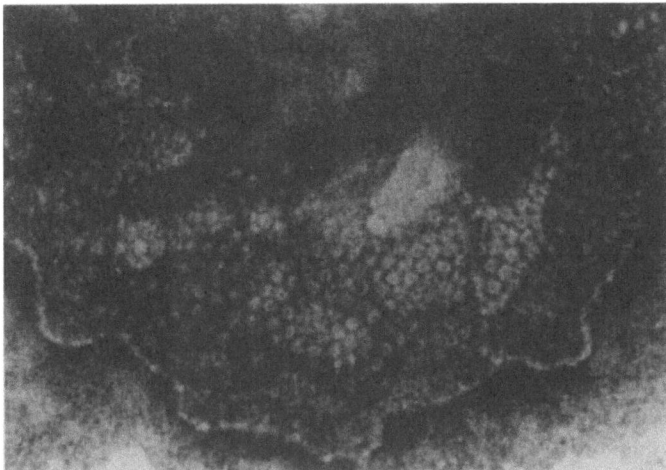

**Figure 23.4.** Negative staining transmission electron microscopy of eiseniapore-treated sheep erythrocyte membranes and untreated membranes. The images clearly show many pore-like structures with an outer diameter of 10 nm. Note that each of the pores contains a dark inner-channel of 3 nm. The image has a honeycomb appearance [from S. Lange et al., Biochemical characteristics of Eiseniapore, a pore-forming protein in the coelomic fluid of earthworms, Eur. J. Biochem. 262 (1999) 547–556] [from E.L. Cooper et al. Digging for innate immunity since Darwin and Metchnikoff. Bioessays, 24 (2002), 319–333.]

eiseniapore mediated lytic activity; 2) membrane glycoproteins, as receptors for eiseniapore, are not required as long as specific sphingolipids are present. Still further analyses by electron microscopy revealed interesting characteristics concerning eiseniapore and erythrocyte membranes.[27] Several ring shaped structures (pores) were found to be composed of a central channel that contained outer (10 nm) and inner (3 nm) diameters. This channel complex consists of six monomers as deduced from the molecular mass of 228 kDa. Functional evidence of pore formation by eiseniapore was revealed since protection of lysis by carbohydrates occurred at an effective diameter above 3 nm (Fig. 23.4). Channel formation complex suggest two conclusions: 1) the existence of a channel or a "barrel stave model" (closed beta-pleated sheet) for vesicle and cell interaction of eiseniapore; 2) a plausible explanation for the mechanism by which components of the earthworm humoral immune system directly destroys nonself cells.

## 6. Lytic Molecules that React Against Cancer Cells

### 6.1. Earthworm Effectors can Bind to and Lyse K562 Targets

We were among the first invertebrate immunobiologists to demonstrate cytotoxicity using leukocyte effectors co-cultured with a variety of targets including allogeneic cells.[86,87] Results of more rigorous analyses follow, especially the all-important question concerning the nature of viability in reactions involving autogeneic cells alone or in mixtures as allogeneic cells.[22] Information would then give us other relevant, newer parameters against which to clarify possible mechanisms of cytotoxicity, especially after contact, *i.e.* initial binding between effectors and targets. In the following experiments, we defined autogeneic (self) cells as leukocytes derived from a single earthworm and those that are allogeneic (nonself) as derived from at least two different earthworms of the same genus and species. In the remaining descriptions, two main leukocytes will be highlighted: small coelomocytes (SC) and large coelomocytes (LC), revealing for the first time their different behavior when co-cultured together or with tumor cell targets.

## 6.2. Viability of Earthworm Effectors When Co-Cultivated with Tumor Cells

After cultivation, we observed a higher incorporation of [$^3$H]-thymidine in autogeneic cultures accompanied by a significantly greater number of cells in S, G2, or M phases than in allogeneic cultures. The disparity and apparent preferential killing resulting in varied efficacy of cytotoxicity has been interpreted in several ways. First, [$^3$H]-thymidine incorporation results suggest that autogeneic coelomocytes are in a better overall state of cellular health than allogeneic populations. Second, when viewed with respect to histocompatibility, autogeneic coelomocytes *"face to face"* are not confronted by potentially deleterious *nonself* alloantigens as when allogeneic cells are *"face to face"*. Putative allogeneic effectors are vulnerable to reactions against themselves even if the alloantigens are weak or the effectors may lack receptors that can recognize minor, nonself, and histocompatibility differences. One controversial explanation suggests a primitive "allogeneic inhibition" as has been observed in mammalian bone marrow transplantation systems.[22] We suggest two other more plausible interpretations: 1) recognition of/binding to and lysing of foreign cells in a NK cell-like reaction may reflect innate immunity; 2) cytotoxicity is a response that is more advanced than phagocytosis.[88]

## 6.3. What Characterizes the Membrane Contact?

Small coelomocyte lyse classical NK-sensitive cells like K562, and NK-resistant targets recalling the nonspecific, innate character of the response.[23] Cell contact is essential as revealed by the intricacies of lysis after binding of effectors to target cell membranes and an actual tearing away of target membranes by voracious effectors. We have observed intimate contacts and holes (pores), reminiscent of those effected by eiseniapore, between effectors and targets by scanning electron microscopy. By contrast to this apparently exclusive cell mediated mechanism, lysis of targets will also occur in cell free coelomic fluid and this suggests humoral mechanisms as the more immediate mediators of lysis. Lysis is only one component of immunodefense and lysins are dependent upon one or more of the coelomocyte types, as mentioned earlier, for their synthesis and secretion.

The complexity of the earthworm cellular armamentarium as revealed by light and electron microscopy as well as by flow cytometry is significantly diverse. For critical future research, it will be essential to determine the molecular mechanisms of what appears to be a pathway involving sphingomyelin in the lytic process of eukaryotic cells such as tumors as revealed in the following results. Mediators of cytotoxic reactions would then suggest effectors of primordial NK cell activity.[87] Invertebrate lytic systems spontaneously kill various targets without evidence that they possess clonal-specific immunocytes (the adaptive system) such as cytotoxic lymphocytes (CTLs), suggesting that an NK-like activity is present.[22] To support this view, dissociation of phagocytosis from NK-like killing has been defined in earthworms wherein cellular and humoral inflammatory responses and transplant rejection have been consistently observed.

## 7. Lysis of Cancer: Molecules that Require Further Definition

### 7.1. An International Effort

In this portion of the review, the work of investigators representing a perhaps-different approach to the problem of earthworm immunity will be presented in some details. The previous paragraphs were more concerned with the possible applications of lytic components from the earthworm as vital to the immune response armamentarium. This next series of works from Japan, Korea, China and Croatia present a different picture, one aimed at finding some more clinical relevance. This activity from the decade of the 1990s provides possible clinical approaches. From earthworms, one factor (lombricine) inhibits growth of spontaneous tumors in SHN mice whereas another, the "killer" glycolipoprotein extract called G-90 retards murine tumor growth *in vivo*.[89,90] However, all other biological activities associated with coelomic fluid are retained. These include hemolysis, agglutination, mitogenicity, and bacteriostatic activity and cytotoxicity in vitro. There are even claims that earthworms are used by the pharmaceutical industry in China[8] and elsewhere to prepare certain medicinals (lumbrokinase) for treatment and prevention of thrombus and embolic diseases.[61,90]

### 7.2. Lombricine: Inhibits Growth of Spontaneous Mammary Tumors in SHN Mice

In 1991, Nagasawa et al. studied the effects of lombricine extracted and purified from earthworm (*Lumbricus terrestris*) skin on the growth of palpable sizes (approximately 5 mm) of spontaneous mammary tumors in SHN mice. In Experiment 1, daily subcutaneous injections of lombricine (0.3 mg/0.05 ml olive oil) inhibited markedly the growth of tumors associated with the retardation of the growth of preneoplastic mammary hyperplastic alveolar nodules. In 1H-NMR spectra, the experimental mice had lower serum levels of lactic acid and glucose than the control. On the other hand, urine of the former group contained higher levels of allantoin, creatine and creatinine than that of the latter. In Experiment 2, lombricine given as diet at the concentration of 120 mg/kg also inhibited the growth of tumors, though to a lesser degree than the injection. The treatment had little effect on 1H-NMR spectra of either serum or urine and normal and preneoplastic mammary gland growth. All results indicate that the inhibition by lombricine of the growth of mammary tumors is at least partly due to the maintenance of homeostasis of the body including the regulation of the excess uptake of glucose as a source of energy and nutrition.[89]

### 7.3. G-90: A glycoliproprotein

*7.3.1. A Glycoliproprotein that Slows Murine Tumor Growth in Vivo*

According to Hrzenjak et al. 1992 earthworms (*Eisenia foetida* and *Lumbricus rubellus*) possess immunological recognition and memory as well as high regenerative abilities.

Using coelomic fluid as a source of biologically active compounds, a biologically active glycolipoprotein extract from a whole earthworm tissue homogenate was isolated and named G-90.[91] G-90 forms precipitation arcs in gel with different animal and human sera. It alters murine cell growth rate in vitro in serum in a dose dependent manner and slows murine tumor growth in vivo G-90 does not contain mutagens or carcinogens. Later this same group separated G-90 into seven fractions by gel-filtration.[92] Radioimmunoassays revealed that each of the fractions, except the lightest one, is crossreactive with porcine anti-insulin antibodies. Also, all fractions, except the heaviest and the lightest one, stimulate mammalian normal and transformed cell proliferation in serum-free conditions *in vitro*. The intensity of stimulation depends on cell type. Stimulation is completely abolished if the medium is supplemented with fetal calf serum.

### 7.3.2. Lysis of Fibrin Clots from Venous Blood of Patients with Malignant Tumors.

According to Hrzenjak et al, 1998 u-PA is secreted by the most malignant tumors. As a response to u-PA synthesis surroundings cells synthesize inhibitors of plasminogen activators for tissue protection. Plaminogen activators were found in earthworm tissue from the tissue homogenate of earthworm *Eisenia foetida* from which the glycolipoprotein mixture named G-90 was isolated. It contains two serine proteases (P I, P II) with fibrinolytic and anticoagulative activities. The fibrinolytic activity of G-90, P I and P II was tested in an in vitro euglobulinic test applied to fibrin clot from blood plasma of patients suffered from malignant tumors. G-90 and above-mentions proteases applied in this study showed euglobulinic time proportionally with the concentrations of added substances. The influence of G-90 and the fibrinolysis rate does not depend only on its concentration, but depends too on histological type of tissue (organ) where the malignant tumors are located. Enzyme P I and P II do not show this activity.

The glycolipoprotein mixture G-90 was shown to contain two serine proteases (P I, P II) with fibrinolytic and anticoagulative activities.[61,90] The fibrinolytic activity of G-90, P I and P II was tested in an *in vitro* euglobulinic test applied to fibrin clot from blood plasma of patients suffering from malignant tumors. G-90 and proteases showed euglobulinic time proportionally with the concentrations of added substances. The influence of G-90 on the fibrinolysis rate does not depend only on its concentration, but depends as well on histological tissue type (organ) where malignant tumors are located. Both P I and P II exhibit fibrinolytic and anticoagulative activities, but the activity of P I is much higher. P I in concentration of $10^5$ ng/ml of plasma shortened the physiological time of fibrin clot lysis by 54% and completely inhibited blood clotting at a concentration of $10^3$ ng/ml of venous blood.

### 7.3.3. Action on Venous Blood of Dogs with Cardiopathies and Malignant Tumors

According to Popovic, et al, 2001, the stability of homeostasis is important to keep a balance between coagulation and fibrinolysis. A disorder of homeostasis leads to different physiological changes and causes different diseases such as cardiopathies and malignant tumors. Cardiopathies are characterized by a hypercoagulation. In the malignant tumors, besides the hypercoagulation due to plasminogen activators (PA) formed inside the tumor, a disorder of homeostasis leads also to acceleration of the fibrinolysis. The variety of internal and external factors in both cases determines the deviation of time for the clot formation, as well as the lyses of blood and fibrin clots. In this study the venous blood as well as the blood and the fibrin clots, derived from healthy dogs, the dogs with cardiopathies and with malignant tumors, were examined for the time of coagulation and fibrinolysis by adding different substances.[93] In these experiments we used a glycolipoprotein extract from earthworm tissue homogenate (G-90) and the proteolytic enzymes P I and P II, isolated from G-90. The efficacy of the tested substances was comparable with the clinically administered anticoagulants. The most significant differences in clotting time among the three tested groups of dogs were obtained by application of the original G-90. The results suggest a possibility that G-90, along with the fibrinolytic enzymes and other biologically active factors, also contains a factor that decelerates the formation of clot in a specific medium, such as the blood from the dogs with malignant tumors.

## 7.4. Eisenin I: A Serine Protease Related to Cell Apoptosis

### 7.4.1. Earthworm Extract is Cytotoxic to Tumor Cell Lines

According to Xie et al, 2003, the extract of earthworm *Eisenia fetida* (EE), isolated by acetone sedimentation and gel filtration, was found to be cytotoxic to several tumor cell lines. This earthworm extract, which is rich in trace elements, is mainly composed of protein with the content of $60.43 \pm 2.36\%$ (n = 5), and the main active components (pI < 7.2) were identified as glycoproteins or glycopeptides. By fibrin plate assay, EE was found to be fibrinolytic and plasminogen-activating.

*In vitro*, the anti-tumor activity of EE was identified by MTT assay and SRB assay and the concentrations of EE required for 50% growth inhibition of human tumor cell strains (HCT-116, SY5Y, MGc803 and Hela) were between 60 and 120 mg/L. *In vivo*, EE was found to significantly prolong the life span of the ascites tumor (S180)-bearing mice by 135.3% and 123.5% when the concentrations (i.p.) were 28 mg/kg and 36 mg/kg, separately. And, EE was even much more effective than cyclophosphamide of which the result of treatment was 76.5%.

The anti-tumor mechanism, *in vitro*, of EE was studied by several methods which included phase-contrast microscopy, fluorescence microscopy, transmission electron microscopy, ordinary agarose gel electrophoresis and flow cytometry. When the effect concentration was not higher than 80 mg/L (taking HCT-116 as the main tested cells), apoptosis was the main mechanism of cell death induced by EE. With different microscopes, obvious apoptotic changes of cell morphology, distributing and content of nucleic acid constitutes and ultrastructure of tested cells were observed in most affected cells. The DNA ladder was assayed by agarose gel electrophoresis. And, when the effect concentration were 40 mg/L and 80 mg/L, the percentage of apoptotic HCT-116 cells, which was assayed by flow cytometry, was 18.9% and 46.1%, respectively. However, cytolysis was also observed in most of effected cells which were treated with EE ($\geq$120 mg/L).[94]

### 7.4.2. Eisenin I, a Possible Apoptosis Related Serine Protease.

Based on the above studies, an apoptosis related protein named Eisenin I was purified from EE by hydrophobic interaction chromatography and ion exchange chromatography.

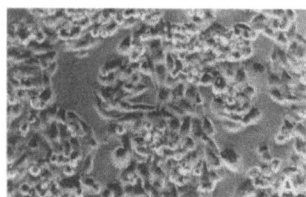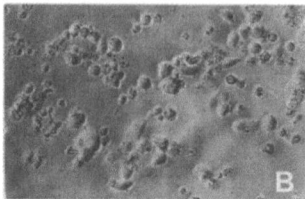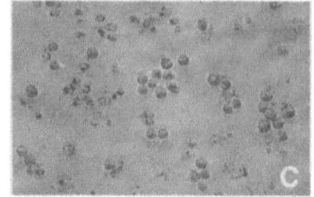

**Figure 23.5.** Morphology of cultured human tumor cells, HCT-116, photographed by phase-contrast microscopy, in 48 h experiment tested for cytotoxicity of Eisenin I; (200 mg/L) and 24 h experiment tested for cytotoxicity of EE (60 mg/L). (a) Control cells without any treatment in 48 h experiment; (b) Cells treated with Eisenin I; (c) Cells treated with EE. Compared with the flattened lineament of normal control cells, b shows that the treatment with 200 mg/L of Eisenin I induced apoptosis with the typical morphology symptoms, such as round contour, restricting and wrinkled cell membrane and apoptotic bodies; however, some necrotic cells were also observed in c, with swollen cell membrane or even membrane fragments. [from Xie et al, identification and partial characterization of antitumor related Serine Protease Eisenin I. Progress in Biochemistry and Biophysics, 2003, in press].

The molecular weight of Eisenin I was about 28 kD assayed by SDS-PAGE. Eisenin I was identified to be a glycopeptide by Schiff′s staining, and the glucide content was about 2.4%. Several coterminal bands could be observed by PAGE of natural Eisenin I and several coterminal peaks of Eisenin I were also detected with MALDI-TOF-MS while the molecular weight of three main peaks are 24645, 25052 and 25281, separately. The amino acid composition of Eisenin I (pI < 3.8) is specialized with high content of acidic amino acids and low content of alkaline amino acids. By fibrin plate assay, Eisenin I was identified to be a plasmin and also a plasminogen activator, and the fibrinolytic activity was inhibited by PMSF (an inhibitor of serine proteases).

*In vitro*, the anti-tumor activity of Eisenin I was not only identified by phase-contrast microscopy observation of apoptotic cells (Fig. 23.5), but also studied further by the localization of fluorescent antibodies.[95] The mechanism of the apoptotic effect still requires further definition.

The complete sequence of mRNA and that of amino acids of Eisenin I were deduced from cDNA. The theory molecular weight is 24691.1. With the complete sequence, this protein was proved to be a serine protease. Now other related studies are being developed.[96]

## 8. Other Clinical Applications of Earthworm Lytic Molecules

Earthworm lytic molecules are antimicrobial and may prove useful as antibacterial agents and prophylactic molecules, an idea that is not farfetched since the discovery of antibiotics was serendipitous. Lysenin and eiseniapore depend to some extent on intracellular lipid trafficking mechanisms. In fact, trafficking dysfunction lead to disease development, such as Tangier disease and Niemann-Pick disease type C, or contribute to the pathogenesis of diseases such as Alzheimer disease and atherosclerosis. Lysenin reacts specifically with fibroblast membranes from patients with Niemann-Pick disease, a rather curious finding, but one that may have some clinical relevance (Table 23.4).[73] Thus, specific binding of lysenin to sphingomyelin on cellular membranes may prove to be a useful tool to probe the molecular motion and function of sphingomyelin in biological membranes, especially in an effort to explain the mechanism of lysis in earthworms.

These results stress the need for concerted analyses of various lytic pathways that may be mediated by the earthworm immunodefense system.

## 9. What Have We Learned from Earthworms: A Model or Source of Clinical Applications?

### 9.1. Novel Approaches to the Treatment of Thrombosis

Thromboembolic diseases remain the main cause of death in Western societies despite current antithrombotic treatments. Recent advances by Gresele and Agnelli, 2001 in the molecular bases of hemostasis have highlighted new targets for novel antiplatelet or anticoagulant agents. Considering antiplatelet agents, selective antagonists of specific receptors (von Willebrand factor, collagen or thrombin receptors) are effective in thrombosis models. For example, direct ADP receptor antagonists, and nitric-oxide-releasing aspirin are in Phase I–II clinical trials. Concerning anticoagulants, inhibitors of tissue-factor-induced clotting activation, selective inhibitors of thrombin and factor Xa, and components of the anticoagulant protein C system (recombinant activated human protein C or human soluble thrombomodulin) have been studied. Some of these agents have had promising results in Phase III studies. Several achievements are anticipated from the development of new antithrombotics, including a further reduction of cardiovascular mortality and unwanted bleeding, and easier patient management.[97]

### 9.2. Thrombolytic Agents

In their review article dealing with thrombolysis during acute myocardial infarction, Anderson and Willerson state, "urokinase is a true enzyme ... [that] exists in both double-chain and single-chain forms."[98] This statement obscures the fundamental clinically important difference between these two forms, a subject never made clear in this otherwise informative review. Urokinase is a true enzyme. The single chain form is in fact, a proenzyme that is inert and stable in plasma (where it is naturally present) and does not form complexes with plasma inhibitors.[99] Moreover, unlike urokinase, the proenzyme is relatively fibrin specific both *in vitro* and *in vivo*.[100] Because of its particular mode of action, which complements that of tissue plasminogen activator, it is synergistic with tissue plasminogen activator as well as with urokinase.[101] The proenzyme is rapidly incorporated by platelets, a finding that may explain the exceptionally low rate of reocclusion (1.5 percent) after coronary thrombolysis reported with its use.[102,103] These properties distinguish the proenzyme from urokinase as well as from other thrombolytic agents: tissue plasminogen activator, streptokinase, and an isolated plasminogen-streptokinase activator complex. It is currently available in Japan and is expected to be approved in Europe within two years. Currently clinical trials in the United States are taking place. It is a novel second-generation activator that should not be confused with first-generation urokinase.

Earthworms, belonging to oligochaete annelids, became a model for comparative immunologists in the early sixties with the publication of results from transplantation experiments. These results proved the existence of self/nonself recognition in earthworms, paving the way for more extensive studies on the earthworm immune mechanisms that evolved to prevent the invasion of pathogens. Within the last 35 years numerous papers on proteolytic, hemolytic, antibacterial, and cytolytic properties of earthworm coelomic fluids were published.[79,104] Moreover, annelids represent one of the animal groups traditionally studied by biologists to seek new molecules with potentially therapeutic use. Though the earthworm has been used as a drug for various diseases in China and the Far East since a few thousand of years ago, more detailed scientific studies have been performed only recently.[45,105] Lumbrokinase, a fibrinolytic enzyme isolated from the earthworm *Lumbricus rubellus*, has been suggested for therapeutic use as a potent thrombolytic agent and recently is commercially available as a novel orally administered fibrinolytic agent for the prevention and treatment of cardiac and cerebrovascular diseases.[106]

Two points deserve consideration with respect to lysis induced by earthworm factors: 1) source of targets *i.e.* intrafamilial allogeneic or xenogeneic targets; those derived from mammals, *e.g.* K562 and erythrocytes; 2) nature of cell death. Regardless of the target, in either instance, cell death may be due to necrosis or apoptosis.[107–109] This clarification would strengthen proposals concerning evolutionary views of apoptosis as have been deciphered for *Caenorhabditis elegans* and *Drosophila*. We may even reveal more informed clues concerning the absence of proven cancer in earthworms and most other invertebrates (with the possible exception of *Drosophila*) (Fig. 23.3). Assuming that cancer poses a natural threat despite the presence of an efficient innate immune system, we can question whether natural cell killing capacity of earthworms is so efficient that any potentially malignant cells never escape its surveillance. Products of the earthworm innate immune system could be tested in clinical trials against tumors as has been reported for the didemnins, a family of seven amino acid, cyclic depsipeptides isolated from deuterostome marine invertebrates (tunicates, Didemnidae).[110,111] Although not clear, didemnin effects may be mediated by high-abundance of a low-affinity cytosolic receptor similar to that for cyclosporine, FK-506, and rapamycin, cytosolic proteins named the immunophilins.[111]

## 10. Acknowledgements

We express appreciation to our most recent students, Eric Jeng with the UCLA Student Research Program for assistance in preparing the manuscript.

## 11. References

1. J. W. Reynolds, W.M. Reynolds, Earthworms in Medicine, *American Journal of Nursing*, 72, 1273–1274. 1972.
2. J. Stevenson, *Oligochaeta*. Claredon Press Oxford, 1930, pp. 658–659.
3. G.E. Gates, The Earthworms of Rangoon, *J. Burma Res. Soc.* 25, 196–221 (1926).
4. H.S. Bristowe, Insects and other invertebrates for human consumption in Siam, *Trans. Entomol. Soc.* 80, 387–404 (1932).
5. L.G.K Carr, Interesting animal foods, medicines and omens of the eastern Indians, with comparisons to ancient European practices, *J. Wash. Acad. Sci.* 41, 229–235 (1951).
6. S. F. Price, Kentucky Folklore, *J Am Folklore.* 14, 30–38 (1901).
7. The Eu Yan Sang Heritage An Anthology of Chinese Herbs and Medicines. Eu Yan Sang International Holdings Pte Ltd, 269A South Bridge Road, Singapore 05818.
8. R. Fan, The new medicine "lumbrokinase capsules" for treatment and prevention of thrombus and imbolicdiseases, pp. 1–7 [Institute of Biophysics, Academia Sinica, Beijing Bio Pharmaceuticals Co. Ltd. 15

Datun Road, Beijing 100101 CHINA] (Personal Communication through Dr. Z. Zhang, National Laboratory of Molecular Virology and Genetic Engineering, Beijing CHINA) (1996).
9. D.S. Linthicum, E.A. Stein, D.H. Marks, and E.L. Cooper, Electron microscopic observations of normal coelomocytes from the earthworm *Lumbricus terrestris*, *Cell and Tissue Res.*, 185, 315–330 (1977).
10. E. A. Stein, R.R. Avtalion, and E.L. Cooper, The coelomocytes of the earthworm *Lumbricus terrestris*: morphology and phagocytic properties, *J. Morphol.* 153, 467–477 (1977).
11. E.A. Stein, A. Wojdani, and E.L. Cooper, Agglutinins in the earthworm *Lumbricus terrestris*: naturally occurring and induced, *Dev. Comp. Immunol.*, 6, 407–421 (1982).
12. E.A. Stein, S. Younai, and E.L. Cooper, Bacterial agglutinins of the earthworm *Lumbricus terrestris*, *Comp. Biochem. Physiol.*, 84: 409–415.
13. E.A. Stein and E.L. Cooper,. In vitro agglutinin production by earthworm leukocytes, *Dev. Comp. Immunol.* 12, 531–547 (1988).
14. E.A. Stein, A. Morovati, P. Rahimian, and E.L. Cooper, Lipid agglutinins from coelomic fluid of the earthworm *Lumbricus terrestris*, *Comp. Biochem. Physiol.*, 94B, 703–707 (1989).
15. E.A. Stein, S. Younai, E.L. Cooper, Separation and partial purification of agglutinins from coelomic fluid of the earthworm, *Lumbricus terrestris*, *Comp. Biochem. Physiol.* 97, 701–705 (1990).
16. E.A. Stein, and E.L. Cooper, Cytochemical observations of coelomocytes from the earthworm, *Lumbricus terrestris*, *J. Histochem*, 10, 657–678 (1978).
17. E.A. Stein, and E.L. Cooper, The role of opsonins in phagocytosis by coelomocytes of the earthworm, *Lumbricus terrestris*, *Dev. Comp. Immunol.* 5, 15–25 (1981).
18. E.A. Stein, and E.L. Cooper, Carbohydrate and glycoprotein inhibitors of naturally occurring and induced agglutinins from the earthworm, *Lumbricus terrestris*, *Comp. Biol. Chem.* 76B, 197–206 (1983).
19. A. Wojdani, E.A. Stein, E.L. Cooper, and L.J. Alfred, Agglutinins and proteins in the earthworm, *Lumbricus terrestris*, before and after injection of erythrocytes, carbohydrates and other materials, *Dev. Comp. Immunol.* 6, 613–624 (1982).
20. P. Roch, E.L. Cooper, and D.P. Eskinazi, Serological evidences for a membrane structure related to human $\beta_2$-microglobulin expressed by certain earthworm leukocytes, *Eur. J. Immunol.* 13, 1037–1042 (1983).
21. A.H. Saad, and E.L. Cooper, Evidence for a Thy-1 like molecule expressed on earthworm leukocytes. *Zool. Sci.* 7, 217–222 (1990).
22. E.L. Cooper, A. Cossarizza, M.M. Suzuki, S. Salvioli, M. Capri, D. Quaglino, C. Franceschi, Autogeneic but not allogeneic earthworm effector coelomocytes kill the mammalian tumor target K562, *Cell. Immunol.* 166, 113–122 (1995).
23. A. Cossarizza, E.L. Cooper, M.M. Suzuki, S. Salvioli, M. Capri, M., G. Gri, D. Quaglino, C. Franceschi, Earthworm leukocytes that are not phagocytic and cross-react with several human epitopes can kill human tumor cell lines, *Exp. Cell Res.* 224, 174–182 (1996).
24. D. Quaglino, E.L. Cooper, S. Salvioli, M. Capri, M.M. Suzuki, I. Pasquali-Ronchetti, C. Franceschi, A. Cossarizza, Earthworm coelomocytes *in vitro*: cellular features and "granuloma" formation during cytotoxic activity against the mammalian tumor cell target K562, *Eu. J. Cell Biol.* 70, 278–288 (1996).
25. E. Kauschke, P. Pagliara, L. Stabili, E.L. Cooper, Characterization of proteolytic activity in coelomic fluid of *Lumbricus terrestris* L. (Annelida, Lumbricidae), *Comp. Biochem. Physiol.* 116B, 235–242 (1997).
26. S. Lange, F. Nussler, E. Kauschke, G. Lutsch, E.L. Cooper, A. Herrmann, Interaction of earthworm hemolysin with lipid membranes requires sphingolipids, *J. Biol. Chem.* 272, 20884–20892 (1997).
27. S. Lange, E. Kauschke, W. Mohrig, E.L. Cooper, Biochemical characteristics of eiseniapore, a pore forming protein in the coelomic fluid of earthworms, *Eur. J. Biochem.* 263, 1–11 (1999).
28. Ph. Roch, P. Ville, E.L. Cooper, Characterization of a 14 kDa plant-related serine protease inhibitor and regulation of cytotoxic activity in earthworm coelomic fluid, *Dev. Comp. Immunol.* 22, 1–12 (1998).
29. I. Eue, E. Kauschke, W. Mohrig, E.L. Cooper, Isolation and characterization of earthworm hemolysins and agglutinins, *Dev. Comp. Immunol.* 22, 13–25 (1998).
30. V. Vetvicka, P. Sima, E.L. Cooper, M. Bilej, Ph. Roch, *Immunology of Annelids* (CRC Press, Boca Raton, 1994), p. 300.
31. E. Kauschke, and W. Mohrig, Cytotoxic activity in the coelomic fluid of the annelid *Eisenia foetida.*, *J. Comp. Physiol. B.* 157, 77–83 (1987).
32. F. Hirigoyenberry, F. Lassalle, M. Lassegues, Antibacterial activity of *Eisenia fetida* andrei coelomic fluid: transcription and translation regulation of lysozyme and proteins evidenced after bacterial infestation., *Comp. Biochem. Physiol.* 95B, 71–75 (1987).

33. A. Wojdani, E.A. Stein, L.J Alfred, E. L. Cooper, Mitogenic effect of earthworm (*Lumbricus terrestris*) coelomic fluid on mouse and human lymphocytes. *Immunobiology* 166, 157–167 (1984).
34. R. Hanusova, M Bilej, L. Brys, P. De Baetselier, A. Beschin, Identification of a coelomic mitogenic factor in *Eisenia foetida* earthworm, *Immunol. Lett.* 65, 203–311 (1990).
35. M. Lassègues, A. Milochau, F. Doignon, L. Du Pasquier, P. Valembois, Sequence and expression of an *Eisenia foetida* derived cDNA clone that encodes the 40 kDa fetidin antibacterial protein,. *Eur. J. Biochem.* 246, 756–762 (1997).
36. Ph. Roch, E.L Cooper, E.L. Invertebrate immune responses: cells and molecular products, *Adv. Comp. Env. Physiol.* 23, 116–145 (1996).
37. P. Valembois, Ph. Roch, D. Boiledieu, in: *Phylogeny of Immunological Memory*, edited by M.J. Manning (Elsevier Biomedical Press, North-Holland, 1980), pp. 47–55.
38. P. Valembois, Ph Roch, L. Du Pasquier, in: *Aspects of developmental and comparative immunology*, edited by J. B. Solomon (Pergamon Press, Oxford, 1980), pp. 23–30.
39. E. L. Cooper, and Ph. Roch, Earthworm leukocyte interactions during early stages of graft rejection, *J. Exp. Zool.* 232, 67–72 (1984).
40. F. Lassalle, M. Lassegues, Ph. Roch, Protein analysis of earthworm coelomic fluid. IV. Evidence, activity, induction and purification of *Eisenia fetida andrei* lysozyme, *Comp. Biochem. Physiol.* 91, 187–192 (1988).
41. P. Valembois, J. Seymour, Ph. Roch, Evidence and cellular localization of an oxidative activity in the coelomic fluid of the earthworm *Eisenia fetida andrei*, *J. Invert. Pathol.* 57, 177–183 (1991).
42. Ph. Roch, P. Valembois, N. Davant, M. Lassegues, Protein analysis of earthworm coelomic fluid. Isolation and biochemical characterization of the *Eisenia fetida andrei* factor (EFAF), *Comp. Biochem. Physiol.* 69, 829–836 (1981).
43. A. Milochau, M. Lassegues, P. Valembois, Purification, characterization and activities of two hemolytic and antibacterial proteins from coelomic fluid of the annelid *Eisenia fetida andrei*, *Biochem. Biophys. Acta.* 1337, 123–132 (1997).
44. H. Mihara, H. Sumi, T. Yoneta, H. Mizumoto, R. Ikeda, M. Seiki, M. Maruyama, A novel fibrinolytic enzyme extracted from the earthworm, *Lumbricus rubellus*, *Jpn. J. Physiol* 41, 461–472 (1991).
45. H. Mihara, M. Maruyama, H. Sumi, Novel thrombolytic therapy discovered from traditional oriental medicine using the earthworm, *Southeast Asia J. Trop. Med. Public Health*. 23,131–140 (1992).
46. U. Rester, I. Bode, M. Moser, M.A.A. Parry, R. Huber, E. Auerswald, Structure of the complex of the antistasin-type inhibitor bdellastasin with trypsin and modeling of the bdellastasin-microplasmin system, *J. Mol. Biol.* 293, 93–106 (1999).
47. N. Nakajima, H. Mihara, H. Sumi, Characterization of potent fibrinolytic enzymes in earthworm, *Lumbricus rubellus*, *Biosci. Biotechol. Biochem.* 10, 1726–1730 (1993).
48. N. Nakajima, K. Ishihara, M. Sugimoto, H. Sumi, K. Mikuni, H. Hamada, Chemical modification of earthworm fibrinolytic enzyme with human serum albumin fragment and characterization of the protease as a therapeutic enzyme, *Biosci. Biotechnol. Biochem.* 60, 293–300 (1996).
49. Z.X. Zhang, F.F. Wang, Effects of crude extract of earthworm on promoting blood circulation to removing stasis, *Chung Kuo Chung His I Chieh Ho Tsa Chih.* 12, 741–743, 710 (1992).
50. K.M. Woo, W. Yi, Y.J. Sohn, C.S. Chang, M.S. Kang, D.B. Ha, C.H. Chung, Purification and characterization of a poly-L-lysine-activated serine endoprotease from *Lumbricus rubellus. Comp. Biochem. Physiol. B. Biochem. Mol. Biol.* 1, 71–80 (1994).
51. G.H. Ryu, S. Park, M. Kim, D.K. Han, Y.H. Kim, B. Min, Antithrombogenicity of lumbrokinase-immobilized polyurethane, *J. Biomed. Mater. Res.* 28, 1069–1077 (1994).
52. G.H. Ryu, Surface characteristic and properties of lumbrokinase-immobilized polyurethane, *J. Biomed. Mater. Res.* 29, 403–409 (1995).
53. Y.S. Kim, M.K. Pyo, K.M Park, B.S. Hahn, K.Y. Yang, H.S. Yun-Choi, Dose dependency of earthworm powder on antithrombotic and fibrinolytic effects, *Arch. Pharm. Res.* 4, 374–377 (1998).
54. Y. Xiong, S.C. Yang, X.Y. Liu, L.Y. Li, B.G. Ru, Purification and determination of partial sequence of Earthworm Fibrinolytic enzyme, *Chinese Biochemical Journal*, 13, 292–296(1997).
55. B.D. Xing, S.M. Yin, B.G. Ru, Purification and characterization of the fibrinolytic enzyme (eFE-D) from earthworm *Eisenia fetida, Acta Biochim. Biophys. Sinica.* 29, 609–612 (1997).
56. J.S. Yang, B.G. Ru, Purification and Characterization of an SDS-Activated Fibrinolytic Enzyme from *Eisenia fetida, Comp. Biochem. Physiol.* 118B, 623–631 (1997).

57. J.S. Yang, L.Y. Li, B.G. Ru, Purification of a Plasminogen Activator from *Eisenia fetida*, *Chinese Journal of Biochemistry and Molecular Biology*. 14, 156–163 (1998a).
58. J.S. Yang, L.Y. Li, B.G. Ru, Characterization of a Plasminogen Activator from *Eisenia fetida*, *Chinese Journal of Biochemistry and Molecular Biology*. 14, 156–169 (1998b).
59. J.S. Yang, L.Y. Li, B.G. Ru, Degradation of Benzoyl-L-argiflifle Ethyl Ester (BAEE) by a Plasminogen Activator from *Eisenia fetida* (e-PA), *Chinese Journal of Biochemistry and Molecular Biology*. 14, 412–416 (1998c).
60. J.S. Yang, L.Y. Li, B.G. Ru, Degradation of N-Acetyl-L-tyrosine Ethyl Ester (ATEE) by a Plasminogen Activator from *Eisenia fetida* (e-PA) *Chinese Journal of Biochemistry and Molecular Biology*. 14, 417–421 (1998d).
61. J.S. Yang, L.Y. Li, B.G. Ru, The enzymology properties and the CD spectra of the active centers of the small subunit of a plasminogen activator from *Eisenia fetida* (e-PA), *Chinese Journal of Biochemistry and Molecular Biology*. 14, 721–725 (1998e).
62. T. Hrzenjak, M. Popovic, L. Tiska-Rudman, Fibrinolytic activity of earthworm's extract (G-90) on lysis of fibrin clots originated from the venous blood of patients with malignant tumor, *Pathol. Onco. Res.* 4, 206–211 (1998).
63. Ph. Roch, L. Stabili, P. Pagliara, Purification of three serine proteases from the coelomic cell of earthworms (*Eisenia fetida*), *Comp. Biochem. Physiol.* 98, 597–602 (1991).
64. C. Leipner, L. Tuckova, J. Rejneck, J. Langner, Serine proteases in coelomic fluid of annelids *Eisenia fetida* and *Lumbricus terrestris*, *Comp. Biochem. Physiol.* 105, 637–641 (1993).
65. A. Polanowski, and T. Wilusz, Serine proteinase inhibitors from insect hemolymph, *Acta. Biochem. Polinica.* 43, 445–454 (1996).
66. Ph. Roch, Protein analysis of earthworm coelomic fluid. I. Polymorphic system of the natural hemolysin of *Eisenia fetida andrei*. *Dev. Comp. Immunol.* 3, 599–608 (1979).
67. Ph. Roch, C. Canicatti, P. Valembois, Interactions between earthworm hemolysins and sheep red blood cell membranes. *Biochem. Biophys. Acta.* 983, 193–198 (1989).
68. P. Valembois, Ph. Roch, M. Lassegues, N. Davant, Bacteriostatic activity of a chloragogen cell secretion, *Pedobiologia.* 24, 191–195 (1982).
69. P. Valembois, Ph. Roch, M. Lassegues, Evidence of plasma clotting system in earthworms. *J. Invert. Pathol.* 15, 221–228 (1988).
70. P. Valembois, M. Lassegues, Ph. Roch, J. Vaillier, Scanning electron-microscopic study of the involvement of coelomic cells in earthworm antibacterial defense, *Cell Tissue Res.* 240, 479–484 (1985).
71. J.H. Cho, C.B. Park, Y.G. Yoon, S.C. Kim, Lumbricin I a novel proline-rich antimicrobial peptide from the earthworm: purification, cDNA cloning and molecular characterization. *Biochim. Biophy. Acta.* 1408, 67–76 (1998).
72. Y. Sekizawa, T. Kubo, H. Kobayashi, T. Nakajima, S. Natori, Molecular cloning of cDNA for lysenin, a novel protein in the earthworm, *Eisenia foetida* that causes contraction of rat vascular smooth muscle, *Gene.* 191, 97–102 (1997).
73. A. Yamaji, Y. Sekizawa, K. Emoto, H. Sakuraba, K. Inoue, H. Kobayashi, M. Umeda, Lysenin, a novel sphingomyelin-specific binding protein, *J. Biol. Chem.* 273, 300–5306, (1988).
74. N. Ohta, S. Shioda, Y. Sekizawa, Y. Nakai, H. Kobayashi, Sites of expression of mRNA for lysenin, a protein isolated from the coelomic fluid of the earthworm *Eisenia foetida*, *Cell Tissue Res.* 302, 263–270 (2000).
75. H. Kobayashi, Y. Sekizawa, M. Aizu, M. Umeda, Lethal and non-lethal responses of spermatozoa from a wide variety of vertebrates and invertebrates to lysenin, a protein from the coelomic fluid of the earthworm *Eisenia foetida*, *J. Exp. Zool.* 286, 538–549 (2000).
76. H. Kobayashi, M. Ohtomi, Y. Sekizawa, N. Ohta, Toxicity of coelomic fluid of the earthworm *Eisenia foetida* to vertebrates but not invertebrates: probable role of sphingomyelin, *Comp. Biochem. Physiol.* 128C, 401–411 (2001).
77. A. Beschin, M. Bilej, E. Torreele, P. De Baetselier, On the existence of cytokines in invertebrates, *Cell. Mol. Life Sci.* 58, 801–814 (2001).
78. M. Bilej, L. Brys, A. Beschin, R. Lucas, E. Vercauteren, R. Hanusova, P. De Baetselier, Identification of a cytolytic protein in the coelomic fluid of *Eisenia foetida* earthworms, *Immunol. Lett.* 45, 123–128 (1995).
79. A. Beschin, M. Bilej, F. Hanssens, J. Raymakers, E. Van Dyck, H. Revets, L. Brys, J. Gomez, P. De Baetselier, M. Timermans, Identification and cloning of a glucan- and lipopolysaccharide-binding protein from *Eisenia*

foetida earthworm involved in the activation of prophenoloxidase cascade, *J. Biol. Chem.* 273, 24948–24954 (1998).
80. M. Bilej, P. Rossmann, M. Sinkora, R. Hanusova, A Beschin, G. Raes, P. De Baetselier, Cellular expression of the cytolytic factor in earthworms *Eisenia foetida. Immunol. Lett.* 60, 23–29 (1998).
81. M. Bilej, P. De Baetselier, Beschin, A. Antimicrobial defense of earthworm, *Folia Microbiol.* 45, 283–300 (2000).
82. M. Bilej, P. De Baetselier, E. Van Dijck, B. Stijlemans, A. Colige, A. Beschin, Distinct carbohydrate recognition domains of an earthworm defense molecule recognize Gram negative and Gram-positive bacteria, *J. Biol. Chem.* 276, 45840–45847 (2001).
83. A. Beschin, M. Bilej, L. Brys, E. Torreele, R. Lucas, S. Magez, P. De Baetselier, Convergent evolution of cytokines, *Nature.* 400, 627–628 (1999).
84. A. Bloc, R. Lucas, E. Van Dijck, M. Bilej, Y. Dunant, P. De Baetselier, A. Beschin, An invertebrate defense molecule activates membrane conductance in mammalian cells by means of its lectin-like domain, *Dev. Comp. Immunol.* 26, 35–43 (2002).
85. E. Olivares, E. Fontt, A. Beschin, E. Van Dijck, V. Vercruysse, M. Bilej, R. Lucas, P. De Baetselier, B. Vray, *Trypanosoma cruzi* is lysed by coelomic cytolytic factor-1, an invertebrate analogue of TNF, and induces phenoloxidase activity in the coelomic fluid of *Eisenia foetida, Dev. Comp. Immunol.* 26, 27–34 (2002).
86. E.L. Cooper, Phylogeny of cytotoxicity *Endeavor* 4, 160–165 (1981).
87. E.L. Cooper, in: *Immunology of Annelids*, edited by V. Vetvicka *et al.* (CRC Press, Boca Raton 1994), pp. 1–12.
88. E.L. Cooper, in: Invertebrate Immunology, edited by B. Rinkevich and W.E.G. Müller (Springer-Verlag, Heidelberg, 1996), pp. 10–45.
89. H. Nagasawa, K. Sawaki, Y. Fujii, M. Kobayashi, T. Segawa, R. Suzuki, H. Inatomi, Inhibition by lombricine from earthworm *(Lumbricus terrestris)* of the growth of spontaneous mammary tumors in SHN mice. *Anticancer Research.* 11, 1061–1064 (1991).
90. T.M. Hrzenjak, M. Popovic, T. Bozic, M. Grdisa, D. Kobrehel, L. Tiska-Rudman, Fibrinolytic and anticoagulative activities from the earthworm *Eisenia foetida, Comp. Biochem. Physiol.* 119B, 825–832 (1998b).
91. T. Hrzenjak, M. Hrzenjak, V. Kasuba, P. Efenberger-Marinculic, S. Levanat, A new source of biologically active compounds earthworm tissue (*Eisenia foetida, Lumbricus rubellus*), *Comp. Biochem. Physiol.* 102A, 441–447 (1992).
92. M. Hrzenjak, D. Kobrehel, S. Levanat, M. Jurin, T. Hrzenjak, Mitogenicity of the earthworm's (*Eisenia foetida*) insulin-like proteins, *Comp. Biochem. Physiol.* 104B, 723–729 (1993).
93. M. Popovic, T.M. Hrcenjak, T. Babic, J. Kos, M. Grdisa, Effect of earthworm (G-90) extract on formation and lysis of clots originated from venous blood of dogs with cardiopathies and with malignant tumors, *Pathol Oncol Res.* 7, 197–202 (2001).
94. J.B. Xie, N. Weng, W.G. He, M.M. Yu, B.G. Ru, Antitumor activity and partial characterization of the extract from earthworm *Eisenia fetida. Chinese Journal of Biochemistry and Molecular Biology*, in press (2003).
95. J.B. Xie, N. Weng, W.G. He, M.M. Yu, B.G. Ru, Purification, identification and partial characterization of antitumor related Serine Protease Eisenin I. *Progress in Biochemistry and Biophysics*, in press (2003).
96. J.B. Xie, N. Weng, B.G. Ru, Cloning and analysis of antitumor related Serine Protease Eisenin I mRNA sequence, *Progress in Biochemistry and Biophysics* (in preparation).
97. PGresele, and G. Agnelli, Novel approaches to the treatment of thrombosis, *Trends in Pharmacological Sciences* 23, 25–32 (2001).
98. H.V. Anderson, J.T. Willerson, Thrombolysis in acute myocardial infarction, *N. Engl. J. Med.* 329, 703–709 (1993).
99. R. Pannell, and V. Gurewich, Pro-urokinase: A study of its stability in plasma and of a mechanism for its selective fibrinolytic effect, *Blood.* 67, 1215–1223 (1986).
100. V. Gurewich, R. Pannell, S. Louie, P. Kelley, R.L. Suddith, R. Greenlee, Effective and fibrin-specific clot lysis by a zymogen precursor form of urokinase (pro-urokinase): a study in vitro and in two animal species, *J. Clin. Invest.* 73, 1731–1739 (1984).
101. R. Pannell, J. Black, V. Gurewich, The complementary modes of action of tissue-type plasminogen activator and pro-urokinase by which their synergistic effect on clot lysis may be explained, *J Clin Invest.* 81, 853–859 (1988).

102. V. Gurewich, M. Johnstone, J.P. Loza, R. Pannell, Pro-urokinase and prekallikrein are both associated with platelets: implication for the intrinsic pathway of fibrinolysis and for therapeutic thrombolysis, *FEBS Lett.* 318, 317–321 (1993).
103. V.Gurewich, Thrombolytic agents, *N. Engl. J. Med.* 330, 291 (1994).
104. M. Bilej, in: *Immunology of annelids*, edited by V. Vetvicka *et al.* (CRC Press, Boca Raton, 1994) pp. 167–200.
105. H. Mihara, Fibrinolytic enzymes extracted from the earthworm *Lumbricus rubellus*: a possible thrombolytic agent, *Nippon Seirigaku Zasshi.* 53, 231–243 (1991).
106. Daedo Pharmaceutical Co. Ltd., Yongshim Capsule, 1990, pp. 1–2.
107. E.W. Skowronski, R.N. Kolesnick, D.R. Green, Fas-mediated apoptosis and sphingomyelinase signal transduction: the role of ceramide as a second messenger for apoptosis, *Death Different* 3, 171–176 (1996).
108. S. Rowan, D.E. Fisher, Mechanisms of apoptotic cell death, *Leukemia* 11, 457–465 (1997).
109. S. Bourteele, A. Hausser, H. Doppler, J. Horn-Muller, C. Roopke, G. Schwarzmann, K. Pfizenmaier, G. Muller, Tumor necrosis factor induces ceramide oscillations and negativity controls sphingolipid synthases by caspases in apoptotic Kym-1 cells, *J. Biol. Chem.* 273, 31245–31251 (1998).
110. K.L. Rinehart Jr., J.B. Gloer, G.R. Wilson, R.G. Hughes Jr., L.H. Li, H.E. Renis, J.P. McGovern, Antiviral and antitumor compounds from tunicates, *Fed. Proc.* 42, 87–90 (1983).
111. D.W. Montgomery, G.K. Shen, E.D. Ulrich, C.F. Zukoski, Immunomodulation by didemnins. Invertebrate marine natural products, *Ann. N.Y. Acad. Sci.* 712, 301–314 (1994).

# 24

# What Can We Learn from Marine Invertebrates to be Used as Complementary Antibiotics?

## Philippe Roch

## 1. The Place for Anti-Infectious Therapies

### 1.1. The Need for Novel Alternative Antibiotics

Several biotechnology start-ups are engaged in developing novel antibiotics using antimicrobial peptides as templates. The pioneer was Magainin Pharmaceuticals from Philadelphia (PA USA) in the early 1990s with Pexiganan®, derived from frog skin magainin, but with little success. Another North-American company, Intrabiotics Pharmaceuticals from Mountain View (CA) developed Iseganan®, a synthetic analogue to pig protegrin. Micrologix Biotech from Vancouver (BC Canada) developed several linear antimicrobial peptides, particularly against resistant *Staphylococcus aureus*. EntoMed from Strasbourg (France) was the first company to use insect (a non revealed tropical butterfly) antimicrobial peptides to fight fungal infections due to *Candida* and *Aspergillus* in immunologically compromised patients. Anti Gram+ bacterial activity, particularly *Staphylococcus*, have also been analyzed. Recently, SelectBiotics from Montpellier (France) has become interested in the peptides produced by several bacterial species to protect themselves from other microorganisms.

It was during the campaign of British vessel *HMS Challenger* (1872–1876) led by Charles Wyville Thomson, that life in the abyss (-5,200 m) was definitively revealed. However, due to the hostile environment, it was only in the early 1960's that the extended species diversity was observed. On February 15, 1977, the first community living without light on the deep sea hydrothermal vents was reported by the expedition led by John Corliss from the University of Oregon. The surprising fauna discovered there included bacteria, mollusks (*Bathimodiolus*), annelids (*Riftia pachyptila*) and crustaceans (shrimp, *Rimicaris exoculata*) (Gaill, 1993). More recently, rich fauna associated with whale bones and with sunken wood was reported from the deep-sea bottom, including annelids and mollusks

---

**Philippe Roch** • Laboratoire DRIM, Université de Montpellier 2, case courrier 080, Place Eugène Bataillon, 34095 Montpellier cedex 5, France. E-mail: proch@univ-montp2.fr

*Complementary and Alternative Approaches to Biomedicine*, edited by Edwin L. Cooper and Nobuo Yamaguchi. Kluwer Academic/Plenum Publishers, 2004.

(Distel et al., 2000). These creatures have developed particular metabolisms based on sulfur released from lipid degradation.

Due to their life in extreme environments, marine invertebrates may have also developed specific molecules which can be of interest. DNA polymerase is one example that has been isolated from *Thermophilus aquaticus* (Taq polymerase) collected from hydrothermal vents and capable of duplicating any DNA up to 75°C, being stable for long exposure to 95°C (Innis et al., 1988). The drawback of such unusual capacity is the high rate of error and synthesis limit at 5 kb (Hamilton et al., 2001). Fortunately, better capacities are discovered periodically: the bacteria *Pyrococcus abyssi* (Erauso et al., 1993) possesses a DNA polymerase (Isis DNA Polymerase™, Qbiogene Molecular Biology) characterized by a half-life of 5 h at 100°C with an error rate of $0.66 \ 10^{-6}$ mutations/nucleotide/duplication (Dietrich et al., 2002).

The long delay between lab discovery and commercial availability makes urgent the need for new families of antibiotics, not only to prevent already existing microorganisms but also to act on the new ones that we have selected.

## 1.2. Biological Basis of Microbial Viability

In all organisms, three reactions mediated by enzymes occurred continuously: DNA duplication to double genetic material, transcription of DNA into RNA and RNA translation into proteins. Numerous antibiotics act on transcription or translation. For instance, molecules from the rifamycin family blocked RNA synthesis and molecules from the macrolid family inhibited protein synthesis. Other antibiotics such as the β-lactamates (penicillins, cephalosporins...) bound to enzymes involved in bacterial cell wall synthesis. All the present antibiotics have been derived from a restricted number of molecules with only slight differences introduced into the same frameworks. No new family has been discovered for more than 20 years.

During that time, more and more bacteria have developed resistance mechanisms against nearly all the known antibiotics. They belong to three categories: modification of membrane permeability, rejection of these toxic molecules and mutation of the target proteins. β-lactamate resistance is of particular interest as the resistance can suddenly appear by exchange of plasmid containing β-lactamases with a resistant strain. Systematization of antibiotics used to prevent putative infection, and use of irrelevant high doses, as during non-serious viral infections, have selected multi-drug resistant bacteria. Such resistances are particularly dangerous in immunologically compromised patients and for people suffering from sepsis due to long term hospitalization.

## 2. New Interest in Antimicrobial Peptides

## 2.1. What Are Antimicrobial Peptides

Antimicrobial peptides are among the most important effectors of innate immunity. They are quite universal as found in plants (Terras et al., 1993; Osborn et al., 1995), invertebrates (Steiner et al. 1981) and vertebrates, including birds (Evans et al. 1994) and humans (see Levy, 1996; Cole and Ganz, 2000 for reviews). They represent elements of an ancestral immune system that predates lymphocytes and immunoglobulins. Even if they have been mostly studied in insects in terms of molecular structure and gene expression, few quantitative data are available. In the pupae of the fruit fly, *Drosophila melanogaster*, normal concentrations of peptides are undetectable. Following injury, concentrations can be significantly different according to the peptide family, ranging from less than 2 µM for defensin to 10 µM for metchnikowin, 20 µM for cecropin, 40 µM for drosocin, and up to 100 µM for drosomycin (Bulet, 1999). In humans, minimal daily production of the only defensin has been estimated at 10 mg/kg, mainly secreted by epithelial cells at the host-environment interface (Ganz et al., 1992). Meanwhile, neutrophils from mice lacked appreciable defensin content, revealing that not all of the species possess such peptides (Eisenhauer and Lehrer, 1992). In invertebrates, only a ridiculously low number of species have been investigated and surprising discoveries might occur.

## 2.2. Mechanism of Action and Regulation

Several results revealed that antimicrobial peptides may enhance the activity of classical antibiotics. For instance, a sub-inhibitory concentration of cecropin increased the sensitivity of *Staphylococcus epidemidis* to benzylpenicillin (40 folds), fusidic acid (4 folds) and novobiocin (2 folds) (Moore et al., 1996). Other immune-related molecules such as lysozyme, also act in synergy with antimicrobial peptides such as rat defensins in bactericidal activity against *Pseudomonas aeruginosa* (Kohashi et al., 1992). In fact, at least in the larvae of the silkworm, *Bombyx mori*, the elicitor specificity for lysozyme induction is identical to that for cecropin, suggesting a common mechanism for recognition of bacteria and signal transduction giving rise to the simultaneous synthesis of cecropin and lysozyme (Morishima et al., 1995).

Regulation of antimicrobial peptide synthesis has been investigated extensively, particularly in *Drosophila* where numerous mutants are available and the knock-out technology has allowed rapid breakthroughs. Several signal transduction pathways have been found to control expression of the antimicrobial peptides. The Toll signaling pathway directs expression of the gene encoding the antifungal peptide drosomycin (Lemaitre et al., 1997). The IMD (immune deficiency) pathway controls drosocin and diptericin (Hoffmann and Reichhart, 1997). In addition, three distinct pathways (Toll, IMD and 18-wheeler) control cecropin, attacin and defensin genes (Williams et al., 1997; Engstrom, 1999) whereas only two (Toll and IMD) control the expression of metchnikowin (Levashina et al., 1998). The Toll pathway exhibits structural and functional similarities closely related to the pathway which triggers immune genes in mammals (Lemaitre, 1999). Consequently, at least the Toll pathway appeared very early in evolution and is maintained in more recent animals, even when acquired immunity developed.

Until recently, the development of antibacterial agents was based on classical strategies, including combinatorial libraries generated by synthetic chemistry targeting DNA replication, cell division, bacterial envelopes, protein secretion, bacteria adhesion and uptake of antibiotic by macrophages (Desnottes, 1996). Antimicrobial peptides function in both phagocytic and extracellular killing of microbes, attacking multiple targets to penetrate and disrupt the microbial surfaces. The latter aspect made them of interest for the development of novel pharmaceuticals for the treatment of infections both as primary and as sequelae of other diseases.

## 2.3. Basic Effectors of Innate Immunity

In the early 1980's, the group of Hans Boman at Stockholm University, isolated a peptide named cecropin possessing strong antibacterial activity from diapausing pupae of the moth *Hyalophora cecropia* (Steiner et al., 1981). Activity was induced by a bacterial challenge suggesting its implication in immune defense. To date, more than 200 antimicrobial peptides have been isolated from insects (see Bulet et al., 1999 for review). Invertebrates do not possess adaptive (acquired) immunity. Their anti infectious defense mainly relies on phagocytosis and destruction of pathogens by several mechanisms including actions of antimicrobial peptides. Even if less specific than the adaptive response, innate immunity acts during the first minutes following the pathogen's entry and is obviously efficient as it has been maintained in evolution for hundreds of thousands of years. Nowadays, the role of antimicrobial peptides in host defenses has become quite obvious. Evidence includes induction-stimulation by bacterial products, kinetics of such neo-synthesis and protection against experimental infections in animal models including transgenic animals.

Insects interest many people, even non-biologists, as an extremely prolific phylum with more than 2 millions species, that colonize all environments. Still, only about 200 species have been tested for antimicrobial peptides and not one of them possesses the same molecule. But diversity of invertebrates is even more extended and one can consider that in evolutionary terms, some species may have developed specific systems, processes or molecules. This might be particularly the case for marine organisms living in a high osmolarity medium associated with minimal sources of light, high pressure and toxic chemicals at great depths.

## 3. Common Features of Antimicrobial Peptides

### 3.1. Structure of Antimicrobial Peptides

Antimicrobial peptides constitute multiple families of gene-encoded molecules comprising between 12 and 50 amino acids, with at least 2 excess positive charges and about 50% hydrophobic amino acids. They are found throughout the animal and plant kingdoms. Although not admitted by all authors, they may be grouped into at least 3 structural classes, (i) linear peptides frequently folded into helix(s), (ii) peptides stabilized by disulfide bond(s), (iii) peptides containing an elevated percentage of a particular amino acid (mainly proline or glycine).

In classical medicine, there has been a tendency to emphasize the role of the humoral-cellular immune system in defense against infections. Meanwhile, it is clear that the adaptive immune system requires a significant delay to be efficient. In contrast, antimicrobial peptides are induced quite rapidly after introducing a pathogen. For instance, in the mosquito *Aedes aegypti*, induction of cecropin B occurred between 2 and 6 h post *in vivo* inoculation with the Gram- *E. coli* (Chalk et al., 1994). Similarly, a 19 amino acid O-glycosylated peptide purified from *Drosophila* inhibits the *in vitro* growth of *E. coli* in less than 3 h (Bulet et al., 1993). The coleoptericin isolated from the large tenebrionid beetle, *Zophobas atratus*, is even more potent since it can kill *E. coli in vitro* in less than 30 min (Bulet et al., 1991).

A single polypeptide chain, translated from a single cDNA, derived from a single gene constitutes antimicrobial peptides. Consequently, and even if the genes include one

(Yamano et al., 1998) or several (Mitta et al., 2000b) introns, synthesis of peptides constitutes a relatively simple process. As a result, there may be ways to manipulate them rather simply.

## 3.2. Activity of Antimicrobial Peptides

Activity of antimicrobial peptides is a direct result of physical mechanisms of action involving electrostatic binding. As positively charged, the peptides strongly bind to the negative charges of the membrane disrupting the integrity of the phospholipid bilayer. Two slightly different modes of action have been reported. In both models, the peptides initially interact with the polyanionic surface lipopolysaccharides (LPS) and competitively displace the divalent cations that bridge and partly neutralize the LPS (Hancock ad Scott, 2000). In the **carpet** model, the peptides are inserted into the membrane and are oriented parallel to it. When a critical number of molecules is inserted, structural organization of the bilayer is disrupted, and may be visualized as surface blebbing. In the **barrel** model, inserted peptides move perpendicular to the membrane. Association of multiple molecules forms transmembrane channels. The lifetime of such channels is short (Wu et al., 1999) and, when they collapse, some of the peptides are left in the membrane where they contribute to its disorganization. In fact, the two modes of action can be observed simultaneously but the importance of each of the two models varies from peptide to peptide. In addition, due to the disruption of the outer membrane, free peptides penetrate the cell and interact with cytoplasmic polyanions, such as DNA (Zhang et al., 1999).

Classical antibiotics neutralize reactions crucial for survival or development, as those that are mediated by enzymes. To avoid such an effect, bacteria mutate and most of the time, a mutation of only one gene is sufficient. Consequently, resistance occurs very rapidly. To avoid the physical mechanism of action by an antimicrobial peptide, bacteria must deeply reorganize the membrane that involves one hundred genes, a very unlikely situation. Meanwhile, one case has been reported concerning the mutation of the PhoP/PhoQ system, but it increases the minimum inhibitory concentration (MIC) to cationic peptides by only 2 to 4-fold (Steinberg et al., 1997).

## 3.3. What are Defensins?

Defensins are abundant constituents of cytoplasmic granules of several mammalian neutrophils and macrophages. In human, the three major defensins make up about 5% of total cellular proteins (Ganz, 1987). Consequently, when defensin-rich granules fuse to a phagosome, the intra-phagolysosome defensin concentration is likely to reach several mg/ml. Only a small volume of the phagolysosome allows such elevated concentrations, exceeding those employed in most *in vitro* experiments. Numerous molecules and actions inside restricted organelles underlined the power of the defense system based on antimicrobial peptides.

## 4. The Molluscan Antimicrobial Peptides

Applying the technique used for insects, several small proteins of 4 kDa have been purified from mussel plasma and hemocyte granules (Charlet et al., 1996; Hubert et al.,

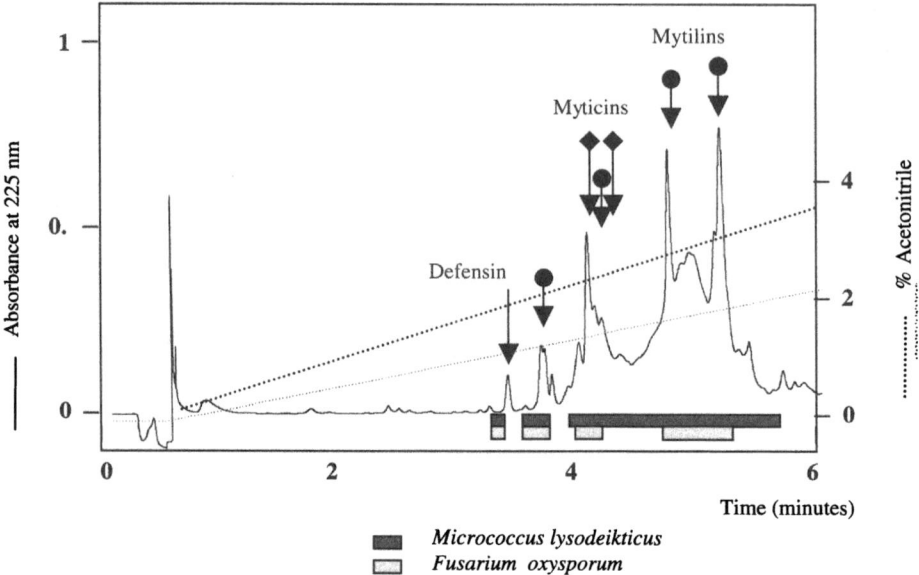

**Figure 24.1.** Reverse phase HPLC of acid extract from hemocyte granule of the Mediterranean mussel, *Mytilus galloprovincialis*. Antimicrobial activity was tested against the Gram+ bacteria *Micrococcus lysodeikticus* and the fungus *Fusarium oxysporum*. Note the multiple active fractions with identical or different activities

1996). Acid extracts were separated to homogeneity in reverse phase HPLC. Routinely, antimicrobial activities were checked all along the elution profiles and consisted of *in vitro* growth inhibition of Gram + *Micrococcus lysodeikticus* and fungus *Fusarium oxysporum*. About 10 active fractions were located, revealing the presence of several molecules. Most fractions shared both anti bacterial and anti fungal activities (Fig. 24.1).

A complete sequence of the peptides showed that they are of similar size with 36–40 amino acids. Based on primary amino acid sequence homologies, mussel peptides have been arranged into 3 families: defensins (Mitta et al., 1999b), mytilins (Mitta et al., 2000a) and myticins (Mitta et al., 1999a) (Fig. 24.2). In *M. galloprovincialis*, all the peptides possess 8 cysteines arranged in 3 specific conserved arrays (Mitta et al., 2000d). Meanwhile, *M. edulis* defensins, referred to as peptides A and B, contain only 6 cysteines (Charlet et al., 1996). Slight differences in other amino acids determined several isoforms. Genes encoding defensin B and mytilin B have been cloned and sequenced, revealing that both genes share the same organization including 4 exons and 3 large introns (Mitta et al., 2000b).

From comparisons of mussel and insect defensins (Cornet et al., 1995), two antibacterial peptides which share the common cystine-stabilized-$\alpha\beta$ (CS$\alpha\beta$) structural motif, the mussel defensin is apparently more compact than the insect's (Yang et al., 2000). Their structures differ mainly by the presence of a fourth disulfide bond, possibly rendering the mussel defensin more stable in a high osmolarity medium, such as sea water (Fig. 24.3). In addition, loop 1 of the mussel defensin is shorter due to deletion of seven residues at the N-terminus. A similar disulfide pattern both in mollusks and in some scorpions argues for a common ancestral molecule. Moreover, a phylogenetic analysis of defensin is indicative of a slow evolutionary process in mussels, scorpions and dragonflies, but a more rapid evolutionary

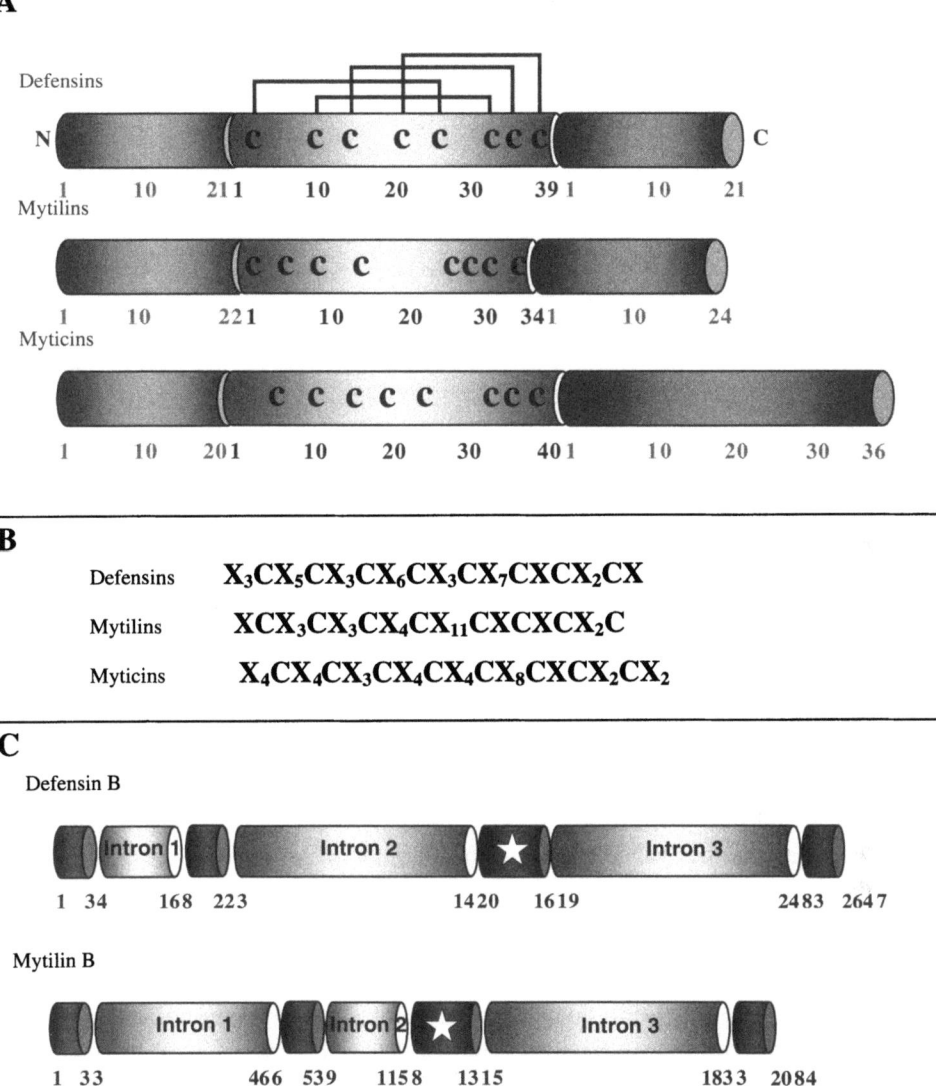

**Figure 24.2.** Structure of *Mytilus galloprovincialis* antimicrobial peptides. **A**, amino acid organization of the pro-peptides. **B**, consensus cysteine patterns defining the 3 families. **C**, defensin and mytilin gene organizations. Active peptides are entirely coded by exon 3 (stars).

process in neopteran insects (Charlet et al., 1996). Rapid evolution of defensins, duplication of defensin genes and multiple abundant defensin peptides in several species, suggest also that structural variations constitute important features of the defensin system.

In mussels, expression of antimicrobial peptide genes is strictly restricted to hemocytes; there have been exceptions. Using polyclonal antibodies, cross-reacting defensins

## A

| | Defensin | Q | M | K | I | H | B |
|---|---|---|---|---|---|---|---|
| Micrococcus lysodeikticus | 0.5 | 9 | 16 | 12 | 13 | 17 | 28 |
| Staphylococcus epidermidis | 3.1 | 30 | 34 | 43 | 38 | 49 | 43 |
| Bacillus megaterium | 0.8 | 24 | 34 | 29 | 46 | 49 | 51 |
| Fusarium oxysporum | 5 | 13 | 13 | 17 | 15 | 35 | 30 |

## B

| | Native 2 Arg + | Native open | 3 + | 6 + | Neutral |
|---|---|---|---|---|---|
| Bacillus megaterium | 28 | >75 | 13 | 6 | >75 |

**Figure 24.3.** Dissection of the defensin molecule. **A**, activities (μM) related to β-sheets and loop 3. **B**, essential role of cyclization and positive charges. Partially from Yang et al., 2000

and mytilins have been detected inside enterocytes that surround the digestive lumen (Mitta et al., 1999b; 2000c). Similar cross-reactivities have been reported in the Paneth cells of mammals (Ouellette and Selsted, 1996) and in insects (Lehane et al., 1997). Although the cross-reacting molecules have not been identified, one can hypothesize that antimicrobial peptides are secreted into the gut to kill potential pathogens. Inside hemocytes, defensins are predominantly located in small granules of a granulocyte subtype and in large clear granules of another granulocyte subtype (Mitta et al., 1999b). Mytilins appeared to be restricted to only one granulocyte subtype, but in two structures: dense granules of 0.5–1.5 μm and multivesicular structures of 3–8 μm (Mitta et al., 2000a). Confocal microscopic observations using double labeling with anti-defensin and anti-mytilin antibodies revealed that the two peptides could be detected (i) in different hemocyte subtypes, (ii) in the same granulocyte but in different intra cellular compartments, and (iii) even in the same compartment (Mitta et al., 2000c). Different hemocytes can be positive for defensins (16%) or mytilins (37%), but both immune reactivities often appeared within the same hemocyte (32%), stored in different compartments or in the same compartment. It is important that 15% of the circulating hemocytes did not stain for defensins or mytilins.

All the mussel peptides inhibited the *in vitro* growth of Gram+ bacteria (*M. lysodeikticus*, *Bacillus megaterium*, *Staphylococcus aureus*, *Listeria monocytogenes*, *Enterococcus faecalis*). Only mytilins are also active against Gram- bacteria, such as *Enterobacter aerogenes* and several *Vibrios* (Table 24.1). This activity strongly depends upon the isoform: mytilin E is not active against Gram- bacteria. Antifungal activity is also dependent upon

Table 24.1. Activity spectrum of mussel antimicrobial peptides (MIC and MBC in µM)

| | Defensin | Mytilins | | | | | Myticins | |
|---|---|---|---|---|---|---|---|---|
| | A | B | C | D | E | | A | B |
| Gram positive bacteria (MBC) | | | | | | | | |
| Micrococcus lysodeikticus | 0.5 | 0.2 | 0.7 | 0.2 | 0.7 | | 2.2 | 1.0 |
| Bacillus megaterium | 0.8 | nt | nt | nt | nt | | 2.2 | 1.0 |
| Staphylococcus aureus | 0.6 | >6 | 0.7 | 3 | >6 | | >20 | >20 |
| Listeria monocytogenes | nt | 0.3 | 0.7 | 0.7 | 1.0 | | >20 | >20 |
| Enterococcus faecalis | nt | 0.2 | 3.0 | nt | 1.0 | | >20 | >20 |
| Enterococcus viridans | nt | nt | nt | nt | nt | | 9.0 | 2.0 |
| Gram negative bacteria (MBC) | | | | | | | | |
| Escherichia coli D31 | >18 | 0.7 | 0.3 | 0.3 | >6 | | >20 | 120 |
| Salmonella newport | nt | nt | >11 | nt | >6 | | >20 | 120 |
| Salmonella typhimurium | nt | nt | 19 | nt | >6 | | >20 | >20 |
| Brucella suis S1 | nt | >11 | >11 | >6 | >6 | | >20 | >20 |
| Pseudomonas aerogenes | nt | >11 | >11 | nt | >6 | | >20 | >20 |
| Vibrio alginolyticus | >18 | 1.4 | 1.4 | 0.8 | nt | | >20 | nt |
| Vibrio vulnificus | nt | 1.4 | 1.4 | nt | nt | | >20 | >20 |
| Vibrio splendidus | 15 | 0.2 | 1.4 | nt | nt | | >20 | >20 |
| Fungus (MIC) | | | | | | | | |
| Fusarium oxysporum | 5.0 | 0.7 | >6 | 0.7 | >6 | | >20 | 5.0 |
| Protozoa (MIC) | | | | | | | | |
| Perkinsus marinus | >20 | nt | >20 | nt | nt | | >20 | >20 |
| Trypanosoma brucei brucei | 5.0 | 1.0 | nt | nt | nt | | nt | nt |

isoforms: mytilins B and D inhibited *in vitro* growth of *F. oxysporum*, mytilins C and E exert no effect. Similarly, myticin B, not myticin A is active. In vitro growth inhibition is due to bacteria killing and all the mussel peptides possess bactericidal effect, but with significant differences between isoforms. Defensin A and mytilin C kill all *M. lysodeikticus* in less than 3 min after contact. In contrast, myticin A required 2 h contact and mytilin E more than 6 h, revealing different kinetics of activity (Mitta et al., 1999a). The different families of mussel antimicrobial peptides, and inside a family, the different isoforms, possess complementary activities, both in terms of target specificity and kinetics, thus extending the target spectrum.

Both defensins and mytilins are toxic for the parasite protozoans, *Trypanosoma brucei brucei* and *Leishmania major*. For instance, 50% of the *T.b. brucei* were killed by 1 µM mytilin B after 3 h of cultivation at 30°C. This antiprotozoa activity, directed towards mammalian parasites, is extremely interesting with respect to the development of new products and medical strategies/therapies.

Antiviral activity is also interesting. Tested in the *in vitro* system using human Hela Magic cells, 20 µM of mytilin B prevented the infection due to HIV-1 Bru by 96% whereas 10 µM were still responsible for 65% inhibition. Similar results have been obtained with defensins. Antiviral activity appeared to be mediated by peptide binding onto the virus as demonstrated by differential incubations. In a totally different system using crustacean White Spot Syndrome Virus (WSSV) and shrimp, *Palaemon serratus*, mytilin B was able to delay the mortality when incubated with the virus before injection.

## 5. Dissecting the Mollusk Defensin

Based on the 3D structure of mussel defensin A (Yang et al., 2000), a set of peptides was designed and chemically synthesized. Dilutions of the purified peptides were further tested for growth inhibition of the Gram+ bacteria *M. lysoteikticus*. Peptides corresponding to the α-helix or to the α-helix prolonged by the N-terminal turn (loop 1) and by the short sequence corresponding to loop 2, did not exhibit measurable activity although the latter peptide represents almost 50% of the complete defensin amino acid sequence (Fig. 24.3). However, the 9-mer peptide corresponding to loop 3 had a MIC of 28 µM (i.e., about 2.5% of the synthetic defensin). This peptide was active only when a cysteine bond was created, cycling the peptide. Identical but linear peptides did not show any measurable activity. Inhibitory activity was not simply due to the basic or cyclic character, since the loop 2 possessing both criteria was inactive.

To study the influence of positive charges on antibacterial activity several peptides derived from loop 3 were designed (Fig. 24.3). Replacing the 2 arginines with 2 isosteric but non ionisable citrullines neutralized the charges of loop 3. As a consequence, the pI was lowered from 9.0 to 8.0 and the bacteriostatic activity was almost completely lost. In contrast, increasing the positive charges by replacing naturally occurring residues with lysins correlated with higher activity. MIC ranged from 28 µM (native 2 positive charges) to 13 µM (3 positive charges), 8 µM (4 positive charges) and even 6 µM (6 positive charges). One can hypothesize that more positive charges resulted in stronger binding of the peptide to the negatively charged prokaryote membrane, as reported for magainin (Matsuzaki et al., 1997). In that case, electrostatic attractions play a crucial role in the binding process and consequently in the antimicrobial activity.

## 6. The Future of Alternative Antimicrobial Approaches

Cationic antimicrobial peptides represent one of the most original anti infectious molecules discovered in the last 25 years. They possess many of the desirable features for a novel antibiotic class: broad spectrum of activity, kill bacteria rapidly, are unaffected by classical antibiotic resistance mutations, do not easily select resistant variants, show synergy with classical antibiotics and other innate immunity processes, neutralize endotoxins and are active in animal models (Hancock and Lehrer, 1998). Despite all of these advantages, they are in the early stage of their pharmaceutical experimentation and many aspects remain to be solved, including demonstration of their therapeutic capacities. Meanwhile, originality of their mode of action, coupled with their stability, small size, resistance to degradation and facility to be synthesized at low cost, will ultimately render these molecules a top choice for 21st Century antibiotics.

## 7. Acknowledgements

Speculative development of mussel antimicrobial peptides is based on experimental results obtained mainly from Guillaume Mitta, Bernard Romestand (DRIM Laboratory), Claude Granier (Institut Biotechnologie Pharmacologie, University Montpellier 1, France),

Alain Beschin (Flanders Interuniversity Institute for Biotechnology, St Genesius-Rode, Belgium) and Eric Bernard (Centre de Recherche en Biochimie des Macromolécules, Montpellier, France). Edwin L. Cooper is warmly acknowledged for improving the language of the present manuscript and for promoting CAM.

## 8. References

Bulet, P., 1999, Les peptides antimicrobiens de la drosophile, *Medecine/Sciences* **15**:23–29.
Bulet, P., Cociancich, S., Dimarq, J-L., Lambert, J., Reichhart, J-M., Hoffmann, D., Hétru, C., and Hoffmann, J.A., 1991, Isolation from a coleopteran insect of a novel inducible antibacterial peptide and of new members of the insect defensin family, *J. Biol. Chem.* **266**:24520–24525.
Bulet, P., Dimarcq, J-L., Hétru, C., Lagueux, M., Charlet, M., Hegy, G., Van Dorsselaer, A., and Hoffmann, J.A., 1993, A novel inducible antibacterial peptide of *Drosophila* carries an O-glycosylated substitution, *J. Biol. Chem.* **268**:14893–14897.
Bulet, P., Hétru, C., Dimarcq, J-L., and Hoffmann, D., 1999, Antimicrobial peptides in insects: structure and function, *Develop. Comp. Immunol.* **23**:329–344.
Chalk, R., Townson, H., Natori, S., Desmond, H., and Ham, P.J., 1994, Purification of an insect defensin from the mosquito, *Aedes aegypti*, *Insect Biochem. Molec. Biol.* **24**:403–410.
Charlet, M., Chernysh, S., Philippe, H., Hétru, C., Hoffmann, J.A., and Bulet, P., 1996, Innate immunity: Isolation of several cysteine-rich antimicrobial peptides from the blodd of a mollusk, *Mytilus edulis*. *J. Biol. Chem.* **271**:21808–21813.
Cole, A.M., and Ganz T., 2000, Human antimicrobial peptides: analysis and application, *Biotechniques* **29**:822–831.
Cornet, B., Bonmatin, J-M., Hétru, C., Hoffmann, J.A., Ptak, M., and Vovelle F., 1995, Refined three-dimensional solution structure of insect defensin A, *Structure* **3**:435–448.
Desnottes, J-F., 1996, New targets and strategies for the development of antibacterial agents, *TIBTECH.* **14**:134–140.
Dietrich, J., Schmitt, P., Zeiger, M., Preve, B., Rolland, J-L., Chaabihi, H., and Gueguen, Y., 2002, PCR performance of the highly thermostable proof-reading B-type DNA polymerase from *Pyrococcus abyssi*, *FEMS Microbiol. Letters* **217**:89–94.
Distel, D.L., Baco, A.R., Chuang, E., Morill, W., Cavanaugh, C., and Smith, C.R., 2000, Do mussels take wooden steps to deep-sea vents, *Nature* **403**:725–726.
Eisenhauer, P.B., and Lehrer, R.I., 1992, Mouse neutrophils lack defensins, *Infect. Immun.* **60**:3446–3447.
Engström, Y., 1999, Induction and regulation of antimicrobial peptides in *Drosophila*, *Develop. Comp. Immunol.* **23**:345–358.
Erauso, G., Reysenbach, A.L., Godfroy, A., Meunier, J.R., Crump, B., Partensky, F., Baross, J.A., Marteisson, V., Barbier, G., Pace, N.R., and Prieur, D., 1993, *Pyrococcus abyssi* sp. nov., a new hyperthermophilic archeon isolated from a deep-sea hydrothermal vent, *Arch. Microbiol.* **160**:338–349.
Evans, E.W., Beach, G.G., Wunderlich, J., and Harmon, B.G., 1994, Isolation of antimicrobial peptides from avian heterophils, *J. Leuk. Biol.* **56**:661–667.
Gaill, F., 1993, Aspects of life development at deep sea hydrothermal vents, *FASEB J.* **6**:558–65.
Ganz, T., 1987, Extracellular release of antimicrobial defensins by human polymorphonuclear leukocytes, *Infect. Immun.* **55**:568–571.
Ganz, T., Oren, A., and Lehrer, R.I., 1992, Defensins: microbicidal and cytotoxic peptides of mammalian host defense cells, *Med. Microbiol. Immunol.* **181**:99–105.
Hamilton, S.C., Farchaus, J.W., and Davis, M.C., 2001, DNA polymerases as engines for biotechnology, *Biotechniques* **31**:370–383.
Hancock, R.E., and Scott, M.G., 2000, The role of antimicrobial peptides in animal defenses, *Proc. Nat. Acad. Sci. USA*, **97**:8856–8861.
Hancock, R.E., and Lehrer, R., 1998, Cationic peptides: a new source of antibiotics, *Trends Biotech.* **16**:82–88.
Hoffmann, J.A., and Riechhart, J-M., 1997, *Drosophila* immunity, *Tr. Cell Biol.* **7**:309–316.
Hubert, F., Noël, T., and Roch, Ph., 1996, A member of the arthropod defensin family from edible Mediterranean mussels, *Mytilus galloprovincialis*, *Eur. J. Biochem.* **240**:302–306.

Innis, M.A., Myambo, K.B., Gelfand, D.H., and Brow, M.A., 1988, DNA sequencing with *Thermus aquaticus* DNA polymerase and direct sequencing of polymerase chain reaction-amplified DNA, *Proc. Nat. Acad. Sci. USA*, **85**:9436-9440.

Kohashi, O., Ono, T., Ohki, K., Soejima, T., Moriya, T., Umeda, A., Meno, Y., Amako, K., Funakosi, S., Masuda, M., and Fujii, N., 1992, Bactericidal activities of rat defensins and synthetic rabbit defensins on *Staphylococci*, *Klebsiella pneumoniae*, *Pseudomonas aeruginosa*, *Salmonella typhimurium* and *Escherichia coli*, *Microbiol. Immunol.* **36**:369-380.

Lehane, M.J., Wu, D., and Lehane, S.M., 1997, Midgut-specific immune molecules are produced by the blood-sucking insect *Stomoxys calcitrans*, *Proc. Nat. Acad. Sci. USA*, **94**:11502-11507.

Lemaitre, B., 1999, La drosophile: un modèle pour l'étude de la réponse immunitaire innée, *Médecine/Sciences* **15**:15-22.

Lemaitre, B., Reichhart, J-M., and Hoffmann, J.A., 1997, Drosophila host defense: differential induction of antimicrobial peptide genes after infection by various classes of microorganisms, *Proc. Nat. Acad. Sci. USA*, **94**:14614-14619.

Levashina, E.A., Ohresser, S., Lemaitre, B., and Imler J-L., 1998, Two distinct pathways can control expression of the gene encoding the *Drosophila* antimicrobial peptide metchnikowin, *J. Mol. Biol.* **278**:515-527.

Levy, O., 1996, Antibiotic proteins of polymorphonuclear leukocytes, *Eur. J. Haematol.* **56**:263-277.

Matsuzaki, K., Nakamura, A., Murase, O., Sugishita, K-I., Fujii, N., and Miyajima, K., 1997, Modulation of magainin 2-lipid bilayer interactions by peptide charges., *Biochemistry-USA*, **36**:2104-2111.

Mitta, G., Hubert, F., Dyrynda, E.A., Boudry, P., and Roch, Ph., 2000b, Mytilin B and MGD2, two antimicrobial peptides of marine mussels: gene structure and expression analysis, *Develop. Comp. Immunol.* **24**: 381-393.

Mitta, G., Hubert, F., Noël, T., and Roch, Ph., 1999a, Myticin, a novel cysteine-rich antimicrobial peptide isolated from hemocytes and plasma of the mussel, *Mytilus galloprovincialis*, *Eur. J. Biochem.* **265**:71-78.

Mitta, G., Vandenbulcke, F., Hubert, F., and Roch, Ph., 1999b, Mussel defensins are synthesised and processed in granulocytes then released into plasma after bacterial challenge, *J. Cell Sci.* **112**:4233-4242.

Mitta, G., Vandenbulcke, F., Hubert, F., Salzet, M., and Roch, Ph., 2000a, Involvement of mytilins in mussel antimicrobial defense, *J. Biol. Chem.* **275**:12954-12962.

Mitta, G., Vandenbulcke, F., and Roch, Ph., 2000b, Original involvement of antimicrobial peptides in mussel innate immunity, *FEBS Letters* **486**:185-190.

Mitta, G., Vandenbulcke, F., Noël, T., Romestand, B., Beauvillain, J-C., Salzet, M., and Roch, Ph., 2000c, Differential distribution and defence involvement of antimicrobial peptides in mussel, *J. Cell Sci.* **113**: 2759-2769.

Moore, A.J., Beazley, W.D., Bibby, M.C., and Devine, D.A., 1996, Antimicrobial activity of cecropins, *J. Antimicrob. Chemother.* **37**:1077-1089.

Morishima, I., Horiba, T., Iketani, M. Nishioka, E., and Yamano, Y., 1995, Parallel induction of cecropin and lysozyme in larvae of the silkworm, *Bombyx mori*, *Develop. Comp. Immunol.* **19**:357-363.

Osborn, R.W., De Samblanx, G.W., Thevissen, K., Goderis, I., Torrekens, S., Van Leuven, F., Attenborough, S., Rees, S.B., and Broekaert, W.F., 1995, Isolation and characterization of plant defensins from seeds of Asteraceae, Fabaceae, Hippocastanaceae and Saxifragaceae, *FEBS Letters* **368**:257-262.

Ouellette, A.J., and Selsted, M.E., 1996, Paneth cell defensins: Endogenous peptide components of intestinal host defense, *FASEB J.* **10**:1280-1289.

Steinberg, D.A., Hurst, M.A., Fujii, C.A., Kung, A.H.C., Ho, J.F., Cheng, F.C, Loury, D.J., and Fiddes, J.C., 1997, Protegrin-1: a broad-spectrum, rapidly microbicidal peptide with *in vivo* activity, *Antimicrob. Agents Chemother.* **41**:1738-1742.

Steiner, H., Hultmark, D., Engstrom, A., Bennich, H., and Boman, H.G., 1981, Sequence and specificity of two antibacterial proteins involved in insect immunity, *Nature* **292**:246-248.

Terras, F.R.G., Torrekens, S., Van Leuven, F., Osborn, R.W., Vanderleyden, J., Cammue, B.P.A., and Broekaert, W.F., 1993, A new family of basic cysteine-rich plant antifungal proteins from Brassicaceae species, *FEBS Letters* **316**:233-240.

Williams, M.J., Rodriguez, A., Kimbrell, D.A., and Eldon, E.D., 1997, The *18-wheeler* mutation reveals complex antibacterial gene regulation in *Drosophila* host defense, *EMBO J.* **16**:6120-6130.

Wu, M., Maier, E., Benz, R., and Hancock, R.E.W., 1999, Mechanism of interaction of different classes of cationic antimicrobial peptides with planar bilayers and with the cytoplasmic membrane of *Escherichia coli*, *Biochemistry-USA* **38**:7235-7242.

Yamano, Y., Matsumoto, M., Sasahara, K., Sakamoto, E., and Morishima, I., 1998, Structures of genes for cecropin A and an inducible nuclear protein that binds to the promoter region of the genes from the silkworm, *Bombyx mori*, *Biosci. Biotechnol. Biochem.* **62**:237–241.

Yang, Y-S, Mitta, G., Chavanieu, A., Calas, B., Sanchez, J-F., Roch, Ph., and Aumelas, A., 2000, Solution structure and activity of the synthetic four disulfide bond Mediterranean mussel defensin, MGD-1, *Biochemistry-USA* **39**:14436–14447.

Zhang, L., Benz, R., and Hancock, R.E.W., 1999, Influence of proline residues on the antibacterial and synergistic activities of alpha-helical peptides, *Biochemistry-USA* **38**:8102–8111.

# 25

# Nervous, Endocrine, Immune Systems As a Target for Complementary and Alternative Medicine

Shinji Kasahara and Edwin L. Cooper

## Introduction

Homeostasis is a fundamental process in life conducted under various stressors. Human health in a highly variable environment is dependent upon the proper balance of physiological processes. According to emerging views, homeostasis may be achieved by the coordinated activities of the three major integrative systems: the nervous, endocrine, and immune systems. The immune system has for a long time been considered as independent, but recent numerous studies have revealed it to be in a delicate state of balance with the neuro-endocrine system. In this chapter, the author reviews how this reciprocal modulation between neuro-endocrine and immune systems is maintained in relation to stress and diseases. Possible animal models for the basic study of neuro-endocrine-immune system are also presented. Lastly, there is a discussion of how: 1) the development of these animal models is important; 2) complementary and alternative medicine (CAM) could contribute to the alleviation of stress-induced diseases.

## Overview on Neuroendocrineimmunology

### Reciprocal interactions between neuroendocrine system and immune system

A. HPA axis, glucocorticoid, and sympathetic nerve system

Two major components of neuroendocrineimmune regulation in vertebrates are known: hypothalamic-pituitary-adrenal (HPA) axis and the sympathetic nervous system (SNS)[1,2]. In a healthy condition, these two systems are operating within a delicate state of balance. They maintain homeostasis even under highly challenging conditions. The main components of the HPA axis are: 1) the paraventricular nucleus (PVN) in the hypothalamus

---

Shinji Kasahara and Edwin L. Cooper • Laboratory of Comparative Neuroimmunology, Department of Neurobiology, David Geffen School of Medicine at UCLA, 10833 Le Conte Avenue, Box 951763, Los Angeles, California 90095-1763. E-mail: shinkoro@u.washington.edu.

*Complementary and Alternative Approaches to Biomedicine*, edited by Edwin L. Cooper and Nobuo Yamaguchi. Kluwer Academic/Plenum Publishers, 2004.

of the brain; 2) the anterior pituitary gland located at the base of the brain; 3) the adrenal glands. Corticotropin releasing hormone (CRH) is secreted from the PVN of the hypothalamus into the hypophyseal portal blood supply and induces adrenocorticotropin hormone (ACTH) expression in the anterior pituitary gland. ACTH then circulates in the bloodstream to the adrenal glands and induces the expression and release of glucocorticoids.

Early studies have demonstrated that adrenal glucocorticoids, the end products of HPA activation, influence immune functions[3,4]. For example, glucocorticoids modulate: 1) trafficking, maturation, and differentiation of immune cells; 2) expression of adhesion molecules; 3) expression of chemoattractants and cell migration; 4) production of inflammatory molecules[5,6]. Most notably, glucocorticoids modulate the transcription of many cytokines. They suppress the proinflammatory cytokines such as interleukin (IL-) $1^{7-11}$, IL-$2^{12}$, IL-$6^{13-15}$, IL-$8^{16}$, IL-$11^{17}$, IL-$12^{18-20}$, TNF-alpha[21], IFN-gamma[12-22], and GM-CSF[23] while upregulating the anti-inflammatory cytokines such as IL-$4^{24,25}$ and IL-$10^{26,27}$. These cytokine modulations can lead to systemic immunosuppression. Glucocorticoids also induce immunosuppression by repressing NFκB, which is a major regulative factor of cytokines and other immune responses[28,29]. It has been reported that glucocorticoids induce the expression of the inhibitory protein IκB via glucocorticoid receptor (GR). IκB then sequesters NFκB in the cytoplasm and inhibits its translocation to the nucleus and gene activation[30,31]. NFκB suppression by GR is known only in certain cell types, particularly monocytes and lymphocytes. There is another mechanism that physical interaction or cross-talk between NFκB and GR prevents gene expression[30,32,33]. The sympathetic nervous system (SNS) is the other major route, which affects immune responses. Cells in SNS make synapse-like contacts with lymphocytes in primary and secondary lymphoid tissue[34]. The local release of sympathetic hormones (e.g. norepinephrine and neuropeptide Y) can modulate the functions of immunocompetent cells[35,36]. These immunocompetent cells possess adrenergic receptors, which are involved in the regulation of the immune responses. The existence of these receptors depends on cell type. For instance, helper type 1 T cells (Th1) and B cells have ß2-adrenergic receptors, but helper type 2 (Th2) cells do not have them. Activation of ß2-adrenergic receptors is associated with increased intracellular cAMP and inhibition of cell function[37]. For example, activation of ß2-adrenergic receptors is caused by stress, leading to the suppression of natural killer (NK) activity and increases tumor colonization in rats[38]. There is other evidence suggesting that the adrenergic system is involved in immune suppression. Temporary ablation of SNS fibers with the selective denervation of peripheral noradrenergic nerves using 6-hydroxydopamine (6-OHDA) augmented initial immune responses against *Listeria monocytogenes*[39], suggesting that norepinephrine inhibits host resistance. In relation to clinical aspects, altered sympathetic function is involved in impaired immune function associated with depression[40]. Thus, sympathetic activation is involved in the immunomodulatory effects of stress[35].

B. Cytokines and the central nervous system (CNS)

Cytokines are multifunctional pleiotropic proteins that play essential roles in cellular communication and activation. Cytokines have been classified into two types: proinflammatory (Th1-type, stimulatory) or anti-inflammatory (Th2-type, inhibitory) depending on the eventual state of balance in the immune system[41]. Cytokines have originally been thought to be functioning only in the immune system. Recent studies, however, have revealed that cytokines are also involved in a variety of physiological and pathological processes, in which peripheral- (PNS) and the CNS are implicated. Thus, the current understanding is that cytokines are both immune- and neuro-modulators.

There are several direct and indirect actions through which cytokines might exert their effect in the CNS[42-44]. **First**, cytokines themselves are present in the brain and other related neuronal cells (e.g., glia, neurons, and macrophages) and play a role in both neuronal cell survival[45] and death[46,47]. It has been suggested that cytokine-mediated neuronal cell death is partially responsible for several neuro-degenerative disorders, such as Alzheimer's disease, multiple sclerosis, stroke, nerve trauma, and neuro-AIDS. There have also been claims that cytokines present in the CNS might influence cognition, learning, and memory. **Second**, peripheral cytokines can act as hormones and stimulate the CNS through several pathways. They may cross the brain blood barrier (BBB) in small amounts either through leaky points or by active transport [e.g., organum vasculosum lamina terminalis (OVLT) or median eminence] [48,49]. Then, these peripheral cytokines can reach the CNS and act directly. **Third**, the stimulation of basic CNS functions including behavioral patterns, fever, and sleep require the indirect effect of cytokines. Evidence of the constitutive expression of cytokines and their receptor genes suggests that some cytokines may be responsible for normal functions of the brain. Especially, IL-1 family is well known to be involved in the modulation of neural functions including sleep[50], feeding[51], and ovulation[52]. IL-1-mediated signals can be transduced from the blood stream into neurons without the need to cross the BBB. Peripheral administration of cytokines may also activate genes that are involved in the regulation of transcription of cytokines within the brain[53]. **Fourth**, cytokines induce indirect secondary effects that are the result of their action on other targets. For example, peripheral LPS induces TNF-$\alpha$ mRNA expression initially in perivascular cells, meningeal cells, and neurons in different brain regions[54], suggesting that peripheral cytokines are involved in the synthesis of cytokines in the brain. The appearance of TNF-$\alpha$ and IL-1$\beta$ in the CNS can cause somnogenic, lethargic, pyrogenic, and anorectic consequences[55]. Peripheral cytokines can also stimulate HPA axis leading to the glucocorticoid-induced immuno suppression[56-58]. In addition, when endothelial cells of blood vessels are activated by cytokines, these cells release second messengers, such as NO and prostanoids, that can influence the CNS through the vagal route. Moreover, circulating cytokines can activate the endothelium to produce more IL-1, leading to indirect signal amplification in the CNS[59,60].

On the other hand, cytokine synthesis is under the control of the CNS and the PNS. Neurotransmitters are critically involved in regulating cytokine balance[36,61,62]. The cytokine-CNS interactions are therefore bidirectional: cytokines and other immune-related molecules can modulate the survival and death of neuronal cells, while the neurotransmitter and neuropeptide plays a pivotal role in influencing immune responses.

### Stress and the immune system

Today, stress is used as a common term that summarizes the effects of internal or external stimuli which affect physical or mental wellbeing. Stress responses require behavioral, psychological, and physiological integrations (adjustments) with the induction of necessary counteracting reactions. This complex equilibrium is recognized as homeostasis, with which humans maintain their survival in the face of psychosocial or environmental stimuli. Since homeostasis is constantly challenged, all life forms have developed mechanisms which are essential for their survival and longevity[63,64]. Therefore, the origin of stress responses can even be tracked back to invertebrates[65]. For example, both in humans and invertebrates, the immune and nervous systems appear to utilize similar signal molecules, which are involved in similar activities[66]. Thus, stress responses are fundamental through phylogeny.

Regarding medical implications, psychological stress as well as chemical and physical stressors, can affect immune responses. It can also have influence on the onset and the development of immunological diseases. Today, there are controversies about how to make a distinction between transient stress and stress that exists over longer periods. Current consensus is that acute stress might facilitate some aspects of immune function, while intense and/or long-term stressors impair certain immune functions. Rats subjected to restraint displayed stronger antigen-specific delayed-type hypersensitivity (DTH) responses than do control animals, when the stress was given either with initial antigen exposure or on re-exposure[66]. In humans, it was found that natural killer-cell (NK) numbers and function were temporally increased when subjects were exposed to a laboratory stressor[67] or did tandem parachute jumping[68]. More long-lived stressors, however, suppress the same measures of immune function. For example, rats exposed to daily restraint for weeks showed significant declines in DTH responses[69]. In humans, reduced NK activity was found in subjects who had histories of prolonged life stress during and after a laboratory speech stress paradigm[70]. It has also been reported that chronic stress further reduced the induction of tumor cell apoptosis, in which nitric oxide pathways are involved[71,72]. In addition, intensity of stress also causes different outcomes. The expression of the nuclear transcription factors AP-1 and NFκB in peripheral blood lymphocytes was decreased in women experiencing distress associated with breast biopsy[73]. This result suggests potential immunosuppressive effects of profound stress at molecular mechanism levels.

Taken together, the effects caused by stress can be beneficial for some types of immune responses but deleterious for others[74,75]. In addition to this view, recently there are emerging scientific/biomedical areas such as neuroimmunology and psychoneuroimmunology. The interdisciplinary character of stress and its applied research brought about these disciplines, indicating that stress may represent a modulator of the immune responses and can be a risk of immunological diseases, although its consequences depends on a multitude of factors[76]. Inflammation, infection, autoimmune processes, and even the onset and development of malignant tumors may be associated with stress-induced events[2].

## Modulatory effects of stress on the immune system

As described above, stress could either enhance or reduce immune functions depending on the duration and the intensity[76]. Stress has been demonstrated in both humans and animals to alter the pathogenesis of infectious diseases, tumor development, and autoimmune disorders[77]. The stress hormone-dependent translocation of immune cells is one of the key mechanisms in stress-influenced immunological diseases[74]. It has been reported that acute stress-promoted immune responses are dependent upon the HPA axis[78], while the mechanisms of chronic stress-induced immune modulations remain controversial.

Acute stress induces immune reactions. For example, mediators such as CRH and substance P stimulate mast cells to degranulate. Mast cell-derived granules contain histamine and proinflammatory cytokines, which activate antigenic inflammatory processes (Fig. 25.1). These cascading reactions can further amplify the inflammatory process initiated

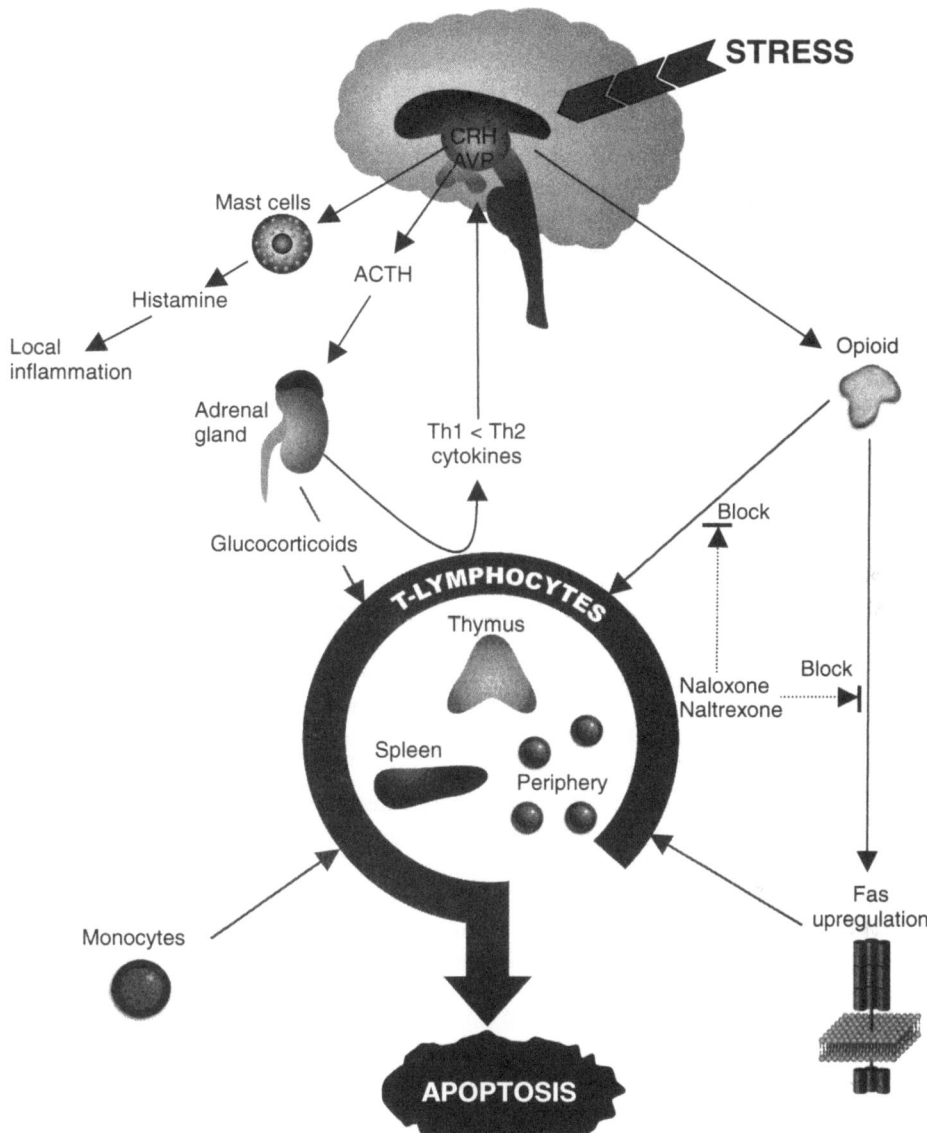

**Figure 25.1.** Stress-induced apoptotic pathways of T-lymphocytes (the red arrows represent the signals initiating apoptosis). Stress induces CRH and AVP activation leading to the glucocorticoid release through the HPA axis. CRH itself can cause mast cell degranulations, leading to local inflammations. Glucocorticoids initiate apoptosis of T-cells in thymus, spleen, and periphery, and also induce the Th2 cytokine predominance, which in turn affects hypothalamic functions. Stress-induced apoptosis in immature T cells is through the HPA axis, while in mature T cells it is mediated by endogenous opioids [96,101,102]. *In vivo* administration of morphine could also induce apoptosis in immature thymocytes, and these reactions by opioids can be reversed by naltrexone or naloxone[102]. Chronic stress causes Fas expression in T-lymphocytes, which also results in the initiation of apoptosis [96]. Monocytes play an important role in the induction of T cell apoptosis [103]. CRH: corticotropic releasing hormone; AVP: arginine vasopressin; ACTH: adrenocorticotropin hormone; Th1/Th2: helper type 1 or 2 T lymphocytes

by the nervous system[75,79]. In addition, the sympathetic hyperactivity induces vasodilation, plasma extravasation, cellular permeability, and chemotaxis of immune cells. This can be seen in an acutely activated stress response, such as in acute trauma or pain[75]. Moreover, sympathetic efferent nerves are involved in the function of primary afferent fibers, which can also cause inflammatory responses. Activated sensory afferent neurons induce the release of neuropeptides such as substance P, calcitonin gene related peptide (CGRP), and bradykinin. It has been known that these neuropeptides upregulate vasodilation, microvascular permeability, and plasma extravasation[80-82]. Thus, acute stress initiates a sympathetic arousal through norepinephrine (catecholamine) pathways, and this alteration may induce local and systemic inflammation[62,75].

Although there is a general understanding that acute stress enhances part of the immune responses, chronic stress often causes immunosuppression. Stress induced-immunosuppression is in part due to the reduction of lymphocytes[83-85]. Experimental models have demonstrated this phenomenon in animals subjected to physical restraint[86-88]. In addition, this stress-induced lymphopenia has also been observed in humans: 1) persons under various psychological stress[89,90]; 2) surgical patients[91-93]; 3) over exercised athletes[94,95]. Chronic stress causes lymphocysis through apoptotic mechanisms[87], for which endogenous opioid-mediated Fas expression is playing a critical role (Fig. 25.1)[87,96].

Still, stress responses cannot simply be distinguished. While the release of norepinephrine from peripheral sympathetic nerve terminals induces inflammatory responses, it also stimulates type 2 cytokines such as IL-10 downregulating the Th1 response[62]. This Th1/Th2 shift brings immune responses from its cell-mediated arm towards humoral predominant responses which includes anti-inflammatory properties[97]. This process resembles the effects which can be caused by cortisol release[79,98]. In addition, under stress, catecholamines, glucocorticoids, and histamine may selectively suppress cellular immunity via down-regulation of Th1, thereby facilitating humoral immune responses[79]. In short, stress may not only induce anti-inflammatory properties by suppressing the Th1 responses via CNS and PNS[62,79], but also facilitate local pro-inflammatory activities through CRH-mast cell-histamine axis[79,99] (Fig. 25.1). Thus, stress carries both properties of anti- and pro-inflammation within itself. The outcome of pro-inflammatory responses or overall imbalances in activated stress responses may lead to the onset or exacerbation of infectious, neoplastic, and autoimmune diseases[79,100].

### Diseases and neuroendocrineimmune system
A. Dysregulation of neuroendocrineimmune system and its consequences

Inbalances of hormonal and neuronal mechanisms can be a risk for immunological and inflammatory diseases that are related to stress. For example, disturbances of the HPA axis or glucocorticoid responses lead to enhanced susceptibility to infection, inflammation and autoimmune diseases. Overflow of the HPA axis-related responses often causes excessive release of circulating glucocorticoids and then induce immune suppression, leading to enhanced susceptibility to infection. On the other hand, the consequence of the HPA axis understimulation is lower circulating levels of glucocorticoids and susceptibility to inflammation. Dissociation of neuroendocrineimmune system may also occur at the molecular level. In addition, neuroimmune mechanisms include SNS activity through which direct innervation of immune organs is induced[100]. In all, dysregulation of balances between: 1)

HPA/SNS pathways; 2) central/peripheral, or neuronal/hormonal stress responses, may have implications in the exacerbation of neuroimmune diseases[100]. These stress induced-neuroimmune modulations may be of clinical significance for certain infections, major injury, sepsis, autoimmunity, chronic pain, and neoplastic diseases[62].

Cytokines are also playing critical roles in neuroimmune diseases. The balance between the pro-inflammatory and anti-inflammatory cytokines are important for neurotoxic and neuroprotective mechanisms. For example, the neurodegenerative processes are closely related to the shift towards pro-inflammatory cytokines such as IL-1 or TNF-α, while anti-inflammatory cytokines in the CNS inhibit degeneration protecting cell viability[104]. A variety of diseases in which the CNS is involved exhibit altered cytokine expressions in the brain. These disorders include: viral or bacterial infections[105,106], Alzheimer's disease[107], multiple sclerosis[108], stroke[109], ischemia[110], and other encephalopathies[105,111]. Experimental data using rats showed that low doses of LPS induced a two-four fold TNF-α increase in the brain[112]. Other studies using transgenic mice which showed cytokine overexpressions in the brain revealed that cytokine dysregulation in the CNS was an important factor in the pathogenesis of neurotoxic and neurodegenerative disorders[111].

B. Stress-induced diseases: infection, cancer, and auto-immune disorders

The impact of stress on immune functions has been linked to increased susceptibility to illness. There is evidence that stress can alter the onset and course of bacterial[113] or viral[114] infections. Impaired antiviral immune responses, for example, have been observed in caregivers of Alzheimer's patients[115,116] or in students with stress associated with medical school exams[117]. These modified immune states could initiate reactivation of latent viral infection and impaired responses to vaccination[118,119]. Nonhuman primates subjected to maternal separation stress have shown significant and lasting alterations in immune functions[120] leading to increased vulnerability to parasitic infection[121]. Thus, stress may be of particular significance in potentially deadly infectious diseases such as AIDS[114]. In addition, stress-induced modulation of IL-1β pathways is associated with AIDS-related dementia[122]. A recent study in HIV-infected men has revealed that severe stress, especially when combined with depression, significantly decreased $CD4^+$ and $CD8^+$ T lymphocyte numbers over the course of two years[123]. In similar analyses using nonhuman primates, psychological stress has been demonstrated to be associated with shortened survival time in animals with simian immunodeficiency virus infection[124]. Thus, in patients (or animal models) already diagnosed with disease, psychological stress could result in poorer disease outcome and turn out to be deleterious in specific cases[114]. In addition, there is epidemiologic data that chronic stress following imbalances in stress response pathway (found in childhood traumatic events, for example) not only increase the risk of being infected with potential pathogens, but also the odds of acquiring a sexually transmitted disease such as AIDS later in life[125].

Stress could dramatically reduce the resistance to various pathogens such as herpes infection[126]. Thus, prolonged outcome of stress-induced immunodeficiency could also lead to increased vulnerability to oncogenic viruses. These infections in individuals with crippled immune functions caused by stress, could dramatically increase the risk of cancer. Moreover, stress itself has been shown to increase the likelihood of cell transformation. Patients newly admitted to a psychiatric clinic possessed reduced levels of DNA repair capability in their lymphocytes[127]. This report suggests an important potential implication of stressors in cancer, since damaged DNA accumulation is a major cause of cell transformation. Another investigation has revealed that methyltransferase, which is an enzyme essential for DNA repair, is dramatically decreased in stressed rats[128]. In addition, it has been shown that swim stress in rats induced chromatid exchange[129]. Studies focusing on other indices have shown that stress significantly reduced the organism's ability to combat malignant tumor growth[2]. Psychological stress suppressed NK function, rendering rats more susceptible to tumor colonization[130]. Taken together, stress could increase the risk of cancer by facilitating oncogenic viral infection, genomic changes, and decreased immune responses to tumors.

There are contradictory results on how stresses contribute to autoimmune disorders[131]. One supportive claim is that overstimulation or imbalance of the HPA axis is an important factor for the development of autoimmune disorders[131], although the physiology of stress in these diseases appears to be particularly complex and a precise impact is not easy to define. Nevertheless, stress has been shown to be of importance in autoimmune disorders. Experimental disease models have demonstrated that abnormalities of stress responses are playing critical roles in autoimmune arthritis and other related inflammatory processes[1,131]. It has also been studied that daily stress/stressors exacerbate rheumatoid arthritis (RA)[132], suggesting that stress seems to play a critical role in this disease through not only peripheral but also CNS pathways[133]. Multiple sclerosis is also sometimes classified as one of the autoimmune disorders, which are associated with stress responses[131]. Other diseases, such as atopic dermatitis, psoriasis, celiac disease, and ulcerative colitis also appear to be related to stress physiology, although their definite etiology remains unclear[131]. Psychological stress has also been shown to slow the process of wound healing[134]. Mast cells, which are perivascularly located close to sympathetic and sensory nerve endings, are involved in the pathophysiology of these various disseases[135]. Acute stress initiates degranulation of mast cells via CRH pathway, leading to a local inflammation[135]. Hyperactivity of the HPA axis and following profound stimulations of adrenocortical cell function and excessive glucocorticoid release are related to murine colitis[136]. HPA overshoot is also seen in patients with multiple sclerosis in the beginning of the disease and then high plasma cortisol levels appear when the disease is established[131]. It seems that the stress-induced hyperactivity or dysregulation of the HPA axis may represent one of the crucial pathophysiological components in autoimmune disorders including multiple sclerosis.

## Possible Animal Models for Neuroimmune Study

### Developing new animal models

Recently, there is evidence in invertebrates and vertebrates that the immune system interacts with the nervous system in both phylogenetical groups[137]. Components previously thought to be involved in either neural or immune systems are found playing overlapping roles and therefore their functions are reciprocal[138,139]. In the clinical arena, for example, research in both humans and experimental mammals show alcohol modulation of neuroimmune responses such as suppression of IL-6 and IL-10, of cytotoxicity of NK cells, induction of sleep deprivation[139–142], enhancement of thymocyte mitogenic responses[142]. Despite these findings, there are still large gaps in our understanding of the neuroimmune interaction and behaviour. Therefore, it would be beneficial to develop appropriate in vivo experimental model systems to evaluate the contribution of neurons in immune reactions. Animal models, especially invertebrates and lower vertebrates can be useful models since they are generally easy to handle, socially non-controversial, genetically manipulatable (e.g. *drosophila*, *C. elegans*, and zebrafish), and inexpensive. These models could also contribute directly or indirectly to the development of future alternative and complementary medicine. Here, three models are presented in which the author has been involved: *Hydra* (cnideria), and *Tilapia* and guppy (fish).

### *Hydra (Cnidaria)*
A. *Hydra*–cnidarians in which nervous systems evolved

The nervous system evolved within cnidarians. The freshwater polyp *Hydra*, a cnidarian, belongs to the most basal animal phylum having a nervous system, which can represent a model for basic neuroimmune study. *Hydra* polyps consist of only two epithelia: the ectoderm and the endoderm surrounding a gastric cavity[143]. The nervous system in *Hydra* is simple: a nerve net of sensory neurons and ganglion cells[144]. *Hydra* neurons differentiate continuously from multipotent stem cells and are constantly intercalated into the nerve net at a rate appropriate to the rate of epithelial cell division[145,146]. It is still unclear how *Hydra* or any other cnidarian polyp defend themselves against infectious microorganisms. A few investigations indicate that epithelium can function as the protective skin and an integral part of the *Hydra* innate immune defense against microbial infection[147]. At the molecular level, however, nothing is known about the components of the *Hydra* innate immune system. In contrast, molecules which are involved in functioning and regulation of the *Hydra* nervous system are well described[144].

B. Enhanced antibacterial activity in *Hydra magnipapillata* lacking nerve cells

When investigating the antibacterial activity in *Hydra*, the author's group unexpectedly observed a strong correlation between the number of neurons present and the level of the antibacterial response[148]. Two strains of *H. magnipapillata* were used for the experiments. Strain 105 is wildtype, while strain *sf-1* is a mutant containing a temperature-sensitive interstitial cell lineage[149]. *H. magnipapillata* strain *sf-1* cultured at the permissive temperature (18°C) contains high levels of nerve cells (about 0.14–0.16 nv/epi). Strain *sf-1* cultured at the non-permissive temperature (25°C) loses its neurons within 8–10 days of temperature treatment while it loses interstitial cells and nematoblasts within 4–5 days[150]. Fig. 25.2 shows a typical epithelial *H. magnipapillata* strain *sf-1* polyp (Fig. 25.2(B)) as well as a mutant polyp kept at the permissive temperature (Fig. 25.2(A)). For assessing antimicrobial activity, a modified radial diffusion assay was performed for both polyp groups[151]. Protein extracts were prepared and added to gels containing *E. coli* XL blue. After incubation at 37°C for 16 h, zones of bacterial clearing became readily apparent (Fig. 25.2). After 4 days of culture at non-permissive temperature, interstitial cells and nematoblasts disappeared, but there was no change in antimicrobial response. This result suggests that the presence or absence of interstitial cells and nematoblasts has no influence on the immune response. Temperature treatment for longer periods of time (8–12 days), however, reduced nerve cells and resulted in drastically enhanced zones of bacterial clearing, about three times that of control tissue (Fig. 25.3). The data indicated that in the absence of neurons *Hydra* epithelial

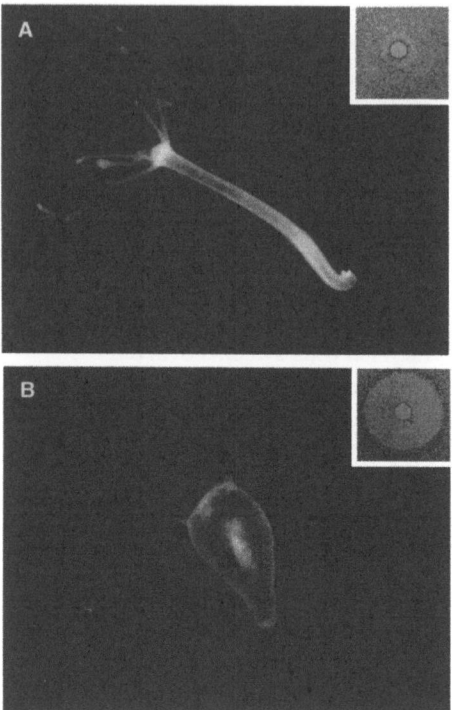

**Figure 25.2.** *H. magnipapillata* strain *sf-1* polyps cultured at permissive (A: ×16) and nonpermissive (B: ×40) temperatures for 12 days. Note the typical morphology of the epithelial polyp. Insets: corresponding radial diffusion assays against E. coli indicating increased antibacterial activity in polyps lacking neurons. Reprinted from: Kasahara S and Bosch TCG (2002) Enhanced antibacterial activity in *Hydra* polyps lacking nerve cells. *Dev. Comp. Immunol.* 27: 79–85; with permission from Elsevier

tissue had enhanced antimicrobial activity against *E. coli* XL blue. We then confirmed that this augmented activity in *sf-1* was due only to the absence of neurons and not to temperature treatment. For this, we conducted control experiments using three other strains. Temperature treatment itself did not affect antimicrobial activities of these strains. We also challenged *B. subtilis* strain DSM 34 as gram-positive bacteria and obtained similar results (Fig. 25.4) although the difference between control and nerve-free tissue was smaller than in experiments using *E. coli* XL blue.

The results revealed two novel facts concerning innate immunity in *Hydra*. The findings indicated for the first time that *Hydra* tissue possesses antimicrobial activity against Gram-negative and Gram-positive bacteria. It is most likely that this protection is in part due to the synthesis of potent antimicrobial peptides. Second and most surprisingly, antibacterial activity was positively correlated with the absence of nerve cells, the fewer the nerve cells, the stronger the response was (Fig. 25.3, 25.4). The data raise the possibility that *Hydra* neurons may actively be involved in the regulation of the antibacterial response and negatively control synthesis of antibacterial peptides. Our observation suggests that neuro-immune interaction can be traced back in evolution from vertebrates to *Hydra*, a member of an animal phylum in which the nervous system first evolved.

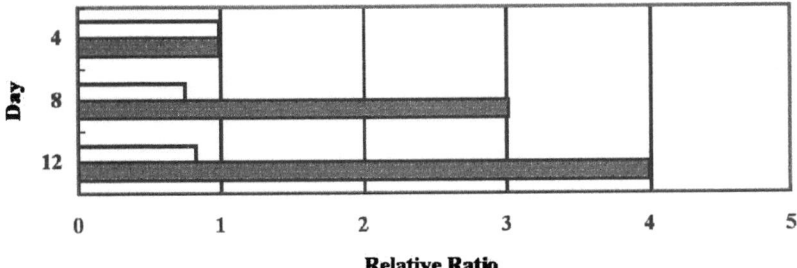

**Figure 25.3.** Antimicrobial activities against *E. coli* in normal and nerve free *H. magnipapillata* strain *sf-1*. Radial diffusion assay with Hydra extracts prepared 4, 8, and 12 days after temperature shift. More than 30 polyps were used for each group. The relative ratio of the zone (18°C, 4 days) was calculated as one. White bar, normal *sf-1* polyps at permissive temperature. Grey bar, polyps cultured at 25°C for 4, 8 and 12 days. Reprinted from: Kasahara S and Bosch TCG (2002) Enhanced antibacterial activity in *Hydra* polyps lacking nerve cells. *Dev. Comp. Immunol.* 27: 79–85; with permission from Elsevier

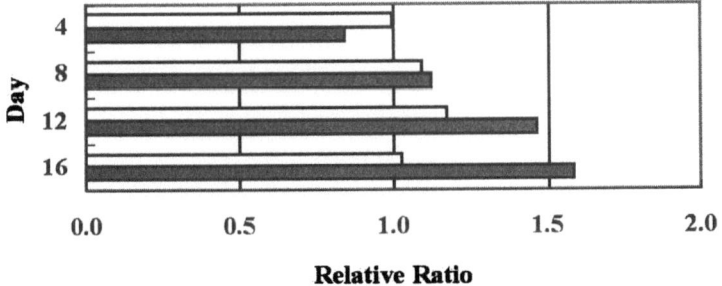

**Figure 25.4.** Antimicrobial activities against *B. subtilis* in normal and nerve-free *H. magnipapillata* strain *sf-1*. Radial diffusion assay with Hydra extracts prepared 4, 8, 12, and 16 days after temperature shift. More than 30 polyps were used for each group. The relative ratio of the zone (18°C, 4 days) was calculated as one. White bar, normal *sf-1* polyps at permissive temperature. Grey bar, polyps cultured at 25°C up to 16 days. Reprinted from: Kasahara S and Bosch TCG (2002) Enhanced antibacterial activity in Hydra polyps lacking nerve cells

### *Tilapia (Fish)*

As in other vertebrates, teleost fish have shown neuroimmune interactions similar to those of humans. Glucocorticoids together with catecholamines and opioid peptides are the major actors in the two main pathways through which the brain regulates the stress response: the hypothalamic-pituitary-adrenal (HPA) axis and the hypothalamic-sympathetic nervous system-chromaffin cell axis. As described in the previous sections in this chapter, in the former pathway, hypothalamic factors including CRH, control hypophyseal secretion of the proopiomelanocortin (POMC)-derived ACTH that in turn stimulates release of glucocorticoids by steroidogenic cells. In the latter pathway, the hypothalamus controls release of catecholamines by chromaffin cells through cholinergic fibers of the sympathetic nervous system and ACTH[152]. In teleosts, steroidogenic cells and chromaffin cells are located in the interrenal, a homologue of the adrenals of mammals. Interestingly, the interrenal is adjacent to the head kidney (pronephros), the principal hematopoietic organ in adult fish[152]. This characteristic strengthens the reason that fish are appropriate models to study neuroimmune interactions.

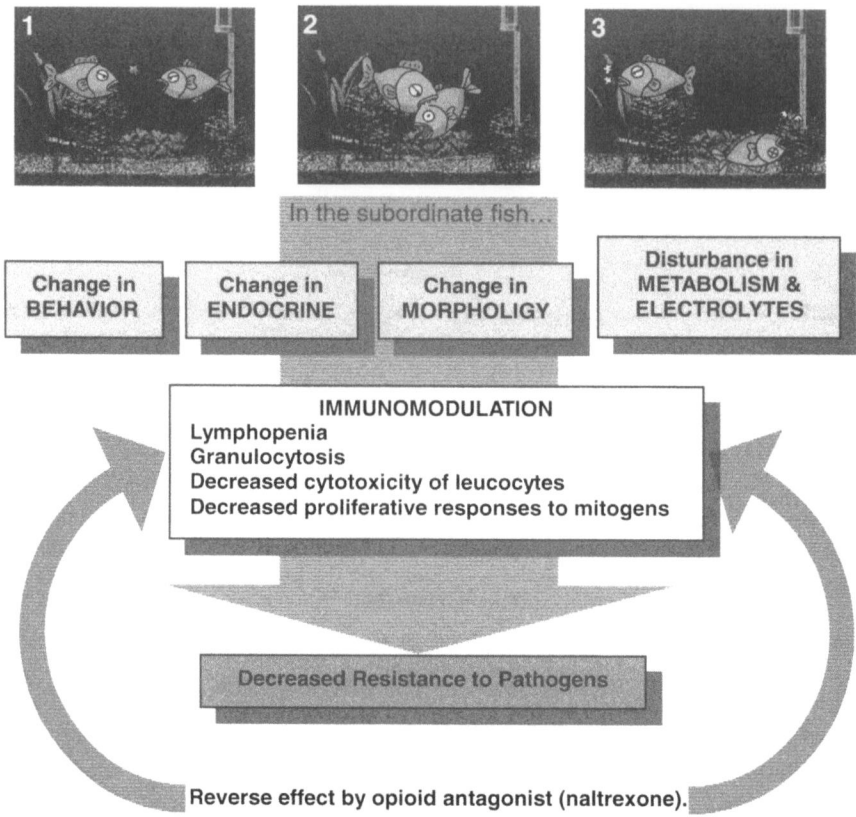

**Figure 25.5.** Social subordination paradigm: encounters between dominant (α) and subordinate (β) fish, the resulting physiological chances in the subordinate fish, and how these modulate the immune system [88]

Fasal et al. have previously shown that stress caused by social confrontation between aggressive fish (*Tilapia*) produces a suppression of several immunological parameters (such as nonspecific cytotoxicity and mitogen-stimulated proliferation in pronephric leukocytes) in the subordinate fish (Fig. 25.5)[88]. Injection of naltrexone, the opioid antagonist, blocked alterations of these immune parameters. It has been therefore demonstrated indirectly that this immunosuppression is in part mediated by the endogenous opioid system. Naltrexone-mediated reversal of immunosuppression may be limited to the populations of the cytotoxic and T-cell lineages. The proliferation response to LPS was unaffected by naltrexone. It was also demonstrated that serum from subordinate (immunosuppressed) fish is immunosuppressive in normal fish–an effect that can also be reversed by naltrexone. These results support a link between the neuroendocrine and immune systems in fish, the lowest vertebrate.

### *Guppy (Fish)*

Guppies can be a model for the study of the relationship between behavior and immune responses. Rarely are the evolutionary origins of mate preferences known, but recently, the preference of female guppies (*Poecilia reticulata*) for males with carotenoid-based sexual colouration has been linked to a sensory bias that may have originally evolved for detecting carotenoid-rich fruits[153]. We were intrigued by the hypothesis that this mate preference may be related to the immune system-enhancing effects of carotenoids[154]. If carotenoids enhance immune responses in guppies, this could explain both the origin of the attraction to orange fruit and the maintenance of the female preference for orange males. Since the orange coloration of males is carotenoid-based[155], females may be able to visually assess the strength of a male's immune system[156] (Fig. 25.6).

We performed tissue grafting[157] for assaying the rejection levels, which can represent the activity of cell-mediated component of the adaptive immune system. Individual scales were swapped between pairs of unrelated fish, creating reciprocal allografts (In all instances, control autograft scales showed no evidence of rejection, except for initial transient inflammatory responses). Transplanted scales were inspected under a microscope and scored blindly on a five-point rejection scale (based on swelling, scale opacity and changes in melanocytes) once per day for 10 days. Five days later, the same pairs of fish were subject to a second set of allografts and scored again. We used 48 mature males and 24 mature females. They were fed one of two diets that were identical except for the concentration of carotenoids ($0.79 \pm 1.27$ µg/g for the L diet versus $2081.0 \pm 229.6$ µg/g for the H diet). Both diets were nutritionally complete, with vitamin A supplied in palmitate form.

We found that the H diet males exhibited a significantly stronger immune response than did the L diet males, to the second allograft but not to the first[158]. For females, the difference between diet groups was less pronounced, perhaps because only males were raised on these diets from birth. These results suggest that carotenoids boost the adaptive immune system of guppies and may explain why carotenoids appear to be mobilized from the skin of guppies infected with a monogenean parasite[159]. We conclude that, attraction to orange may be beneficial for female guppies in the wild because they could: 1) intake carotenoids for their own immune system; 2) avoid sexually-transmitted diseases from males who have utilized their carotenoids stored in the skin to combat infections[159]; 3) choose healthy mates and pass on good genes for disease resistance to their offspring.

## Conclusion and Future Directions

The integration between the neuroendocrine and immune system provides a finely tuned regulatory system required for health. A profound understanding of how CNS and neuroendocrine systems regulate the immune system would be essential. Investigations at the systemic, anatomical, cellular, and molecular levels will inform not only the pathogenesis and treatment of infectious disease, neoplastic disorders, and inflammatory/autoimmune conditions and but also conditions predisposing to susceptibility and resistance to these illnesses.

Molecules such as cytokines, which were initially discovered as chemical messengers regulating immune responses, have overlapping roles with neuroendocrine system. It is of importance in current medicine that there is an increasing body of evidence indicating roles of cytokines as modulators of CNS function and behavior. This has exciting implications in future understanding and treatment of psychiatric disorders, such as major depression, bipolar disorder, and schizophrenia, whose etiology has been linked to dysfunction of the immune system. Thus, investigating how cytokines have bidirectional regulations and

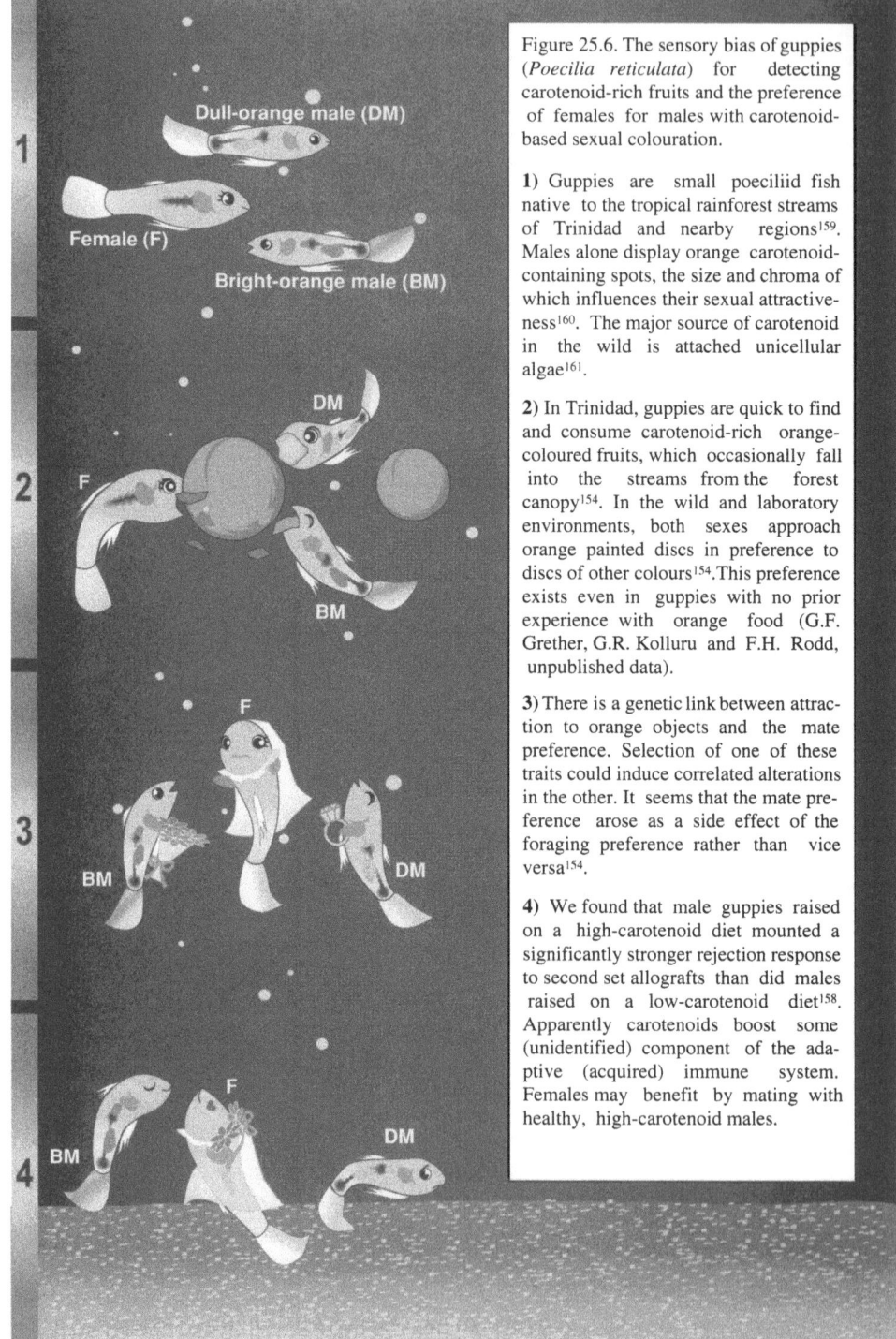

Figure 25.6. The sensory bias of guppies (*Poecilia reticulata*) for detecting carotenoid-rich fruits and the preference of females for males with carotenoid-based sexual colouration.

1) Guppies are small poeciliid fish native to the tropical rainforest streams of Trinidad and nearby regions[159]. Males alone display orange carotenoid-containing spots, the size and chroma of which influences their sexual attractiveness[160]. The major source of carotenoid in the wild is attached unicellular algae[161].

2) In Trinidad, guppies are quick to find and consume carotenoid-rich orange-coloured fruits, which occasionally fall into the streams from the forest canopy[154]. In the wild and laboratory environments, both sexes approach orange painted discs in preference to discs of other colours[154]. This preference exists even in guppies with no prior experience with orange food (G.F. Grether, G.R. Kolluru and F.H. Rodd, unpublished data).

3) There is a genetic link between attraction to orange objects and the mate preference. Selection of one of these traits could induce correlated alterations in the other. It seems that the mate preference arose as a side effect of the foraging preference rather than vice versa[154].

4) We found that male guppies raised on a high-carotenoid diet mounted a significantly stronger rejection response to second set allografts than did males raised on a low-carotenoid diet[158]. Apparently carotenoids boost some (unidentified) component of the adaptive (acquired) immune system. Females may benefit by mating with healthy, high-carotenoid males.

functions with the CNS may provide new insights in brain disorders, ultimately contributing to the development of novel strategies including CAM to treat these diseases.

The relationship between stress and immunological diseases also represents an important aspect of extended modern medicine. For example, stress can induce lymphocyte apoptosis, which may play a prominent role in the pathogenesis of various diseases. Increasing evidence suggests that a compromised immune system may be a major contributor to cancer development and can impact the outcome of cancer therapies. Importantly, studies of breast cancer patients revealed that alleviation of psychological distress extends life expectancy[162], an effect thought to be mediated by having good quality social support[163]. There is similar evidnece that cognitive behavioral interventions designed to alleviate the distress associated with HIV infection were effective in improving the immunological indices of disease course[164]. In this context, complementary and alternative medicine can be a useful aid for the alleviation of stress. An increasing number of studies on how to palliate malignant psychological or physical consequences of stressors has been conducted (yoga, meditation, music therapy, Kampo-medicine, holistic healing method, etc). In addition, it will be essential to develop animal model systems, which can be easily handled at molecular, cellular, and behavioral levels. These models could contribute to the fundamental understanding of neuroendocrineimmune system and eventually may provide the insight of how to use CAM in human patients.

## Acknowledgement

The author appreciates the help of Ms. Atsue Kamata, a graphic designer (Kanagawa, Japan), for the illustrations, and Ms. Mami Kasahara (Kanagawa, Japan) for editing the manuscript.

## References

1. A.B. Negrao, P.A. Deuster, P.W. Gold, A. Singh and G.P. Chrousos, *Biomed. Pharmacother.* **54**, 122 (2000).
2. R. McCarty and P.E. Gold, *Psychosom. Med.* **58**, 590 (1996).
3. R. Ader, D. Felten and N. Cohen, *Psychoneuroimmunology*. Academic Press, New York. (1991).
4. A. Munck, P.M. Guyre and N.J. Holbrook, *Endocr. Rev.* **5**, 25 (1984).
5. I.M. Adcock, *Pulmon. Pharmacol. Ther.* **13**, 115 (2000).
6. P.J. Barnes, *Clin. Sci.* **94**, 557 (1998).
7. S.W. Lee, A.P. Tsou, H. Chan, J. Thomas, K. Petrie, E.M. Eugui and A.C. Allison, *Proc. Natl. Acad. Sci. USA* **85**, 1204 (1988).
8. E. Groujon, S. Layé, P. Parnet and R. Dantzer, *Psychoneuroendocr.* **22**, S75 (1997).
9. M. Schmidt, H.G. Pauels, N. Lügering, A. Lügering, W. Domschke and T. Kucharzik, *J. Immunol.* **163**, 3484 (1999).
10. J.E. Kunicka, M.A. Talle, G.H. Denhardt, M.Brown, L.A. Prince and G. Goldstein, *Cell. Immunol.* **149**, 39 (1993).
11. F. Colotta, F. Re, M. Muzio, R. Bertini, N. Polentarutti, M. Sironi, J.G. Giri, S.K. Dower, J.E. Sims and A. Mantovani, *Science* **261**, 472 (1993).
12. J.A. Moynihan, T.A. Callahan, S.P. Kelley and L.M. Campbell, *Cell. Immunol.* **184**, 58 (1998).
13. A. Ray and K.E. Prefontaine, *Proc. Natl. Acad. Sci. USA*. **91**, 752 (1994).
14. K. De Bosscher, M.L. Schmitz, W. Vanden Berghe, S. Plaisance, W. Fiers and G. Haegeman, *Proc. Natl. Acad. Sci. USA*. **94**, 13504 (1997).

15. W. Vanden Berghe, L. Vermeulen, G. De Wilde, K. De Bosscher, E. Boone and G. Haegeman, *Biochem. Pharmacol.* **60**, 1185 (2000).
16. M.M.J. Chang, M. Juarez, D.M. Hyde and R. Wu, *Am. J. Physiol. Lung Cell. Mol. Physiol.* **280**, 107 (2001).
17. J. Wang, Z. Zhu, R. Nolfo and J.A. Elias, *Am. J. Physiol. Lung Cell. Mol. Physiol.* **276**, 175 (1999).
18. C.Y. Wu, K. Wang, J.F. McDyer and R.A. Seder, *J. Immunol.* **161**, 2723 (1998).
19. D. Franchimont, J. Galon, M. Gadina, R. Visconti, Y.J. Zhou, M. Aringer, D.M. Frucht, G.P. Chrousos and J.J O'Shea, *J. Immunol.* **164**, 1768 (2000).
20. I.J. Elenkov and G.P. Chrousos, *Trends Endocrinol. Metab.* **10**, 359 (1999).
21. J.H. Steer, D.T.S. Ma, L. Dusci, G. Garas, K.E. Pederson and D.A. Joyce, *Ann. Rheu. Dis.* **57**, 732 (1998).
22. C.M. Verhoef, J.A.G. van Roon, M.E. Vianen, F.P.J.G. Lafeber and J.W.J. Bijlsma, *Ann. Rheu. Dis.* **58**, 49 (1999).
23. K.K. Adkins, T.D. Levan, R.L. Miesfeld and J.W. Bloom, *Am. J. Physiol. Lung Cell. Mol. Physiol.* **275**, L372 (1998).
24. R.A. Daynes and B.A. Araneo, *Eur. J. Immunol.* **19**, 2319 (1989).
25. R.A. Daynes, B.A. Araneo, T.A. Dowell, K. Huang and D. Dudley, *J. Exp. Med.* **171**, 979 (1990).
26. E.R. de Kloet, E. Vreugdenhil, M.S. Oitzl and M. Joëls, *Endocr. Rev.* **19**, 269 (1998).
27. J. Visser, A. van Boxel-Dezaire, D. Methorst, T. Brunt, E.R. de Kloet and L. Nagelkerken, *Blood* **91**, 4255 (1998).
28. A.S. Baldwin Jr., *Annu. Rev. Immunol.* **14**, 649 (1996).
29. S. Ghosh, M.J. May and E.B. Kopp, *Annu. Rev. Immunol.* **16**, 225 (1998).
30. B. van der Burg, J. Liden, S. Okret, F. Delaunay, S. Wissink, P.T. van der Saag and J. Å. Gustafsson, *Trends Endocinol. Metab.* **8**, 152 (1997).
31. J. Ramdas and J.M. Harmon, *Endocrinology* **139**, 3813 (1998).
32. K.P. Ray, S. Farrow, M. Daly, F. Talabot and N. Searle, *Biochem. J.* **328**, 707 (1997).
33. J. Liden, I. Rafter, M. Truss, J. Å. Gustafsson and S. Okret, *Biochem. Biophys. Res. Commun.* **273**, 1008 (2000).
34. D.L. Felten, S.Y. Felten, D.L. Bellinger, S.L. Carlson, K.D. Ackerman, K.S. Madden, J.A. Olschowki and S. Livnat, *Immunol. Rev.* **100**, 225 (1987).
35. E.M. Friedman and M.R. Irwin, *Ann. N.Y. Acad. Sci.* **771**, 396 (1995).
36. K.S. Madden, V.M. Sanders and D.L. Felten, *Annu. Rev. Pharmacol. Toxicol.* **35**, 417 (1995).
37. V.M. Sanders, R.A. Baker, D.S. Ramer-Quinn, D.J. Kasprowicz, B.A. Fuchs and N.E. Street, *J. Immunol.* **158**, 4200 (1997).
38. G. Shakhar, S. Ben-Eliyahu, J.D. Ashwell, F.W. Lu, M.S. Vacchio, S. Cohen, S. Line, S.B. Manuck, B.S. Rabin, E.R. Heise and J.R. Kaplan, *J. Immunol.* **160**, 3251 (1998).
39. P.A. Rice, G.W. Boehm, J.A. Moynihan, D.L. Bellinger and S.Y. Stevens, *J. Neuroimmunol.* **114**, 19 (2001).
40. M. Irwin, in: *Psychopharmacology*, edited by F.E. Bloom and D.J. Kupfer, Raven Press, New York. (1995).
41. T.R. Mosmann, H. Cherwinski, M.W. Bond, M.A. Giedlin and R.L. Coffman, *J. Immunol.* **136**, 2348 (1986).
42. H.O. Besedovsky, A. del Rey, I. Klusman, H. Furukawa, G. Monge Arditi and A. Kabiersch, *J. Steroid Biochem. Mol. Biol.* **40**, 613 (1991).
43. C.D. Breder, C.A. Dinarello and C.B. Saper, *Science* **240**, 321 (1988).
44. C.R. Plata-Salaman, Y. Oomura and Y. Kai, *Brain Res.* **448**, 106 (1988).
45. D.E. Brenneman, M. Schultzberg, T. Bartfai and I. Gozes, *J. Neurochem.* **58**, 454 (1992).
46. E.N. Benveniste, *Cytokine Growth Factor Rev.* **9**, 259 (1998).
47. S.J. Hopkins and N.J. Rothwell, *Trends Neurosci.* **18**, 83 (1995).
48. W.A. Banks and A.J. Kastin, *J. Neuroimmunol.* **79**, 22 (1997).
49. L.R. Watkins, S.F. Maier and L.E. Goehler, *Life Sci.* **57**, 1011 (1995).
50. J.M. Krueger, S. Takahashi, L. Kapas, S. Bredow, R. Roky, J. Fang, R. Floyd, K.B. Renegar, N. Guha-Thakurta, S. Novitsky and F. Obal, *Adv. Neuroimmunol.* **5**, 171 (1995).
51. C.R. Plata-Salaman, *Semin. Oncol.* **25**, 64 (1998).
52. J.G. Cannon and C.A. Dinarello, *Science* **227**, 1247 (1985).
53. A.H. Swiergiel, A.J. Dunn and E.A. Stone, *Brain Res. Bull.* **41**, 61 (1996).
54. C.D. Breder, M. Tsujimoto, Y. Terano, D.W. Scott and C.B. Saper, *J. Comp. Neurol.* **337**, 543 (1993).
55. A.J. Dunn and A.H. Swiergiel, *Ann. N. Y. Acad. Sci.* **840**, 577 (1998).
56. A.J. Dunn, *Brain Res. Bull.* **29**, 807 (1992).
57. V.S. Palamarchouk, J. Zhang, G. Zhou, A.H. Swiergiel and A.J. Dunn, *Brain Res. Bull.* **51**, 319 (2000).

58. R.H. Straub, J. Westermann, J. Scholmerich and W. Falk, *Immunol. Today.* **19**, 409 (1998).
59. M. Hashimoto, Y. Ishikawa, S. Yokota, F. Goto, T. Bando, Y. Sakakibara and M. Iriki, *Brain Res.* **540**, 217 (1991).
60. M.L. Wong, P.B. Bongiorno, P.W. Gold and J. Licinio, *Neuroimmunomodulation* **2**, 141 (1995).
61. A. del Rey, H.O. Besedovsky, E. Sorkin, M. da Prada and S. Arrenbrecht, *Cell. Immunol.* **63**, 329 (1981).
62. I.J. Elenkov, R.L. Wilder, G.P. Chrousos and E.S. Vizi, *Pharmacol. Rev.* **52**, 595 (2000).
63. G.P. Chrousos, P.W. Gold, *JAMA.* **4**, 1244 (1992).
64. G.L. Fricchione, G.B. Stefano, *Adv. Neuroimmunol.* **4**, 13 (1994).
65. E.L. Cooper, *Animal Biol.* **1**, 169 (1992).
66. F.S. Dhabhar, *Ann. N.Y. Acad. Sci.* **840**, 359 (1998).
67. M.R. Larson, R.Ader, J.A. Moynihan, *Psychosom. Med.* **63**, 493 (2001).
68. M. Schedlowski, R. Jacobs, G. Stratmann, S. Richter, A. Hadicke, U. Tewes, T.O. Wagner and R.E. Schmidt, *J. Clin. Immunol.* **13**, 119 (1993).
69. F.S. Dhabhar, *Ann. N.Y. Acad. Sci.* **917**, 876 (2000).
70. J.L. Pike, T.L. Smith, R.L. Hauger, P.M. Nicassio, T.L. Patterson, J. McClintick, C. Costlow and M.R. Irwin, *Psychosom. Med.* **59**, 447 (1997).
71. P. Secchiero, A. Gonelli, C. Celeghini, P. Mirandola, L. Guidotti, G. Visani, S. Capitani and G. Zauli, *Blood* **98**, 2220 (2001).
72. J.P. Kolb, V. Roman, F. Mentz, H. Zhao, D. Rouillard, N. Dugas, B. Dugas and F. Sigaux, *Leuk. Lymphoma* **40**, 243 (2001).
73. M. Nagabhushan, H.L. Mathews, L. Witek-Janusek, A.P. Kohm, Y. Tang, V.M. Sanders and S.B. Jones, *Brain Behav. Immun.* **15**, 78 (2001).
74. B.S. McEwen, *Brain Res.* **886**, 172 (2000).
75. S. Lutgendorf, H. Logan, H.L. Kirchner, N. Rothrock, S. Svengalis, K. Iverson, D. Lubaroff, *Psychosom. Med.* **62**, 524 (2000).
76. R. Ader and N. Cohen, *Ann. Rev. Psychol.* **44**, 53 (1993).
77. B.S. Rabin, S. Cohen, R. Ganguli, D.T. Lysle and J.E. Cunnick, *Crit. Rev. Immunol.* **9**, 279 (1989).
78. F.S. Dhabhar and B.S. McEwen, *Proc. Natl. Acad. Sci. USA.* **96**, 1059 (1999).
79. I.J. Elenkov, G.P. Chrousos, *Bailliers Best Pract. Res. Clin. Endocrinol. Metab.* **13**, 583 (1999).
80. P. Holzer, *Neuroscience* **24**, 739 (1988).
81. T.J. Coderre, A.I. Basbaum, J.D. Levine, *J. Neurophysiol.* **62**, 48 (1989).
82. S.D. Brain, T.J. Williams, *Br. J. Pharmacol.* **86**, 855 (1985).
83. F. Berthiaume, C.L. Aparicio, J. Eungdamrong and M.L. Yarmush, *Tissue Eng.* **5**, 499 (1999).
84. C.M. Pariante, B. Carpiniello, M.G. Orru, R. Sitzia, A. Piras, A.M. Farci, G.S. Del Giacco, G. Piludu and A.H. Miller, *Psychother. Psychosom.* **66**, 199 (1997).
85. E.P. Zorrilla, L. Luborsky, J.R. McKay, R. Rosenthal, A. Houldin, A. Tax, R. McCorkle, D.A. Seligman and K. Schmidt, *Brain Behav. Immun.* **15**, 199 (2001).
86. D.A. Padgett, P.T. Marucha and J.F. Sheridan, *Brain Behav. Immun.* **12**, 64 (1998).
87. D. Yin, R.A. Mufson, R. Wang and Y. Shi, *Nature* **397**, 218 (1999).
88. M. Faisal, F. Chiappelli, I.I. Ahmed, E.L. Cooper, H. Weiner, *Brain Behav. Immun.* **3**, 223 (1989).
89. J.P. Capitanio and N.W. Lerche, *AIDS* **5**, 1103 (1991).
90. S.G. Zakowski, C.G. McAllister, M. Deal and A. Baum, *Health Psychol.* **11**, 223 (1992).
91. A. Galinowski, *Encephale.* **19**, 147 (1993).
92. H. Iwagaki, Y. Morimoto, M. Kodera and N. Tanaka, *Rinsho Byori* **48**, 505 (2000).
93. P. Kunes and J. Krejsek, *Cas. Lek. Cesk.* **139**, 361 (2000).
94. B.K. Pedersen, H. Bruunsgaard, M. Klokker, M. Kappel, D.A. MacLean, H.B. Nielsen, T. Rohde, H. Ullum and M. Zacho, *Int. J. Sports Med.* **18**, S2 (1997).
95. B.K. Pedersen, H. Bruunsgaard, M. Jensen, K. Krzywkowski and K. Ostrowski, *Proc. Nutr. Soc.* **58**, 733 (1999).
96. D. Yin, D. Tuthill, R.A. Mufson and Y. Shi, *J. Exp. Med.* **191**, 1423 (2000).
97. K. Bendtzen, M.B. Hansen, C. Ross, L.K. Poulsen, M. Svenson, *Stem Cells.* **13**, 206 (1995).
98. B.M. Jones, *BMC Complement. Altern. Med.* **1**, 8 (2001).
99. M. Kohno, Y. Kawahito, Y. Tsubouchi, A. Hashiramoto, R. Yamada, K.I. Inoue, Y. Kusaka, T. Kubo, I.J. Elenkov, G.P. Chrousos, M. Kondo, H. Sano, *J. Clin. Endocrinol.* **86**, 4344 (2001).
100. Sternberg EM. *Horm Res.* **43**, 159 (1995).

101. B.S. McEwen, C.A. Biron, K.W. Brunson, K. Bulloch, W.H. Chambers, F.S. Dhabhar, R.H. Goldfarb, R.P. Kitson, A.H. Miller, R.L. Spencer and J.M. Weiss, *Brain Res. Rev.* **23**, 79 (1997).
102. D.O. Freier and B.A. Fuchs, *J. Pharmacol. Exp. Ther.* **265**, 81 (1993).
103. K. Kono, A. Takahashi, H. Iizuka, H. Fujii, T. Sekikawa and Y. Matsumoto, *Br. J. Surg.* **88**, 1110 (2001).
104. N.J. Rothwell, *J. Physiol.* **514**, 3 (1999).
105. I.L. Campbell, M. Eddleston, P. Kemper, M.B. Oldstone and M.V. Hobbs, *J. Virol.* **68**, 2383 (1994).
106. A. Waage, A. Halstensen, R. Shalaby, P. Brandtzaeg, P. Kierulf and T. Espevik, *J. Exp. Med.* **170**, 1859 (1989).
107. J. Bauer, S. Strauss, U. Schreiter-Gasser, U. Ganter, P. Schlegel, I. Witt, B. Yolk and M. Berger, *FEBS Lett.* **285**, 111 (1991).
108. J.E. Merrill, *J. Immunother.* **12**, 167 (1992).
109. B.D. Klein, H.S. White and K.S. Callahan, *Neurochem. Int.* **36**, 441 (2000).
110. A.C. Yu and L.T. Lau, *Neurochem. Int.* **36**, 369 (2000).
111. I.L. Campbell, A.K. Stalder, Y. Akwa, A. Pagenstecher and V.C. Asensio, *Neuroimmunomodulation.* **5**, 126 (1998).
112. C. Sacoccio, J. Dornand and G. Barbanel, *Neuroreport.* **9**, 309 (1998).
113. I.G. Rojas, D.A. Padgett, J.F. Sheridan, P.T. Marucha, *Brain Behav. Immun.* **16**, 74 (2002).
114. H. Land, S. Hudson, *Soc. Sci. Med.* **54**, 147 (2002).
115. R. Glaser and J.K. Kiecolt-Glaser, *Ann. Behav. Med.* **19**, 78 (1997).
116. J.T. Cacioppo, K.M. Poehlmann, J.K. Kiecolt-Glaser, W.B. Malarkey, M.H. Burleson, G.G. Berntson and R. Glaser, *Health Psychol.* **17**, 182 (1998).
117. R. Glaser, J. Rice, J. Sheridan, R. Fertel, J. Stout, C. Speicher, D. Pinsky, M. Kotur, A. Post and M. Beck, *Brain Behav. Immun.* **1**, 7 (1987).
118. R. Glaser, J. Sheridan, W.B. Malarkey, R.C. MacCallum and J.K. Kiecolt-Glaser, *Psychosom. Med.* **62**, 804 (2000).
119. J.K. Kiecolt-Glaser, R. Glaser, S. Gravenstein, W.B. Malarkey and J. Sheridan, *Proc. Natl. Acad. Sci. USA.* **93**, 3043 (1996).
120. G.R. Lubach, C.L. Coe and W.B. Ershler, *Brain Behav. Immun.* **9**, 31 (1995).
121. M.T. Bailey and C.L. Coe, *Dev. Psychobiol.* **35**, 146 (1999).
122. C. Rachal Pugh, M. Fleshner, L.R. Watkins, S.F. Maier, J.W. Rudy, *Neurosci. Biobehav. Rev.* **25**, 29 (2001).
123. J. Leserman, J.M. Petitto, D.O, Perkins, J.D. Folds, R.N. Golden and D.L. Evans, *Arch. Gen. Psychiatry.* **54**, 279 (1997).
124. J.P. Capitanio and N.W. Lerche, *Psychosom. Med.* **60**, 235 (1998).
125. S.E. Hobfoll, A. Bansal, R. Schurg, S. Young, C.A. Pierce, I. Hobfoll and R. Johnson, *J. Consult. Clin. Psychol.* **70**, 252 (2002).
126. R.M. DeLano and S.R. Mallery, *J. Neuroimmunol.* **89**, 51 (1998).
127. J.K. Kiecolt-Glaser, L. McGuire, T.F. Robles and R. Glaser, *Psychosom. Med.* **64**, 15 (2002).
128. Y. Shi, S. Devadas, K.M. Greeneltch, D. Yin, R. Allan Mufson and J.N. Zhou. *Brain Behav. Immun.* **17**, S18 (2003).
129. H.K. Fischman, R.W. Pero and D.D. Kelly, *Int. J. Neurosci.* 84, 219 (1996).
130. S. Ben-Eliyahu, R. Yirmiya, J.C. Liebeskind, A.N. Taylor and R.P. Gale, *Brain Behav. Immun.* **5**, 193 (1991).
131. D. Michelson, L. Stone, E. Galliven, M.A. Magiakou, G.P. Chrousos, E.M. Sternberg, P.W. Gold, *J. Clin. Endocrinol. Metab.* **79**, 848 (1994).
132. G. Affleck, S. Urrows, H. Tennen, P. Higgins, D. Pav and Aloisi R, *Ann. Behav. Med.* **19**, 161 (1997).
133. D.E. Yocum, W.L. Castro and M. Cornett, *Rheum. Dis. Clin. North Am.* **26**, 145 (2000).
134. J.K. Kiecolt-Glaser, G.G. Page, P.T. Marucha, R.C. MacCallum and R. Glaser, *Perspect. Psychoneuroimmunol.* **53**, 1209 (1998).
135. T.C. Theoharides, L.K. Singh, W. Boucher, X. Pang, R. Letourneau, E. Webster and G. Chrousos, *Endocrinol.* **139**, 403 (1998).
136. D. Franchimont, G. Bouma, J. Galon, G.W. Wolkersdorfer and A. Haidan, *Gastroenterol.* **119**, 1560 (2000).
137. M. Salzet, D. Vieau and R. Day, *Trends Neurosci.* **23**, 550 (2000).
138. E.J. Goetzl, J.K. Voice, S. Shen, G. Dorsan, Y. Kong, K.M. West, C.F. Morrison and A.J. Harmar, *Proc. Natl. Acad. Sci. USA.* **98**, 13854 (2001).
139. S. Kasahara, H. Wago and E.L. Cooper, *Int. J. Immunopathol. Pharmacol.* **15**, 1 (2002).
140. M. Sander, M. Irwin, P. Sinha, E. Naumann, J. Kox and D. Spies, *Intensive Care Med.* **28**, 285 (2002).

141. M. Irwin and C. Miller, *Alcohol Clin. Exp. Res.* **24**, 560 (2000).
142. M. Irwin, J.C. Gillin, J. Dang, J. Weissman, E. Phillips and C.L. Ehlers, *Biol. Psychiatry.* **51**, 632 (2002).
143. T.C.G. Bosch, in: *Cellular and molecular basis of regeneration*, edited by P. Ferretti and J. Géraudie, Wiley, Sussex (1998).
144. C.J.P. Grimmelikhuijzen, L.I. Leviev and K. Carstensen, *Int. Rev. Cytol.* **167**, 37 (1996).
145. T.C.G. Bosch and C.N. David, *Dev. Biol.* **121**, 182 (1987).
146. H.R. Bode, *Trends Genet.* **8**, 279 (1992).
147. T.C.G. Bosch and C.N. David, *J. Exp. Zool.* **238**, 225 (1986).
148. S. Kasahara and T.C. Bosch, *Dev. Comp. Immunol.* **27**, 79 (2003).
149. B.A. Marcum, T. Fujisawa and T. Sugiyama, in: *Developmental and cellular biology of coelenterates*, edited by P. Tardent and R. Tardent, Elsevier/North Holland, Amsterdam (1989).
150. T.C.G. Bosch, R. Rollbühler, B. Scheider and C.N. David, *Roux's Arch. Dev. Biol.* **200**, 269 (1991).
151. R.I. Lehrer, M. Rosenmann, S.S.S.L. Harwig, R. Jackson and P. Eisenhauer, *J. Immunol. Meth.* **137**, 167 (1991).
152. S.E. Wendelaar-Bonga, *Physiol. Rev.* **77**, 591 (1997).
153. G.F. Grether, K.A. Hughes and F.H. Rodd, *Science* **296**, 847 (2002).
154. F.H. Rodd, K.A. Hughes, G.F. Grether, C.T. Baril, *Proc. Roy. Soc.* **269**, 475 (2002).
155. A.P. Møller, C. Baird, J.D. Blount, D.C. Houston, P. Ninni, N. Saino and P.F. Surai, *Poult. Avian Biol. Rev.* **11**, 137 (2000).
156. G.F. Grether, J. Hudon and D.F. Millie, *Proc. Roy. Soc.* **266**, 1317 (1999).
157. E.L. Cooper, *Transplantation* **2**, 2 (1964).
158. G.F. Grether, S. Kasahara, G.R. Kolluru and E.L. Cooper, *Proc. Roy. Soc.* in press.
159. J.A. Endler, *Evol. Biol.* **11**, 319 (1978).
160. G.F. Grether, *Evolution* **54**, 1712 (2000).
161. G.F. Grether, D.F. Millie, M.J. Bryant, D.N. Reznick and W. Mayea, *Ecology* **82**, 1546 (2001).
162. D. piegel, J.R. Bloom, H.C. Kraemer and E. Gottheil, *Lancet* **2**, 888 (1989).
163. J.M. Turner-Cobb, S.E. Sephton, C. Koopman, J. Blake-Mortimer and D. Spiegel, *Psychosom. Med.* **62**, 337 (2000).
164. S. Cruess, M. Antoni, D. Cruess, M.A. Fletcher, G. Ironson, M. Kumar, S. Lutgendorf, A. Hayes, N. Klimas and N. Schneiderman, *Psychosom. Med.* **62**, 828 (2000).

# VI

**Education and Philosophy**

# 26

# Scientific Thinking: Its History, Methods, and Advantages

## C.S. Wallis

## 1. Introduction

Philosophers and scientists often dismiss complementary and alternative medicine as "pseudoscience." In applying the pseudoscience moniker they often intend to suggest that the beliefs and practices of complementary and alternative medicine are as irrational as a fear of flying, as effective as a rabbit's foot, and as pretentiously deceptive as a mongrel in a dog show. Many people in western society have come to think of science as the deity of reason and pseudoscience is its doppelganger. In this paper I wish to offer a less bipolar view of science and pseudoscience wherein one sees science and pseudoscience as sometimes overlapping partitions on a continuum between utopic rational inquiry and chaotic irrationality. My thesis does not represent a diminution of science, nor an exultation of pseudoscience. Rather, I portray pseudoscientific beliefs and practices as resulting from human, but nevertheless unacceptable, hypotheses, evidence standards, and innate judgment mechanisms. Likewise, science while striving towards laudable intellectual standards should not be thought of as an epistemic authority before which one must unthinkingly genuflect.

I proceed by first sketching the history and basics of scientific theories and scientific method. As part of my explication of scientific method, I look more closely at the placebo effect. Western science often remains suspicious of complementary and alternative medicine, dismissing claims of its efficacy as placebo effect. I suggest that placebo effect is an umbrella for a large number of different phenomena in scientific experimental results. I then discuss common methodological problems associated with pseudoscience. My thesis will be that science is not the exclusive practitioner of its methodological principles nor does science prove completely immune to the sorts of difficulties associated with pseudoscience.

---

**Charles S. Wallis, Ph.D.** • Department of Philosophy, California State University, Long Beach, 1250 Bellflower Boulevard, McIntosh Humanities Building (MHB) 917, Long Beach, CA 90840-2408
*Complementary and Alternative Approaches to Biomedicine*, edited by Edwin L. Cooper and Nobuo Yamaguchi. Kluwer Academic/Plenum Publishers, 2004.

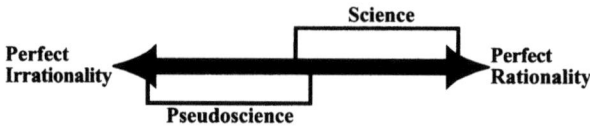

Figure 26.1.

## 2. Scientific Theories

Science is a way of thinking much more than it is a body of facts. The next three sections outline that thought process. Specifically, these sections delineate the nature of scientific theories and the evidence that scientists gather in support of a given scientific theory. Understanding more precisely the product and processes of science serves to better illustrate its similarities and differences with respect to other methods of inquiry like pseudoscience.

The semantic view of scientific theories current in philosophy of science supposes that scientific theories are collections of models.[1] That is, scientific theories are structures that mirror structural relations in the real world. A model fits the world like a glove to a hand. More accurately, a model fits the world because it is similar to the world in certain respects and to certain degrees. Theories are groups of models sharing a core insight which has been modified to introduce or maximize similarity in some respect. For example, consider the ideal pendulum law: $\mathbf{P} = 2\pi\sqrt{l/g^2}$

If one interprets the ideal pendulum law literally, real pendulums would falsify the law. Real pendulums have varying degrees of frictional resistance in their arm pivot and wind resistance against their weights and arms. Frictional resistance in the arm pivot and/or air resistance in against the arm and bob alter the relationship of arm length to periodicity

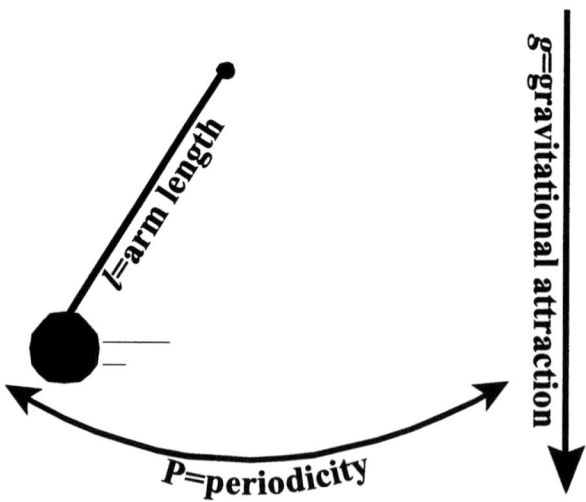

Figure 26.2.

making the ideal pendulum law literally false. However, frictional and wind resistance prove negligible, i.e., introduce only small amounts of error, in many applications. When these forces prove negligible, pendulum periodicity varies inversely with the square root of the arm length divided by gravitational acceleration. Thus, one can idealize from friction and wind resistance so as to quantify a real relationship between arm length and periodicity in a manner that proves predictively adequate. Likewise, one can alter the ideal pendulum law so as to introduce a constant for friction. Important modifications of a law to cover common or important applications of the law become part of the cluster of models used to understand and interact with the class of phenomena they describe. Alas, one cannot, and would not want to, generate modified laws for every possible instance that deviates from the ideal pendulum law. For instance, modifying the law to cover elastic pendulum arms, magnetic fields and metal pendulum bobs, moving arm pivots, etc. proves both unnecessary and scientifically uninteresting absent the need for such modifications.

A theoretical hypothesis, then, is simply a claim that a given model is similar to the world in the required respects or to an appropriate degree. If such a similarity relation holds, then scientists deem the hypothesis true; absent such similarity, scientists reject the hypothesis as false. The ideal pendulum law is a model. It has a standard application to pendulums where frictional resistance at the arm pivot and air resistance are negligible. In specifying conditions for the application of the model together with an acceptable margin of error one formulates theoretical hypothesis for how the model relates to the world.

The process of creating a model and specifying a theoretical hypothesis does not differ dramatically from every day problem solving. One might have a model; Bill pilfered your cookies and consumed them. You have a theoretical hypothesis; this model accurately represents the relationship between Bill and my cookies at some point in time between last night when I went to bed and this morning. Your model either fits the world in accordance with the theoretical hypothesis or it does not. Similarly, the auto mechanic has a model; your car's alternator is no longer charging your battery. She has a theoretical hypothesis; this relationship is holding right now in your car. Her model either fits the world in the ways specified by her hypothesis, or it does not.

## 3. Scientific Methods: Types of Studies

Scientists perform experiments to assess the fit of a theory with the world. In western science, experimental methodology has become relatively uniform. In the medical, biological, and social sciences most studies are designed to gather statistical assessments of a theory's fit with the world. Historically, Arbuthnott[2] published the first test of a statistical hypothesis, when arguing that only "divine providence" could explain the equal birth rates for men and women.[3] The philosopher John Stuart Mill outlined a set of methods for causal inference[4] at approximately the same time as Jules Gavarret[5] seems to have first used probable error to test significance in experimental biology, though John Venn[6] was one of the first to use the terms "test" and "significant."[7] Wood and Stratton are thought to be the first to employ such tests for experiments in agriculture.[8]

Dr. John Arbuthnott
1637–1735

Sir Ronald Alymer Fisher 1890–1962

Egon Pearson 1895–1980

Jerzy Neyman 1894–1981

In the works of R.N. Fisher[9] as well as Jerzy Neyman and Egon Pearson[10–12] the modern outline of contemporary statistical hypothesis testing emerged. At the dawn of the 20$^{th}$ century Karl Pearson (Egon's father) and others had developed the field of applied statistics to the point where statistical analysis consisted primarily of collecting large data sets for the purpose of determining the characteristics of a given population. For instance, Karl Pearson is credited with coining the term "standard deviation". Fisher determined the exact distributions and introduced numerous key concepts and methods such as the analysis of variance, the concept of maximum likelihood, and hypothesis testing using statistical methods. The later idea was similar to work by the great French mathematician and scientist Jules Poincare'[13–16] and the philosopher Karl Popper[17]. The shared core idea of their view consists in the claim that one cannot verify scientific hypotheses. Rather scientific hypotheses are subject only to rejection or revision based upon constant new empirical tests. Both Fisher and William Gossett, who corresponded regularly, developed techniques for the analysis of smaller samples.

Statistical assessments of a theory's fit with the world can occur in several formats. The distinction between descriptive and experimental studies constitutes perhaps the most important difference. Descriptive studies attempt to test the fit of the theory by carefully considering distributions within the actual population. Descriptive studies can take either a prospective or a retrospective form. Prospective studies follow subjects from the beginning of the study into the future. Retrospective studies look backwards in time from the study. In contrast, experimental studies always proceed prospectively (forward-looking from the initiation of the experiment) and involve carefully manipulating the conditions during the time of the study.

All of the above studies assess the fit of the world in the same basic manner. Imagine two Earths, identical in every way except one: On one Earth there is no suspected causal factor (call this the control planet), on the other Earth the suspected causal factor is omnipresent (call this the experimental manipulation planet). To determine if a substance or treatment is a causal factor, one simply compares the rates of the effects in the two worlds. If the causal factor world differs from the control world at a rate not explicable by chance alone, then the substance or treatment is a causal factor. One has excellent evidence for a

Types of Studies

|              | Manipulate Conditions | Prospective | Retrospective |
|---|---|---|---|
| Descriptive  | NO  | YES | YES |
| Experimental | YES | YES | NO  |

| Control |  |
|---|---|
|  |  |

Figure 26.3.

causal hypothesis in this scenario since the only difference between the two worlds is the presence of the suspected causal factor.

Of course, there is only one real world and gathering complete data from an entire population is almost always prohibitively expensive and difficult, if not impossible. So, one must design one's studies to replicate as closely as possible the data from each of the above-described ideal worlds. First, to overcome expense and logistics, scientists gather data only from subsets of the studied population, i.e., samples. From these samples, researchers make statistical inferences regarding the population as a whole. Samples can be generated either randomly or non-randomly. In a random sample each individual in the population is equally likely to have been included in the study sample. A random sample of sufficient size provides researchers with a sample that is more likely to reflect the general make-up of the target population in that differences between individuals within the sample are more likely to reflect the distribution of such differences in the population. Non-random samples are selected in ways in which some non-random constraints act to determine sample membership. Consider a jar of marbles with white marbles on top and green marbles on the bottom. If one selects a sample of marbles from the top of the jar, then one's sample will not reflect the presence of green marbles. Thus, the bias introduced by selecting marbles only from the top of the jar makes the sample a non-random sample. On the other hand, if one selects marbles from random depths in the jar, no such bias is introduced in sampling and one's random sample is much more representative of the marble population.

Once one has selected a sample, one can proceed to gather data from the samples in one of two manners: (1) through experimental manipulation of experimental controls or (2) through data manipulation of descriptive data.

If medical researchers adopt a descriptive methodology, they can elect to test a theory using a retrospective or a prospective data gathering process. In conducting a retrospective descriptive study, one selects as large a sample as possible from the population. For instance, one might look at the effect of moderate exercise on the development of coronary heart disease by selecting a large group of people within the age range normally associated with the onset of coronary heart disease. One then looks back at their exercise history comparing comparable groups of exercisers to groups of non-exercisers. In descriptive studies there is no control for individual variation by random selection of the study sample, so one must carefully control for biasing factors by selecting a pool from the original sample that matches the control and causal factor groups as closely as possible. Usually researchers select this pool by eliminating other known causal factors. For example, in looking at the effect of exercise on heart disease one would like the exercise and no exercise groups to have the same approximate age distribution, number of smokers, individuals with family histories of heart disease, and so on.

Retrospective studies have the advantage that they can yield relatively quick assessments of theoretical fit. Coronary heart disease develops over a long period of time. In order to perform a prospective descriptive study or an experimental study, experimenters must follow the study subjects over that long period of time in order to gather data. In retrospective studies the time period has already elapsed and data as to outcomes is already available. The primary disadvantage of retrospective studies consists in the difficulty in collecting accurate and complete information regarding the suspected causal factor as well as regarding the potential influence of other known causal factors. In the case of a heart disease study, personal recollections of exercise patterns, dietary and other habits can prove unreliable or nonexistent. The problem of accurately recording the subject's history during the period studied can result in spurious assessments of theoretical fit.

Prospective studies suffer much less dramatically from above-mentioned problem of data collection. In a prospective study researchers select a large sample from the population and record each subject's data as the study unfolds. Based upon the subject's choices, researchers can later classify them into the control or causal factor group for the purposes of evaluating the fit of the theory.

Prospective studies often occur over long periods of time, and results may not be known for decades. However, if the subjects of the study are properly monitored, the results of the study are much more conclusive than in a retrospective study. Prospective and retrospective studies are widespread research tools in part because they share two additional advantages. First, performing an experiment lasting decades and occurring outside the laboratory setting often proves very difficult. Second, medical researchers regularly wish to evaluate the causal effects of a substance or treatment when the suspected effect is deleterious to the individual. In such cases ethical considerations prevent experimental studies. Imagine, for example, that one wishes to determine whether smoking is a causal factor in the development of lung cancer. One cannot experiment upon subjects by requiring them to expose themselves to a potentially dangerous treatment or substance. One cannot simply say to subjects, "I think that this might kill you. Take it and let's see." Such a practice would be a thousandfold worse than the practice of asking another to taste the milk to see if it has gone bad.

When ethical concerns and experimental logistics do not prove prohibitive, medical researchers perform experiments. Experimental methodology shares the same general ideal of comparing two populations that differ from each other only by the presence or absence of the suspected causal factor. In particular, scientists gather a great deal of evidence regarding the fit of scientific theories using null hypothesis testing. The methodology of null hypothesis testing consists in comparing two samples or a sample and a hypothetical standard in order to determine how much significance one ought to legitimately attach to observed differences between them. Ironically, the hypothesis that is often tested in null hypothesis testing is the hypothesis that a given causal factor has no effect or that there is no relationship between two samples. The hypothesis that a given causal factor has no effect or that there is no relationship between two samples is called the null hypothesis, and is often referred to as $H_0$. The null hypothesis is the mutually exclusive and jointly exhaustive counterpart to one's

Elements of Null Hypothesis Testing

| Role of Hypothesis | Content of Hypothesis | Prior Known Probability | Experimental Goal |
|---|---|---|---|
| Hypothesis Tested | The Null or No Effect Hypothesis | Probability Known to Equal 0 | Falsify |
| Target Hypothesis | The Hypothesis that some Agent has some Causal Effect | Prior Probability not Precisely Known | Gain Evidentiary support through falsification of HT |

hypothesis of interest. For example, suppose that one's hypothesis of interest, $H_1$, is that ginkgo biloba causes measurable improvements in the memories of young adults. The null hypothesis, $H_0$, would then consist in the assertion that ginkgo biloba causes no measurable improvements in memories of young adults. Experimenters test the null hypothesis because the probability that the causal factor will alter the outcome given this (null) hypothesis is quite easy to calculate. It is zero! That is, the ginkgo biloba sample and the no-ginkgo sample should differ by no more than chance (i.e., the sampling error). One gathers evidence for the target hypothesis (not the null) by gaining negative evidence for (i.e., falsifying) the null hypothesis.

Several Applications of Null Hypothesis Testing

| Type of Hypothesis Tested | Null Hypothesis | Method of Data Evaluation |
|---|---|---|
| S Has A Causal Effect | S Has No Causal Effect | Evaluate Sample S Differences From Control Sample For Statistically Significant Difference |
| S Has A High Causal Effectiveness | S Has No Causal Effect | Evaluate Degree of Statistical Difference Between Sample S and Control Sample |
| S Has A Greater Causal Effectiveness Than Alternative A | Alternative A Has A Greater Causal Effectiveness Than S | Evaluate Sample Differences For Statistically Significant Difference Between Sample A and Sample S OR Sample S and Control Sample |
| S Has Fewer/Less Severe Side Effects Than Alternative A | Alternative A Has Fewer/Less Severe Effects Than S | Evaluate Sample Differences In Rates of Side Effects/Severity of Side Effects For Statistically Significant Difference Between Sample A and Sample S OR Between Sample S and Control Sample To Compare To Known Rates For A |

In western medicine experimenters employ null hypothesis testing to investigate a number of properties. As in the above ginkgo biloba example, scientists may wish to test to see if a substance or treatment has any **causal effect**. For example, doctors may wish to discover if Tamoxifen can cause breast cancer remission. Often researchers in western biomedical science want to test the **effectiveness** of a known or suspected causal factor. How strong of a causal factor is it? For instance, in the late 1980s the findings of a study suggested that oat bran was very effective in reducing blood serum cholesterol. Were this finding to have been born out by follow-up studies, oat bran could have had medical uses as a cholesterol reduction tool. Medical scientists may also want to **compare a substance's or treatment's effectiveness** to another substance or treatment already in use. Doctors may want to compare, to take a case, the efficacy of surgery and chemotherapy to chemotherapy alone for breast cancer. Likewise, scientists often compare the two treatments or substances in terms of their cost-to-benefit ratio. In such cases diverse factors like comparative side effects, side effects severity, and effectiveness may be assessed through the formulation of the null hypothesis.

## 4. Scientific Methods: Experimental Blinds and Placebo Controls

In addition to the basic structure of null hypothesis testing, medical researchers almost invariably employ a double or single blind design and a "placebo control" group their experimental studies. In a **single blinded** study knowledge regarding a subject's role in an experiment is kept from the subject or the experimenter. In a **double blind** study the subject's role in the experiment is kept secret from both the subject and the experimenter. Blinded studies prevent subject and/or experimenter bias from tainting the experiment's results. Contamination of experimental data can happen via numerous routes. Biases or expectations can influence how subjects report to experimenters or how experimenters record subject reports and other data. Such misreporting and inaccurate recording can result from conscious intent or happen at an unconscious level. For example, as many high school drug dealers know, people who are given innocuous or inert substances and told that they are taking LSD will often report experiences consistent with their expectations. Gould [18] chronicles how the racist biases of such scientists as Samuel Morton tainted their research into comparative intelligence of peoples of various ethnicity. Similarly, Watson [19] criticized structuralist psychologists like Titchener and Wundt for training their subjects so as to create reporting biases.

While blinding helps to mitigate experimenter and subject biases, not all experiments can be blinded. For example, one would have great difficulty concealing from a subject that he or she had in fact taken LSD or that she or he were in the vigorous exercise sample in an experiment. Likewise, not all data collection can be performed by blinded experimenters. Experimenters would likely know the group to which a subject belonged when studying the effects of electric shock on subject response time.

A placebo is an innocuous or inert substance. The use of placebo control groups in experiments dates back to at least, 1907 when WHR Rivers employed a placebo control group in his study of the effects of alcohol. The word "placebo" ("I will please" in Latin) dates back much farther, finding its way into English through an incorrect translation of Psalm 116. The erroneous translation exchanged the phrase, "I will please the Lord" for the more correct phrase "I will walk before the Lord". Since in medieval Catholicism the mistranslated verse opened the Vespers for the Dead, and since Catholics occasionally hired professional mourners to sing vespers, the phrase "to sing placebos" acquired the derogatory connotation of a sycophant. One finds such a usage, for instance, in Chaucer. In the 1800s "placebo" connoted medicine given "more to please than to benefit the patient" ([20] p. 272). Today, the term "placebo" has two standard uses. Within the context of scientific and medical research, it refers to an innocuous or inert substance or treatment that the experimental subject would find indistinguishable from the substance or treatment under investigation. Outside the context of modern scientific research, "placebo" often connotes substandard or fraudulent practices used by unscrupulous charlatans.

William Halse Rivers Rivers (1864–1922)

The two researchers most responsible for the widespread contemporary use of placebo control groups and blinds in scientific studies are Harry Gold and H. K. Beecher. The first use of the term "blind or test experiments" occurs in Jevons' *The Principles of Science: A Treatise on Logic and the Scientific Method* in 1874[21]. Usage of the term appears sporadically into the 1930's, at which time Harry Gold employs the term more systematically in advocating blinded experiments. One account alleges that Gold derived his usage from advertisements for Old Gold cigarettes in which consumers were challenged to take a blindfolded taste test.[20] Gold and several colleagues at the Medical College of Cornell University urged the adoption of experimental blinding as a means to introduce greater rigor and impartiality into clinical tests of drugs. They also advocated the use of placebo pills in such research. As drug researchers, Gold and colleagues found themselves suspicious of much of the scientific tests of pharmaceuticals conducted at that time. Beecher's 1955 article, "The Powerful Placebo," represents the first scientific attempt to quantify the placebo effect.[22] Beecher based his analysis upon 15 different trials with different diseases. Based upon his analysis, Beecher concluded that placebo alone provided satisfactory relief for 35% of the 1082 patients in the 15 studies.

H.K. Beecher (1904–1976)

When western medical researchers use the placebo effect as a blanket dismissal of complementary and alternative medicine they are often thinking of fraudulent practices or false reporting by subjects of either symptoms or relief thereof. However, research on the placebo effect suggests that the placebo effect serves as an umbrella for a number of phenomena. Some occurrences of the placebo effect constitute the medically irrelevant or even dangerous brummagems intimated by medical practitioners. Other classes of the placebo effect stem from an eclectic assortment of inconsequent non-causal factors or extraneous and/or nonspecific causal factors. In some cases placebos seem to actually belie their moniker and cause beneficial physiological changes.

One can partially account for the placebo effect in many scientific studies as the result of spurious non-causal factors.[23,24] Inaccuracy or error in data collection constitute one non-causal factor that can partially account for placebo effects. In addition to self-delusory reports of improvement, subjects may provide incorrect data out of politeness or to please the experimenter. Likewise, scaling biases or intra-subjective reports may taint the results.

An evaluation scale that ranges from 1=same to 10=dramatically improved, will likely bias reports towards improvement when compared with a scale ranging from ~10=dramatically worse to 10=dramatically better. An absence of careful controls for underlying natural history or the prevalence of symptoms under study can also lead researchers to overestimate the rate of seeming placebo induced changes. For instance, a placebo may seem effective at inducing sleep if researchers fail to control for drowsiness resulting from the fact that 71% of Americans fail to sleep 8 hrs during the week.[25]

Spurious and/or nonspecific causal factors can contribute to placebo effects.[24] Disorders that tend to vary in their severity and duration, like pain, mood disorders, and prostrate problems, can exhibit spontaneous improvement. Regression to the mean likewise can explain improvement in ailments in which symptoms vary in their intensity. Subjects may enter a study or seek treatment when symptoms reach the more extreme levels of a normal range of values. In such cases, improvement can occur without any effective treatment as the symptoms return to the mean. For example, a person may try a new diet when holiday binging pushes their weight to the higher end of their normal range. The person may see the diet as effective when such variation in weight will often reverse itself. Subjects may also experience improvement as a side effect of the additional treatment and/or improved medical care provided as part of the study. Some of a depressed patient's listlessness may disappear when blood tests diagnose his or her anemia. Similarly, studies intended as placebo-controlled actually follow an additive design in which the treatment and placebo group receive additional supportive therapies. Such additive therapies are quite common in HIV and cancer studies, where patients will often attribute improvements to alternative therapies that they underwent in addition to traditional medical treatment.

Finally, many researchers attribute therapeutic causal effects to placebos in such diverse subjectively assessed illnesses as migraines, back pain, post-surgical pain, rheumatoid arthritis, angina, and depression.[26–29] Placebos have elicited significant alterations in blood pressure, skin temperature, cholesterol level, and heart rate. Studies have likewise reported improvement in warts and contact dermatitis.[28] More recently, a study of depressed subjects by Leuchter *et al* recorded significant changes in prefrontal cordance using quantitative EEG in placebo responders (those who showed improvement) that was not seen in placebo non-responders or either drug group.[30]

## 5. Science and Pseudoscience

If the double blind, placebo controlled experiment represents scientific method at it's best, how does one define pseudoscience. There are a number of features generally thought to be indicative of pseudoscience. However, as I suggest in the introduction to this paper pseudoscience is not an isolated human activity definable in terms of necessary and sufficient conditions. A given pseudoscientific belief or practice will usually exemplify some subset of the larger class of features I discuss. Moreover, science and scientists are not totally immune from the difficulties I describe. Indeed, as I note, many of the reasoning processes associated with pseudoscience are general patterns of reasoning common to the scientist and layman alike. Science distinguishes itself from pseudoscience in its explicit attempt to overcome and minimize the foibles less well recognized and regulated in pseudoscience and daily life.

Pseudoscience represents itself as scientific, empirical, and/or objective. However, pseudoscience displays varying degrees of indifference to facts indicative of an unprincipled and/or flawed criteria of evidence and evidential support. Furthermore, pseudoscience does not progress. That is, it does not vigorously test its theories and revise or abandon them in light of new evidence. Finally, pseudoscientific hypotheses and research generally do not uncover new facts about the world. Often pseudoscience is parasitic upon mainstream science for empirical discoveries.

A classic example of the non-progressive and parasitic nature of some pseudoscience is creation science. Creation science yields no new data. Instead, as indicated by Judge Overton in McLean v Arkanas:

The proof in support of creation science consisted almost entirely of efforts to discredit the theory of evolution through a rehash of data and theories which have been before the scientific community for decades. The arguments asserted by creationists are not based upon new scientific evidence or laboratory data which has been ignored by the scientific community. (Overton Opinion, McLean v Arkansas, 1981)

In short, creation science consists in offering alternative explanations for scientific data or attempting to discredit that data.

Science evolves through both the corroboration **and** the falsification of hypotheses. For this reason science strives to formulate theories that make predictions which experimentation can potentially reveal as nonveridical. In more extreme cases, pseudoscientific practitioners construct their claims as to render them impervious to falsification. James Randi once tried to test the claim of a Russian clinic that their "energized" water had medicinal properties. Since the operators of the clinic claimed to be able to sense the special properties of their water, Randi proposed a simple detection test. He asked them to leave the room while he placed several containers around the room. Some of the containers contained regular tap water, while others contained the "energized" water. When the Russians returned, they were asked to find the "energized" water. Eventually he was told that the thoughts of the Russians while they looked for the "energized" water had "energized" all of the water. Similarly, the popular psychic medium John Edward claims to reunite people with dead loved ones by communicating with those dead loved ones. One way one might test the accuracy of his claims would be to see how often he is right about the details he relays from the departed. However, Mr. Edwards has said that due to the difficulty of his work his proclamations often prove wrong. Since he distances himself from any claims about the high reliability of his psychic pronouncements, it becomes nearly impossible to refute, or even test, his claimed psychic ability based upon the accuracy of his psychic pronouncements.

The above-discussed operators of the Russian clinic revised their claims to prevent refutation. Another approach to theorizing common in pseudoscience consists in belief perseverance. One dictate guiding science is to revise one's hypotheses in light of new information. Practitioners of pseudoscience will often fail to revise their beliefs in light of new evidence, or even in the face of a clear cut refutation. For instance, I once had a student who professed belief in what is often called the "Chinese Zodiac" in the United States. According to her view, people have specific personalities and physical or mental gifts according to their birth year. I pointed out to her that her belief had the consequence that nearly everyone in our class would have had the same personality and physical or mental gifts. Though she acknowledged that our class was hardly homogenous, she insisted that her belief was clearly true nonetheless. In the 1980's James Randi exposed a faith healer Peter Popoff on national television by catching his wife transmitting information to his earpiece that he then portrayed as being sent to him from God. Popoff ministries declared bankruptcy shortly thereafter. Nevertheless, Popoff currently has a show on television. He claims to heal people who send him money by sending them his "miracle healing water."

Like many of the tendencies attributed to pseudoscience, belief perseverance is a failing of human reasoning generally. Once people form a belief, they often will continue to hold that belief even in the face of massive counter evidence. Psychologists have repeatedly demonstrated this phenomenon using what is often called the "debriefing paradigm." In these experiments subjects form beliefs based upon "evidence" provided by the experimenters. The researchers later debrief the subjects, informing them that they have been utterly deceived. Despite the refutation of the basis of their new beliefs, subjects tend to maintain their new beliefs, even going so far as to defend their new beliefs.[31]

It would be false to suppose that western science is bereft of belief perseverance or irrefutable theories. Western science actively eschews such practices, but instances of such theories can be found in the history of science. Ptolemaic astronomy was the scientific precursor of Copernican astronomy. Ptolemaic astronomy was also remarkably accurate in its predictions of the positions of various the planets and stars. However, this accuracy was not the result of a valid model of the solar system and the heavens. Ptolemaic astronomy supposed that the Earth was the center of the universe and all celestial bodies revolved around the Earth. Rather, Ptolemaic astronomy was virtually irrefutable before the advent of telescopes precisely because Ptolemaic astronomers had several mechanisms to modify the Ptolemaic model in inconsequential ways in light of new evidence. The philosopher and mathematician Imre Lakatos (1980) deemed the formulation of such *ad hoc* hypotheses one of the hallmarks of legitimate scientific research. However, Lakatos found that excessive use of *ad hoc* hypotheses to shield theories from refutation to be the hallmark of degenerative science and pseudoscience.[32]

Similarly, prior to the advent of the oxygen theory of combustion, the accepted view of combustion was that it involved the release of phlogiston, the fire substance, from the burned objects, thereby rendering them "dephlogisticated." Lavoisier mounted an extensive experimental refutation of the phlogiston theory in favor of the oxygen theory. Despite the massive evidence generated by Lavoisier, phlogiston theory's most intelligent and able defender, Joseph Priestley, never rejected the theory. At one point Priestley even introduced the *ad hoc* hypothesis that phlogiston had negative mass. Priestley persevered in his belief in phlogiston until his death, long after the general scientific community had abandoned it.

Like the formulation of pseudoscientific theories, the inadequate or inappropriate use of evidence to support pseudoscientific claims varies in severity, and can take a number of forms. The most stereotypical, though not necessarily the most common, manifestation of indifference to evidence by pseudoscience practitioners consists in the extensive propagandistic employment of rhetoric, misrepresentation, and common informal fallacies in advocating pseudoscientific positions. Pseudoscientific literature is rife with fallacious arguments such as appealing to unqualified authority, appealing to emotion, and appeal to ignorance. In the case of ginkgo biloba, many herbal supplements misrepresent the facts by suggesting that studies have shown that ginkgo biloba improves memory in the general population. In fact, most research involving gingko studies elderly subjects, usually with some form of memory impairment or Alzheimer's disease. One summary of the ginkgo literature to date concludes that

> collectively, the behavioral literature reviewed cannot be used conclusively to document or to refute the efficacy of ginkgo in improving cognitive functions. At best, the effects seem quite modest. In particular, it is questionable whether effects of ginkgo, if present, are equal to those obtained by the administration of acetylcholinesterase inhibitors, hearing an arousing story, or ingesting glucose. ([33], p. 2)

Likewise, one often finds the use of fallacious reasoning in pseudoscience. A pseudoscience practitioner may, for instance, dismiss conventional scientific objections to their view by saying, "science doesn't know everything." Such a statement constitutes an appeal to ignorance fallacy, since the supposed limits of scientific knowledge neither refute specific scientific findings, nor make pseudoscientific views more likely to be true. Pseudoscientists appeal to unqualified authority to support their views when they cite one another or celebrity believers as evidence for their view.

Though the scientific community reviles scientific misconduct, it does exist. Not all cases of scientific fraud have as high a profile in the general populace as the claim by Martin Fleischmann and Stanley Pons that they had discovered a means for achieving cold fusion. Nevertheless, the National Institutes of Health receives nearly 200 complaints of scientific misconduct each year and finds misconduct in about 20% of those cases. The NSF has reported approximately 50 cases being investigated each year.[34] Such prestigious journals as Science and Nature have published articles later discovered to be fraudulent.[35,36]. As late as 1997, though many organizations have codes covering scientific misconduct, Denmark was the only country to implement a national code covering scientific misconduct.

Though deliberate misrepresentation and use of fallacious reasoning proliferates in pseudoscience, more subtle, less consciously malicious instances of inadequate or inappropriate use of evidence prove more ubiquitous. For instance, pseudoscience practitioners often do not regularly and rigorously subject their beliefs to meaningful tests. I often ask students who profess to place some credence in astrology to pick their horoscope for the previous day from a random unidentified list. Not only are these students surprised that their success rate falls squarely in the range predicted by chance alone, but they confess that it never occurred to them to test their belief in this fashion. Likewise, many people will start taking a supplement or undergoing a therapy on the basis of a friend's recommendation or a single ad or article without further research.

Pseudoscientific practitioners who do look for evidence often assign inordinate weight to anecdotal evidence.[37,38] Diet fads often provide excellent examples of the misuse of anecdotal evidence. In Los Angeles I often see signs for Ephedra-based herbal weight loss formulas that read, "Lost 40 pounds in 1month." Such testimonials typically exhaust the "evidence" for the product's efficacy and safety. Believers in the paranormal often offer the experiences of themselves and their friends as evidence of paranormal phenomena. Anecdotal evidence suffers from multiple difficulties. First, though psychologically influential, individual occurrences provide inadequate sample sizes for the purposes of statistical inference. Such evidence occurs in circumstances lacking controls to eliminate alternative causal factors. The evidence is also more likely to be tainted by the person's biases and the limitations of human memory. When someone testifies Echinacea cleared up their cold much quicker. They have not ruled out other possible causal factors that might explain their experience. Did they get more rest, eat better, drink more fluids, etc.? Was the cold virus itself less pernicious than other viruses they have battled? Likewise, their memory may be flawed or imprecise. Have they made careful note of the lengths of sicknesses in the past, or is their claim based upon their impression? What criteria did they use to determine when they had become well?

Flawed or inadequate experimental designs rank highest among the deficiencies in evidential criteria pervading research in pseudoscience. For example, pseudoscientific research often exhibits confirmation bias. Carl Jung, to take a case, claimed to have analyzed 40,000 dreams to gather what he portrayed as striking evidence for this psychological theory that dreams and the unconscious contained archetypes–primordial images inherited from our ancestors.[39] However, Jung's analysis consisted entirely of his own interpretation of the dreams as they were told to him. Though perhaps more forgivable in Jung's time, such methodology lacks mechanisms to prevent experimenter bias from tainting the evidence or its interpretation. Similarly, Gold *et al* note the following difficulties in assessing the efficacy of ginkgo: (1) A great deal of the research on ginkgo is published in non-English and/or obscure journals and other media. (2) The bulk of the experimental studies involve older subjects with varying degrees of mental impairment, making generalization to the population at large difficult. (3) Investigators often fail to administer both pre- **and** post-ginkgo psychological tests to better specify the nature and extent of ginkgo's effects. (4) Experiments exhibit dramatic differences in the subject sample size and in control over experimental conditions. (5) Many studies fail to use double-blind placebo controlled methodology.[33]

In addition to flawed or inadequate experimental designs followers of pseudoscience often fail to seek overall consistency with the scientific body of knowledge, with all available evidence or elements of a given theory, and amongst individual practitioners. A graduate student once showed me a book on personality theory. The book proclaimed that there were sixteen distinct personalities. Each personality had its associated cluster of distinct characteristics. It immediately occurred to us that a person need not manifest all and only the characteristics of one of the sixteen specified groupings. Later in the book, in response to such concerns, the authors assert that a person could have a particular personality type, even when they manifest none of the characteristics of that personality type. The astrological constellations upon which astrology demarcates its signs no longer occupy the positions at the times supposed by astrological theory. This astronomical fact often surprises astrology buffs. Nevertheless, a common response to this fact is that, "it doesn't matter."

Pseudoscientists invent vocabularies in which many terms lack definitions or have imprecise or ambiguous definitions. For instance, Healing Touch International describes their practice as using "... gentle, non-invasive touch to influence and support the human energy system within and surrounding the body." No further clarification of the term "human energy system" is offered on their web site.[40] According to the American Chiropractic College, "a subluxation is a complex of functional and/or structural and or pathological articular changes that compromise neural integrity and may influence organ system and general health." This particular definition does not indicate that chiropractors attribute subluxations exclusively to the spine. The *Basic Chiropractic Procedural Manual* published by The American Chiropractic Association lists the 18 different types of subluxations generated at the 1972 chiropractic consensus conference.[41]

Pseudoscientific explanations tend towards narrative rather than appeal to underlying general laws or principles. Pseudoscientists often explain something by telling a story or by imagining a reasonable scenario. For instance, here are several accounts of the functioning of ginkgo biloba:

> The ginkgo is the oldest living tree species, geological records indicate this plant has been growing on earth for 150–200 million years. Chinese monks are credited with keeping the tree in existence, as a sacred herb. It was first brought to Europe in the 1700's and it is now a commonly prescribed drug in France and Germany.... Ginkgo works by increasing blood flow to the brain and throughout the body's network of blood vessels that supply blood and oxygen to the organ systems.
> Herbal Information Center 2003

> The Ginkgo Biloba tree is the oldest known tree species. Today, worldwide studies reveal Ginkgo's role in promoting health by increasing blood circulation in the brain.
> Discount Vitamins and Herbs 2003

> Ginkgo Biloba has been used in traditional Chinese medicine for more than 4,000 years.... Ginkgo Biloba enhanced circulation in the brain includes improved short and long term memory, increased reaction time and improved mental clarity.
> VitaDigest 2003

None of the above accounts specifically misleads or deceives. However, none of the above accounts indicate that the bioflavonoids in ginkgo increase blood flow by acting as anticoagulants (inhibit blood platelet aggregation and clot formation) and vasodilators (stimulates blood vessel dilation). Such underlying principles are quite relevant to the decision to use ginkgo, since they are also some of the properties of common (much less expensive) aspirin. Ginkgo is likewise touted as an antioxidant, which by absorbing oxygen molecules, may dampen the creation of free radicals, highly reactive oxygen molecules that may injure neurons and cause age-related changes in the brain. However, antioxidants are quite common and include vitamin C.

The tendency towards explanation by narrative stems from a stable aspect of human reasoning. People will often increase their subjective estimate of the likelihood of an event after being told a story about how the event could have occurred. Researchers often refer to the phenomena as "The Othello Effect."[42]

The above accounts of ginkgo biloba also illustrate a final tendency in pseudoscience towards unwarranted deference to antediluvian ideas. The fact that ginkgo has been used for over 4,000 years does not in itself establish it efficacy or it superiority to other substances. "Ancient wisdom," though not necessarily to be dismissed out of hand, is often the beliefs of people who knew less about the body, the world, and the universe than my seven year old son.

## 6. Conclusion

Western scientific theories and methods represent a critical and conscious attempt spanning centuries to formulate and test our understanding of the world. Science and scientific method takes as its goal the creation theories supported by the strongest, most objective evidence possible, and the elimination of misleading reasoning practices and innate dispositions. Western science has achieved these goals to a high degree. However, science does not exist in a rarified intellectual vacuum. Science remains a human activity and scientists remain "merely human." The sorts of theories and reasoning patterns characteristic of pseudoscience, including fraud, can be found in science. No one should accept seeming reports of scientific findings uncritically. For instance, Sparks and Pellechia (1997) found people assign significantly greater credence to reports of UFOs if the reports contain reference to a scientific authority.[43] One should hold scientific findings as well as all claims representing themselves as scientific, empirical, and/or objective to the same critical standards.

## 7. References

1. R. Giere, *Understanding Scientific Reasoning*. 4th ed. 1996, New York, NY: Holt Rinehart & Winston.
2. J. Arbuthnott, An Argument for Divine Providence, Taken From the Constant Regularity Observ'd in the Births of Both Sexes. 1710.
3. I. Hacking, *Logic of Statistical Inference*. 1965, Cambridge: Cambridge University Press.
4. J. Mill, *A System of Logic: Ratiocinative and Inductive*. 1843.
5. J. Gavarret, Principes Généraux de Statistique Médicale. 1840.
6. J. Venn, Comment. *Journal of the American Statistical Association*, 1888. **82**: p. 130–131.
7. L. Hogben, *Statistical Theory: The Relationship of Probability, Credibility, and Error; An Examination of the Contemporary Crisis in Statistical Theory From a Behaviourist Viewpoint*. 1968, New York, NY: W. W. Norton & Co., Inc.
8. E. Beaven, Discussion on Dr. Neyman's Paper. *Journal of the Royal Statistical Society*, 1935. **Supplement 2**: p. 159–161.
9. R. Fisher, *Statistical Methods for Research Workers*. 1925, London: Oliver and Boyd.
10. J. Neyman and E. Pearson, On the Problem of the Most Efficient Tests of Statistical Hypotheses. *Philosophical Transactions of the Royal Society A*, 1933a. **231**: p. 289–337.
11. J. Neyman and E. Pearson, The Testing of Statistical Hypotheses in Relation to Probabilities A Priori. *Proceedings of Cambridge Philosophical Society*, 1933b. **20**: p. 492–510.
12. J. Neyman and E. Pearson, The Use of the Concept of Power in Agricultural Experimentation. *Journal of the Indian Society of Agricultural Statistics*, 1933c. **9**: p. 9–17.
13. J. Poincaré, *La Science et L'hypothèse (Science and Hypothesis)*. 1905, Paris: Flammarion.
14. J. Poincaré, *La valeur de la Science (The Value of Science)*. 1907, Paris: Flammarion.
15. J. Poincaré, *Science and Méthode (Science and Method)*. 1914, Paris: Flammarion.
16. J. Poincaré, *Dernières Pensées (Mathematics and Science: Last Essays)*. 1963, Paris: Flammarion.
17. K. Popper, *Logik der Forschung (The Logic of Scientific Discovery)*. 1935, Vienna: Julius Springer Verlag.
18. S. Gould, *The Mismeasure of Man*. 1981, New York, NY: W.W. Norton and Company.
19. J. Watson, Psychology as the Behaviorist Views it. *Psychological Review*, 1913. **20**: p. 158–177.
20. A. Shapiro and E. Shapiro, *The Powerful Placebo: From Ancient Priest to Modern Physician*. 1997, Baltimore: Johns Hopkins University Press.
21. W. Jevons, *The Principles of Science: A Treatise on Logic and the Scientific Method*. 1874, New Yrok, NY: Macmillian.
22. H. Beecher, The Powerful Placebo. *Journal of the American Medical Association*, 1955. **159**: p. 1602–1606.
23. J. Dodes, The Mysterious Placebo. *Skeptical Inquirer*, 1997. **21**: p. 44–45.
24. G. Kienle and H. Kiene, The Powerful Placebo Effect: Fact or Fiction? *Journal of Clinical Epidemiology*, 1997. **50**: p. 1311–1318.
25. National_Sleep_Foundation, *2002 Sleep in America Poll*. 2002, National Sleep Foundation: Washington, D.C.
26. C. Hart, The Mysterious Placebo Effect. *Modern Drug Discovery*, 1999. **2**: p. 30–40.
27. D. Price and H. Fields, *The Contribution of Desire and Expectation to Placebo Analgesia: Implications for New Research Strategies*, in *The Placebo Effect: An Interdisciplinary Exploration*, A. Harrington, Editor. 1997, Harvard University Press: Cambridge. p. 117–137.

28. I. Kirsch, *Specifying Nonspecifics: Psychological Mechanisms of Placebo Effects*, in *The Placebo Effect: An Interdisciplinary Exploration*, A. Harrington, Editor. 1997, Harvard University Press: Cambrdige, MA. p. 166–186.
29. D. Cherkin, et al., A Review of the Evidence for the Effectiveness, Safety, and Cost of Acupuncture, Massage Therapy, and Spinal Manipulation for Back Pain. *Annals of Internal Medicine*, 2003. **138**: p. 898–906.
30. A. Leuchter, et al., Changes in Brain Function of Depressed Subjects During Treatment With Placebo. *American Journal of Psychiatry*, 2002. **159**: p. 122–129.
31. L. Ross, M. Lepper, and M. Hubbard, Perseverance in Self-Perception and Social Perception: Biased Attributional Processes in the Debriefing Paradigm. *Journal of Personality and Social Psychology*, 1975. **32**: p. 880–892.
32. I. Lakatos, *The Methodology of Scientific Research Programmes: Volume 1 : Philosophical Papers*, ed. J. Worrall and G. Currie. Vol. 1. 1980, Cambridge: Cambridge University Press.
33. P. Gold, L. Cahill, and G. Wenk, Ginkgo Biloba: A Cognitive Enhancer? *Psychological Science in the Public Interest*, 2002. **3**(1): p. 2–11.
34. D. Buzzelli, The Definition of Misconduct in Science: A View from NSF. *Science*, 1993. **259**: p. 584–585, 647–648.
35. E. Staff, Conduct Unbecoming. *New Scientist*, 2002. **176**: p. 3.
36. E. Staff, Learning From Scientific Misconduct. *Nature Medicine*, 1997. **3**: p. 1175.
37. D. Radner and M. Radner, *Science and Unreason*. 1982, Belmont, CA: Wadsworth Publishing Co.
38. T. Gilovich, *How We Know What Isn't So*. 1993, New York, NY: Free Press.
39. C. Jung, *The Collected Works of C. G. Jung*, ed. G. Adler and M. Fordham. 2000, Princeton, NJ: Princeton University Press.
40. Healing_Touch_International. 2003, Healing Touch International.
41. R. Schafer, ed. *Basic Chiropractic Procedural Manual*. Fourth ed. 1984, American Chiropractic Association: Arlington, VA.
42. M. Piattelli-Palmarini, *Inevitable Illusions : How Mistakes of Reason Rule Our Minds*. 1996, New York: John Wiley & Sons.
43. G. Sparks and M. Pellechia, The Effect of News Stories About UFOs onRreaders' UFO Beliefs: The Role of Confirming or Disconfirming Testimony From a Scientist. *Communication Reports*, 1997. **10**(2): p. 165–172.

# 27

# Glycome: A Medical Paradigm

## Arnold Loel

## 1. Origins

The folk cures of early society have become the Complementary and Alternative Medicine of today. Basic science has unlocked many secrets of early herbal compounds and therapies. Computer science has aided in the opening of a new frontier called Glycobiology, the study of saccharides (sugars). This body of knowledge, known as the Glycome, may be the last frontier in medicine's quest for health.

Healthy organisms show resistance to disease and infection. Research constantly seeking to explain how cells interact in defending and protecting us against disease. Improving the body's cell-to-cell communication is the focus of this paper. Studying the past sheds light on our path to the future. Tracing the development of life on earth, observing how cells functions, and researching the nutrients required for cell-to-cell communication may be the key to achieving the goal of good health.

Scientists estimate that our solar system was created 4.5 billion years ago. Approximately 1.1 billion years into Earth's history elements came together to form reproducible organic structures. Primitive single cell organisms such as algae, bacteria, and fungi have a common thread, an outer membrane composed of saccharides (sugar)-protein molecules. These life forms evolved into multi-cell creatures. The progeny of these multi cell species adapted to grouping and specializing, eventually developing into complex organisms. Over time these life forms became the organisms that inherit the Earth today. Earthworms are some of the earth's early life forms. They still nourish themselves on available organisms and remain healthy throughout their life. Earthworms are ubiquitous and their diets have not changed much in 350 million years. Most of the world's fertile soil was created by the activities of earthworms.

## 2. Quantitation and Codes

Numbers, the universal language of man, have been associated with technologic advancement. The past 4000 years has yielded significant advances in the use of digits.

---

**Dr. Arnold Loel** • Founder—The Institute for Glyconutritional Study, Practicing Dentist, 2080 Century Park East, Los Angeles, CA 90067. E mail: glycodoc@adelphia.net

*Complementary and Alternative Approaches to Biomedicine*, edited by Edwin L. Cooper and Nobuo Yamaguchi. Kluwer Academic/Plenum Publishers, 2004.

Newtonian physics brought order to the known sciences and fostered the beginning of a new society evidenced by developing order in what outwardly appeared chaotic random events. Devices such as the telescope, compass, clock, and lineal measurements here on Earth and in the solar system became possible. In the 1700's, Captain Cook and other sea captains recognized that disease and death was associated with nutrition. In the 20th century the Quantum Theory began to replace Newtonian thinking, slowly our concepts of matter and energy are changing. Computers allow for massive information storage and retrieval. Recently the sequence arrangement of the four-nucleotide parts in the DNA of the human genome was defined aided by computer science advancements. The resulting paradigm shift can be seen in many areas of science. Genes (DNA) carry information for replicating cells. Bruce Lipton, PhD.[8] believes genes do not act as switches to turn cell functions on and off. Proteins in conjunction with saccharides (sugars) appear to perform these functions. In the past the order of cell function was thought of as similar to the order of the Newtonian Universe. We lack the ability to visualize the multitude of changes that take place within various structures of cells. With the advent of the Quantum theory and its application to cell biology scientists, like Lipton, are changing their focus away from the nucleolus of the cell, where the DNA resides, to what may be the master brain of an individual cell, the enclosing glycoprotein membrane. Cells respond to signaling codes, thought to be saccharides. Codes of adjacent structures modify the outside surface membrane of a cell which in turn affects the inside surface membrane. A cell's resistance or openness to invaders is dependent upon the signaling qualities of the protein/saccharide molecule. Due to the minute size and complex structure of saccharides, understanding their exact shapes and relationships requires advanced technology in lasers, spectrometry, and computers. Some of the step-by-step transformations that occur within the glycoprotein molecule only last nano seconds and there are hundreds of transformations.

## 3. Glycobiology

Glycobiology research slowly reveals the amazing partnership of protein and saccharide that was billions of years in the making. Apparently glyconutrients (foods that contain saccharides) provide essential saccharides (sugars) that combine with proteins to make up the cell's membrane. The symbiotic relationship of protein and saccharide started with the development of the first single cell. The protein-saccharide combinations also form the basis for cell-to-cell communications such as: see me, defend me, feed me, repair me, cleanse me, and replace me. There are many more signals that cell membranes respond to and generate. Medicine's awareness of the Glycome, one of the newest bodies of knowledge, and Glycobiology are only now becoming known due to the complexity of the functions of the Glycome and the study of Glycobiology. The progressive development of more powerful computers increased our ability to unveil the very complex aspects of the glyco protein relationship.

## 4. Proteins: The Building Blocks

Long chains of amino acids make up protein molecules. These chains are so long that their survival depends on the folding of these long chains into compact global shaped units.

Without the folding, life may not be possible. The saccharides cause folding of the amino acids within the protein molecule. The process of combining proteins and saccharides is accomplished by several structures within a healthy cell. Saccharides (sugars) from circulating plasma filter across the cell membrane into the cell. They are transformed into glyco building blocks in secretory compartments. In a process known as glycosyltransferases they become oligosaccharides and are bound to protein or lipid scaffolding in the endoplasmic reticulum and Golgi apparatus (a worm like structure observable in normal cells). The new combination glycoprotein moves into position within the cell membrane. We do not know exactly how this happens, what saccharide combinations are needed, or, what directs the assembly process. However, if specific sugars are missing, the process appears unaffected and the resulting glyco protein molecule will not be able to perform expected healthy functions.

To illustrate this concept, think of a child drawing the representation of a person. Children universally draw stick figures. A protein molecule drawn in the same vein would be viewed as various lumps, amino acids, linked together like a long rope with knots. Imagine the rope with the same amino acid at each end. The ends being of the same energy level would repel each other. If the polarity of one end could be changed with the addition of a saccharide carrying the opposite charge, those same amino acids could be attracted towards each other and movement of the rope would occur. Movement of the protein can be created by a change in energy field of the protein/saccharide combination. Artists have portrayed saccharides as towers of sticks extending from the main body of the protein. Artist can not duplicate the thousands of molecules that make up the surface of a single cell. The relationship of the saccharide to the protein is so very intimate and minute that a single powerful computer of today is not able to process the exact structural configuration in less than 1000 years running 24 hours a day. It has been suggested that by linking 30,000 computers the numbers could be crunched in a year or two.

## 5. Cell Membranes and Cell Communications

Outside surface protein/saccharide combinations connect via lipid/protein to an inside surface protein-saccharide. Some biologists, including Dr. John Axford,[10] have suggested that the signaling substance is saccharide based. Signals from other cells or nearby substances such as pathogens cause a change in the surface saccharide which in turn passes information into the interior saccharide where a structural change results in attached or adjacent proteins to perform some built-in function such as emitting cytotoxin or opening spaces in the membrane for invasion of virus, bacteria, drugs, or nutrients.

## 6. The Role of Nutrituion

Linus Pauling[2] postulated that as life progresses up the evolutionary tree the number and complexity of internal processes occurring within an organism increases through a process Darwin called, natural selection. Coupling evolution with our progress in cyber science, Pauling made some interesting assumptions: Our environment must provide a continuum of nutrients for healthy life to survive. Captain Cook also examined this point. A sea captain, 1728–1779[1], his experiences at sea lead him to examine a theory that a lack of nutrients could cause disease. He would begin a voyage with 1500 able seamen knowing that 800 would die during their 9 months at sea. Their daily diet consisted of salted pork, one quart of oats and a loaf of bread. Cook was derided by the medical community of his day for asserting the lack of fresh fruit and vegetables would lead to disease and death. Of interest, the Royal Naval Academy ordered limes to be carried on British ships in 1836, 57 years after Cook's death. Ships' logs from many captains of the 1700's tell tales of landing sick and dying seamen on uninhabited lands where adequate fresh fruit and wild pigs were available. A week or two of fresh nutrients revived life into morbid seamen. The results of eating fresh food were that 95 percent of sick seamen recovered and only 5 percent of the crew died.

## 7. Vitamin C: An Early Model

We know their disease as scurvy, a Vitamin C deficiency[2]. Today, in our advanced society, people are dying of similar causes while the medical community looks for drugs to reverse the symptoms. A growing number of health care practitioners understand that our foods are deficient of many essential nutrients. For this reason food supplements have become a way of life for many people. It is estimated that more than 50 percent of the health care dollars go to Alternative and Complementary therapies including food supplementation. Another assumption postulated by Linus Pauling: "An organism with the fewest processes to perform will survive over a rival if the individual can procure the process from outside its body, i.e. from the environment. To prove this postulate, Pauling placed a population of fruit flies that produced its own Vitamin C with a population of fruit flies that needed to obtain Vitamin C from the environment. The Vitamin C producers died out as long as Vitamin C was available from their food[3]. Humans do not produce their own Vitamin C. They must obtain it daily from their food sources. In Pauling's view Vitamin C must have been plentiful in the food sources of man some two to five million years ago when our ancestors began to stand up on two legs. Knowing that sheep produced their own vitamin C, Pauling measured the quantity of vitamin C a 175-pound sheep would produce in a day and applied that amount to human needs. In his estimation the average human needs 2000 mg per day. Vitamin C is water-soluble and cannot be stored. We use 100 to 200 milligrams per hour. It would be best to take small amounts of a vitamin C supplement throughout the day as though it were food.

## 8. Modern Day Diseases and Stress

In modern society, as in the past, disease is of concern but accepted as a condition of life. Our population lives longer with medical advances and we are tracking disease more carefully. A longer life span does not mean a more productive and comfortable life. Heart Disease, still the number one killer in the USA[4]. Diabetes, referred to as the "accelerated aging disease" is up 700 percent since 1960[5] and obesity affects one in five adults and one in ten children and rising. Cancer will affect one in three women and one in two men. Multiple Sclerosis, Autism, and AIDS are not under control. Fortunately the paradigm of the past is changing due to our ever expanding knowledge and the inventiveness of researchers. We are able to view disease in a new light. It is now understood that symptoms are the language of the body expressing imbalance. Biological stressors within our cells precipitate symptoms. Biological Stressors are the real issue, not the disease that conventional medicine often confuses with a symptom or array of symptoms that are described in Greek or Latin to be treated with a miraculous drug. Biologic Stressors take many forms: Nutrient insufficiency, pathogenic microbes of bacteria, virus, yeast, and parasites, environmental toxins, metabolic imbalance, physical and emotional stress, pharmaceutical drugs, recreational drugs, structural misalignments and the interaction of these factors with genetics.

## 9. Glycobiology and a Paradigm Shift

This paradigm shift has to do with our understanding of cell-to-cell communications. Glycobiology, the new frontier in medicine, has been enhanced by our expanding knowledge and information storage capability[6]. The Australian Patent office, in October, 2001, approved a patent for food supplements based on the premise that freeze dried food products containing essential glyconutrients taken daily have an influence on the reduction and cure of disease, including: color blindness, diabetes, Downs Syndrome, fibromyalgia, chronic fatigue syndrome, arthritis, heart disease, ADD, ADHD, asthma, and many more mal-functions of the human body. The food supplements mentioned in the patent included the following nutrients: Dietary Carbohydrates—8 essential saccharides, Antioxidants—mature freeze-dried fruits and vegetables, Phytosterol—Dioscorea from the Mexican Yam, and Glycinated vitamins and minerals. After reading the patent claims one can conclude that whatever nutrients are in these products allows cells to function more normally. The premise being that health is the normal condition and disease involves pathologic function of cells.

Do you know the most important nutrient for your body? The answer according to Sam Caster, a food supplement distributor, is the one you are missing, of course[9] . My assumptions, like Pauling, are that nutrients, especially glyconutrients were plentiful during our development many millions of years ago. We don't produce them within our bodies and we rely on dietary sources to replenish these nutrients daily. Using the Linus Pauling postulate, "the fewer biologic functions performed by an organism the more likely its genetic stream will survive over a contemporary that performs those functions within its own cells" I infer that individuals with disease are not getting the nutrients required to maintain healthy cells. Just as the sailors on Cook's ships. When people ingest needed nutrients their cells return to optimal function and disease symptoms disappear.

The journal *Science*, March 23, 2001 devoted the entire issue to Carbohydrates and Glycobiology. The pharmaceutical manufactures have taken notice of the potential of glyconutrients to carry drugs to specific targets, such as cancer cells; open pathways for drugs to penetrate cells infected by virus DNA, and direct white cells to foreign substances and trauma. The exact method of communication is not completely understood. Many theories have been contemplated but none have been completely documented. Under investigation are the data codes and sequence of changes that take place during the signaling process. Inroads are being made. So many factors are involved it will take years for investigators to understand the process completely. Current research confirms that glyconutrients provide the essential saccharides that combine with proteins to make up the cell's outer membrane. It is thought that deficiencies in the glyco portion of the glycoprotein molecule interrupts to normal flow of signaling information and results in abnormal cell function. For example, a protein folding defect in the astrocyte cell, found in the vertebral neural synapses has been identified as a factor in Lou Gehrig's disease. The affect, muscles in the chest wall malfunction resulting in distressed breathing . Similar disturbances in the brain may be the cause of Alzheimer.

## 10. Cell-Cell Communication

Our survival depends on cell-to-cell communications. Using mathematical probabilities determined that carbohydrates, not proteins, provide the signaling symbols for communication between the 5 trillion cells that make up our bodies. The mathematical possibilities point to saccharides as the code signals and receptors that cells use to communicate their needs. The four-nucleotide arrangement in the DNA alphabet can combine to produce 256 different four-unit structures. The 20 amino acids in protein can yield 16,000 four-unit configurations. Simple sugars in the body can assemble into more than 15,000,000 four-unit component arrangements[7]. Saccharides (sugars) stand out because they alone are capable of offering the massive divergence of combinations that are required by the over 200 different kinds of cells that make up our bodies.

Research scientist, Dr. John Axford has ventured to predict that the study of Glycobiology is the next and maybe the last frontier in medicine[10]. Saccharides working in conjunction with protein are the alphabet in the language cells developed when they banded together for the benefit of longevity and comfort. Currently there are 8 essential saccharides that are known to be needed to preserve health: D-Galactose (Gal), D-Glucose (Glc),—acetyl-D-Galactoseamine (GalNAc), D-Mannose (Man), L-Fucose (Fuc), N-acetly-D-glucosamine (GlcNAc), D-Xylose (Xyl), and N-acetylneuraminic acid (NeuAc).More may be identified as research progresses.

## 11. Predictions

Tomorrow's medicine will be the result of applying basic science to the therapies of Complementary and Alternative Medicine. As our view of cell function expands so does our understanding of Complementary and Alternative Medicine. Computers are speeding up medical advances. In time, prevention through the use of Glyco-nutritional supplements will substantially reduce the need for medical cures. New drugs will enhance the paradigm of cell-to-cell communication through glycosylation. The Alternative Medicine of today will evolve into a new order of understanding which will complement both Complementary Medicine and Traditional Medicine by providing glyconutrient substrates necessary to build healthy cell communications in the world of tomorrow.

## 12. Acknowledgements

Special thank you to my editors, Liz Stevens, Joyce Grunauer, and Vicky Loel-grant writer. And to Nancy Silverman, Ready Radio.com for formatting a camera ready copy.

## 13. References

1. History of the Earth. http://www.bbc.co.uk/history/discovery/explore
2. Linus Pauling, Research Notebooks volume V Special Collection, Subject Index. Linus Pauling Institute, Oregon State University, Lecture series Los Angeles, California 1978, A Dr. Manuel Charaskin project, University of Alabama.
3. http://www.lbl.gov/Science-Articles/Archive/pauling
4. American Heart Association Dec. 31, 1998 http://www.oralchelation.com/heart/aha5.htm5. http://www.mercola.com/2000/aug/27/diabetes_increasing.htm
6. Sugars that Heal, Emil I. Mondoa, M.D. and Mindy Kitel, Ballantine Books.ISBN 0-345-44106-0
7. American Scientific, June 2002 pa 40-45 Sweet Medicine
8. Bruce Lipton, The Biology of Belief, Video 2000 Jenny Myers Productions. www.brucelipton.com
9. Sam Caster, Mannatech lecture, September 2002, Pasadena, California. And *www.wallstreetreproter.com* August 22, 2003 interview
10. John Axford, B Sc, M.D. FRCP St. George's Hospital University of London, London England. Lecture Dallas, TX., April 2002

# Abstracts

VII

**1. Michael Irwin:** *Exploring the scientific basis of complementary and alternative medicine: Plenary.* Norman Cousins Professor, Neuropsychiatric Institute, University of California, Los Angeles, CA, USA. Member, Advisory Council, National Center for Complementary and Alternative Medicine

Complementary and alternative medicine (CAM) practices are most generally considered to be those that are not yet an integral part of conventional medicine. Five general domains of CAM research will be reviewed with recognition of overlap across these categories: biologically based systems, manipulative or body-based systems, mind-body medicine, alternative medical systems, and energy therapies. Increasingly people have turned to complementary and alternative (CAM) approaches when mainstream medicine does not meet all their expectations and needs. As CAM practices are found to be safe and effective, patients will use them to complement conventional care and healthcare practices will fully adopt these approaches.
**Keywords**: complementary medicine; alternative medicine

**2. Gerd G. Uhlenbruck[1] and Edwin L. Cooper:[2]** *Compliments to complementary medicine: Plenary.* [1]Institute Immunobiology, University of Cologne, GERMANY, [2]Department of Neurobiology, University of California, Los Angeles, CA, USA

Complementary Medicine (CM) represents a supporting strategy for treatment of patients (in and outside the hospital) in collaboration with University Medical Schools and is not an alternative treatment to the one in a recommended University Clinic. Alternative aspects are only discussed within the field of CM, for instance various combinations of Echinacea or of vitamin mixtures and micronutrients. All methods and treatments suggested by CM must have been scientifically evaluated in highly qualified Institutes or Clinics and must have been published in recommended high-level Journals, for instance those listed in Current Contents. For those who are engaged in CM, quality control examinations and seminars in orthomolecular medicine are suggested as well as the yearly participation in Congresses or Meetings dealing with CM. In any case CM should be of help for patients or those who fear to become patients by leading them to a healthy lifestyle. A special license for using CM within the framework of therapeutic efforts seems important. Psychoneuroimmunological strategies must be ensured in order to cope with the consequences of a severe disease.
**Keywords**: complementary medicine; alternative medicine

**3. Michael Irwin, J. L. Pike and M. N. Oxman:** *Effects of Tai Chi Chih on varicella-zoster virus specific immunity and health functioning in older adults.* Cousins Center for Psychoneuroimmunonology, UCLA Neuropsychiatric Institute, Los Angeles, CA, USA

We determined if Tai Chi Chih (TCC) affects varicella zoster virus (VZV) specific immunity and health functioning in older adults who, on average, show impairments of health and are at risk for shingles. In a randomized controlled trial, older adults (n = 36) were assigned to 15 weeks of TCC instruction or wait list control. VZV-specific immunity increased 50% from baseline to post-intervention in the TCC group ($P < 0.05$) but unchanged in the wait list control group. In older adults who had low health functioning, TCC led to improvements in Medical Outcome SF-36 scale scores for role-physical, general health, vitality, role-emotion, and mental health (all $P$'s $< 0.01$). Health functioning was unchanged in the wait list control group. Administration of TCC for 15 weeks led to an increase in VZV-specific CMI comparable in magnitude to that observed in adult

recipients of investigational varicella-zoster vaccine. Gains in health functioning occurred in participants who received TCC, and were most marked in older adults with the greatest impairments of health status at entry.
**Keywords:** psychoneuroimmunology; Tai Chi Chih; immunity; aging; health; functioning; mind-body; shingles; varicella-zoster virus

**4. Bruce S. Rabin:** *Safety and effectiveness of health strategies to eliminate behaviors detrimental to health.* University of Pittsburgh Medical Center, Pittsburgh, PA, USA

Conventional medicine, in providing the highest standard of health care in the history of mankind, has contributed immeasurably to the well being of most people in the United States and to many elsewhere in the world. However, many, despite well-documented risks, continue to lead unhealthy lifestyles that contribute to avoidable diseases that result in a decreased quality and shortened life span. As part of its commitment to help people achieve a high quality of health and life, the UPMC Health System (UPMCHS) has established the UPMC Health Enhancement Program (HEP). The HEP is based on a growing body of medical research demonstrating the safety and effectiveness of health strategies that eliminate behaviors detrimental to health and enhance behaviors that aid in coping with psychologic stressors prevalent in our environment. Capitalizing upon the public's search for and utilization of additional approaches to their healthcare, HEP offers evidence based options both for healthy individuals to maintain their health, and also for those individuals with disease to improve their health. To enhance accessibility by the public, HEP programs are offered throughout our region, placing programs close to where people reside. Further, HEP is promoting the development of an "Enhanced Healing Environment" for hospitalized patients. The HEP has been accepted across the entire demographic spectrum.
**Keywords**: preventive medicine; behavioral medicine; stress coping

**5. Zhang Shulan, Lin Bei, Cai Wei, et al.:** *Expression of matrix metalloproteinase in cervical squamous cell carcinoma tissue and its significance.* Department of Obstetrics and Gynecology, the Second Clinical Hospital, China Medical University, Shenyang, China

To investigate the relationship between the expression of matrix metalloproteinases (MMP-2, MMP-9) and tumorigenesis, development and metastasis of cervical squamous cell carcinoma. 36 cases with cervical squamous cell carcinoma were enrolled, chronic cervical inflammation, cervical intraepithelial neoplasia and normal cervices comprised the control group. The expression and distribution of MMP-2, MMP-9 protein were detected by immunohistochemical SP method. Their active proteins and mRNA were analyzed by gelatin zymography and RT-PCR respectively. Immunohistochemical staining showed that MMP-2 and MMP-9 proteins mostly localized in the plasma, and secondarily in the membrane. The positive expression rates of MMP-2 and MMP-9 was 77.78 and 66.67% respectively, in the carcinoma tissues. The presence of MMP-2 and MMP-9 proteins showed no relationship between pathologic grade, clinical stage and the size of the carcinoma. In CIN, the positive rate for these two proteins was 10% and 20% respectively. However, they were absent in normal tissues. MMP-2 and MMP-9 correlate well with the occurrence, invasion and metastasis of cervical squamous cell carcinoma. The two indicators can be used to identify the metastasis trend of cervical carcinoma in clinic.
**Keywords:** cervix neoplasms: carcinoma, squamous cell; metalloproteinase; controlled clinical trials; neoplastic metastasis

**6. Zhang Shulan, Zhao Chang, Qing Lin Pei, Li Yan and Gao Hong:** *Expression of surviving gene and its relation with the expression of Bcl-2 and Bax protein in epithelial ovarian cancer.* Department of Obstetrics and Gynecology, The Second Affiliated Hospital, China Medical University, Shenyang, China

To study the expression of apoptosis related gene surviving and its relation with expression of Bcl-2, Bax protein in epithelial ovarian cancer. Expression of surviving mRNA was evaluated by reverse transcriptase polymerase chain reaction (RT-PCR) in 35 cases of epithelial ovarian cancer, 10 cases of borderline cancer, 10 cases of benign tumors and 10 cases of normal tissue. Expression of Bcl-2 and Bax was detected by immunohistochemistry streptomecin-avidin-biotin-peroxidase complex (SABC) method, and correlation between them was analyzed. Expression of surviving gene was detected in a significantly greater proportion in epithelial ovarian cancer and borderline cancer than benign tumors and normal tissue. There was no relationship between surviving gene expression and FIGO stage, histologic grade, pathological type and lymphatic metastasis. Expression of Bcl-2 and Bax protein was positively and negatively correlated with expression of surviving gene respectively. The surviving genes may play an important role in pathogenesis of ovarian cancer, apoptosis related gene Bcl-2 may have a synergic role and Bax had an antagonistic role with surviving in formation and progression of ovarian cancer.

**Keywords:** surviving gene; apoptosis; ovarian cancer

# Index

# Index

## A
Abstracts, 455–457
Acetoxy-valerenic acid, 51
N-Acetylcysteine, as multiple sclerosis treatment, 95–96
N-Acetyl-D-glucosamine, 450
N-Acetyl-L-tyrosine ethyl ester, 370
N-Acetylneuraminic acid, 450
Actinia, 342
Acupressure
    as childbirth-related pain treatment, 201
    prevalence of use in Japan, 16
Acupuncture
    as atopic dermatitis treatment
        effect on eosinophils, 231
        effect on granulocytes, 229, 234
        effect on leukocytes, 230–231, 234
        effect on lipid levels, 231, 232
        during steroid therapy withdrawal, 231, 233, 234, 235, 236
    health insurance coverage for, 10, 40
    in HIV/AIDS patients, 107, 108, 109
    in Japan, 10, 15, 16
    meridians in, 200
    as pain treatment, during childbirth, 200–201
    as pain treatment, efficacy evaluation of, 217–227
        adequacy of treatment protocol in, 219–220
        BRITS method of, 219
        credibility of controls in, 224–225
        with decommissioned electrical unit, 224
        with double-blind trials, 218–219
        with dummy needling, 223–224
        experimental controls in, 221–222
        guidelines for, 225
        placebo controls in, 221, 222–225
        with pseudo-acupuncture, 223
        randomization in, 220
        sample size in, 221
        with sham acupuncture, 222–223
        with superficial needling, 223
    traditional Chinese medicine basis of, 217, 220
Acute stress disorder, 182
Adaptive immunity, effect of music on, 273–274
Adhesion molecules, 406
Adrenal glands, 405–406
Adrenergic system, in immunosuppression, 406
Adrenocorticotropin hormone (ACTH)
    effect on glucocorticoids, 406
    immunomodulatory function of, 266
    in stress response, 415
*Aedes aegypti*, 394
*Aegle marmelos*, 319–320
Aeroplysinin, 350–351
*Aesculus hippocastanum* (horse chestnut), 49, 50, 54
*Agaricus blazei* Murrill, 16, 19

Agglutinins, in earthworms, 375
Aggregation factors, in sponges, 333
Agoraphobia, 182
AIDS (acquired immunodeficiency syndrome). *See* HIV/AIDS
*Alcyonium aurantium*, 327
*Alcyonium dioscoridis*, 327, 331
*Alcyonium domuncula*, 330
*Alcyonium durum*, 331, 334
*Alcyonium lyncurium*, 327
*Alismatis rhizoma*, 307
*Allium sativum* (garlic), 319
Allorecognition, in sponges (Porifera), 335, 336
Aloe, 58
Alzheimer's disease, 31–33, 382, 407, 411, 450
*Amani semen. See* Ekki-youketsu-fusei-zai (EYFZ)
American Chiropractic College, 440
American Medical Association, 253
American Society of Oncology, 13
Amino acids
    as protein components, 446–447, 450
    *Spirulina* content of, 312
Analysis of variance, 430
"Ancient wisdom," 441
Anemia, malnutrition-related, 311
*Angelicae. See also* Ekki-youketsu-fusei-zai (EYFZ)
    use in Kampo medicine, 306
Angiogenesis, genistein-related inhibition of, 149–150
Animals, dietary supplements for, 10
Annelids. *See also* Earthworms; Leeches
    from deep-sea bottom, 391–392
*Anopheles*, *Azadirachta indica* (neem)-mediated killing of, 112
Anoxia, superantigen induction by, 87
Antibiotics
    microbial protein synthesis-inhibiting activity of, 392
    novel alternative, 391–392
Antibodies, pathogenic, detection of, 90–91
Anticancer lytic molecules, in earthworms, 377–382
Anticarcinogenic activity
    of antioxidants, 96–97, 314
    of earthworm lytic molecules, 377–382
    of ekki-youketsu-fusei-zai (EYFZ), 173–178
    of genistein, 121–165
    of mushrooms, 19–20
Antimicrobial molecules/peptides, 392–400
    action mechanisms of, 393
    antiviral activity of, 399
    common features of, 394–395
    definition of, 392–393
    in earthworms, 363, 371–377
    as innate immunity effectors, 392, 394
    molluscan, 395–400
    structure of, 394–395

Antioxidants
  anticarcinogenic activity of, 96–97
  β-carotene, 312–313, 314
  fermented papaya preparation (FFP), 65–70
  *gingko biloba*, 441
  immune function-enhancing activity of, 97–99
  as multiple sclerosis therapy, 95–96
  papaya-derived gastroprotective effects of, 65–70
  pycnogenol, 23
  vitamin C, 314
  vitamin E, 314
Anxiety
  as stress cause, 265
  stress-related, 266
Anxiety disorders
  central nervous system in, 182–184
  classification of, 182
  treatment of, 181
Anxiolytic compounds, natural, efficacy of, 181–191
  in behavioral models, 184–185
  behavioral tests of, 186–188
  flavonoids-related, 185–189
  *in vitro* tests of, 188
Anxiolytic drugs, 181
  interaction with gamma-aminobutytic acid$_A$ receptors, 182–184
*Aphrocallistes vastus*, 337, 340, 341
Apigenin, 186, 188
*Aplysia*, 327
*Aplysina aerophoba*, 350–351
Apoptosis
  earthworm serine proteases-induced, 381–382
  genistein-induced, 125, 149, 151–152
  in lymphocytes, stress-induced, 410, 419
  in sponges (Porifera), 336
  staphylococcal enterotoxin B-induced, 81
  stress-induced, 409, 410, 419
  vitamin-mediated protection against, 98
Arabinofuranosyladenine, 328, 351
Arbuthnott, J., 429
Arginine vasopressin, in T-cell apoptosis, 409
Aricept® (donepesil), 31, 33
Aristotle, 325, 327, 337
Aromatherapy
  for childbirth-related pain management, 203
  prevalence of use in Japan, 15, 16
*Artemia vulgaris*, 200
Arteriosclerosis, heat shock protein-related, 88
Arthritic peptide antibody, 90
Artichoke, 49, 58
*Ascetta primordialis*, 337
*Ascidia*, 342
Ascorbic acid. *See* Vitamin C
Asia, herbal drugs/remedies market in, 47
Aspergillosis, antimicrobial peptide treatment of, 391

*Astragali*. *See also* Ekki-youketsu-fusei-zai (EYFZ)
  use in Kampo medicine, 306
Astrology, 437–438, 439, 440
Astronauts, 64–65
Astronomy, Ptolemaic, 438
(2-5)A synthase, 332, 334–335
Atherosclerosis, 382
  infection-related, 86–87, 88
*Atractylodis lanceae*, 306, 307
*Atropa* (belladonna), 50, 58
Attacin, 393
*Aurantii nobilis*, 306
Australia, relevance of traditional medicine use in, 58
Autism, 448
  as complex disease model, 91
  maternal infection as risk factor for, 85–86
  music therapy for, 265, 273
Autoimmune diseases
  environmental factors in, 75–76
  genetic susceptibility to, 85
  heat shock protein-related, 87–88, 89–90
  infection-related, 85–86
  neurological, 91–92
  stress-related, 4–8, 410–411
  xenobiotics-related, 85
Autonomic nervous system, effect of music therapy on, 274, 275
Axford, John, 450
Ayurvedic medicine, 13, 57, 318–319
*Azadirachta indica* (neem), 112–113. *See also* Praneem polyherbal formulation
  therapeutic applications of, 312, 318–319

B
*Bacillus megaterium*
  antimicrobial peptide-related inhibition of, 398, 399
  fetidin-related inhibition of, 372
Bacteria. *See also* specific bacteria
  antibiotic resistance in, 395
    mechanisms of, 392
  intestinal
    role in gastrointestinal function, 78–80
    role in immune function, 82–83
  superantigens of, 86, 87
Bacterial infections. *See also* specific bacterial infections
  as atherosclerotic cardiovascular disease cause, 88
  neuronal cytokines in, 411
Bacteriocytes, 334
*Basic Chiropractic Procedural Manual* (American Chiropractic Association), 440
*Bathimodiolus*, 391
B cells, β2-adrenergic receptors of, 406
Bcl-2, in sponges (Porifera), 336
Beecher, H. K., 435

# Index

Belgium, prevalence of traditional medicine use in, 58
Belief perseverance, in pseudoscience, 437–438
Belladonna *(Atropa)*, 50, 58
Bellastatin antistatin-type inhibitor, 365–366
*Ben Cao Gang Mu (Compendium of Materia Medica)*, 364
Benzalkonium chloride, 111
Benzodiazepine-binding site ligands, role in anxiolytic compounds' activity, 183– 184, 185–186, 187, 188, 189
Benzodiazepines, anxioselective, 184
Benzoyl-L-arginine ethyl ester, 370
Benzylpenicillin, *Staphylococcus aureus*' sensitivity to, 393
Betel, 314
B43-genistein immunoconjugate, 151–152, 154
Bias
   in descriptive studies, 431–432
   in sampling, 431
   scaling, 435–436
   in scientific experiments, 434
*Biconiosporella corniculata*, 351
Bilobalides, 51
Biofeedback, for childbirth-related pain management, 198
Biogenic amines, i stress, 266
Biological response modifiers, 97
Biomarkers, for disease, 76
   for cardiovascular disease, 90–91
   for celiac disease, 80–81
   for Crohn's disease, 81–82
   for multiple sclerosis, 95
   for ulcerative colitis, 81, 82
Birth rate, 429
Black gram *(Vigna mungo)*, genistein and daidzen content of, 122
Black pepper *(Piper nigrum)*, 319
Bladder stones, earthworm extracts treatment of, 359–360
Blindness, vitamin A deficiency-related, 311, 314
Blinds, experimental, 434, 435
Blood flow measurement, 20–22
Boman, Hans, 394
Bone marrow transplants, natural killer cells in, 97
Bonny Method Guided Imagery in Music, 255
Bradykinin, in stress, 410
Brain, cytokine synthesis in, 407
Brain cancer, genistein-related prevention of, 126
Brain function, role of cytokines in, 407
Breast cancer
   dietary factors in, 96
   genistein-related prevention of, 123, 124
     *in vitro*, 124, 127–133, 148, 150, 151, 152–153
     *in vivo*, 125, 154
Breast cancer patients, stress reduction in, 419

Breast implants, silicone, 97
Brown rice, pre-germinated, as atopic dermatitis treatment, 285–289
*Bufo japonicus formosus*, earthworm coelomic fluid toxicity in, 374
*Bupleuri*, 306
Butyrophilin, 92

## C

*Cacospongia*, 327
Calcitonin gene-related peptide, in stress, 410
*Callyspongia diffusa*, 333
CAM. *See* Complementary and alternative medicine (CAM)
Canada, prevalence of traditional medicine use in, 58
Cancer. *See also* Anticarcinogenic activity; specific types of cancer
   carotene-related prevention of, 314
   Chinese traditional medicine treatment of, 167–179
   complementary and alternative medicine research in, 14, 19–20
   dietary factors in, 96
   nutrition-based prevention of, 318
   stress associated with, 267
   stress-related susceptibility to, 266–267, 411–412
Cancer patients
   *Agaricus blazei* Murrill use by, 16
   music therapy for, 254–255
Candidiasis
   antimicrobial peptide treatment of, 391
   dietary inhibition of, as atopic dermatitis treatment, 282–283
   intestinal, 79
     as atopic dermatitis cause, 303
     as immunity impairment cause, 83
   Praneem polyherbal formulation treatment of, 115, 116, 117, 118
Carbohydrates
   complex, 96
   effect on intestinal *Candida* growth, 283
   role in cellular communication, 450
Cardiomyopathy, fibrinolysis associated with, 380–381
Cardiovascular disease
   atherosclerotic, infection-related, 86–87
   heat shock protein-related, 89–90
   laboratory tests for early detection of, 90–91
   as mortality cause, 448
   superantigens in, 87–90
Cardiovascular disease patients, music therapy for, 255–256
L-Carnitine, as multiple sclerosis treatment, 95–96
Carotene(s), 314
$\alpha$-Carotene, anticarcinogenic activity of, 96

β-Carotene
　anticarcinogenic activity of, 96
　antioxidant activity of, 312–313, 314
　dietary sources of, 314
　*Spirulina* content of, 312–313, 314
Carotenoids, 318, 416–417, 418
Caspases, in sponges (Porifera), 336
*Cassia* (senna), 50, 54, 58
Caster, Sam, 449
Catecholamines, 273
　in stress, 266, 410
Cathepsin, 334
Caulophyllum, 201
Causal effects, null hypothesis testing of, 434
Causal factors, 430–431
　in descriptive studies, 431–432
　effectiveness of, 434
　spurious and/or nonspecific, 436
Causal inference, 429
CD26, 91
CD69, 86
CD69 antibodies, 91
Cecropin, 393, 394
Cecropin B, 394
Celiac disease, 77, 80–81
　autoantigens to, 81
　biomarkers of, 80–81, 82
　stress-induced, 412
Cell-cell communication, role of saccharides in, 447, 449
Cell cycle, genistein-related inhibition of, 125, 148, 149, 150
Central nervous system
　in anxiety disorders, 182–184
　effect of cytokines on, 406–407
Cerebrospinal fluid analysis, for multiple sclerosis diagnosis, 94
*Cervi parvum cornu.* See Ekki-youketsu-fusei-zai (EYFZ)
*Challenger* (HMS), 391
Chamomile, 49
Charlemagne, 349
Chemoattractants, 406
Chemotherapeutic agents, interaction with genistein, 150–151
Chen-Yuan Lee, 36
Cherokee Indians, 360
Chick peas *(Cicer arietinum)*, genistein and daidzein content of, 122
Childbirth
　"natural," 197
　"painless," 197
　pain treatment during, 193–206
　　with complementary and alternative medicine, 196–204

　　with conventional treatments, 195–196
　　with music therapy, 256
China, Kampo medicine use in, 27
Chinabark, 58
China Medical College, Chinese medicine programs, 37–38, 39
Chinese traditional medicine, 57
　acupuncture in, 217, 220
　as cancer treatment, 167–179
　modernization of
　　in China, 35
　　in Taiwan, 35–42
　in Taiwan, 35–42
　　current status of, 36–37, 39–40
　　history of, 36, 37–39
　　during Japanese occupation, 36
　　modernization of, 35–42
　　National Research Institute of Chinese Medicine and, 36
　　public insurance coverage for, 40
　　use of earthworms in, 360–361, 364
Chinese zodiac, 437–438
Ching Dynasty, 36
Chiropractic, prevalence of use
　in Japan, 15
　in the United States, 16
*Chlamydia*, heat shock protein-60 in, 88
*Chlamydia pneumoniae*
　as atherosclerotic cardiovascular disease cause, 88
　peptides from, as encephalomyelitis cause, 92
*Chlamydia trachomatis*, Praneem polyherbal formulation-related inhibition of, 115, 118
*Chlamys*, 330
Chlordiazepoxide, 183–184
Chocolate, as atopic dermatitis aggravant, 282
Cholesterol-lowering drugs, multiple sclerosis patients' use of, 95
Choline acetyltransferase, 31–33
Chromatography
　high-performance liquid, 45
　thin-layer, 44–45
Chronic fatigue immune dysfunction, 97
Chronic obstructive pulmonary disease patients, music therapy for, 257
Chrysin, 185, 186, 187–188
Chuen-Chi Shen, 36
*Cicer arietinum* (chick peas), genistein and daidzen content of, 122
*Cimicifugae*, 306
*Cinnamomi cortex*, 307
Clinical trials, in herbal medicine, 62–63
*Cnidii*, 306
Coelomic cytolytic factor (CCF-1), 363, 373, 375

Coenzyme Q$_{10}$
 anticarcinogenic activity of, 96
 as multiple sclerosis treatment, 95–96
Coffee, as atopic dermatitis aggravant, 282
Coix seed, 11–12
Cold fusion, 439
Coleoptericin, 394
Colon-adenocarcinoma cells, effect of herbal anti-tumor preparation on, 174–176, 177, 178
Colon cancer, dietary factors in, 96
Comatose patients, music therapy for, 256–257
Complementary and alternative medicine (CAM)
 adverse/side effects of, 23
 classification of, 9–10
 definition of, 9–10
 international prevalence of use, 15, 16
 non-holistic approach in, 59–60
 as placebo effect, 427, 435
 popularity of, 9
 as pseudoscience, 427
 relationship with allopathic medicine, 57–58
 research in, 11
  in Taiwan, 36–37
 role of, 75–76
Complementary-determining regions (CDRs), 331, 335
Condyloma acuminatum, 11–12
Contraception, Praneem polyherbal formulation as, 114
Controls, experimental, 430, 431
 in acupuncture efficacy evaluation, 221
 in descriptive studies, 431–432
 in herbal medicine clinical trials, 62
Cook, James (Captain), 446, 447–448, 449
*Coptidis*, 306, 307
Corliss, John, 391
Cornell University Medical College, 435
Corticotropin-releasing hormone, in stress, 406, 408, 409, 415
Cortisol, music-induced modulation of, 266
Coxsackie virus, as atherosclerotic cardiovascular disease cause, 88
Creation science, 437
Crohn's disease, 77
 biomarkers of, 81–82
 differentiated from ulcerative colitis, 81–82
Crossover design, in acupuncture efficacy evaluation, 218
Cross-reactivity, 90
*Cryptotethya crypta*, 351
Ctenophora, 326–327
*Curcuma longa*, therapeutic applications of, 312, 319–320
 oral cancer treatment, 314
Curcumin, 151

Curcuminoids, 319
Cyclobutane pyrimidine dimers, 23
Cytochrome P-450 enzymes, in ulcerative colitis, 83, 84
Cytokines
 anti-inflammatory, 411
 *Azadirachta indica* (neem)-related increase in, 113
 central nervous system effects of, 406–407, 417, 419
 classification of, 406
 definition of, 406
 glucocorticoid-related modulation of, 406
 macrophage production of, effect of herbal anti-tumor preparation on, 168–171
 neuro-communication effects of, 86
 peripheral nervous system effects of, 406–407
 pro-inflammatory, 406, 408, 409, 411
 in sponges (Porifera), 335
 in stress, 410
 in stress-related neuroimmune disease, 411
 in T-cell apoptosis, 409
Cytomegalovirus infections
 as atherosclerotic cardiovascular disease cause, 88
 as autism cause, 85

D
Daidzein, 122
 chemical structure of, 122
 excretion of, 123, 124
Daidzin, 122
Darwin, Charles, 447
Data collection
 inaccuracy or errors in, 435
 methods in, 431
Davidigenin, 212
Deafness, iodine deficiency-related, 311
"Debriefing paradigm," 438
Deep breathing, during singing, 258, 259
Defensins, 393
 definition of, 395
 in mussels, 396–400
  synthetic derivatives of, 400
Dementia
 AIDS-related, 411
 music therapy for, 265, 273
Demyelination, 91–92, 93, 94
 in multiple sclerosis, 92, 93, 95
Dendritic cells, 77–78
Dentistry, complementary and alternative medicine use in, 207–214
Depression
 disability cost of, 182
 immunosuppression associated with, 406
 stress-related, 266

Dermatitis, atopic, 281–296
    acupuncture treatment of, 229–237
        effect on eosinophils, 231
        effect on granulocytes, 229, 231, 234
        effect on leukocytes, 230–231, 234
        effect on lipid levels, 231, 232
        during steroid therapy withdrawal, 231, 233, 234, 235, 236
    definition of, 281
    Japanese traditional diet treatment of, 281–296, 301, 303, 309
        allergic approaches in, 282
        effect on unsaturated fatty acid balance, 289–293
        non-allergic approaches in, 282
        pre-germinated brown-based, 285–289
    Kampo medicine treatment of, 297–310, 303–309
        with diet therapy, 301, 303, 309
        efficacy evaluation of, 304–309
    medical treatment of, 281
    onset age of, 229
    remission/exacerbation patterns in, 297–298, 299, 300, 301, 302
    steroid hormone treatment of, 229
        disease-exacerbating effects of, 299–301, 302
        granulocytosis associated with, 230, 231, 234, 235, 326
        as lipid peroxide source, 231, 232, 235
        as oxidized cholesterol cause, 229–230, 231, 232, 234–236
        withdrawal syndrome associated with, 230, 231, 234, 235, 236
    stress as risk factor for, 412
Dermatitis herpetiformis, 80
Dermatologic disorders, *Azadirachta indica* (neem) treatment of, 319
δ6-Desaturase, as atropic dermatitis treatment, 282
Descriptive studies, 430, 431–432
Diabetes mellitus, 320, 448
Diaries, diet, 283, 284
Diazepam, 181, 183–184
    action mechanism of, 182, 183, 189
    adverse effects of, 181, 189
    anxiolytic effects of, 181, 182, 183–184, 189
    myorelaxant effects of, 187–188
    sedative effects of, 187–188, 189
Dietary factors, in disease, 96
*Dietary Guidelines for Americans*, 96
Dietary supplements
    herbal preparations as, 44
    most popular, 16
    prevalence of use in Japan, 15, 16
Diet diaries, 283, 284
Digitalis, 58
Di Long, 364
6,3'-Dinitroflavone, 186

Dipeptidylpeptidase IV (DPP IV), 86, 91, 92
Diptericin, 393
Diseases, symptoms of, 448–449
DNA damage
    as cancer cause, 411–412
    gastrointestinal, effect of papaya-derived antioxidant on, 66–70
    ultraviolet light-induced, 23
DNA duplication, 392
DNA polymerase, 392
DNA transcription, 392
Donepesil (Ariceptr®), 31, 33
Double blinded studies, 434
    in acupuncture efficacy evaluation, 218–219
Dragonflies, antimicrobial peptides in, 396–397
Dream analysis, 440
Drosocin, 393
Drosomycin, 393
*Drosophila*, antimicrobial peptides in, 376, 393, 394
Dynorphin, immunomodulatory function of, 266
Dysmenorrhea, 19

E
Earthworms, 359–389. *See also Eisenia; Lumbricus*
    as anticoagulant source, 370–371
    as antimicrobial molecule source, 363, 371–377
    coelomic fluid in, 363, 372, 375
    diet of, 445
    fetidins in, 363, 371–372, 373, 375
    humoral products in, 362–363
    immune system of, 361–363
    lumbrokinase in, 364–368
    lysins of, 361–376, 363
    lytic molecules in, 370–371, 377–384
        anticarcinogenic activity of, 377–382
        as thrombosis treatment, 361, 364–371, 383–384
    medicinal use of
        in Chinese traditional medicine, 360–361, 364
        history of, 359–361
        as thrombosis treatment, 361, 364–371, 383–384
    plasminogen activators in, 360–361, 364, 369–371, 380
    serine proteases in, 363, 370–371
        anticarcinogenic activity of, 381–382
East-West Integrative Medicine, 35, 41–42
Echinacea, 49
Edward, John, 437
Eicosapentaenoic acid, 290, 291, 293, 294
*Eisenia*
    agglutinins in, 375
    lysins in, 375
*Eisenia foetida*
    coelomic fluid in, 363, 372, 375
    fetidins of, 371–372
    fibrinolytic enzymes from, 363, 368–369

immunological memory and recognition in, 379
  lysins of, 371–375
  lytic factors in, 363
  plasminogen activator from, 369–371
  serine proteases of, 363, 371
Eiseniapore, 363, 374, 376–377, 378
Eisenin I, 363, 381–382
Ekki-youketsu-fusei-zai (EYFZ), 167–178
  anticarcinogenic activity of, 173–178
    on natural killer cell activity, 173, 176, 177–178
  lymphocyte activation by, 171–173, 177
  macrophage activation by, 168–171, 177
  preparation of, 167–168
Elevated plus-maze test, 184
Embryo transfer, 10
Encephalomyelitis, peptide-related, 92
Encephalopathy, neuronal cytokines in, 411
Endometrial cancer, genistein-related prevention of, 133
Endometriosis, 19
β-Endorphin, 266
Endotoxins, immunosuppression-mediating effects of, 82–83
*Enterobacter aerogenes*, antimicrobial peptide-related inhibition of, 398
*Enterococcus faecalis*, antimicrobial peptide-related inhibition of, 398, 399
Enteropathy, gluten-sensitive. *See* Celiac disease
EntoMed, 391
Environmental factors, in disease, 75–76
Ephedra
  adverse effects of, 23
  as Kampo medicine component, 307
  as weight loss formula component, 439
Ephedrine, 23
Epidermal growth factor-genistein conjugate, 152, 154
Epidural analgesia, as childbirth pain treatment, 195–196
Epigallocathecin
  as multiple sclerosis treatment, 95–96
  synergistic interaction with genistein, 151
Epinephrine, 406
  music-induced modulation of, 266
Epstein-Barr virus, retroviral superantigen trans-activation by, 87
Equol, 122
*Escherichia coli*
  antimicrobial peptide-mediated inhibition of, 394
  heat shock protein-60 in, 88
  intestinal overgrowth of, 82
  multidrug-resistant, 115, 118
Escins, 50
Ethics, in experimental studies, 432
Ethnopharmacognosy research, 62
*Ethya aurantium*, 327, 328

*Eugenia jambolina*, 320
*Euplectella aspergillum*, 339, 341
Europe, quality requirements for herbal medicinal products in, 43–56
  for analysis test techniques, 45
  as basis for safety and efficacy, 44–45
  for identification tests, 44–45, 52
  for labeling, 46, 55
  for solvent quality, 46, 55
  for specific extract types, 45–46, 54
  for stability evaluation, 45, 53
European Agency for the Evaluation of Medicinal Products, 45
*European Pharmacopoeia*, herbal medicinal products quality guidelines of, 43, 44–46, 48–55
European Union, herbal medicinal products markets in, 47
*Eu Yan Sang Heritage: An Anthology of Chinese Herbs and Medicine*, 360
Evening primrose oil, as atopic dermatitis treatment, 282
Evidence-based medicine, 109
Evoked responses, for multiple sclerosis diagnosis, 94
Evolution, 447
Experimental design/methodology, 429–434
  blind designs, 434
  in pseudoscience, 440
Experimental studies, 430, 432–434
  null hypothesis (Ho) testing in, 432–434
Experiments, 429

F

*Farrea occa*, 339, 341
Fas, stress-induced expression of, 409, 410
Fatty acids, polyunsaturated
  *Spirulina* content of, 312–313
Fatty acids, polyunsaturated. *See also* Omega-3 fatty acids; Omega-6 fatty acids
  effect on atopic dermatitis, 281–296
Fenugreek *(Trigonella foenum-graecum)*, 320
Fermented papaya preparation (FFP), antioxidant properties of, 65–70
Fertilization, *in vitro*, 10
Fetidins, 363, 371–372, 373, 375
Fiber, dietary, 96
Fight-or-flight response, 182
Fish, neuroendocrineimmunological studies in, 415–417, 418
Fisher, Ronald Alymer, 430
Fish oils. *See* Omega-3 fatty acids; Omega-6 fatty acids
Flavones, 122, 185
Flavonoids
  anxiolytic activity of, 185–189
  of *Ginkgo biloba*, 51

Fleischmann, Martin, 439
Folate/folic acid
  deficiency of, 311
  as multiple sclerosis treatment, 95–96
Food allergies, 79, 293
*Forkhead/T-box* genes, 343
France
  herbal drugs/remedies market in, 47
  prevalence of traditional medicine use in, 58
*Frangula*, 54
Free radicals, 314
  in multiple sclerosis, 95
Frontal Alpha Wave Pulsed Photic Synchronization (FAPPS), 22–23
L-Fucose, 450
Fungi. *See also* Mushrooms; Yeast
  in association with sponges (Porifera), 350, 351
  as intestinal microflora, 79–80
*Fusarium oxysporum*, antimicrobial peptide-related inhibition of, 396, 398–399
Fusidic acid, *Staphylococcus aureus'* sensitivity to, 393

G
D-Galactose, 450
Gamma-aminobutyric acid, 182
Gamma-aminobutyic acidA receptors
  interaction with
    anxiolytic drugs, 182–184
    natural anxiolytic compounds, 185–186
  structure of, 183
*Gardeniae fructus*, 306, 307
Garlic *(Allium sativum)*, 319
Gastric mucosa, effect of papaya-derived antioxidants on, 65–70
Gastritis, atrophic, antioxidant supplementation in, 66–70
Gastrointestinal cancer, genistein-related prevention of, 133–134
Gastrointestinal disease
  intestinal flora in, 78–80
  intestinal permeability in, 78
  mucosal immune system in, 77–78
  xenobiotic and drug metabolism in, 83–84
Gastrointestinal tract
  antigen-lymphocyte receptor interactions in, 86
  host-microbial relationships in, 82–83
  mucosal immune system of, 77–80
    role of immunoglobulin A in, 77, 78–80
Gavarret, Jules, 429
Generalized anxiety disorder, 182
Genistein
  anticarcinogenic activity of, 121–165
    action mechanism of, 124, 149–150
    in active specific immunotherapy, 152–153, 154
  in interaction with chemotherapeutic drugs, 150–151
  in interaction with polyphenol food supplements, 151
  in passive immunotherapy, 151–152, 154
  chemical structure of, 122
*Geodia cydonium*, 325, 330, 331, 333–336, 340, 346
*Geodia rovinjensis*, 331, 334, 343
Germany, herbal drugs/remedies market in, 47, 48
Ginger *(Zingiber)*, 306, 307, 319
*Gingko biloba*, 49, 51, 441
  antioxidant activity of, 441
  efficacy assessment of, 440
  memory-enhancing effect of, 438–439, 441
  as multiple sclerosis treatment, 95–96
Ginkgolides, 51
Ginseng, 306
Gliadin, 81, 82
Gliadin antibodies, 91
Glucocorticoids
  adrenal, immunomodulatory activity of, 406
  in stress, 409, 410, 415
D-Glucose, 450
Glucosinolates, 318
Glutathione
  anticarcinogenic activity of, 96
  as multiple sclerosis treatment, 95–96
*Glycine max* (soybeans), genistein and daidzein content of, 122
Glycobiology, 445, 446, 449–450
Glycome, 445, 446
Glyconutrients, 446, 449–450
$\beta$2-Glycoprotein antibody, 90
Glycoproteins, formation of, 447
Glycosyltransferase, 447
*Glycyrrhizae*, 306, 307
G-90, 370–371
  anticarcinogenic activity of, 379–381
  anticoagulant activity of, 381
Gold, Harry, 435
Gossett, William, 430
Granulocyte-macrocyte colony-stimulating factor, glucocorticoid-related suppression of, 406
Green tea, as multiple sclerosis treatment, 95–96
Growth hormone, music-induced modulation of, 266
Guided imagery, surgical patients' use of, 257
Guisquiamo, 58
Guppies, neuroendocrineimmunological studies in, 416–417, 418
Gut-associated lymphoid tissue (GALT), 86
Gynecological procedures. *See also* Childbirth
  music therapy during, 256
*Gypsum fibrosum*, 307

# Index

## H

Haptonomy, 199
Hawthorn, 49
Healing Touch International, 440
Health insurance coverage
  for complementary and alternative
    medicine, 10
  for Kampo medicine, 10, 28
Heart disease. *See* Cardiovascular disease
Heat shock protein-60, 87–88
  in atherosclerosis, 87
Heat shock protein-65, 88
Heat shock protein-60 antibody, 90
Heat shock proteins, 86, 87, 89–90
  as autoimmune disease cause, 87–88, 89–90
  definition of, 87
  immune reactivity to, 87–88
Heavy metals, superantigen induction by, 87
*Helicobacter pylori*, 87
Hemolysin, from earthworms, 363, 372
Hemorheology, 20–22
Hepatitis, kava kava-related, 181
Herbal drugs/remedies. *See also* Kampo medicine; specific herbs
  adverse effects of, 60
  as anxiolytic agents, 181–191
  for childbirth-related pain management, 203
  composition of
    effectors, 44, 45, 50, 54
    markers, 44, 45, 50, 51
  definition of, 43, 48, 203
  drug interactions of, 63
  efficacy/safety evaluation of, 60–62
    clinical trials in, 62–63
    literature reviews in, 61–62
    need for scientific research in, 60–61
    WHO guidelines for, 60, 61–62
  European quality requirements for, 43–56
    for analysis test techniques, 45
    for identification tests, 44–45, 52
    for labeling, 46, 55
    for solvent quality, 46, 55
    for specific extract types, 45–46, 54
    for stability evaluation, 45, 53
    guidelines for rationale use of, 63
  HIV/AIDS patients' use of, 107
  interaction with antiretroviral drugs, 107
  molecular research in, 62–63
  non-holistic use of, 59–60
  prevalence of use of
    in Japan, 15, 16
    in the United States, 16
    in industrialized countries, 58
  relationship with traditional medicine, 57–58
  world market for, 47
Herpes infections, genital, Praneem polyherbal formulation treatment of, 116–117
Herpes simplex virus-2. *See also* HIV/AIDS
  Praneem polyherbal formulation-related inhibition of, 114–115, 117
H1 factor, 375
H2 factor, 375
H3 factor, 375
Highly-active antiretroviral (HAART) therapy, interaction with complementary and alternative medicine, 107, 108
*Hippospongia*, 350
*Hippospongia equina*, 344
Hirustatin, 365–366
Histamine, in stress, 410
Histocompatibility, in sponges (Porifera), 335
HIV. *See* HIV/AIDS
HIV/AIDS, 448
  $CD8^+$ cell count in, stress-related decrease in, 411
  $CD4^+$ T cell count in, 106
    stress-related decrease in, 411
  complementary and alternative medicine treatment for, 106–110
  interaction with highly-active antiretroviral (HAART) therapy, 107, 108
  pandemic of, 111
  stress-related susceptibility to, 411
HIV/AIDS patients, stress reduction in, 419
*HMS Challenger*, 391
*Hoelen*, 307
Holistic approach, in complementary and alternative medicine, 59–60, 61
Homeopathy
  as childbirth-related pain treatment, 201
  as HIV/AIDS treatment, 107–108
  prevalence of use in Japan, 15
Homeostasis, 307, 405–406, 407
  intestinal, 77
Hong-Yen Hsu, 35
Horse chestnut *(Aesculus hippocastanum)*, 49, 50, 54
Hospitals, music therapy in, 253, 254, 260
Human herpesvirus-6, as multiple sclerosis cause, 92–93
Human immunodeficiency virus-1. *See also*
  Praneem polyherbal formulation-related inhibition of, 113, 115, 116, 118
Human immunodeficiency virus infection. *See* HIV/AIDS
Human papilloma virus, 118
Humoral products, of earthworms, 362–363
*Hyalonema*, 325
*Hyalophora cecropia*, 394
*Hydra*, antibacterial activity in, 413–415
Hydrogen donor emissions, measurement of, 20, 21

Hydrotherapy
 for childbirth-related pain management, 203
 hot springs, 239–251
  age factors in, 243, 244, 245, 246, 248–249
  effect on $CD^+$ cells, 239, 243–244, 246, 247
  effect on cytokine-producing cells, 239, 243, 244–245
  effect on granulocytes, 241, 246, 247, 248, 249
  effect on interferon-$\gamma$, 244, 246, 250
  effect on interleukin-4, 244, 246, 250
  effect on interleukin-1$\beta$, 244, 246, 250
  effect on leukocytes, 241, 242, 243, 244
  effect on lymphocytes, 241, 243, 246, 247, 248, 249, 250
  effect on monocytes, 247, 248
  effect on white blood cells, 240, 241, 243, 245, 247
  historical analysis of, 239–240
3-Hydroxyanthranilate 3,4-dioxygenase, 344
8-Hydroxydeoxyguanidine, antioxidant-induced reduction in, 66, 68–69, 70
8-Hydroxyguanosine, 98
Hydroxy-valerenic acid, 51
Hyodo, I., 13
L-Hyoscyamine, 50
*Hypericum perforatum*. See St. John's wort
Hypersensitivity, delayed-type, 83, 408
Hypnosis
 adverse effects of, 198
 for childbirth-related pain management, 197–198
Hypothalamic-pituitary-adrenal axis
 interaction with sympathetic nervous system, 405–406, 410–411
 in stress, 266, 409, 410–411, 412
  in stress-related disease, 410
   sympathetic nervous system interactions in, 410–411
 in stress-related disease, 410
Hypotheses
 pseudoscientific, 437
 theoretical, 429
Hypotheses testing, 437
 of null hypothesis ($H_o$), 432–434
 statistical, 430

I
Immersion therapy, 239–240
Immune complex, 90
Immune deficiency pathway, in antimicrobial peptide synthesis, 393
Immune system. *See also* Adaptive immunity; Innate immunity
 aging-related impairment of, 229
 of earthworms, 361–363
 effect of hot springs hydrotherapy on, 240–250
 effect of music therapy on, 265–278
 intestinal mucosal, 77–80
  role of immunoglobulin A in, 77, 78–80
 of sponges (Porifera), 333–337
 stress-related enhancement of, 410
Immunoglobulin A
 celiac disease-related deficiency in, 80
 as multiple sclerosis biomarker, 95
 secretory
  in intestinal mucosal immunity, 77, 78
  music-enhanced secretion of, 268, 269, 274
Immunoglobulin A antigliadin antibodies, as celiac disease biomarker, 80–81, 82
Immunoglobulin G, 91
 in intestinal immunity, 79, 80
 as multiple sclerosis biomarker, 94, 95
Immunoglobulin G oligoclonal band, 94
Immunoglobulin-like domains, in sponges (Porifera), 331, 335
Immunoglobulin M
 in gastrointestinal immunity, 79, 80
 as multiple sclerosis biomarker, 95
Immunosuppression
 as disease risk factor, 410–412
 endotoxin-mediated, 82–83
 glucocorticoids-related, 406
 music therapy-related reversal of, 267
 stress-related, 408, 410
Immunotherapy
 active specific, genistein use in, 152–153, 154
 passive, genistein use in, 151–152, 154
India, nutritional supplements development in, 311–321
Indinavir, 107
Infant soy formula, 122
Infections
 as atherosclerotic cardiovascular disease cause, 86–87, 88
 as autoimmune disease cause, 85–86
 endogenous superantigen induction in, 86
 maternal, as behavioral disorder cause, 85–86
 stress-related, 410–411, 411–412
 superantigen induction by, 87
Inflammatory bowel disease
 definition of, 81
 etiology of, 81
 pathophysiology of, 77–78
Inflammatory molecules, 406
Inflammatory processes, stress-induced, 408, 409, 410–411
*In vitro* fertilization, 10
Innate immunity
 in earthworms, 361–363
 effect of music on, 273–274
 in invertebrates, 361, 362

Insects, antimicrobial peptides in, 393, 394, 396–397, 398
Insomnia, music therapy for, 265, 273
Intelligence, comparative, 434
Intelligence quotient, effect of Mozart's music on, 274
Interferon-α
  macrophage production of, 168
  retroviral superantigen trans-activation by, 87
Interferon-β, macrophage production of, 168
Interferon-γ
  glucocorticoid-related suppression of, 406
  herbal anti-tumor preparation-induced expression of, 171–173, 174, 177, 178
  intestinal mucosal production of, 78
  as multiple sclerosis biomarker, 95
Interleukin-1
  glucocorticoid-related suppression of, 406
  music-related increase in, 271–272
  role in neuronal functions, 407
Interleukin-1α, macrophage production of, 168
Interleukin-1β
central nervous system effects of, 407
macrophage production of, 168
in stress-related AIDS dementia, 411
Interleukin-2
  glucocorticoid-related suppression of, 406
  herbal anti-tumor preparation-induced expression of, 171–173, 174
Interleukin-4
  glucocorticoid-related upregulation of, 406
  herbal anti-tumor preparation-induced expression of, 171–173, 174
Interleukin-6
  alcohol-related suppression of, 412
  glucocorticoid-related suppression of, 406
  macrophage production of, 168
Interleukin-10
  alcohol-related suppression of, 412
  glucocorticoid-related upregulation of, 406
  in intestinal immune function, 77
  in stress, 410
Interleukin-11, glucocorticoid-related suppression of, 406
Interleukin-12
  glucocorticoid-related suppression of, 406
  intestinal mucosal production of, 77–78
  macrophage production of, 168
    herbal anti-tumor preparation-related enhancement of, 169–171, 177
International Research Center for Traditional Medicine (Japan), 13
Intestinal barrier function test, 79, 80
Intrabiotics Pharmaceuticals, 391
Invertebrates
  antimicrobial peptides in, 392–394
  innate immunity in, 361, 362
  marine
    DNA polymerase from, 392
    sulfur-based metabolism in, 391–392
    neuroendocrineimmunological studies in, 413–415
    stress responses in, 407–408
Iodine deficiency, 311
  *Spirulina*-based prevention of, 317
*Ircinia fasciculata*, 350
Iron deficiency, 311
  *Spirulina*-based prevention of, 317
Ischemia
  neuronal cytokines in, 411
  pathophysiology of, 87
Isegananr®, 391
Isoflavones, 122
Italy, herbal drugs/remedies market in, 47
Ivy, 49

J
Japan
  complementary and alternative medicine in, 9–25
    government's attitudes towards, 13–14
    history of, 11–13
    medical societies' acceptance of, 13–14
    medical societies relevant to, 12–13
    prevalence of, 16
    public health insurance coverage for, 10
    types of, 10
  herbal drugs/remedies market in, 47
Japanese Medical Association, 1314
Japanese Society for Complementary and Alternative Medicine, 12
Japanese Society of Internal Medicine, 13–14
Japanese traditional diet therapy, for atopic dermatitis, 281–296, 301, 303, 309
  allergic approaches in, 282
  effect on unsaturated fatty acid balance, 289–293
  non-allergic approaches in, 282
  pre-germinated rice brown-based, 285–289
Japanese traditional medicine, 281
Japan Medical Congress, 13–14
Japan Society of Clinical Oncology, 13, 14
Jaundice
  *Azadirachta indica* (neem) treatment of, 319
  earthworms as treatment for, 360
*Journal of the American Medical Association*, 13–14
Judo, 10
Judo-Orthopedics, 10
Jung, Carl, 440

K
Kampo medicine, 27–34
  as Alzheimer's disease treatment, 31–33

Kampo medicine (cont.)
    as atopic dermatitis treatment, 297–310, 303–309, 306–307
        with diet therapy, 301, 303, 309
        efficacy evaluation of, 304–309
    basic studies of, 31–33
    in dentistry, 207–214
    effect on intestinal immunity, 30–31
    health insurance coverage for, 10, 28
    herbal formulations in, 27
        bakumondo-to, 210–213
        eppi-ka-jutsu-to, 307
        gorei-san, 307
        hochu-ekki-to, 306
        juzen-taiho-to, 28, 29, 30
        kami-untan-to, 31–33
        keishi-bukuryo-gan, 307
        manufacturing guidelines for, 28
        oren-gedoku-to, 307
        syo-saiko-to, 27
        unsei-in, 306
    history of, 27
    nonclinical basic studies of, 28, 29
    oketsu concept of, 303
    prevalence of use, 15, 27
Kanazawa University, complementary and alternative medicine research group, 12
Kanazawa University Graduate School of Medical Science, Complementary and Alternative Medicine Department, 13, 15, 16–23
    curricula of, 18–19
    research conducted by, 19–23
    role of, 18
    staff of, 17, 18
Kava kava (Piper methylsticum), 49, 50
    adverse effects of, 23, 181
Kava-pyrones, 50
Kidney beans (Phaseolus vulgaris), genistein and daidzein content of, 122
Kudzu (Pueraria lobata), genistein and daidzein content of, 122
Kwong-Wen Shen, 36

L
$\beta$-Lactamate antibiotics, bacterial resistance to, 392
Lakatos, Imre, 438
Lamaze method, of childbirth, 197
Lamina properia, 77–78
Lavoisier, Antoine, 438
Laws, scientific, 428–429
LDL (low-density lipoprotein), 87
LDL (low-density lipoprotein) antibody, 90
Leboyer's method, of childbirth, 197
Lecithin, as multiple sclerosis treatment, 95–96

Leeches, as bdellastatin antistatin-type inhibitor source, 365–366
*Leishmania major*, antimicrobial peptide-related inhibition of, 399
*Lentinus edodes*, 20, 21
Leukemia, genistein-related prevention of
    *in vitro*, 134–137, 149
    *in vivo*, 134
Leukocytes
    in atopic dermatitis, 230–231, 234
    in earthworms, 361
    immune functions of, 361
Leukotriene $B_4$, as atopic dermatitis risk factor, 303
Life, origin and evolution of, 445
Light/dark box test, 184, 186
Lignans, 122
Linablue, 314
Linné, Carolus (Linnaeus, Carl), 327, 334
Linoleic acid
    as atopic dermatitis risk factor, 303
    deficiency of, 281
$\gamma$-Linoleic acid, *Spirulina* content of, 313
$\gamma$-Lipids, *Spirulina* content of, 312, 313
(n-3) Lipids. *See* Omega-3 fatty acids
(n-6) Lipids. *See* Omega-6 fatty acids
Lipoic acid, as multiple sclerosis treatment, 95–96
Lipopolysaccharide, 168, 407, 411
Lipton, Bruce, 446
Liquititigenin, 212
*Listeria monocytogenes*, 406
    antimicrobial peptide-related inhibition of, 398, 399
Liver cancer, genistein-related prevention of, 137–138, 150, 151
Lombricine, 379
Lou Gehrig's disease, 450
Low-calorie diet therapy, for atopic dermatitis, 282
Low-density lipoprotein (LDL), 87
Low-density lipoprotein (LDL) antibody, 90
Lumbricin I, 373, 375, 376
*Lumbricus*
    agglutinins in, 375
    lysins in, 375
*Lumbricus rubellus*
    fibrinolytic enzymes in, 364–365
    immunological memory and recognition in, 379
    lumbrokinase in, 364–368
        basic characteristics of, 364
        isolation of, 364–366
        thrombolytic effects of, 364, 366–368
        thrombolytic enzyme proteins from, 364–365, 366, 367–368
*Lumbricus terrestris*
    lombricine in, 379
    serine proteases in, 371

Lumbrokinase, 360–361
  enzyme fractions of, 364–365
  polyurethane-immobilized, 367
Lung cancer, genistein-related prevention of, 138–139, 150
Lupus peptides antibody, 90
Lycopene
  anticarcinogenic activity of, 96
  as multiple sclerosis treatment, 95–96
Lymphocytes. *See also* B cells; Peripheral blood lymphocytes; T cells
  stress-induced apoptosis of, 410, 419
Lymphocyte surface markers, 86
Lymphoma
  B-cell lineage, 151–152
  genistein-related prevention of
    *in vitro*, 138–140
    *in vivo*, 138
Lysenin, 363, 372, 373, 374, 382
Lysins, from earthworms, 363, 371–376, 378
Lysozyme, from earthworms, 362
  synergistic interaction with antimicrobial peptides, 393
Lytic compounds, from earthworms, 362–384
  anticarcinogenic activities of, 377–382
  antimicrobial activities of, 371–377, 382
  lumbrokinase, 364–368
  plasminogen activator, 360–361, 369–371

## M

Macrophages, 334
  in cardiovascular disease, 87
  effect of herbal anti-tumor preparation on, 168–171
Magainin Pharmaceuticals, 391
Magnetic resonance imaging (MRI), for multiple sclerosis diagnosis, 94, 95
Malaria, 319
Malnutrition, nutritional supplements for prevention of, 311–321
D-Mannose, 450
Manual healing, for childbirth-related pain management, 201–202
Massage therapy
  for childbirth-related pain management, 202
  for HIV/AIDS patients, 107
  prevalence of use
    in Japan, 15, 16
    in the United States, 16
Mast cells, stress-related degranulation of, 408, 409, 412
Maximum likelihood, 430
MCFAN, 20–22
Measles, as autism cause, 85
Medicarpin, 212

Medicine
  evidence-based, 109
  traditional Chinese. *See* Chinese traditional medicine, 9
Meditation, music use during, 260
Meephwanoyeekhun thwaykhan, 359
*Megascolides australis*, 361
Megavitamins
  HIV/AIDS patients' use of, 107
  prevalence of use, 16
Melanoma, genistein-related prevention of, 140, 148, 150, 153
Melatonin, 266, 273
Memories, music-associated, 259–260
Mental healing, 16
Mercury, as autoimmunity cause, 85
Metazoa, 325, 326, 330, 338, 339, 343
Metchnikowin, 393
Methyltransferse, 412
Microbial peptides. *See also* Antimicrobial molecules/peptides
  as autoimmune disease cause, 85–86
Microbial viability, biological basis of, 392
*Microciona prolifera*, 333
*Micrococcus galloprovincialis*, 396, 397
*Micrococcus lysodeikticus*, antimicrobial peptide-related inhibition of, 396, 398, 399, 400
Micrologix Biotech, 391
Milk proteins, 78
Milk thistle *(Silybum marianum)*, 49, 50
Mill, John Stuart, 429
Mind-body interventions, 10
  as childbirth-related pain treatment, 196–199
Minelli, Emilio, 63
Minerals, glycinated, 449
Ming Dynasty, 36
Min-Huo Huang, 38
Mistletoe, 49
Models, scientific theories as, 428–429
Molecular mimicry, 90, 93
Mollusks. *See also* Mussels
  from deep-sea bottom, 391–392
*Momordica charantia*, 320
Mongolia, hot springs hydrotherapy in, 239–240
Monocytes, in T-cell apoptosis, 409
*Moringa oleifera*, 314
Morton, Samuel, 434
Motherwort, 203
*Moutan radix*, 307
Moxibustion, prevalence of use in Japan, 15
Mozart, Wolfgang Amadeus, 274, 275
*Mucor rouxii*, 313
Müller, J., 327

Multiple sclerosis, 92–96, 448
  antioxidant therapy for, 95–96
  bacterial and viral infections associated with, 85
  as complex disease model, 91
  cytokines in, 407, 411
  demyelination in, 92, 93, 95
  diagnosis of, 93–95
  effect of stress on, 412
  viral etiology of, 92–93
  xenobiotics-related, 92, 93
Mushrooms, anticarcinogenic activity of, 19–20
Musical associations, 259–260
Music therapy, 253–263, 265–278
  autonomic nervous system response to, 274, 275
  for cancer patients, 254–255
  for cardiovascular disease patients, 255–256
  during childbirth/gynecological procedures, 199, 256
  for chronic obstructive pulmonary disease patients, 257
  for comatose patients, 256–257
  definition of, 253
  effect on neuro-endocrine-immunology, 266–267
  history of, 253–254
  in hospitals, 253, 254, 260
  immune function effects of
    interleukin-1 increase, 271–272, 274
    neutrophil-leukocyte proportion, 268–270
    neutrophil phagocytosis activation, 271, 272, 273, 274
    salivary immunoglobulin A secretion, 268, 269
  for Parkinson's disease patients, 257
  for pediatric patients, 258
  for premature infants, 254, 260
  for stress reduction, 259–260, 265
  for stroke patients, 257
  for surgical patients, 257–258
  for wellness development and maintenance, 253, 260–261
Mussels, antimicrobial peptides in, 395–400
Mutism, iodine deficiency-related, 311
Myelin, immunologic tolerance to, 91
Myosin, 87
Myosin antibody, 90
Myrtle, 49
Myticins, 396, 399
Mytilins, 396, 397–399

N

Nanticoke Indians, 360
National Center for Complementary and Alternative Medicine, 9–10
National Institutes of Health, 439
  Office of Alternative Medicine, 13

National Shikoku Cancer Center, 13
National Taiwan University, 36
Native Americans
  fatty acid deficiencies in, 293
  medicinal use of earthworms by, 360
Natural killer cells
  antioxidant-related stimulation of, 97–98
  in bone marrow transplants, 97
  cytotoxicity of
    alcohol-related decrease in, 412
    herbal anti-tumor preparation-related enhancement of, 173, 176, 177–178
    stress-related decrease in, 406, 408, 412
    stress-related increase in, 408
  in stress response, 97
Natural selection, 447
*Nature*, 439
Neem. See *Azadirachta indica*
*Neisseria gonorrhea*, Praneem polyherbal formulation-related inhibition of, 114, 118
Nerve growth factor, 31, 32
Neuroendocrineimmunology
  basic concepts of, 405–412
  effect of music therapy on, 266–267
Neuroendocrine system, effect of music on, 273
Neurohormones, in stress, 266, 273
  Neuroimmunology, 266
Neuronal cells, cytokine-induced death of, 407
Neuropathy, HIV-related peripheral, acupuncture treatment of, 108
Neuropeptides, stress-induced, 410
Neurotransmitters, cytokine-regulating function of, 407
Neyman, Jerzy, 430
Niemann-Pick disease type C, 382
Nitric oxide, 168, 407
  in multiple sclerosis, 95
*N*-Nitroso-fenfluramine, 23
Nitrous oxide, for childbirth-related pain management, 196
Noise, as immune system impairment cause, 266
  effect of music on, 267
Nonoyxnol-9, 111
Nonrandom samples, 431
Norepinephrine, 406
  music-induced modulation of, 266
North America, herbal drugs/remedies market in, 47
Novobiocin, *Staphylococcus aureus*' sensitivity to, 393
Nuclear factor KB, 406, 408
Nucleic acids, *Spirulina* content of, 312
Null hypothesis ($H_o$) testing, 432–434
Numbers, 445
Nutrition, guidelines for, 317–318

## O

Obesity, 448
Obsessive-compulsive disorder, 182
*Oenothera*, 313
Ohno, Yoshiyuki, 284
Old Gold cigarettes, 435
Oligosaccharides, 447
*Olynthus*, 342
Omega-3 fatty acids
    as atopic dermatitis prophylaxis, 282
        Japanese traditional diet-related, 290–293
    as multiple sclerosis treatment, 95–96
Omega-6 fatty acids, as atopic dermatitis risk factor, 283, 290
    effect of Japanese traditional diet on, 290–293
Oncology. *See* Cancer
Oolong tea, as atopic dermatitis treatment, 282
Ophiogonins, 210
Opioid analgesia, for childbirth-related pain management, 196
Opioid peptides
    immunomodulatory function of, 266
    in stress-related immunosuppression, 416
    in stress response, 415
Oral cancer
    dietary treatment of, 314
    genistein-related prevention of, 141
Oroxylin, 188
Oroxylin A, 185
Osteopathy, prevalence of use in Japan, 15
Osteosarcoma, genistein-related prevention of, 141
Ovary cancer, genistein-related prevention of, 141
Oxidative stress/radicals
    antioxidant-based protection against, 95–96
    superantigen induction by, 87
8-Oxoguanine, 98

## P

Pacifier Activated Lullaby, 254
Paeoniae, 306, 307
*Pagurites oculatus*, 330
Pancreatic cancer, genistein-related prevention of, 142
Panic disorder, 182
Panthothenic acid deficiency, 281
Papaya, fermented, antioxidant properties of, 65–70
Paraventricular nucleus, 405–406
Parkinson's disease patients, music therapy for, 257
Passionflower
Passionflower, anxiolytic effects of, 181, 185
    behavioral tests of, 186–188
Pauling, Linus, 447, 448, 449
Peanuts, genistein and daidzein content, 122
Pearson, Egon, 430
Pearson, Karl, 430
Pediatric patients, music therapy for, 258

Pellagra, 281
Pendulum law, 428–429
*Penicillin chrysogenum*, 350
Perinuclear anti-neutrophil cytoplasmic autoantibodies, as ulcerative colitis biomarkers, 81, 82
Peripheral blood lymphocytes, effect of herbal anti-tumor preparation on, 171–173, 174
Peripheral nervous system, effect of cytokines on, 406–407
Peritonitis, 82
Persicae, 307
Personality theory, 440
Pets, complementary and alternative medicine for, 10
Pexiganan®, 391
Peyer's patches, 77–78
Phagocytosis
    leukocytes in, 361
    music-induced, 271, 272, 273–274
*Phaseolus vulgaris* (kidney beans), genistein and daidzein content of, 122
*Phellinus linteus*, 19–20
*Phellodendri cortex*, 306, 307
Phenolics, 318
*Pheronema carpentari*, 339, 341
Phlogiston, 438
Phobias, 182
Phosphatidyl choline, as multiple sclerosis treatment, 95–96
Phosphatidyl serine, as multiple sclerosis treatment, 95–96
Phycobilins, *Spirulina* content of, 312, 314
Physicians, in East-West Integrative Medicine, 35
Physics, Newtonian, 446
Phytoestrogens, 122, 318
Phytosterols, 449
Pineapple, 49
*Piper methylsticum* (kava kava), 49, 50
    adverse effects of, 23, 181
*Piper nigrum* (black pepper), 319
Pituitary gland, anterior, 405–406
Placebo, definition of, 434–435
Placebo effect, 435–436
    complementary and alternative medicine as, 427, 435
    in herbal medicine clinical trials, 62
Plasmin
    bdellastatin-related inhibition of, 365
    from earthworms, 364, 365
Plasminogen activators
    from earthworms, 360–361, 364, 369–371, 380
    in tumors, 380–381
*Plasmodium falciparum*, *Azadirachta indica* (neem)-mediated killing of, 112

*Poecilia reticulata*, neuroendocrineimmunological studies in, 416–417, 418
Poincare, Jules, 430
Poland, herbal drugs/remedies market in, 47
Polyketide synthases, 346–348
Polyphenol food supplements, interaction with genistein, 151
*Polyporus*, 307
Polysaccharides
  pectic, 31
  *Spirulina* content of, 313
Pons, Stanley, 439
Popoff, Peter, 438
Popper, Karl, 430
*Porphyromonas gingivalis*, heat shock protein-60 in, 88
Posttraumatic stress disorder, 182
"Powerful Placebo, The" (Beecher), 435
Praneem polyherbal formulation, 111–119, 113–118
  antimicrobial activity of, 114–115
  ingredients of, 112–113
  phase II efficacy studies of, 116–118
  toxicity and safety studies of, 116
Premature infants, music therapy for, 254, 260
Priestley, Joseph, 438
Primmorphs, 332, 334–335, 346
*Principles of Science: A Treatise on Logic and the Scientific Method* (Jevons), 435
Proanthocyanidins, 95
Probable error, 429
Prolactin, 273
  immunomodulatory function of, 266
  music-induced modulation of, 266
  in stress, 266
Prospective studies, 430, 432
Prostaglandin E, 77
Prostanoids, 407
Prostate cancer
  genistein-related prevention of
    *in vitro*, 125, 143–146, 148–149, 150
    *in vivo*, 142–143
  genistein-related treatment of, 152
  in combination with radiation, 151
  Prostate-specific antigen (PSA), 153
Protegrin, 391
Proteins
  amino acid composition of, 446–447, 450
  *Spirulina* content of, 312–313
Protein-saccharide combinations, 446–447, 450
Protein tyrosine kinases, genistein-related inhibition of, 124, 153
*Pseudomonas*, as intestinal immunity impairment cause, 83
*Pseudomonas aeruginosa*, 392
Pseudoscience, 436–441

  belief perseverance in, 437–438
  complementary and alternative medicine as, 427
  definition of, 436–437
  fallacious reasoning in, 438–439
  science differentiated from, 436
  vocabularies in, 440
Psoriasis
  dietary factors in, 284
  stress-induced, 412
Psychological disorders, stress therapy for, 265
Psychoneuroimmunology, 266
*Pueraria lobata* (kudzu), genistein and daidzein content of, 122
Pycnogenol
  antioxidant activity of, 23
  as dysmenorrhea and endometriosis treatment, 19
  as multiple sclerosis treatment, 95–96
*Pyrococcus abyssi*, 392

Q
Quality of life, 58, 64–65, 239
Quantum theory, 446
Quercetin, 151
Quinine hydrochloride, as Praneem polyherbal formulation component, 113
Quinolinic acid, 344

R
Radiation therapy, genistein-enhanced cytotoxicity of, 151
Randi, James, 437, 438
Randomization
  in acupuncture efficacy trials, 220
  in herbal medicine clinical trials, 62
Random samples, 431
Raspberry leaf, 203
Reactive oxygen species, chemiluminescence measurement of, 20, 21
Red Cross nurses, music therapy use by, 253–254
Rees, Jonathan, 76
Reflexology, prevalence of use in Japan, 16
Rehmanniae. *See also* Ekki-youketsu-fusei-zai (EYFZ)
  use in Kampo medicine, 306
Relaxation therapy
  for childbirth-related pain management, 198
  HIV/AIDS patients' use of, 107
  prevalence of use in the United States, 16
Renal cancer, genistein-related prevention of, 146, 149–150, 152
Reproductive tract infections
  Praneem polyherbal treatment of, 111–119
  prevalence and incidence of, 111
Resveratrol, as multiple sclerosis treatment, 95–96
Retrospective descriptive studies, 431

Retrospective studies, 430, 432
Retroviruses, superantigens of, 87
*Rhabdocalyptus dawsoni*, 337, 340
Rheumatism, earthworms as treatment for, 360
Rheumatoid arthritis
  bacterial and viral infections associated with, 85
  CD69 in, 86
  dipeptidylpeptidase IV in, 86
  stress-related exacerbation of, 412
Rhythmic auditory stimulation (RAS), 257
Ribonucleic acid (RNA) translation, 392
Rice
  brown, pre-germinated, as atopic dermatitis treatment, 285–289
  nutrients in, 286
Rifamycin, 392
*Riftia pachyptila*, 391
*Rimicaris exoculata*, 391
Rivers, William Halse Rivers, 434, 435
RNA (ribonucleic acid) translation, 392
*Roseobacter*, 347–348
Royal Naval Academy, 448
Rubella, as autism cause, 85

S
*Saccharomyces cervisiae*, 339–340
*Saccharomyces cervisiae* antibodies, as Crohn's disease biomarkers, 81–82
St. John's wort, 49, 54
  interaction with antiretroviral drugs, 107
Saliva, 209–210
  secretory immunoglobulin A content of, 78
Salivary glands, 209–210
Samples, 431
Sample size, in acupuncture efficacy trials, 221
*Sapindus mukerosi*, as Praneem polyherbal formulation component, 113
Saponins, as Praneem polyherbal formulation component, 113
Sarcoma, genistein-related prevention of, 146–147, 163
Saw palmetto, 49
Schizophrenia, 85–86
Schleiden, M.J., 327
Schwann, T., 327
*Science*, 439, 449
Scientific laws, 428–429
Scientific methods, 429–434
  experimental blind design, 434, 435
  flow chart of, 11
  placebo controls, 434–436
Scientific misconduct, 439
Scientific theories, 428–429
  statistical assessment of, 430

Scientific thinking, 427–443
  scientific methods, 429–434
  scientific theories, 428–429, 430
Scorpions, antimicrobial peptides in, 396–397
Scurvy, 98, 281, 447–448
*Scutellaria baicalensis*, flavonoids from, 188
Scutellariae, 306, 307
SelectBiotics, 391
Selenium
  anticarcinogenic activity of, 96
  as multiple sclerosis treatment, 95–96
Senna *(Cassia)*, 50, 54, 58
Sennosides, 50
Serine proteases, from earthworms, 363, 370–371
  anticarcinogenic activity of, 381–382
Serotonin, in stress, 266, 273
Sesquiterpenes, 319
Sexual coloration, carotenoids-related, 416–417, 418
Sexually-transmitted infections
  Praneem polyherbal treatment of, 111–119
  stress-enhanced susceptibility to, 411
Shiitake mushrooms, 21
Silicase, 344
Silicatein, 344
Silicone breast implants, 97
*Silybum marianum* (milk thistle), 50
Singing, deep breathing during, 258, 259
Single blinded studies, 434
Smallpox, earthworms as treatment for, 359
Snake venom, 36, 37
Social phobia, 182
Sophrology, 199
Sorbicillactone A, 350, 351
South America, herbal drugs/remedies market in, 47
Soybeans *(Glycine max)* genistein and daidzein content of, 122
Spain, herbal drugs/remedies market in, 47
Spermatozoa, sphingomyelin content of, 374–375
Sphingomyelin
  lysenin binding to, 374–375, 382
  role in earthworm coelomic fluid toxicity, 374–375
Spiritual healing, HIV/AIDS patients' use of, 107
*Spirulina*, as nutritional supplement, 311–317
  chemical composition of, 312–314
  cultivation of, 315–317
Sponge adhesion molecules (SAM), 331, 335
Sponges (Porifera), 325–357
  bacteria in, 346–348
  bioactive component production in, 344, 346–348, 352
  evolutionary combinatorial chemistry in, 348–349, 351, 352
  gene clusters in, 346, 347, 348
  therapeutic use of, 348–351, 352
  biology of, 325, 327–343

Sponges (Porifera) (*cont.*)
  biochemistry, 344
  endemic species, 343
  evolution, 337–343
  histology and embryology, 325, 327–333, 342, 343
  immunology, 333–337
  stem cells, 329–330
  systemics/taxonomy, 325, 326, 327, 337–343
  cell culture of, 346
  hemostatic extracts from, 325
  mariculture of, 344–346
  paleontology of, 326–327
  taxonomy of, 325, 326
*Spongia*, 350
*Spongia coronata*, 337, 340
*Spongia equina*, 327
*Spongia mollissima*, 327
*Spongia officinalis*, 344
*Spongilla domuncula*, 329, 330, 332, 333, 334, 335, 336, 337, 340, 343, 346, 348
*Spongilla fluviatilis*, 328–329
Staircase test, 184, 186
Standard deviation, 430
*Staphylococcus*
  enterotoxin B in, 81
  heat shock protein-60 in, 88
*Staphylococcus aureus*
  antibiotic-resistant, 391
  antimicrobial peptide-related inhibition of, 398, 399
  sensitivity to benzylpenicillin, 393
  sensitivity to fusidic acid, 393
  sensitivity to novobiocin, 393
*Staphylococcus epidermidis*, 398
  benzylpenicillin sensitivity in, 393
  lymphocyte proliferation suppression by, 83
Statistical analysis, 430
Statistical tests, 429–430
Sterile water blocks, for childbirth-related pain management, 202
Steroid hormone addiction, 235
Steroid hormone therapy, for atopic dermatitis
  disease-exacerbating effects of, 299–301, 302
  granulocytosis associated with, 230, 231, 234, 235, 326
  as lipid peroxide source, 231, 232, 235
  as oxidized cholesterol cause, 229–230, 231, 232, 234–236
  withdrawal from, 231, 233, 234, 235, 236
Stinging nettle root, 49
Stramonium, 58
*Streptococcus*, heat shock protein-60 in, 88
*Streptococcus* group A peptides, as encephalomyelitis cause, 92
*Streptokinase*, 86

*Streptokinase* antigens, 91
Stress
  anti-inflammatory effects of, 410
  definition of, 259, 407
  immune system effects of, 272, 407–412
  immunosuppressive effects of, 272
  natural killer cells in, 97
  physiological effects of, 265
  pro-inflammatory effects of, 409, 410
  types of, 265
Stress management, with music therapy, 259–260
Stressors, biological, 449
Stress response, central nervous system mechanisms of, 182
Stroke, 411
  neuronal cell death in, 407
Stroke patients, music therapy for, 257
*Suberites domuncula*, 330, 332, 333
*Suberites ficus*, 330, 332
Substance P, in stress, 408, 410
Sun-Ten Drug Company, 35
Superantigens, 86
  in cardiovascular disease, 87–90
  definition of, 87
Surgical patients, music therapy for, 257–258
Suzuki, Nobutaka, 11–12
*Sycon raphanus*, 337, 340
Sympathetic nervous system, interaction with hypothalamic-pituitary-adrenal axis, 405–406
  in stress, 410–411
Symptoms, of disease, 448–449
Systemic lupus erythematosus, 85
  dipeptidylpeptidase IV in, 86
Sze-Piao Yang, 39

T
Taiwan, Chinese medicine in, 35–42
  current status of, 36–37, 39–40
  history of, 36, 37–39
  during Japanese occupation, 36
  modernization of, 35–42
  National Research Institute of Chinese Medicine and, 36
  public health insurance coverage for, 40
Tangier disease, 382
T cells
  effect of herbal anti-tumor preparation on, 167, 173, 177, 178
  heat shock protein response in, 88
  stress-related apoptosis in, 409
Th1
  $\beta 2$-adrenergic receptors of, 406
  in stress, 409, 410
Th2, in stress, 409, 410

Tea
  green, as multiple sclerosis treatment, 95–96
  oolong, as atopic dermatitis treatment, 282
Testis cancer, genistein-related prevention of, 147, 151
Tests, experimental, 429
*Tethya*, 327
*Tethya limski*, 328, 343
*Tethya lyncurium*, 327, 328, 332, 345
*Tetia sphaerica*, 327
Th1
  $\beta$2-adrenergic receptors of, 406
  in stress, 409, 410
Th2, in stress, 409, 410
Thalassemia, 365
Thearubigin, 151
*Theonella swinhoei*, 346
Therapeutic touch therapy
  for childbirth-related pain, 201–202
  for HIV/AIDS, 108
*Thermophilus aquaticus*, 392
Thimerosal, 91
Thomson, Charles Wyville, 391
*Thoosa istriaca*, 343
Thrombolytic substances
  from earthworms, 361, 364–371, 383–384
  from leeches, 365–366
Thyroiditis, autoimmune, 85
Thyrotropin, immunomodulatory function of, 266
*Tilapia*, neuroendocrineimmunological studies in, 415–416
Titchener, Edward B., 434
$\alpha$-Tocopherol, anticarcinogenic activity of, 96
$\gamma$-Tocopherol, anticarcinogenic activity of, 96
Toll pathway, in antimicrobial peptide synthesis, 393
Topoisomerase II, genistein-related inhibition of, 124, 151
Touch therapy
  for childbirth-related pain, 201–202
  for HIV/AIDS, 108
Trace elements, s multiple sclerosis treatment, 95–96
Traditional medicine. *See also* Chinese traditional medicine, 9
Transcutaneous electrical nerve stimulation (TENS), for childbirth-related pain management, 202
Transglutaminase, as celiac disease biomarker, 81, 82
*Trigonella foenum-graecum* (fenugreek), 320
Tropomyosin, s ulcerative colitis biomarker, 82
*Trunculariopsis trunculus*, 330
*Trypanosoma brucei*, antimicrobial peptide-related inhibition of, 399
*Trypanosoma cruzi*, coelomic cytolytic factor-1-induced lysis of, 375
Tsumura Pharmaceuticals Company, 207, 210, 211
Tsung-Ming Tu, 36
Tumor growth factor-$\beta$, 77

Tumor necrosis factor, 375
Tumor necrosis factor-$\alpha$
  central nervous system effects of, 407
  glucocorticoid-related suppression of, 406
  macrophage production of, 168
  as multiple sclerosis biomarker, 95
Tumor necrosis factor-$\beta$, s multiple sclerosis biomarker, 95
Turmeric *(Curcuma longa)*, therapeutic applications of, 312, 319–320
  as oral cancer treatment, 314
Tzu-Chi University, 42

U
UFOs (unidentified flying objects), 442
Ulcerative colitis, 77
  biomarkers for, 81, 82
  differentiated from Crohn's disease, 81–82
  stress-induced, 412
  xenobiotics-related, 83–84
Ultraviolet B light, as DNA damage cause, 23
Unani medicine, 57, 318–319
United Kingdom, herbal drugs/remedies market in, 47
United States, prevalence of complementary and alternative medicine use in, 16
University of Milan, World Health Organization Collaborating Center for Traditional Medicine, 63–70
  antioxidant research at, 65–70
  collaboration with Y. Garin Aerospace Center, Moscow, 64–65
  meetings and symposia sponsored by, 63–64
  postgraduate courses at, 64

V
Vaginal discharge, abnormal, Praneem polyherbal formulation treatment of, 117–118
Vaginosis, bacterial, 111
Valerenic acid, 51
Valerian *(Valeriana officinalis)*, 51, 181
Valium. *See* Diazepam
Vanadium, 151
Vascular endothelial growth factor, 149–150
Vasodilators, stress-induced, 410
Vegetarians, nutritional support for, 311–321
Venn, John, 429
Vertebrates, earthworm coelomic fluid toxicity in, 374–375
Veterinarians, complementary and alternative medicine use by, 10
*Vibrio*, antimicrobial peptide-related inhibition of, 398, 399
*Vigna mungo* (black gram), genistein and daidzein content of, 122

Viral infections. *See also* specific viral infections
    as atherosclerosis cardiovascular disease cause, 88
    as autoimmune disease cause, 85
    neuronal cytokines in, 411
    stress-related, 411
Virulence factors, bacterial, 79
Viruses. *See also* specific viruses
    superantigens of, 87
Visualization, music use during, 260
Vitamin(s), glycinated, 449
Vitamin A, as multiple sclerosis treatment, 95–96
Vitamin A deficiency
    *Spirulina*-based prevention of, 317
    as vision loss cause, 314
Vitamin $B_6$, as multiple sclerosis treatment, 95–96
Vitamin $B_{12}$, *Spirulina* content of, 313
Vitamin B complex deficiency, 281
Vitamin $B_6$ deficiency, 281
Vitamin C, 448
    anticarcinogenic activity of, 96, 98
    antioxidant properties of, 314
    as multiple sclerosis treatment, 95–96
    natural killer cell-modifying effects of, 97–98
    recommended dietary allowances of, 98
Vitamin C deficiency, 98, 281
Vitamin D deficiency, 311, 448
Vitamin deficiencies
    disease-associated, 96–97
    as disease risk factor, 96–97
    in the elderly, 96
Vitamin E
    antioxidant properties of, 314
    as multiple sclerosis treatment, 95–96

W

Weight loss formulas, Ephedra-based, 439
World Health Organization (WHO)
    Collaborating Center for Traditional Medicine, 63–70
    Global Burden of Disease Survey, 182
    herbal medicine efficacy/safety evaluation guidelines of, 60, 61–62
Wound healing, stress-related inhibition of, 412
Wundt, Wilhelm, 434

X

Xanthophylls, *Spirulina* content of, 312
Xenobiotics
    as autism cause, 91
    as autoimmune disease cause, 85
    endogenous superantigen-inducing activity of, 86
    as multiple sclerosis cause, 92, 93
    as ulcerative colitis cause, 83–84
Xerostomia, Kampo therapy for, 210–213
D-Xylose, 450

Y

Yang-Ming University, Institute of Traditional Medicine, 39
Yeast
    as intestinal microflora, 79
    therapeutic properties of, 49
*Ye se kun byo*, 359
Yoga, for childbirth-related pain management, 199

Z

*Zingiber* (ginger), 306, 307, 319
*Zizphi fructus. See also* Ekki-youketsu-fusei-zai (EYFZ)
    as Kampo medicine component, 306, 307
Zodiac, Chinese, 437–438
*Zophobas atratus*, 394